T0143185

Green Energy and Technology

For further volumes:
http://www.springer.com/series/8059

Dasarathi Das · S. R. Bharadwaj
Editors

Thoria-based Nuclear Fuels

Thermophysical and Thermodynamic Properties, Fabrication, Reprocessing, and Waste Management

Springer

Editors
Dasarathi Das
Kharghar
New Mumbai
India

S. R. Bharadwaj
Chemistry Division
Bhabha Atomic Research Centre
Trombay
Mumbai
Maharashtra
India

ISSN 1865-3529 ISSN 1865-3537 (electronic)
ISBN 978-1-4471-6956-7 ISBN 978-1-4471-5589-8 (eBook)
DOI 10.1007/978-1-4471-5589-8
Springer London Heidelberg New York Dordrecht

Softcover reprint of the hardcover 1st edition 2013

Printed on acid-free paper

Springer is part of Springer Science+Business Media (www.springer.com)

Foreword

It is now amply clear that nuclear energy can be a sustainable source of clean energy for the whole world if we take into account both the fissile and the fertile material resources availability. If, however, uranium-235 is taken as the only fissionable nuclide from which we can generate nuclear power, the exploitable resources are not adequate to supply fuel for the growing demand of the nuclear power industry beyond a century. The adoption of closed fuel cycle to utilize plutonium-239 which is generated as a transmutation product of uranium-238 in the spent nuclear fuel is, therefore, essential in multiplying the fuel resource base by over 50 times. Thorium-232, the other naturally occurring fertile material can enhance the fissile resources still further several fold. In addition, the use of uranium-233, the fissile product of thorium, offers several advantages in reducing radiotoxic waste burden and in providing a proliferation resistant fuel. Worldwide R&D efforts in establishing the reactor technology with thorium fuel cycle have significantly grown in the recent past mainly because of these factors. India being endowed with a huge reserve of thorium minerals has given the due emphasis on research on thorium fuel cycle and has drawn the roadmap of thorium utilization in her long-term plan for the growth of nuclear energy.

This book gives a comprehensive treatment of various aspects of thoria-based fuels and both the front and the back end of the fuel cycle associated with it. Each chapter is written by experts who have many years of experience in their respective fields. At the same time, the continuity of the theme and coherence among the chapters are maintained. The editors must be complimented for structuring the book very well and make it suitable not only for those who are entering into research in the nuclear fuel cycle area but also for practicing scientists. The authors have paid attention to technological details while presenting the scientific basis from a fundamental standpoint. The fact that the authors have direct research and operating experience in the areas which are covered in their respective chapters gets reflected in the clarity of their presentations. There are only a few in the world who have delved into thorium fuel cycle, fabrication of thoria based fuel, reprocessing of high burn up fuels and their waste management issues. As they have joined hands to prepare this book entitled *Thoria-based Nuclear Fuels: Thermophysical and Thermodynamic Properties, Fabrication, Reprocessing and Waste Management*, the product that has emerged is not only unique but also extremely valuable.

Mumbai, India, April 2013

Srikumar Banerjee
Homi Bhabha Chair Professor
Bhabha Atomic Research Centre
(Former Chairman, Atomic Energy Commission,
& Secretary to the Department of Atomic Energy,
Government of India)

Preface

In nuclear power generation establishing the performance of a fuel calls for collective effort of scientists and technologists for addressing various issues of front and back ends of the fuel cycle. These issues pertain to all the operational aspects for smooth functioning right from fuel resourcing, mining and purification, fuel pin fabrication, reactor irradiation, fuel reprocessing, and management of radioactive wastes. Safety, security, and economy are to be met in the fuel cycle management. Typically nuclear fuels are irradiated in reactors for several decades for deriving power from them. Therefore, a thorough understanding of thermal and thermodynamic properties of the fuel materials is necessary for the evaluation of its performance under these conditions. Reliable data acquisitions are thus carried out for the thermal expansion, thermal conductivity, heat capacity, phase stability of the fuel, and thermodynamic and transport properties of fission products that are formed in the fuel matrix. With the additional knowledge of the transport properties of fission-released oxygen in the case of oxide fuels, the chemical states of the redistributed fission products are evaluated for ensuring containment of the cladded fuel.

Many countries, particularly those having abundant resource of thorium, envisage nuclear power production from this fertile-actinide. The envisaged policy for the thorium-based reactor technology is toward reducing burden to the enriched uranium-based conventional fuels. Thorium-based fuels have the general merits of greater abundance of the element on earth crust, superior physical and nuclear properties, particularly in oxide form, better resistance to weapons proliferation and lesser production of heavier actinides in reactor irradiation. There has been worldwide effort to establish the thorium-based reactor technology and thus a large database on the thermophysical, thermodynamic and transport properties of thoria based fuels, and detailed technical information on the fuel fabrication, reprocessing and waste management exists in the literature. In this context, it is necessary to consolidate the accumulated information at one place, in the form of a book that essentially covers the scientific and technological information on all the stated aspects of thoria-based fuels. With this objective, the chapters in this book are organized accordingly and are written by experts in the respective fields, who collectively contributed to the thorium utilization program in India. The arrangement of chapters has been carefully planned to provide the readers with adequate state-of-art knowledge regarding thoria fuels. We believe that the scientific and technical information in this book will serve as a ready reference to researchers and technologists working in the field of thorium utilization.

Editors of this book sincerely acknowledge the authorities of the Department of Atomic Energy, Government of India, who have appreciated and approved the idea of making such publication that presents a consolidated view of R&D on the thorium-based fuel cycle. We would like to express our gratitude to all the contributors and to the staff of Springer-Verlag for their generous assistance without which this book could not have been successfully published.

Mumbai, India, April 2013

D. Das
S. R. Bharadwaj

Contents

Contents

Contributors

D. Das Chemistry Division, Bhabha Atomic Research Centre, Trombay, Mumbai 400085, India, e-mail: dasd1951@gmail.com

T. R. G. Kutty Formerly with Radio Metallurgy Division, Bhabha Atomic Research Centre, Trombay, Mumbai 400085, India, e-mail: trgovindankutty@gmail.com

J. Banerjee Radiometallurgy Division, Bhabha Atomic Research Centre, Trombay, Mumbai 400085, India, e-mail: joydipta@barc.gov.in

Arun Kumar Radiometallurgy Division, Bhabha Atomic Research Centre, Trombay, Mumbai 400085, India, e-mail: arunk@barc.gov.in

R. Agarwal Product Development Division, Bhabha Atomic Research Centre, Trombay, Mumbai 400085, India, e-mail: arenu@barc.gov.in

S. C. Parida Product Development Division, Bhabha Atomic Research Centre, Trombay, Mumbai 400085, India, e-mail: sureshp@barc.gov.in

S. R. Bharadwaj Chemistry Division, Bhabha Atomic Research Centre, Trombay, Mumbai 400085, India, e-mail: shyamala@barc.gov.in

R. Mishra Chemistry Division, Bhabha Atomic Research Centre, Trombay, Mumbai 400085, India, e-mail: mishrar@barc.gov.in

M. Basu Chemistry Division, Bhabha Atomic Research Centre, Trombay, Mumbai 400085, India, e-mail: deepa@barc.gov.in

S. Kolay Chemistry Division, Bhabha Atomic Research Centre, Trombay, Mumbai 400085, India, e-mail: siddhart@barc.gov.in

A. N. Shirsat Chemistry Division, Bhabha Atomic Research Centre, Trombay, Mumbai 400085, India, e-mail: shirsat@barc.gov.in

S. K. Mukherjee Fuel Chemistry Division, Bhabha Atomic Research Centre, Trombay, Mumbai 400085, India, e-mail: smukerji@barc.gov.in

N. Kumar Fuel Chemistry Division, Bhabha Atomic Research Centre, Trombay, Mumbai 400085, India, e-mail: nkumar@barc.gov.in

R. V. Pai Fuel Chemistry Division, Bhabha Atomic Research Centre, Trombay, Mumbai 400085, India, e-mail: rajeshvp@barc.gov.in

P. V. Achuthan Nuclear Recycle Board, Bhabha Atomic Research Centre, Trombay, Mumbai 400085, India, e-mail: apv@barc.gov.in

A. Ramanujam Formerly with Fuel Reprocessing Division, Bhabha Atomic Research Centre, Trombay, Mumbai 400085, India, e-mail: aiyaswamy.ramanujam@gmail.com

Kanwar Raj Waste Management Division, Bhabha Atomic Research Centre, Trombay, Mumbai 400085, India, e-mail: rajkanwar50@gmail.com

C. P. Kaushik Waste Management Division, Bhabha Atomic Research Centre, Trombay, Mumbai 400085, India, e-mail: cpk@barc.gov.in

R. K. Mishra Waste Management Division, Bhabha Atomic Research Centre, Trombay, Mumbai 400085, India, e-mail: mishrark@barc.gov.in

Introduction

D. Das

This book is introduced with a brief overview of worldwide efforts and recent outlook on the use of thoria-based fuels for power generation. It also summarizes the merits of thoria-based fuels, issues in the thorium fuel cycle, and the past to present accounts of work on the fuels. As preamble to different chapters of the book, reader's attention is drawn toward the need of detailed information on the thermophysical, thermodynamic, and transport properties of the fuels and on the established procedures of the front and back end operations like fabrication, reprocessing, and waste management of the fuel cycle.

Over the last 30 years there has been increased interest in utilizing thorium as nuclear fuel primarily because this actinide element is three times more abundant in the earth's crust as compared to uranium, which is widely used as nuclear fuel. With the economically extractable thorium reserve [1] in excess of 1.2 megaton its usage gives considerable savings in uranium ore and in ^{235}U isotope enrichment units. The most abundant ^{232}Th isotope is not fissile and its application as nuclear material rests on the fact that it absorbs slow neutron in irradiation to produce fissionable daughter nucleus ^{233}U ($^{232}Th + n \rightarrow {}^{233}Th(22\,m) \rightarrow {}^{233}Pa(27d) \rightarrow {}^{233}U(1.5\,x10^{6}y)$). With the higher neutron yield in ^{233}U fission [2, 3] a more efficient breeding cycle compared to the cases of ^{235}U or ^{239}Pu can be set up. The thorium fuel cycle generates fewer long lived heavy actinides, which will be an advantage in its waste management.

For ^{233}U breeding and for maintaining neutron economy in the core of critically run reactor, the Th-based fuels will always contain fissile components, which are usually the ^{235}U or ^{239}Pu isotopes in their chemically compatible forms with the host matrix. Configurationally, the sub-critical blankets mainly of thorium fuel rods surround the seed elements with highly enriched ^{235}U (HEU) or ^{239}Pu. The HEU/Pu seed elements and the Th blankets are spatially separated either within a given assembly, or in between assemblies of fuel rods.

D. Das (✉)
Chemistry Division, Bhabha Atomic Research Centre, Trombay, Mumbai 400085, India
e-mail: dasd1951@gmail.com

D. Das and S. R. Bharadwaj (eds.), *Thoria-based Nuclear Fuels*,
Green Energy and Technology, DOI: 10.1007/978-1-4471-5589-8_1,

The fuel bred out of thorium or a mixture of thorium and depleted uranium inherits resistance to nuclear proliferation due to the presence of the ^{232}U isotope that decays rapidly ($t_{1/2}$ = 73.6 y) to hard gamma active daughters (^{212}Bi, 0.7–1.8 MeV, and ^{208}Tl, 0.72–2.6 MeV) [3]. The spent fuel reprocessing or re-fabrication are not that easy as it needs elaborate remotization engineering for handling the hard gamma active fuel. This engineering blockade is helpful in manufacturing proliferation free thorium blended fuel. The Th-based fuel, (Th, Pu)O_2 is thus superior for Pu-incineration as compared to (U,Pu)O_2. In the absence of advanced remotization, the usual strategy of thorium-based fuel management is to go for once through cycle. Burnt fuel elements are required to be stored over decades to reduce the activity. Attaining economy in power generation using the once through strategy needs a very high discharge burnup (>60 GWd T^{-1}). The fuel containment for achieving high burnup is an issue to be addressed for such fuel.

Many countries, particularly those rich in thorium resources, have focused attention on the research and development of the thorium-based fuels. In the past three decades the Th-based fuel cycles have been studied extensively in Germany, India, Japan, Russia, United Kingdom, and United States of America, and significant experiences have been gained on the performance of the fuel in power generation and breeding. It is generally seen that without making radical change in the configurations of the presently used power reactors like PWR, PHWR, VVER, and HTGR [3–8] or change in their operation strategies, the thorium-based fuels in oxide, alloy, or carbide forms can be used and significantly high burnup (100 GWd ton^{-1} or more) can be attained. Light water breeder reactor (LWBR) concept has successfully emerged [8] by using the fuel assemblies containing the seed of HEU/Pu fissile components and blankets of thorium-based material. A number of other reactor concepts have emerged with the fuel: (a) light water reactors based on the mixed oxides (Th,Pu)O_2 (Pu < 5 %), (Th,U)O_2 ($^{233}U/^{235}U < 5$ %), and ThO_2 matrices in pellet or microsphere forms, (b) high-temperature gas-cooled reactors using SiC and pyrolytic graphite coated fuel particles of dicarbides and oxides of Th/HEU, Th/^{233}U, and Th/Pu in pebble bed and prismatic configurations, (c) light water cooled advanced heavy water reactors (AHWR) with sub-critical core of Th/^{233}U oxide self-sustained by a few seed regions of the conventional mixed oxide fuel (containing Pu/$^{233}U/^{235}U < 4$ %) under an overall negative void coefficient, (d) fast reactors with the mixed oxide cores, (Th,Pu/U)O_2 (Pu/$^{233}U/^{235}U \sim 25$ %), and thoria blankets or with the alloy core, e.g., (Th + Zr + 10 % Pu), and thoria/thoria-urania blanket (e) molten salt reactors with breeding concept, and (f) accelerator driven reactor systems (ADS) employing spallation neutrons for ^{233}U breeding in a sub-critical core of Th. Advanced CANDU reactor (ACR) is designed for operation with slightly enriched fuels (SEU) such as about 2 % enrichment for 21 GWd ton^{-1} burn up, or 4 % for future operation up to 45 GWd ton^{-1}.

In USA, the investigation and utilization of thorium dioxide and thorium dioxide-uranium dioxide (thoria–urania) solid solutions as nuclear fuel materials have been successfully conducted at the Shipping port Light Water Breeder Reactor [8]. Experience with ThO_2 and ThO_2–UO_2 fuels have been carried out at the Elk River (Minnesota) Reactor, the Indian Point (N.Y.) No. 1 Reactor, and the HTGR (High-temperature Gas-cooled Reactor) at Peach Bottom, Pennsylvania, and a commercial HTGR at Fort St. Vrain in Colorado. Recent reviews that take into consideration of pros and cons of going for thorium-based fuels in industrial scale could be seen in [3–9].

India accounting one-fourth of the world's thorium reserve and with about six times more Th than U has aimed at the thorium utilization for large-scale energy production [9, 10]. The utilization program is being implemented through three-stage concept: ^{239}Pu generation from uranium in pressurized heavy water reactors (PHWR), ^{233}U breeding in ^{239}Pu based fast breeder reactors (FBR), and ^{233}U burning for power production. India is also developing advanced heavy water reactor (AHWR) for deriving power directly from thorium through insitu breeding of ^{233}U. AHWR is a new concept in reactor technology. It is a vertical pressure tube type heavy water moderated reactor that has several passive safety features including the core heat removal by natural convective circulation of boiling light water. Currently, a 300 MWe reactor is being developed using the MOX fuels (Th,^{233}U)O_2 and (Th,^{239}Pu) O_2 in composite cluster of 54 pins in circular array with slightly negative void coefficient of reactivity. The fissile content of the pins is kept below 4 wt%. This reactor will derive most of its power from thorium with no external input of ^{233}U in the equilibrium cycle [10].

Many of the Th-utilization schemes for the ^{233}U breeding and power generation involve the uses of the thoria or thoria-based fuels. Usage of the oxide matrix is primarily due to the fact that there is vast experience with oxide fuel in thermal as well as fast reactors. The performance of urania, plutonia, and their solid solutions as reactor fuel is well established. The procedures of the fuel fabrication, storage as spent fuel, reprocessing, and waste management are proven for over so many decades. The fabrication and handling of the oxides are easier than the carbide, nitride, or metallic fuels. The carbide fuel fabrication, for example, needs meticulous control on oxygen and moisture contents of the inert gas as carbide is highly pyrophoric and susceptible to oxidation and hydrolysis. The spent carbide fuel reprocessing is equally problematic as it is difficult to dissolve in nitric acid and the dissolution leaves behind organic complexes. The uses of carbide, nitride, or metallic alloy fuels are generally considered as advanced concepts to cater the strategic need of compact reactor core to achieve high breeding gain, and disposition of weapon grade Pu.

Like the case of the conventionally used oxide fuels, the thoria-based fuels do not pose any difficulty in handling and fabrication in the virgin state. Thoria does not get oxidized or easily hydrolyzed, and as compared to urania it has better chemical stability and desirable thermophysical and radiation resistance properties which ensures better in-pile performance and a more stable waste form. However, as mentioned, the thoria-based spent fuel handling is exceptionally difficult owing to the presence of hard gamma emitting nuclei. Under such situation one adopts extensive burning inside reactor so that the overall economy in the Th-utilization program is met in the once through cycle. Based on the success of attaining high burnup (50–100 GWd ton^{-1}) in the exploratory runs in experimental reactors, the present target is to achieve the same on commercial basis in PWR/PHWR and FBR configurations.

For designing nuclear reactors based on thoria-based fuels it is necessary to have a thorough analysis of the fuel performance using proven simulation code and reliable database of the thermophysical and chemical properties of the fuel in its virgin as well as high burnup states. The wealth of information meanwhile noted from the irradiation studies of thoria-based fuels in the experimental reactors [8] will certainly provide the verification points of the simulation results. An important aspect in the simulation analysis will be the evaluation of the fuel-clad integrity or faultless containment of fuel inside clad. Such evaluation is quite established for the case of the conventionally used urania fuel, but it is not that much as will be called for the commercial implementation of the thoria-based fuels. The input of reliable physical and chemical information of thoria-based fuels in the performance analysis will strengthen the predictability of fuel behavior under the normal course of years' long burning process inside reactor and also under off-normal situations like fuel containment problem due to clad failure out of stress corrosion cracking or loss of coolant accident. In fact, the physico-chemical database of fuel and fission products, and clad are frequently referred while planning the whole fuel cycle program from fuel design and fabrication and reprocessing of irradiated fuel to the management of nuclear waste. For countries like India that has meager resources of natural uranium, the realization of the whole fuel cycle for thoria-based fuels is necessary in order to make use of the fissile isotope ^{233}U bred inside ^{235}U/^{239}Pu fuelled reactors.

The chemistry of the fission products (fps) is principally governed by the matrix within which these are produced. For understanding their chemistry in thoria fuel the available information on the chemical states of fps and their distribution inside urania fuel matrix are useful. The same set of fps with similar yields are formed and settle down inside the two fluorite lattices MO_2 ($M = Th^{+4}, U^{+4}$) with similar crystal radii of the actinide cations. The yields of the fps for different fuels as given in Table 1 of "Thermochemistry of Thoria Based Fuel and Fission Products Interactions" subscribe to the general similarity in the two cases with the exceptions that the thoria-based fuel results in comparatively more gaseous and less metallic fps. Nevertheless, there are some distinctive features in thoria. Chemically, the distinctiveness originates from the rigidly four valency of Th in its compounds in condensed phases. This contrasts with U which is known to acquire higher valencies (four to six) in its oxides and compounds with alkali and alkaline earth fission products.

With increasing the oxygen partial pressure, urania undergoes oxidation from the stoichiometric UO_2 to the hyperstoichiometric composition UO_{2+x} whereas this aspect is absent in thoria. The valency rigidity of thorium results in increasingly less buffering of fission released oxygen and hence development of higher oxygen pressure in the thoria rich $(Th,U)O_2$ fuels during their burnup. For the same reason the oxygen transport in thoria rich matrix is expected to be predominantly by self diffusion unlike the case in urania where the oxygen makes much faster transport through the chemical affinity driven diffusion process [11] to reach clad like zircaloy for its oxidation. The local regulation of fission generated oxygen owing to the impeded transport and poor buffering action in the fuel matrix practically rests on oxidation of the fps.

As against the above-mentioned undesirable features of furthering the fps' oxidations in the irradiated fuel matrix, the thermophysical properties of thoria are superior in many respects to urania. The thermodynamic and kinetic analyses frequently refer to steady state as well as transient thermal profiles in the fuel pin. Thermal conductivity as a function of temperature and fuel composition is an important property in the analyses as it helps in establishing thermal profile at a given power rating. The thermal diffusivity ($\kappa = \lambda/\rho C_v$) derivable from the conductivity (λ), heat capacity (C_v), and density (ρ) is useful in calculating the relaxation time of thermal transients in power ramp when the thermal profile shoots up for a while resulting in augmented thermal stress in the fuel pin and promoting rapid redistribution and release of the gas and volatile fps. The thermal expansion properties help in understanding the fuel dilation relative to the clad and also in the analysis of fuel integrity in presence of thermal stress. The thermal stress developed over temperature differential ΔT is expressed by $E\alpha\Delta T$ [12], where E is modulus of elasticity.

As compared to urania, the fuel dilates less and conducts more heat under a given temperature gradient. On the basis of available data of thermophysical properties of pure thoria and urania a comparative representation of the two material properties is included in Fig. 1 [13]. A look in Fig. 1 indicates that the thermal diffusivity ($\kappa = \lambda/\rho C_v$) of thoria is even higher than urania so that under power ramp the thermal relaxation will be faster in thoria. Dutta et al. [13] has further evaluated thermal profiles of thoria and urania fuel pellets using the Code FAIR-TFC and one of their results is included in Fig. 2. These add to the merits of the thoria-based fuel. On the transport properties of oxygen, and gaseous and volatile fission products in urania and thoria matrices the reported information suggest subtle difference. The distinction in oxygen transport in the two oxide matrices has been indicated already. The combined involvement of vacancy and interstitial in the self diffusion of oxygen is reflected in the reported activation energy for O atom in the two oxide matrices; the energy barrier is higher in thoria (2.8 eV) than in urania (2.6 eV) [11, 14]. The reported value of anion interstitial migration energy (Q_i) is significantly higher in thoria (3.27 eV) than in urania (2.6 eV) and vacancy migration energy (Q_v) is comparable (~0.8–1.0 eV) [11, 14]. The gaseous and volatile species are expected to diffuse using interstitial and vacancy sites in the lattice. Th remaining strictly tetravalent in its oxide, the electronegative species such as I and Te show distinction in their diffusion behaviors in the two oxide matrices. In urania the diffusion is significantly influenced by O/M ratio; the oxygen hyper-stoichiometry augments the diffusion. Thermophysical properties of relevance to the evaluation of performance of thoria as fuel matrix are included in Table 1.

As for the fabrication, reprocessing, and waste management aspects of the thoria-based fuels there are again subtle distinction with urania. The fluorite phase of ThO_2 in highly sintered state is chemically inert and this pose problem in acid

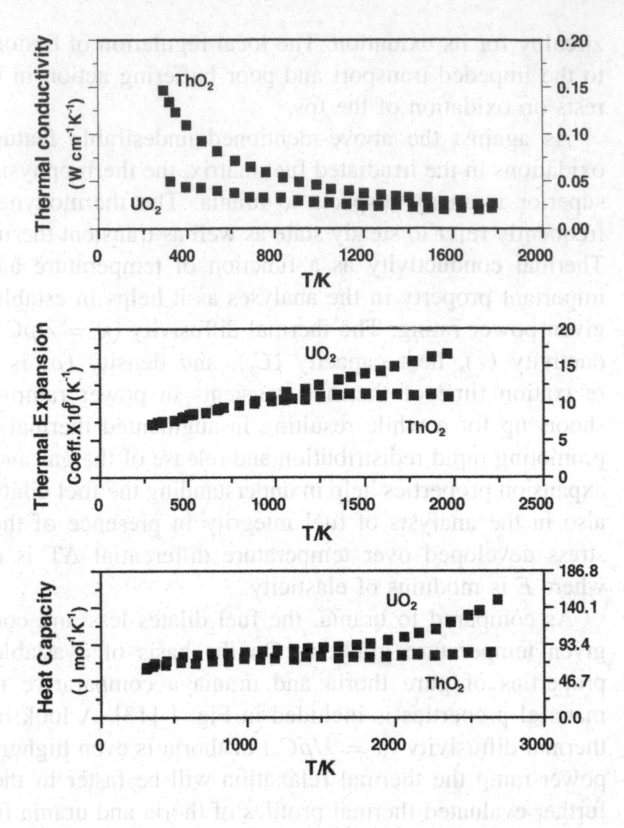

Fig. 1 Thermophysical properties of thoria vis-à-vis urania

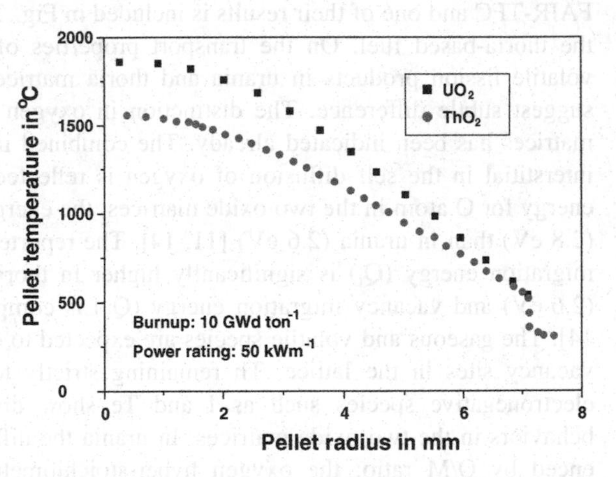

Fig. 2 Thermal profiles in thoria and urania based fuels

dissolution of irradiated fuel in its reprocessing. For improving the dissolution behavior defects are introduced in the fluorite lattice by doping with aliovalent oxides like magnesia, niobium oxide. The doping aids the sintering property also. The highly sintered state can be achieved then at significantly lower temperatures

Table 1 Relevant thermophysical properties of thoria as fuel matrix

Crystal structure	Fluorite (CaF_2 type) structure, space group $Fm\bar{3}m$ (No. 225) [15]
Lattice parameter	559.730(3) pm [15]
Theoretical density	9.9994×10^3 kg m^{-3}
Linear thermal expansion	$\Delta L/L_0 = -0.179 + 5.097 \times 10^{-4}(T/K) + 3.732 \times 10^{-7}(T/K)^2 -7.594 \times 10^{-11}(T/K)^3$ [15]
Thermal conductivity	$1/(A + BT)$, $A = 4.20 \times 10^{-4}$ mKW^{-1}, $B = 2.25 \times 10^{-4}$ mW^{-1} [15]
Zero pressure bulk modulus at 298 K, and its pressure coefficient	196 GPa, and 5 respectively [16–19]
Bulk modulus at different porosity (volume) fractions (f_p, 0.06–0.4) and temperatures (T, 298–1300 K)	196 (1–2.21 f_p) GPa [17] 196 [1.0230–14.05 × 10^{-5}T Exp(−181/T)] GPa [17]
Tensile strength at 298 K	0.082–0.102 GPa [17]
Shear modulus 96.9 (1–2.12 f_p) GPa at 298 K [17, 18], f_p is porosity fraction	Rupture Modulus (MPa) = 440.963 $d^{-0.3578}$ Exp(4.0858 f_p) [17], d = mean dia. of grains in microns, f_p = porosity fraction
Standard enthalpy of formation at 298.15 K	−1226.4 ± 3.5 kJ mol^{-1} [15]
Standard entropy at 298.15 K	65.23 ± 0.20 J $K^{-1}mol^{-1}$ [15]
Standard heat capacity	$C_p^0 = 55.962 + 51.2579 \times 10^{-3}\,T - 36.8022 \times 10^{-5}\,T^2 + 9.2245 \times 10^{-9}\,T^3 - 5.74031 \times 10^5/T^2$ J K^{-1} mol^{-1} (298 ≤ T ≤ 3500 K), [13] and C_p^0(melt) = 61.76 J K^{-1} mol^{-1}(estimated value) [15]
Melting point, heat of fusion	3651 ± 17 K [15], 90 kJ mol^{-1} [15]
Sublimation paths vapor pressures of the sublimates (2400–2800 K)	$ThO_2(s) = ThO(g) + O(g)$, $ThO_2(s) = ThO_2(g)$ log (p_{ThO}/Pa) = −36860/T(K) +13.15 [15] log (p_O/Pa) = −36800/T(K) +12.56 [15] log (p_{ThO2}/Pa) = −35070/T(K) + 12.96 [13]
Self diffusion coefficient of oxygen	log ($D/m^2\,s^{-1}$) = −14362/T–3.35 [11]
Self diffusion coefficient of Th/U	log ($D/m^2\,s^{-1}$) = −32715/T(K) −4.30 [11]
Defects migration energies	Anionic Vacancy [14] (0.78 eV) Cationic vacancy (7.04 eV) [14] Interstitial [14] (3.27 eV) Intertitialcy [14] (0.92 eV)

(~1873 K) and shorter period (~5 h) of thermal programming. Additional research inputs are involved in the fabrication, and reprocessing of thoria-based fuels. Similar inputs are also involved for incorporation of thoria containing nuclear waste inside the glass matrix for immobilization.

This book will cover the essential information on thermophysical, thermodynamic, and transport properties of oxide fuels with particular reference to thoria-based fuels. Besides it will cover the front and back end operations such as the procedures of fuel fabrication and characterization, and reprocessing and waste management of the reactor irradiated fuels. The thermophysical information includes the thermal conductivity/diffusivity and thermal expansion properties, heat capacity, and phase stability of the oxide fuels, oxygen potentials of the fuels, transport property of fission generated oxygen for understanding the possibility of its redistribution in fuel pin and uptake by clad, chemical states of the fission products and their distributions inside the pin, transport and release properties of the fission products xenon, and corrosive volatiles like iodine and tellurium. On fuel fabrication the book will cover fabrication procedures for different fuels such as ThO_2, $(Th,U)O_2$, $(Th,Pu)O_2$, and also for non-oxide fuels for the sake of comparison. Different fabrication routes such as the conventional as well as modified powder-pellet route, sol–gel microsphere pelletization (SGMP) route, pellet impregnation route, and coated agglomerate-particle (CAP) route will be described. The procedures for fuel pin fabrication, compositional, and microstructure characterization of the fabricated pellet/particulate, and also the safety aspects of handling thoria-based fuels will be outlined. On fuel reprocessing, the book will cover radiological problems encountered in irradiated thorium fuel reprocessing to recover ^{233}U/Th and the details of Thorex process steps comprising of fuel decladding, dissolution, solvent extraction by TBP to recover ^{233}U alone or both ^{233}U and thorium, final purification of ^{233}U product and its conversion to oxide. Variations and options in Thorex process to meet the different objectives and the possible areas of improvements in Thorex process and forthcoming developments will be included. The last chapter of the book will present an overview of various types of waste streams, namely, low, intermediate, and high-level radioactive solid and liquid wastes that are generated in the spent fuel reprocessing. Various techniques used in the treatment of the radioactive wastes and safe disposals procedures of the treated wastes adopting different strategies for the LLW/ILW/HLW will be described. Elaboration will be made on the development and characterization of glass matrices used for immobilization of radionuclides present in the high-level waste. Aspects in the development of barium-borosilicate matrix for handling large concentration of sodium, iron, and sometimes sulfate present in the waste will also be covered.

The four chapters that immediately follows deal with the essential features of thermophysical, thermodynamic, and transport properties of thoria-based fuels, while the last three chapters provide the essentials of fabrication, reprocessing, and waste management of the fuel. In all the descriptions special reference has been made with the relevant features of the conventional urania/urania–plutonia fuels in order to bring home the merits and demerits of the thoria-based fuels.

References

1. Jayaram KMV (1985) An overview of world thorium resources, incentives for further exploration and forecast for thorium requirements in the near future. In: IAEA-TECDOC-412 Vienna, p 7
2. IAEA-TECDOC-412 (1985) Summary of the panel discussion. IAEA, Vienna, p 139
3. Greneche D, Szymczak WJ, Buchheit JM, Delpech M, Vasile A, Golfier H (2007) Rethinking the thorium fuel cycle: an industrial point of view. In: Proceedings of ICAPP nice. France, 13–18 May, Paper 7367
4. IAEA TECDOC-1155 (2000) Thorium based fuel options for the generation of electricity: developments in the 1990s. IAEA, Vienna
5. Kazimi MS (2003) Thorium fuel for nuclear energy. Am Sci 91:408–415
6. Benedict M, Pigford TH and Levi HW (1981) Nuclear chemical engineering, 2nd edn. Thorium, McGraw-Hill, pp 283–317. ISBN; 0-07-004531-3. (Chapter 6)
7. Indian Nuclear Society (2001) Conference Proceedings, vol 2
8. IAEA-TECDOC-1450 (2005) Thorium fuel cycle-potential benefits and challenges. IAEA, Vienna
9. Chidambaram R (1990). In: Srinivasan M, Kimura I (eds) Proceedings of the Indo-Japan seminar on thorium utilization, Bombay, p 7
10. Sinha RK, Kakodkar A (2006) Design and development of AHWR—the Indian thorium fuelled innovative nuclear reactor. Nucl Eng Des 236:683–700
11. Matzke HJ (1992) Diffusion processes in nuclear fuels. In: Agarwala RP (ed) Diffusion processes in nuclear materials, North Holland
12. DOE Fundamentals Handbook DOE-HDBK-1017/2-93 (1993)
13. Dutta BK (1999) Evaluation of thermal properties of ThO_2 and $Th_{1-y}U_yO_2$ based on literature survey. In: Venugopal V. et al (eds) Annual progress report, task force on 'Thermodynamic and Transport Properties of Thoria based Fuels', BARC I/007, pp 9–12
14. Shiba K (1992) Diffusion processes in thoria and thorium based oxides. In Agarwala RP (ed) Diffusion processes in nuclear materials, North Holland
15. Bakker K, Cordfunke EHP, Konings RJM, Schram RPC (1997) Critical evaluation of the thermal properties of ThO_2 and $Th_{1-y}U_yO_2$ and a survey of the literature data on $Th_{1-y}Pu_yO_2$. J Nucl Mater 250:1–12
16. Idiri M, Le Bihan T, Heathman, Rebizant J (2004) Behavior of actinide dioxides UO_2 and ThO_2 under pressure. Phys S Rev B70:014113
17. Sah DN (1999) Property models of the ThO_2 fuel for code PROFESS. In: Venugopal V et al (eds) Annual progress report, task force on 'Thermodynamic and Transport Properties of Thoria based Fuels', BARC I/007, pp 13–15
18. Olsen JS, Gerward L, Kanchana V, Vaitheeswaran G (2004) The bulk modulus of ThO_2—an experimental and theoretical study. J Alloys Comp 381:37–40
19. Macedo PM, Capps W, Watchman JB (1964) Elastic constants of single crystal ThO_2 at 25 °C. J Am Ceram Soc 47:651

Thermophysical Properties of Thoria-based Fuels

T. R. Govindan Kutty, Joydipta Banerjee and Arun Kumar

Abstract The behavior of nuclear fuel during irradiation is largely dependent on its thermophysical properties and their change with temperature and burnup. Experimental data on out-of-pile properties such as melting point, density, thermal conductivity, and thermal expansion are required for fuel design, performance modeling, and safety analysis. The variables that influence the out-of-pile properties are fuel composition, temperature, porosity, microstructure, and burnup. Among the above-mentioned properties, thermal conductivity of nuclear fuel is the most important property which influences almost all the processes such as swelling, grain growth, and fission gas release, and limits the linear power. The changes in thermal conductivity occur during irradiation by the formation of fission gas bubbles, porosities, build-up of fission products, and by the change of fuel stoichiometry. Melting point plays a crucial role in determining the power to melt the fuel and decides the operating linear heat rating. The coefficient of thermal expansion (CTE) is needed to calculate stresses occurring in the fuel and cladding on change in temperature. In safety analysis, the values of thermal expansion data are required in determining the gap conductance and the stored energy.

1 Introduction

The behavior of nuclear fuel during irradiation is largely dependent on its physicochemical properties and their change with temperature and burnup [1]. Thermal conductivity is an important parameter to understand the performance of the fuel

T. R. Govindan Kutty (✉)
Formerly at Radiometallurgy Division, Bhabha Atomic Research Centre,
Mumbai 400085, India
e-mail: trgovindankutty@gmail.com

J. Banerjee · Arun Kumar
Radiometallurgy Division, Nuclear Fuels Group, Bhabha Atomic Research Centre,
Mumbai 400085, India

D. Das and S. R. Bharadwaj (eds.), *Thoria-based Nuclear Fuels*,
Green Energy and Technology, DOI: 10.1007/978-1-4471-5589-8_2,

pins under irradiation [2]. It is highly dependent on physical structure, state, chemical composition, and is one of the most important properties for predicting fuel and material performance [3]. If the thermal conductivity is low, the temperature gradient in the radial direction of the fuel pellet is large which results in high temperature at the central part of the fuel pin [2, 3]. The thermal conductivity of nuclear fuel influences almost all important processes such as fission gas release, swelling, grain growth etc. and limits the linear power [4, 5]. The changes in thermal conductivity occur during irradiation by the formation of fission gas bubbles, build-up of fission products, and by the change of oxygen-to-metal ratio (O/M) [6]. Hence, the knowledge of thermal conductivity is needed to evaluate the performance of nuclear fuels. The coefficient of thermal expansion (CTE) values is needed to calculate stresses occurring in the fuel and cladding. If the thermal expansion varies considerably between the fuel and cladding, then stresses will be accumulated during the thermal cycling [7]. This can lead to deformation of the cladding and eventually may result in the breakage of the cladding. Hence, precise evaluation of CTE data of the fuel is needed.

Other important thermophysical properties to be considered are melting point and density. Thorium and uranium oxide fuels used in nuclear reactors have very high melting points, but low density and they suffer from poor thermal conductivity, because in these insulating oxides only phonons (lattice vibrations) conduct heat. Understanding the physics underlying transport phenomena due to electrons and lattice vibrations in actinide systems is an important step toward the design of better fuels [8].

Thermophysical properties of materials depend on various factors, such as microstructure, porosity and its distribution, thermal treatment employed, production technology used, radiation exposure undergone, and other unidentified factors leaving aside the temperature effect [1]. Improving the technology for nuclear reactors through better computer codes and more accurate data of materials property, which can contribute to improved performance as well as economics of future plants by getting rid of currently used large design margins. Accurate representations of thermophysical properties under relevant temperature and neutron fluence conditions are therefore, necessary for evaluating reactor performance under normal operation and accidental conditions [2].

Prior to deploying new fuels and structural materials in a nuclear reactor, its thermophysical properties must be known. The fuel temperature is determined by the thermal conductivity. Such properties of the fuel are not constant during the irradiation period in the reactor, but change with the burnup. Therefore, the evaluation of the thermophysical properties of fuel, including a reliable uncertainty assessment, is required by the nuclear reactor design [3]. A high confidence level on the fuel performance can only be reached from a good interpretation of the irradiation data followed by post-irradiation examinations. A prerequisite for this is to have data on out-of-pile properties such as thermal conductivity or thermal diffusion that allows to understand the influence of parameters such as temperature, temperature gradient, stress, stress gradient, fission rate, and impurities that are effective during the operation. Safety analyses are required by regulatory authorities to prove that the fuel can be burned safely in the reactors. These safety analyses require calculations with safety codes that need the appropriate thermophysical properties of the fuel. These important informations are used by thermohydraulic codes to define operational aspects and to assure the safety, when analyzing various potential accidental scenarios. For each property, the variables that influence the property are to be described, followed by a review of the available data and correlations. Variables considered are temperature, composition, porosity (*p*), burnup (*B*), and oxygen-to-metal (O/M) ratio [9, 10].

Due to its higher thermal conductivity, during normal operation, ThO_2-based fuel will operate with somewhat lower fuel temperatures and release less fission gas than UO_2 fuel at corresponding powers and burnups. This will allow for higher prepressurization and thereby minimizes cladding creep down and fuel cladding mechanical interactions at high burnup and thereby possibly allow for higher burnup use of this material. During an accident such as a large loss-of-coolant accident (LOCA), ThO_2–UO_2 fuel will have less stored energy but a slightly higher internal heat generation rate than UO_2 fuel at similar power levels [11]. As a result, certain parameters for accident evaluation such as the maximum cladding temperature and the timing of fuel rod rupture are expected to be slightly different [8]. These expected differences in behavior between ThO_2–UO_2 fuel rods and UO_2 fuel rods need to be quantified for an objective evaluation of the performance of ThO_2–UO_2 fuel. The mixed ThO_2–UO_2 fuel reduces the amount of total plutonium production by a factor of 3.2 and the ^{239}Pu production by a factor of about 4, when compared with conventional UO_2 fuel irradiated to 45 GWD/t [9]. The plutonium that is produced in the mixed ThO_2–UO_2 fuel is high in ^{238}Pu, producing copious amounts of decay heat and spontaneous neutrons making it proliferation resistant. A mixture of ThO_2 and UO_2 is much more resistant to long-term corrosion in air or oxygenated water than UO_2. Thus, ThO_2–UO_2 is a superior waste form if the spent fuel is slated for direct disposal rather than reprocessing.

2 Melting Point

A very important thermophysical property to be considered for an engineering material, like nuclear fuel, is its melting point. The onset of melting at the centerline of the fuel rod has been widely accepted as an upper limit to the allowable thermal rating of nuclear fuel elements [11, 12]. The melting point must be taken into account when considering a new fuel, as it limits the power that can be extracted from the fuel element. The knowledge of the melting point is also important in the fabrication of chemically homogeneous pellets like thoria–urania since ThO_2 (3,663 K) and UO_2 (3,100 K) have high melting points and relatively low diffusion coefficients at normal sintering temperatures [13].

As a pursuit for the better fuel, it is crucial to understand the underlying transport phenomena due to electrons and lattice vibrations in actinide systems.

According to the Lindemann criterion, solids with large Debye frequencies have high melting points [14]. This is typically found in insulators where atomic bonds are strong due to lack of free electrons. Thorium contains no occupied 5f states while uranium has two unpaired 5f valence electrons, and therefore uranium and thorium possess very different electronic and chemical properties. Since melting point is an important property, it is worth considering from the standpoint of the bonding present in actinide elements [15]. The highest melting points for the actinide metals are for Th and Pa metals. The effect of f-orbitals on the melting point is maximized with Np and Pu; both have very low melting points, which are believed to be a reflection of f-orbital repulsion [16]. Uranium has multiple oxidation states (3+ through 6+) which allow UO_2 to be easily oxidized to U_3O_8 or UO_3, by incorporating interstitial O atoms. In contrast, thorium only exhibits one oxidation state (4+) and hence cannot be oxidized beyond ThO_2 [9, 17]. Among the actinide oxides, only ThO_2 is a white insulating solid and the other AnO_2 solids are all dark and opaque and poorly conducting. While in UO_2 the 5f electrons, which occupy an energy band from 1.37 to 1.50 eV cause a strong visible light absorption resulting in a dark gray color of this oxide. The consequent absence of low electrons in the valence band is the cause of the high transparency of stoichiometric thoria and the low spectral emissivity in the visible range at room temperature [17].

Thorium dioxide exists up to its melting point as a single cubic phase with the fluorite crystal structure, isomorphous, and completely miscible with UO_2. Unlike UO_2, ThO_2 does not dissolve oxygen to a measurable extent. Therefore, it is stable at high temperature in oxidizing atmosphere. On prolonged heating to 1,800–1,900 °C in vacuum, it blackens with loss of oxygen, although the loss is insignificant to be reflected in lattice parameter or in chemical analysis. On reheating in air to 1,200–1,300 °C, the white color is restored.

The melting points of the nuclear fuels are shown to be influenced by the following factors: stoichiometry and composition, irradiation dose, impurities and their contents.

2.1 ThO_2

The melting point of ThO_2 was experimentally measured or estimated by several authors [18–26]. Their results are summarized in Table 1. As it can be seen from the Table, the reported values vary from 3,323 to 3,803 K. Peterson and Curtis [26], in their compilation of data on thorium-based ceramics, arrived at two different values, e.g., 3,573 ± 100 K from the work of Lambertson et al. [21] on ThO_2–UO_2 system and 3,663 K from the work of Benz [22] on Th–ThO_2 system. Lambertson et al. [21] first estimated the melting point of ThO_2 to be between 3,558 and 3,828 K and subsequently arrived at an intermediate value of 3,623 K by extrapolating the melting point data of (Th,U)O_2 compositions corresponding to zero UO_2 content. They further refined their data by introducing some corrections

Table 1 Melting point of ThO_2 determined by various authors

Year	Author	Melting point (*K*)
1929	Ruff et al. [18]	3,323
1932	Wartenberg and Reusch [19]	3,803
1953	Geach and Harper [20]	3,323 ± 25
1953	Lambertson et al. [21]	3,573 ± 100
1969	Benz et al. [22]	3,663 ± 100
1970	Christensen [27]	3,543
1972	Chikalle et al. [23], TECDOC-1319	3,573
1975	Rand [25]	3,643 ± 30
1984	Belle and Berman [12]	3,640 ± 30
1992	ITU [25]	3,640 ± 20
1996	Ronchi and Hiernaut [24]	3,651 ± 17

for the liquidus/solidus curve to effect a curvature correction for the pure ThO_2 end to that of pure UO_2 end of the temperature—composition diagram. Their final recommended data was 3,575 K, which is in good agreement with the data 3,543 K recommended by Christensen [27] from his experimentally measured melting point data on ThO_2–UO_2 system and subsequent extrapolation to zero UO_2 content. Rand [25], however, disagrees with the curvature corrections made by others on the thoria or urania rich side of the temperature composition curve. He justified that the curvature need not be same at both the terminal compositions and the difference could be due to the loss of 'O' from UO_2 in urania-rich side, which is different for thoria-rich side. He recommended a value of 3,643 ± 30 K. Belle and Berman [12] used 3,640 K as the melting point of ThO_2, recommended by Rand [25] in his work on ThO_2. Ronchi and Hiernaut [24] had recently measured the melting temperature of ThO_2 (both stoichiometric and hypostoichiometric) material experimentally by heating a spherical sample by four symmetrically spaced pulsed Nd YAG laser and observing the cooling/heating curve with time. For stoichiometric ThO_2, the measured melting point was found to be 3,651 ± 17 K. The data of Ronchi and Hiernaut [24] reasonably agrees with the data generated by Benz [22] (3,660 ± 100 K) and is also close to that recommended by Rand [25] (3,643 ± 30 K). All these values are markedly different from those of Lambertson et al. [21]. It is also well understood that the curvature difference at the uranium- and thorium-rich side of the temperature versus composition diagram is quite justifiable and was attributed to the loss of oxygen. Hence, the recommended melting temperature of ThO_2 should be taken as 3,651 ± 17 K, and is in fairly good agreement with majority of the previous studies. The value measured at the Institute for Transuranium Elements (ITU) (3,640 ± 20) K, is very close to the value reported by Rand [25].

Measurements of the cooling curves of molten ThO_2 and $ThO_{1.98}$ reveal that the stoichiometric compound melts congruently at 3,651 K, while the hypostoichiometric oxide displays a liquidus at 3,628 K and a solidus transition at 3,450 K. Ronchi et al. [24] conducted pulse-heating experiments on thoria and showed that this compound exhibits a premelting transition at 3,090 K whose features are analogous to those observed in other ionic compounds having fluorite type structures. A class of diatomic compounds which crystallize in the face-centered cubic fluorite lattice (space group Fm3m), and whose component atoms have very different mobilities, exhibit at a temperature corresponding to about 80 % of the absolute melting temperature, a premelting transition. The discovery of this transition in UO_2 [12] gave rise to a number of investigations aimed at defining its nature and possible effects on the high temperature properties of this technologically important material. In ThO_{2-x}, the dependence of the transition temperature, T_c on stoichiometry was found to be weaker than hypostoichiometric urania and exhibits an opposite trend, with T_c decreasing with x.

To conclude, the melting temperature of ThO_2 recommended from this assessment is 3,651 ± 17 K and is in fairly good agreement with majority data available in the literature.

2.2 UO_2

The melting point of UO_2 given in MATPRO [28] is 3,113.15 K. This temperature is based on the measurement made by Brassfield et al. [29] and the equations for the solids and liquids boundaries of the UO_2–PuO_2 phase diagram given by Lyon and Bailey [30]. The recommended values by ORNL [31] for $UO_{2.00}$ are 3,120 ± 30 K and for PuO_2 is 2,701 ± 35 K. This value for UO_2 has been recommended by Rand et al. [32] from their analysis of 14 experimental studies of the melting temperature of UO_2. This recommendation of Rand et al. [32] was accepted internationally and was recommended in the assessment of UO_2 properties by Harding et al. [33] in their 1989 review of material properties for fast reactor safety. The melting point of UO_2, according to Latta et al. [34] is 3,138 ± 15 K, which is considerably higher than that of Christensen (3,073 K) [35]. Belle and Burman [12] recommended the melting temperatures of UO_2 as 3,120 K, with an error of probably ±30 K.

In recent experimental measurements of the heat capacity of liquid UO_2 using laser heating of a UO_2 sphere, Ronchi et al. [36] made several measurements of the freezing temperature of UO_2 on different samples. For specimens in an inert gas atmosphere with up to 0.1 bar of oxygen, they obtained melting points in the interval 3,070 ± 20 K. Higher melting temperatures (3,140 ± 20 K) were obtained for samples in an inert gas atmosphere without oxygen. The variation in melting temperature is in accordance with the expected lower oxygen-to-uranium (O/U) ratio in the latter samples. The melting point of UO_2 drops on variation O/M ratio around stoichiometry: for example, if the melting point of stoichiometric UO_2 is 3,138 K, its value drops to 2,698 K at an O/U ratio of 1.68 and to 2,773 K at an O/U ratio of 2.25. The effect of irradiation is lowering of the melting point of UO_2. At a burnup of 1.5×10^{21} fissions/cm^3, it has been reported [12] that the melting point drops to 2,893 K. Typical values of melting points for UO_2 and PuO_2 obtained by various authors are given in Table 2.

Table 2 Melting point of UO_2 and PuO_2 determined by various authors [31]

	UO_2 (in *K*)	PuO_2 (in *K*)	Remarks
Lyon and Baily [30]	3,113 ± 20	2,663 ± 20	
Aitkens and Evans [146]	3,120 ± 30	2,718 ± 35	
Latta et al. [34]	3,138 ± 15		
Adamson et al. [147]	3,120 ± 30	2,701 ± 35	BU: 0.4 MWD/kg
MATPRO [148]	3,113.15	2,647	BU: 3.2 MWD/kg
Komatsu [149]	3,138 ± 15	2,718	BU: 0.7 MWD/kg
Rand et al. [32]	3,120 ± 30		
Ronchi et al. [36]	3,075 ± 30		
Christensen [37]	3,073		
Belle and Berman [12]	3,120 ± 30		
Belle and Berman [12]	2,893		BU: 1.5×10^{21} fissions/cm^3

BU burnup

2.3 ThO_2–UO_2 System

Phase diagram studies of the ThO_2–UO_2 system have been reported by several authors. Lambertson et al. [21] used a quench technique, Christensen [27, 37] a tungsten filament technique, whereas Latta et al. [34] applied a thermal arrest method. The phase diagram of ThO_2–UO_2 can be constructed with the help of the melting points and enthalpies of fusion of the end members, assuming ideal solid solution behavior in both liquid and solid state. The results of the phase diagram measurements are given in Fig. 1 showing UO_2 and ThO_2 form a continuous series of solid solutions. The measurements of Christensen [27] and Latta et al. [34] show a shallow minimum at 2 and 5 mol% ThO_2, respectively. The melting points calculated for some intermediary composition are shown below in Table 3 [38–49].

There is no direct experimental measurement on the heat of fusion of ThO_2 or ThO_2–UO_2 solid solution. The most probable value is that of Fink for UO_2 as 74.8 ± 1 kJ/mol [50]. The recommended value for ThO_2 is 90.8 kJ/mol [40].

2.4 ThO_2–PuO_2 System

Freshley and Mattys [41, 42] have shown that ThO_2 and PuO_2 form a complete solid solution (Fig. 2) in the whole composition range like ThO_2 and UO_2. A continuous series of solid solution has also been reported by Mulford and Ellinger [44]. They found only a single fluorite structure by X-ray diffraction (XRD) and also showed that the lattice parameter varied linearly with composition.

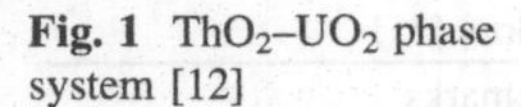

Fig. 1 ThO_2–UO_2 phase system [12]

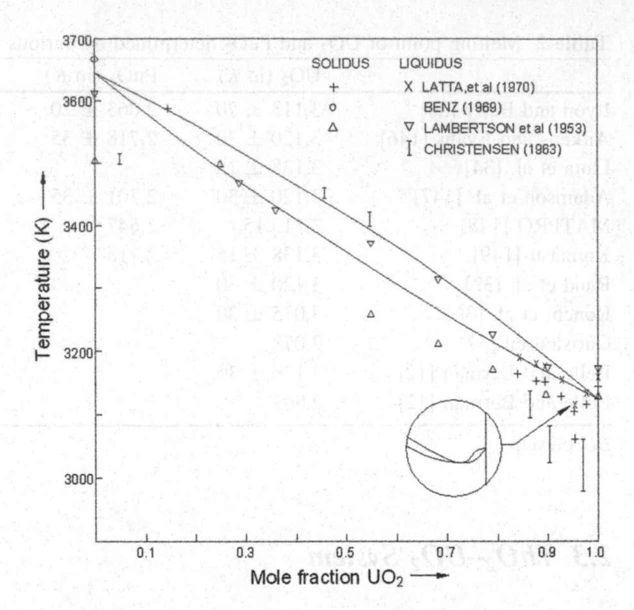

Table 3 Melting temperatures of ThO_2–UO_2 solid solution [12]

Composition (mol% UO_2)	Solidus temperature (*K*)	Liquidus temperature (*K*)
0	3,640	3,640
10	3,580	3,600
20	3,520	3,550
30	3,460	3,510
40	3,410	3,460
50	3,360	3,410
60	3,310	3,360
70	3,260	3,310
80	3,210	3,250
90	3,160	3,190
100	3,120	3,120

Dawson [45] has made magnetic susceptibility measurements on PuO_2 and ThO_2 mixtures and implied that PuO_2 and ThO_2 form solid solutions which follow the Vegard's law. The lattice parameter of fluorite type cubic phase was found to decrease regularly from 0.5597 nm for ThO_2 to 0.5396 nm for pure PuO_2 [43]. Due to the limited amount of experimental data, a more accurate assessment of the phase diagram is not yet possible. The melting point of the ThO_2–PuO_2 solid solutions, containing various amounts of PuO_2, was measured in helium. The melting point of the specimens containing less than 25 % ThO_2 was found to be unchanged as shown in Fig. 3 [42]. The melting point of a PuO_2 sample (whose purity was not specified) used in their study is 2,533 K, which is 130 K lower than the assessed value of Table 2.

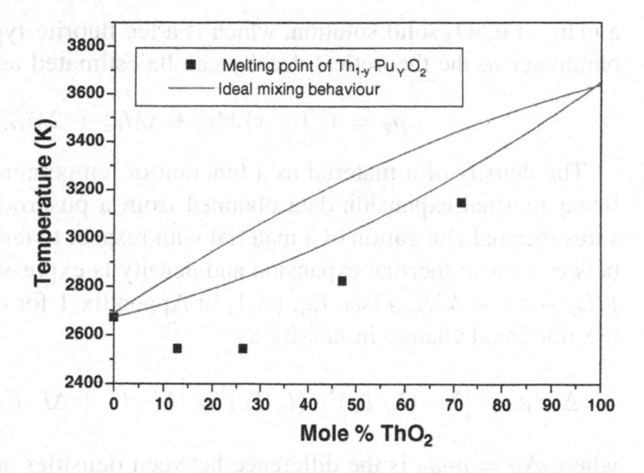

Fig. 2 Phase diagram of ThO_2–PuO_2 system [38]. (permission from Elsevier)

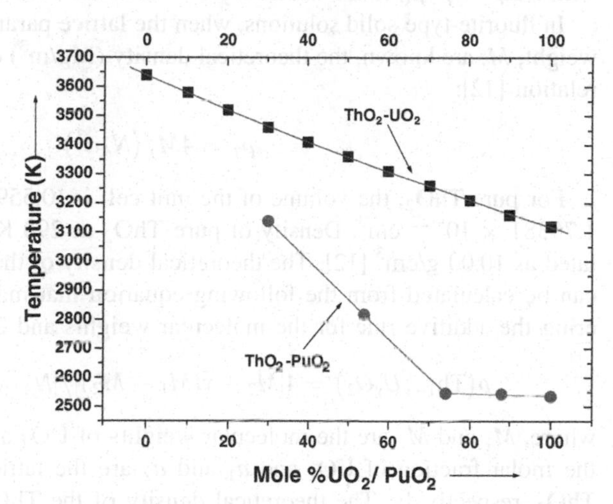

Fig. 3 Melting points of various compositions of ThO_2–UO_2 and ThO_2–PuO_2 system [12, 42]

3 Density

The fuel density, ρ is an important property of the fuel and is a function of the following factors: fuel composition, temperature, amount of porosity, O/M ratio, and burnup. The theoretical densities of the materials (ρ_T) can be calculated from the knowledge of lattice type and values of lattice parameter. Assuming that the elements form a solid solution, the theoretical density of the material can be calculated from the following relation [12]:

$$\rho_T = M_{\text{system}}N/(VN_a), \tag{1}$$

where, M_{system} is the atomic weight of the system, N the number of atoms per unit cell, V the volume of the lattice, and N_a the Avogadro constant. Thus, in the case of a $(Th_{1-x}Pu_x)O_2$ solid solution, which is a fcc fluorite-type structure, with a lattice parameter a, the theoretical density can be estimated as :

$$\rho_T = 4[(1-x)\,M_{\mathrm{Th}} + xM_{\mathrm{Pu}} + 2M_O]/a^3N_a. \quad (2)$$

The density of a material as a function of temperature can be calculated using linear thermal expansion data obtained from a pushrod dilatometer, which measures thermal elongation of a material with respect to temperature (T). The relation between linear thermal expansion and density is expressed [40] as $\rho_0/\rho = (L/L_0)^3$, $L/L_0 \equiv (1 + \Delta L/L_0)$ (see Eq. (A.4) in Appendix 1 for details), so that one writes the fractional change in density as

$$\Delta\rho/\rho_0 \equiv [1-(L/L_0)^3]/(L/L_0)^3 = \left[1-(1+\Delta L/L_0)^3\right]/[1+\Delta L/L_0]^3, \quad (3)$$

where $\Delta\rho = \rho - \rho_0$ is the difference between densities at temperatures T and T_0.

In fluorite-type solid solutions, when the lattice parameter, a, and the molecular weight, M, are known, the theoretical density (Mg/m^3) can be calculated using the relation [12]:

$$\rho_T = 4M/(N_A a^3). \quad (4)$$

For pure ThO_2, the volume of the unit cell is $(0.55975\ nm)^3$ which is equal to $1.75381 \times 10^{-22}\ cm^3$. Density of pure ThO_2 at 298 K may, therefore, be calculated as 10.00 g/cm^3 [12]. The theoretical density of the ThO_2–UO_2 solid solution can be calculated from the following equation that makes use of Eq. (4) considering the additive rule for the molecular weights and cell volumes [39]:

$$\rho(Th_{1-y}U_yO_2) = 4[M_2 + y(M_1 - M_2)]/[N_A(a_2^3 + y(a_1^3 - a_2^3))], \quad (5)$$

where, M_1 and M_2 are the molecular weights of UO_2 and ThO_2, respectively, y is the molar fraction of UO_2 and a_1 and a_2 are the lattice parameters of UO_2 and ThO_2, respectively. The theoretical density of the ThO_2–UO_2 solid solution as a function of UO_2 content at 298 K is shown in Fig. 4.

The recommended equations for the density of solid uranium dioxide are based on the lattice parameter value of 0.54704 nm reported by Gronvold [46] at 293 K and thermal expansion data by Martin [47]. The above lattice parameter values are in good agreement with measurements by Hutchings [49] and are in full agreement with the recommendations of Harding et al. [48]. Assuming the molecular weight of UO_2 is 270.0277, this lattice parameter gives a density at 293 K as 10.956 g/cm^3. Applying the thermal expansion values of Martin [47], the density at 273 K is $10.963 \pm 0.070\ g/cm^3$. The values reported by Benedict et al. [51] and MATPRO for solid UO_2 are 10.970 ± 0.070 and $10.980 \pm 0.020\ g/cm^3$, respectively. Densities of UO_2 and PuO_2 given by various authors are shown in Table 4.

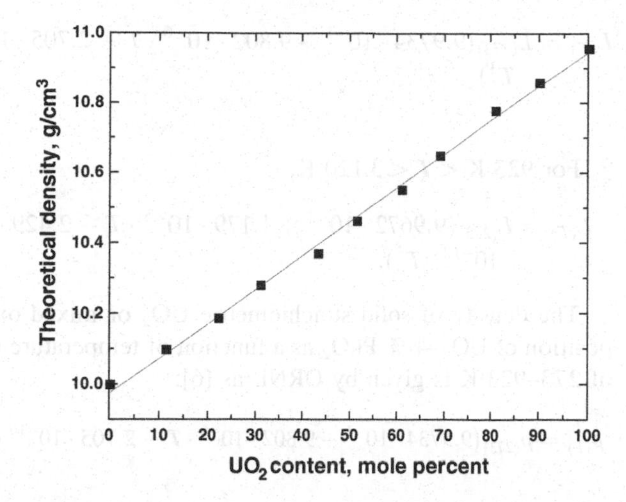

Fig. 4 Theoretical density of the ThO_2–UO_2 solid solution as a function of UO_2 content at 298 K [52]. (permission from Elsevier)

Table 4 Densities of UO_2 and PuO_2

	Density at 273 K (g/cm³)	
	UO_2	PuO_2
Fink [50]	10.963 ± 0.070	
Benedict [51]	10.970 ± 0.070	11.460 ± 0.080
MATPRO [28]	10.980 ± 0.020	

From the data of coefficient of linear expansion, one can evaluate the fractional change in density of UO_2 or ThO_2 as a function of temperature up to their melting point. The following assumptions are considered:

1. There is no volume change on the formation of solid solution or solid solutions are ideal,
2. Densities of the solid solution can be calculated by evaluating the molar volume of the two components at the temperature of interest, and
3. Linear interpolation of the data to the desired temperature.

The density as a function of temperature may be calculated from [40]:

$$\rho_{(T)} = \rho_{(273)} \left[\left(L_{(273)} / L_{(T)} \right) \right]^3, \quad (6)$$

where, $\rho_{(273)}$ is the density at 273 K; $L_{(273)}$ and $L_{(T)}$ are the lengths at 273 K and at temperature T (K), respectively. The ratio of length ($L_{(273)}/L_{(T)}$) can be calculated from the Martin's equation for the thermal expansion of UO_2 [47].

For 273 K < T < 923 K,

$$L_{(T)} = L_{(273)}(9.9734 \cdot 10^{-1} + 9.802 \cdot 10^{-6} \cdot T - 2.705 \cdot 10^{-10} \cdot T^2 + 4.391 \cdot 10^{-13} \cdot T^3) \tag{7}$$

For 923 K < T < 3,120 K,

$$L_{(T)} = L_{(273)}(9.9672 \cdot 10^{-1} + 1.179 \cdot 10^{-5} \cdot T - 2.429 \cdot 10^{-9} \times T^2 + 1.219 \cdot 10^{-12} \cdot T^3). \tag{8}$$

The density of solid stoichiometric UO_2 or mixed oxide (MOX) with a composition of UO_2–4 % PuO_2 as a function of temperature for the temperature range of 273–923 K is given by ORNL as [6]:

$$\rho_{(T)} = \rho_{(273)}(9.9734 \cdot 10^{-1} + 9.802 \cdot 10^{-6} \cdot T - 2.705 \cdot 10^{-10} \cdot T^2 + 4.391 \cdot 10^{-13} \cdot T^3)^{-3} \tag{9}$$

and the density of UO_2 or MOX for the temperature range of 923 K to the melting temperature,

$$\rho_{(T)} = \rho_{(273)}(9.9672 \times 10^{-1} + 1.179 \times 10^{-5} \cdot T - 2.429 \times 10^{-9} \cdot T^2 + 1.219 \times 10^{-12} \cdot T^3)^{-3} \tag{10}$$

The recommended uncertainty in the density value is 1 % in the entire temperature range.

Martin [47] recommends from assessment of the available data on hyperstoichiometric uranium dioxide (UO_{2+x}), using the same equations for the linear thermal expansion of UO_2 and of UO_{2+x} for x in the ranges 0–0.13 and 0.23–0.25. Therefore, Eqs. (9) and (10) are recommended for the density of UO_{2+x} for x in the ranges 0–0.13 and 0.23–0.25. No data on the effect of burnup on density or thermal expansion of UO_2 are currently available. In the absence of data, Eqs. (9) and (10) are recommended for UO_2 during irradiation, in accord with the recommendation of Harding et al. [48]. The density of UO_2 as a function of temperature is shown in Fig. 5.

As mentioned earlier, the phase diagram shows a continuous series of solid solutions between UO_2 and ThO_2. This is supported by the fact that deviations from Vegard's law are within the uncertainties of the lattice parameter measurements. Also, densities of ThO_2–UO_2 solid solution at room temperature can be calculated using additive rule:

$$\rho_{(273)} = 9.99003 + 0.00953 \cdot y \text{ (in g/cm}^3\text{)}, \tag{11}$$

where y is mol% UO_2.

The calculated theoretical densities, obtained from the measured lattice constants, for some ThO_2–UO_2 solid solution are given below in Table 5.

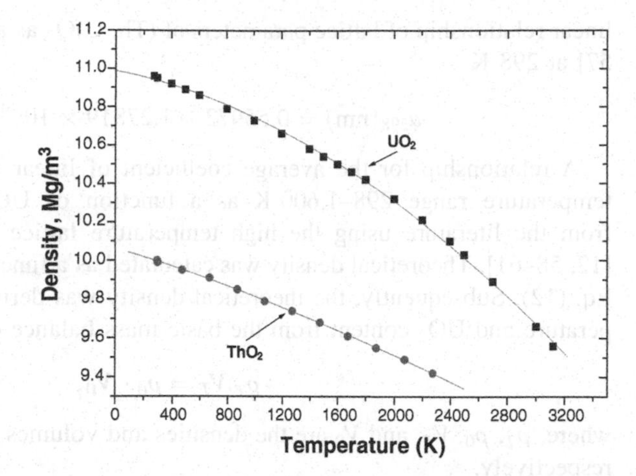

Fig. 5 Density of UO_2 [6, 31] and ThO_2 [40] as a function of temperature

Table 5 Room temperature lattice constants and theoretical densities of ThO_2–UO_2 solid solutions

Composition (mol% UO_2)	Lattice parameter (nm ± 0.00003 nm)	Theoretical density (g/cm^3)
0.0	0.55975	10.00
10.1	0.55846	10.09
20.2	0.55726	10.18
30.1	0.55590	10.28
40.3	0.55475	10.37
50.1	0.55355	10.46
60.1	0.55225	10.55
69.9	0.55098	10.65
80.1	0.54969	10.75
90.0	0.54841	10.85
100	0.54705	10.96

Belle and Berman [12] calculated theoretical density of ThO_2–UO_2 solid solution for different UO_2 content (x) from the lattice parameter data of Cohen and Berman [52]. The following equation shows the relationship between the theoretical density and UO_2 contents.

$$\rho_{(T)}\left(\text{g/cm}^3\right) = 9.9981 + 0.0094 \cdot (x) - 8.7463 \cdot 10^{-6}(x)^2 + 1.1192 \cdot 10^{-7} \cdot (x)^3, \quad (12)$$

where, x is the UO_2 content.

Density of (Th, U)O_2 system has also been calculated as a function of temperature by many authors [12, 40, 53]. Momin and Venketeswarulu [54], Momin and Karkhanwala [55], Kempter and Elliott [56] and Springer et al. [57] reported the densities from the lattice and bulk expansion data. The density of (Th, U)O_2 system as a function of temperature and UO_2 content has been estimated from a linear relationship of lattice parameters of (Th, U)O_2 as a function of UO_2 [12, 38, 57] at 298 K.

$$a_{298}(\mathrm{nm}) = 0.55972 - 1.27819 \times 10^{-4}[\%UO_2] \tag{13}$$

A relationship for the average coefficient of linear thermal expansion in the temperature range 298–1,600 K as a function of UO_2 content was obtained from the literature using the high temperature lattice parameter measurements [12, 58–61]. Theoretical density was calculated as a function of UO_2 content using Eq. (12). Subsequently, the theoretical density was derived as a function of temperature and UO_2 content from the basic mass balance equation, i.e.,

$$\rho_T.V_T = \rho_0 \cdot V_0, \tag{14}$$

where, ρ_T, ρ_0, V_T, and V_0 are the densities and volumes at temperatures T and T_0, respectively.

With the coefficient of thermal expansion, the following equation was derived for the theoretical density [40]:

$$\begin{aligned}\rho_{(T)}(\mathrm{g/cm^3}) = {} & 10.087 - 2.891.10^{-4} \times T - 6.354.10^{-7}(x) \times T \\ & + 9.279.10^{-3}(x) + 5.111.10^{-6} \times (x)^2,\end{aligned} \tag{15}$$

where x is UO_2 content. It is observed that the variation in density obtained from Eq. (15) and that from the literature is within ±0.28 %. The theoretical densities of UO_2 and ThO_2 at different temperatures are given in Table 6.

3.1 Density of Liquid UO_2

The recommended equation for the density of liquid uranium dioxide is based on the in-pile measurements of the vapor pressure, density, and isothermal compressibility of liquid (U, Pu)O_2 by Breitung and Reil [62]. Measurements of density as a function of enthalpy and as a function of temperature were obtained from the melting point to 7,600 K. The equation of Breitung and Reil for the

Table 6 Densities of ThO_2 and UO_2 as a function of temperature [12]

Temperature (*K*)	ThO_2 (g/cm^3)	UO_2 (g/cm^3)
298	10.00	10.96
473	9.95	10.90
673	9.89	10.83
873	9.83	10.76
1,073	9.78	10.69
1,273	9.72	10.62
1,473	9.66	10.55
1,673	9.60	10.48

density of UO_2 and $(U, Pu)O_2$ for mole fractions of Pu ≤ 0.25 is in good agreement with the equation for the density of UO_2 from experiments by Drotning [63], which had been recommended in the 1981 assessment by Fink et al. [64].

The recommended equation for the density of UO_2 as a function of temperature is:

$$\rho = 8.860 - 9.285 \cdot 10^{-4} \cdot (T - 3120), \tag{16}$$

where, density (ρ) is in g/cm^3 and temperature (T) is in K.

No data exists for the volume change of ThO_2 on melting, but some information is available on UO_2. Christensen [65] measured the density of UO_2 between 1,553 and 3,373 K by high-temperature radiography. The density of solid and liquid at 3,073 K were 9.67 and 8.74 g/cm^3, respectively. The accepted value of the density of liquid UO_2 at the melting point is 8.74 ± 0.016 g/cm^3. The volume increase on melting was 10.6 %.

Finally, the burnup also affects the density by the change in the porosity. At low burnup (<15 GWd/t), density increases by the fuel densification process; at the higher burnup, density decreases (porosity increases) due to the fuel swelling [66].

4 Thermal Expansion

As discussed earlier, the performance of a nuclear fuel is highly dependent on its physicochemical properties, especially their variation with temperature. One such property is thermal expansion of the fuel which affects the size of the gap between the fuel outer surface and cladding inner surface. The difference between the coefficients of thermal expansion of the fuel and the cladding determines whether the initial fuel–cladding gap closes or opens when the fuel element is brought to power [1]. If the initial gap is small and the fuel expands more than the cladding, the two come into contact. The resulting pressure at the interface is known as the contact or interfacial pressure. On the other hand, if the cladding expands more than the fuel and the gap is enlarged, heat conduction through the fuel–cladding gap will be low and the fuel temperature will be high because of the thermal resistance of the fuel–cladding gap. If interaction occurs, then differential thermal expansion between the fuel and cladding affects the magnitude of stresses in the cladding and fuel. This will be accounted for by including a thermal expansion strain component in the total strain of the fuel in each of three coordinate directions [1, 2, 4–6]. Fuel melting and excessive thermal expansion of the fuel pellet can occur at high enthalpy levels. The sudden expansion of the pellet can fail the cladding and resulting in molten fuel dispersion that can damage the pressure boundary. Dimensional changes are also affected by the processes such as sintering, bloating, or irradiation-induced changes such as swelling or densification [6]. For cubic crystals, such as ThO_2, UO_2, and ThO_2–UO_2 solid solutions, the expansion or contraction is same in all directions. As long as phase changes do not occur, the dimensional changes are reversible [12].

The pellet–cladding mechanical interaction (PCMI) failures that occurred in the CABRI test reactor (France) and in the NSRR test reactor (Japan) were mainly due to rapid fuel thermal expansion because most of the energy remained in the fuel pellet during the extremely small time scale of the reactivity pulse [67]. With the same amount of energy deposition, thoria fuel will have a higher fuel temperature due to the lower heat capacity [12]. However, this high temperature does not necessarily lead to a larger thermal expansion because of the lower thermal expansion coefficient of thoria fuel. Results have shown that ThO_2–UO_2 fuel will have better performance than UO_2 fuel under Reactivity Initiated Accident (RIA) event conditions due to its lower thermal expansion and flatter power distribution in the fuel pellet [67].

Thermal dilation property of a material has correspondence to lowering in its density with temperature increase. The relative lowering in the density expressed in Eq. (3), leads to the following expression of the density ratio:

$$\rho/\rho_o = \{1 + [1 - (1 + (\Delta L/L_0)^3/(1 + (\Delta L/L_0)^3]\} \quad (17)$$

By Eq. (17), the density ratio can be evaluated from the linear dilation ($\Delta L/L_0$). Conversely, the relative dilation ($\Delta L/L_0$) can be determined using the following relation (see Appendix 1 for details) if density change is known,

$$(\Delta L/L_0) \approx (\rho_0/\rho)^{0.33} - 1 \quad (18)$$

The dilation property in a crystalline solid provides a measure of the extent of defect growth with temperature rise. The thermal expansion $\Delta L/L_0$ of a crystalline solid as per Eshelby theorem [12] can be written as:

$$\Delta L/L_0 = \Delta a/a_0 + 0.33\,\Delta N/N_0, \quad (19)$$

where a_0 is the lattice parameter at the reference temperature, $\Delta a = a_T - a_0$, N_0 is the number of occupied lattice sites at ambient temperature, and ΔN is the number of Schottky defects. The quantity becomes significant only at temperatures >0.6 Tm. By determining both $\Delta L/L_0$ and $\Delta a/a_0$ on the same specimen, it is possible to evaluate the energy of formation of Schottky defects. If, however, the thermally generated defects are of the Frenkel type, then thermal expansion will be same whether measured by XRD or by bulk measurements. The concentration of thermally generated Schottky defects in a cubic crystal can be determined from the above equation [12].

Further, the thermal dilation property is used in the evaluation of thermal stress in solid. In general, the total strain (ε) in a body is the sum of the mechanical strain (ε_M), and the thermal strain (ε_T) [1],

$$\varepsilon = \varepsilon_M + \varepsilon_T \quad (20)$$

$$\text{or, } \varepsilon = \sigma/E + \alpha \cdot \Delta T \quad (21)$$

where E is the Young's modulus and σ is the stress. In Eq. (21), it is assumed that both the elastic modulus, E, and the coefficient of thermal expansion, α, do not vary with temperature. The stresses due to change in temperature or due to temperature gradient are termed as thermal stresses, σ_{thermal} and can be expressed as:

$$\sigma_{\text{thermal}} = \alpha \cdot E \cdot \Delta T \,. \tag{22}$$

UO_2 and ThO_2 have the same isometric structure, and the ionic radii of 8-fold coordinated U^{4+} and Th^{4+} are similar (1.14 and 1.19 nm, respectively) [68]. Yamashita et al. [69] made the comparison for the linear thermal expansion coefficient (α) of actinide dioxides, using their measured values and literature ones at room temperature and at high temperature (1,200 K). They found that at room temperature the α values were almost the same of about $8.5 \times 10^{-6}\,K^{-1}$, but at high temperatures the α values increased with increasing atomic numbers from $10.2 \times 10^{-6}\,K^{-1}$ for ThO_2 to $13.2 \times 10^{-6}\,K^{-1}$ for BkO_2 [69, 70]. The thermal expansion coefficients of ThO_2, UO_2, and PuO_2 reported in the literature are shown in Table 7.

Lattice parameters of ThO_2, UO_2, and PuO_2 reported by various authors at room temperature are shown in Table 8.

It is well known that the lattice parameter (a) of nonstoichiometric UO_{2+x} diminishes with increasing excess oxygen content (x) and that the relation between a and x at room temperature is given as [12]

$$a(\text{pm}) = 547.05 - 9.4\,(x) \tag{23}$$

Values of linear thermal expansion (LTE) at temperature T can be calculated by the relation

$$\text{LTE}_{(T)}(\%) = (a_T - a_0) \times 100/a_0, \tag{24}$$

where, a_T is the lattice parameter at temperature T and a_0 is that at the reference temperature 293 K. The unit cell parameters could be determined as a function of temperature for ThO_2, UO_2, and PuO_2 with help of high temperature X-ray

Table 7 Thermal expansion coefficients of ThO_2, UO_2, and PuO_2 [69]

	ThO_2	UO_2	PuO_2
At 293 K ($\times 10^{-6}$)			
Yamashita et al. [69]	8.43	9.36	9.04
Marples [150]	7.3	9.3	8.4
Fahey et al. [151]	8.21	8.71	8.71
Taylor [71]	7.76	9.01	8.84
TPRC [61]	7.7	9.4	8.1
At 1,200 K ($\times 10^{-6}$)			
Yamashita et al. [69]	10.41	10.76	11.61
Fahey et al. [151]	10.24	12.35	12.14
Taylor [71]	11.00	11.31	12.27
TPRC [61]	10.4	11.6	12.00

Table 8 Lattice parameters of ThO_2, UO_2, and PuO_2 reported by various authors

	ThO_2 (pm)	UO_2 (pm)	PuO_2 (pm)	Year	Remarks
Zachariasen [152]	559.72 ± 0.05			1948	At 298 K
Gronvold [46]		547.04 ± 0.08		1955	At 293 K
Baldock et al. [153]		547.04 ± 0.01		1966	At 298 K
Marples [150]	559.68 ± 0.01	547.05 ± 0.01	539.60 ± 0.01	1976	At 292 K for PuO_2 At 293 K for ThO_2 At 294 K for UO_2
Taylor [71]	559.74	546.80	539.55	1984	At 298 K
Katz et al. [154]	559.7		539.60 ± 0.03	1986	At 298 K
Yamashita et al. [69]	559.74 ± 0.06	547.02 ± 0.04	539.54 ± 0.04	1997	At 298 K

diffractometer. Lattice parameters were measured with an accuracy of ±0.5 pm and are shown in Fig. 6.

Because of their technical importance as nuclear fuel, thermal expansions of ThO_2, UO_2 and PuO_2 have been intensively studied using various techniques (see Table 9). These data were compiled and assessed by Touloukian et al. [61]. They presented the recommended equations of the linear thermal expansion for ThO_2 and PuO_2 and the provisional equation for UO_2. Taylor [71], also, compiled and analyzed thermal expansion data and presented regression equations of lattice parameter as a function of temperature for these actinide dioxides.

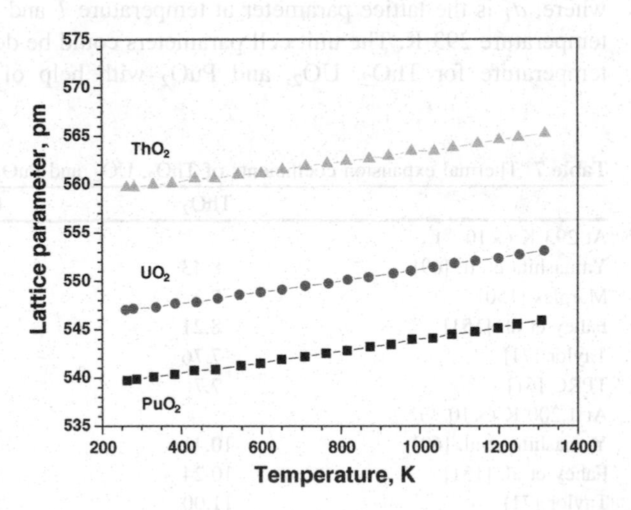

Fig. 6 Lattice parameters of ThO_2, UO_2, and PuO_2 as a function of temperature [69]. (permission from Elsevier)

Table 9 List of the authors worked on thermal expansion measurement on $(Th_{1-y}U_y)O_2$ system and year of publications

Year	Authors	Remarks
2012	Lu et al. [145]	Review based on first principles
2012	Bhagat et al. [85]	ThO_2 based SIMFUEL
2009	Subramaniam et al. [86]	ThO_2 containing 17.9, 41.7 and 52.01 % of $SmO_{1.5}$ in the temperature range 298–2,000 K
2008	Kutty et al. [84]	ThO_2–4 % UO_2, ThO_2–10 % UO_2 ThO_2–20 % UO_2
2006	Kim et al. [40]	Review paper
2006	Mathew et al. [87]	ThO_2–Nd_2O_3 phase with general compositions $Th_{1-x}Nd_xO_{2-x/2}$
2005	Grover et al. [88]	$(Th, Ce, Zr)O_2$
2001	Mathew et al. [89]	ThO_2, $Th_{0.96}Ce_{0.04}O_2$, $Th_{0.92}Ce_{0.08}O_2$
2000	Fink [64]	Review paper
2000	Anthonysamy et al. [78]	$(Th_{0.45}U_{0.55})O_2$, $(Th_{0.87}U_{0.13})O_2$ and $(Th_{0.09}U_{0.91})O_2$
2000	Tyagi and Mathew [77]	ThO_2 and ThO_2–2 wt% UO_2
1997	Bakker et al. [38]	Review paper
1991	Momin et al. [60]	High temperature XRD on ThO_2 and $(Th_{0.8}U_{0.2})O_2$
1991	Momin et al. [60]	ThO_2 and $(Th_{0.8}U_{0.2})O_2$ with 20 wt% Ln_2O_3.
1988	Martin [47]	Review paper
1981	Rodrigues and Sundaram [82]	Review paper
1977	Hirata et al. [75]	ThO_2 having 99.99 % purity
1970	Touloukian et al. [61]	Review paper
1967	Springer et al. [79]	ThO_2–10.09 % UO_2 and ThO_2–20.02 mol% UO_2 in temperature range 293–2,273 K
1967	Turner and Smith [80]	$(U, Th)O_2$ where U: Th = 1:10, 293–1,273 K
1962	Lynch and Beals [81]	<1,173 K
1959	Kempter and Elliot [56]	ThO_2–50.05 % UO_2 in temperature range 293–1,173 K
1959	Powers and Sharpio [83]	$(Th_{0.936}U_{0.064})O_2$, <1,073 K

4.1 ThO_2

The thermal linear expansion of ThO_2 is well established. Touloukian et al. [61] list more than 34 different experimental determinations which are in excellent agreement and recommended the following equation (150–2,000 K):

$$\% \Delta L/Lo = -0.179 + 5.097 \times 10^{-4}(T/K) + 3.732 \times 10^{-7}(T/K)^2 - 7.594 \times 10^{-11}(T/K)^3, \tag{25}$$

Touloukian states that the equation has an accuracy of ±3 % or less. Hoch and Momin [72] showed that their data, obtained by XRD measurements of thermal expansion of ThO_2, is in very good agreement with that obtained by Ohnysty and

Table 10 Recommended values for linear thermal expansion of ThO_2 [12]

Temperature (*K*)	$\%\Delta L/L_0$	Expansion coefficient, $\times 10^{-6}$ K^{-1}	
		Mean	Instantaneous
298.15	0.0	8.43	8.43
400	0.087	8.53	8.63
500	0.174	8.63	8.82
600	0.264	8.73	9.01
700	0.355	8.83	9.20
800	0.448	8.93	9.39
1,000	0.641	9.13	9.77
1,200	0.842	9.33	10.2
1,400	1.05	9.53	10.5
1,600	1.27	9.73	10.9
1,800	1.49	9.93	11.3
2,000	1.72	10.1	11.6
2,500	2.34	10.6	12.5
3,000	3.01	11.1	13.4
3,500	3.72	11.6	14.3

Rose [73] by dilatometry. Their data also agreed well with that of Aronson et al. [74] and Hirata et al. [75]. Hoch and Momin [72] recommended the following equation for ThO_2 (293–2,373 K):

$$\%\,\Delta L/L_0 = -0.2426 + 7.837\times10^{-4}(T/K) + 9.995\times10^{-8}(T/K)^2 \quad (26)$$

The results obtained from the above equations are almost identical. Hoch and Momin concluded from their studies that the lattice defects in ThO_2 were probably of the Frenkel type.

The extrapolation data to higher temperatures other than those covered by measurements need caution [76]. The equation of Touloukian is valid in the range of 150–2,000 K and that of Hoch and Momin [72] should be used for extrapolation. The recommended values of thermal expansion of are shown in Table 10. Thermal expansion curve for ThO_2 is shown in Fig. 7.

4.2 UO_2

Martin [47] had reviewed 15 sets of UO_2 thermal expansion data and provided a relation which has been accepted as the most authenticated data. Fink in her peer review paper [64] on thermophysical properties of UO_2 has also recommended the data suggested by Martin. Recommended percentage linear thermal expansion for UO_2 by Martin [47]:

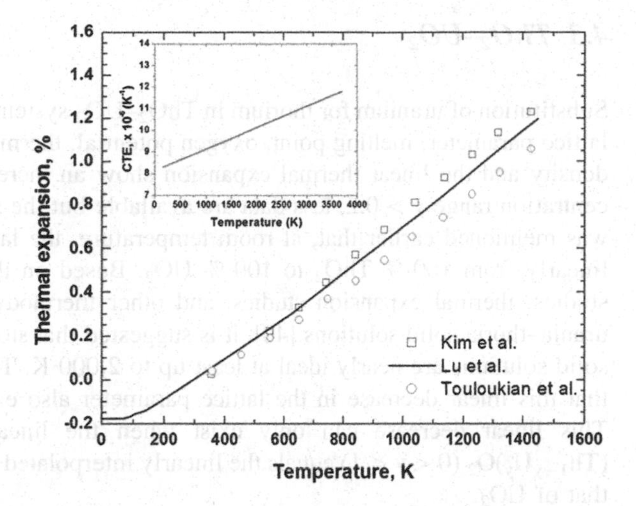

Fig. 7 Temperature dependence of the linear thermal expansion for ThO_2 [145]. The inset is the volume thermal expansion coefficient as a function of temperature [12]

$$(\Delta L/L_0) \times 100 = -0.266 + 9.802 \times 10^{-4} \cdot T - 2.705 \times 10^{-8} \cdot T^2 + 4.391 \times 10^{-11} \cdot T^3, \tag{27}$$

$(273\ \mathrm{K} < T < 923\ \mathrm{K})$.

$$(\Delta L/L_0) \times 100 = -0.328 + 1.179 \times 10^{-3} \cdot T - 2.429 \times 10^{-7} \cdot T^2 + 1.219 \times 10^{-10} \cdot T^3, \tag{28}$$

$(923\ \mathrm{K} < T < 2{,}000\ \mathrm{K})$.

Figure 8 shows a plot of expansion in percentage versus temperature for both ThO_2 and UO_2 and it shows that ThO_2 has a lower expansion than UO_2.

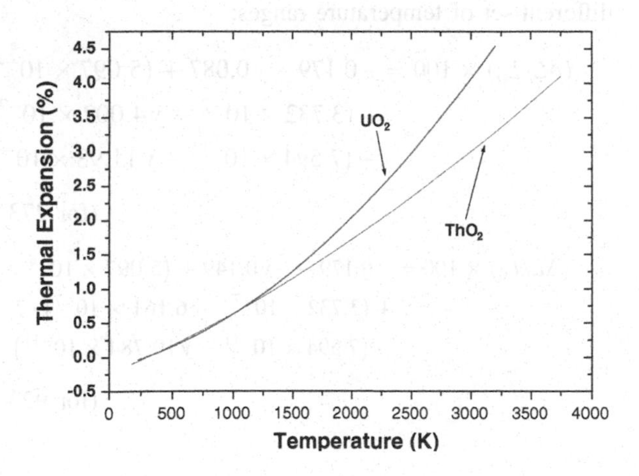

Fig. 8 Comparison of thermal expansion of ThO_2 and UO_2. Data of Belle et al. are re-plotted and shown above [12]

4.3 ThO_2–UO_2

Substitution of uranium for thorium in ThO_2–UO_2 system results in the decrease of lattice parameter, melting point, oxygen potential, thermal conductivity, while the density and the linear thermal expansion show an increase. In the uranium concentration range $y > 0.2$, less data are available but the same trend is expected. It was mentioned earlier that, at room temperature, the lattice parameter decreases linearly from 100 % ThO_2 to 100 % UO_2. Based on the results of calorimetric studies, thermal expansion studies, and other thermodynamic measurements on urania–thoria solid solutions [40], it is suggested that stoichiometric urania–thoria solid solutions are nearly ideal at least up to 2,000 K. Therefore, one can assume that this linear decrease in the lattice parameter also exists at high temperature. This linear decrease can only exist when the linear thermal expansion of $(Th_{1-y}U_y)O_2$ $(0 < y < 1)$ equals the linearly interpolated value of that of ThO_2 and that of UO_2.

The thermal linear expansion of the $(Th_{1-y}U_y)O_2$ solid solution has been studied in much less detail than that of the pure compounds. Konings et al. [39] reviewed the thermophysical properties of ThO_2-based fuels. They concluded from the results of eight studies on $(Th_{1-y}U_y)O_2$, and suggested the equation of the type:

$$\begin{aligned}\Delta L/L_0 = {} & (8.1635\times10^{-4} + 3.8325\times10^{-4}y + 5.2423\times10^{-4}y^2)(T/K - 298.15)\\ & + (1.2144\times10^{-7} + 1.4936\times10^{-8}y + 1.5633\times10^{-7}y^2)(T/K - 298.15)^2,\end{aligned} \tag{29}$$

where, the thermal linear expansion $\Delta L/L_0$ in %, and y is the molar fraction of UO_2.

Bakker et al. [38] have recommended the percentage linear thermal expansion data of $(Th_{1-y}U_y)O_2$ $(0 < y < 1)$ by obtaining the linear interpolation of the values of Touloukian [61] and Martin [47] and obtained the following relations in two different set of temperature ranges:

$$\begin{aligned}(\delta L/L_0)\times 100 = {} & -0.179 - y0.087 + \left(5.097\times10^{-4} + y4.705\times10^{-4}\right)\cdot T\\ & + \left(3.732\times10^{-7} - y4.002\times10^{-7}\right)\cdot T^2\\ & - \left(7.594\times10^{-11} - y11.98\times10^{-11}\right)\cdot T^3\end{aligned} \tag{30}$$

(for $273\text{ K} < T < 923\text{ K}$),

$$\begin{aligned}(\Delta L/L_0)\times 100 = {} & -0.179 - -y0.149 + \left(5.097\times10^{-4} + y6.693\times10^{-4}\right)\cdot T\\ & + \left(3.732\times10^{-7} - y6.161\times10^{-7}\right)\cdot T^2\\ & - \left(7.594\times10^{-11} - y19.784\times10^{-11}\right)\cdot T^3\end{aligned} \tag{31}$$

(for $923\text{ K} < T < 2{,}000\text{ K}$).

Momin et al. [60] measured lattice thermal expansion of (Th, U)O_2 system by X-ray diffraction method. They obtained coefficient of expansion data for pure ThO_2 and $(Th_{0.8}U_{0.2})O_2$ to be 9.5×10^{-6} K^{-1} and 7.1×10^{-6} K^{-1}, respectively, in the temperature range 298–1,600 K. It was observed that the coefficient of thermal expansion of $(Th_{0.8}U_{0.2})O_2$ is lower than either of ThO_2 and UO_2, which is quite unreasonable.

Tyagi et al. [77] found CTE values for ThO_2 and ThO_2–2 wt% UO_2 to be 9.58×10^{-6} and 9.74×10^{-6} K^{-1}, respectively, in the temperature range of 298–1,473 K. The CTE value reported in IAEA-TECDOC [40] for ThO_2 in the temperature range 300–1,473 K is 9.732×10^{-6} K^{-1} and for ThO_2–4 wt% UO_2 it is 9.85×10^{-6} K^{-1}; both these values were found to be in close agreement with those reported by Tyagi et al. The CTE value 10.33×10^{-6} K^{-1} for composition $(Th_{0.87}U_{0.13})O_2$ in the temperature range 298–1,973 K as reported by Anthonysamy et al. [78] matches well with the value obtained in the IAEA study for ThO_2–10 % UO_2 which was found to be 10.21×10^{-6} K^{-1} in the temperature range 300–1,773 K. The average linear thermal expansion coefficients for $(Th_{0.45}U_{0.55})O_2$ and $(Th_{0.09}U_{0.91})O_2$ were measured to be 10.83×10^{-6} K^{-1} and 11.45×10^{-6} K^{-1}, respectively, in the temperature range between 298 and 1,973 K. These data clearly show that thermal expansion coefficients increases with increase in UO_2 content in ThO_2–UO_2 system. Figure 9 shows % thermal expansion plot of some typical ThO_2–UO_2 solid solutions.

The coefficient of expansion data of Momin et al. [60], Springer et al. [79], Turner and Smith [80], Kempter and Elliot [56] and Lynch and Beals [81] show a wide scatter of data points when plotted against composition. Rodriguez and Sundaram [82] in their review article reported an average linear thermal expansion coefficient of 9.67×10^{-6} K^{-1} for ThO_2 (293–2,273 K) and 12.5×10^{-6} K^{-1} for $(Th_{0.8}U_{0.2})O_2$ (1,100–2,400 K). Powers and Shapiro [83] reported the same average linear thermal expansion coefficient value of 9×10^{-6} K^{-1} for both pure

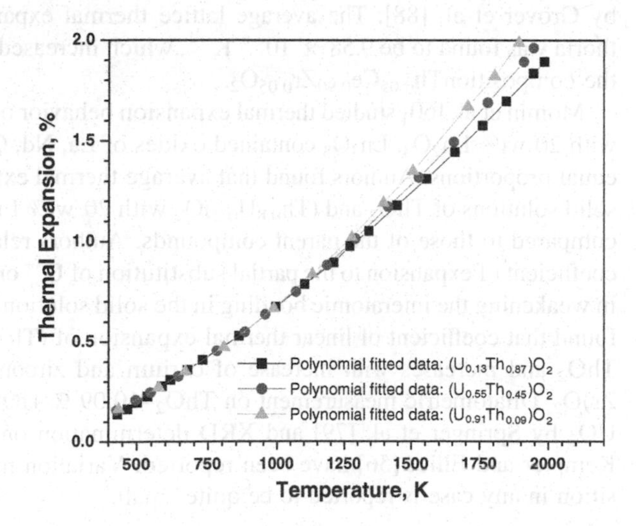

Fig. 9 Percent thermal expansion plot of some typical ThO_2–UO_2 solid solutions [78]. (permission from Elsevier)

UO_2 and $(U_{0.064}Th_{0.936})O_2$. They obtained lower coefficient value (8×10^{-6} K^{-1} up to 1,073 K) for $(Th_{0.8}U_{0.2})O_2$.

Kutty et al. [84] measured thermal expansion of ThO_2, ThO_2–4 % UO_2, and ThO_2–20 % UO_2 pellets fabricated by (Coated Agglomerate Pelletization) CAP route using ThO_2 and U_3O_8 powders as the starting materials. They reported that the thermal expansion of ThO_2–20 % UO_2 pellet was different from that of ThO_2 and ThO_2–4 % UO_2, e.g., it increased more rapidly with increasing temperature in the temperature range of 1,000–1,500 °C which they attributed to the loss of oxygen of (Th, U)O_{2+x} above 1,000 °C. The thermal expansion behavior of polycrystalline samples of ThO_2–3.45 % UO_2 and SIMFUEL corresponding to the burnup of 43,000 MWd/Te has been investigated from room temperature to 1,473 K, and for SIMFUEL corresponding to burnup of 28,000 MWd/Te has been investigated from room temperature to 1,173 K, using a high-temperature X-ray diffraction (HTXRD) by Bhagat et al. [85]. They reported that SIMFUEL has higher thermal expansion than ThO_2–3.45 % UO_2 and this is related to the higher thermal expansion coefficient of dissolved rare earth oxides and also to the lower melting point of SIMFUEL matrix.

The mean linear thermal expansivity for ThO_2–$SmO_{1.5}$ solid solutions containing 17.9, 41.7 and 52.01 % of $SmO_{1.5}$ were determined by Subramanian et al. [86] in the temperature range 298–2,000 K. The mean linear thermal expansion coefficients for ThO_2–$SmO_{1.5}$ solid solution were found to be 10.47, 11.16, and 11.45×10^{-6} K^{-1}, respectively. The synthesis, characterization, and lattice thermal expansion studies of the ThO_2–Nd_2O_3 phase with general compositions $Th_{1-x}Nd_xO_{2-x/2}$ are reported by Mathews et al. [87]. The lattice thermal expansion (293–1,473 K) behavior of the solid solutions has been investigated by high temperature XRD and found to show a gradual increase with increasing content of $NdO_{1.5}$ in $Th_{1-x}Nd_xO_{2-x/2}$ series. The lattice thermal expansion behavior of a number of single-phase compositions of CeO_2–ThO_2–ZrO_2 in the temperature range from 293 to 1,473 K, as investigated by high-temperature XRD are reported by Grover et al. [88]. The average lattice thermal expansion coefficient of pure thoria was found to be 9.58×10^{-6} K^{-1}, which increased to 11.91×10^{-6} K^{-1} in the composition$Th_{0.05}Ce_{0.90}Zr_{0.05}O_2$.

Momin et al. [60] studied thermal expansion behavior of ThO_2 and $(Th_{0.8}U_{0.2})O_2$ with 20 wt% Ln_2O_3. Ln_2O_3 contained oxides of La, Nd, Ce, Y, Sm, Gd, and Eu in equal proportions. Authors found that average thermal expansion coefficient of the solid solutions of ThO_2 and $(Th_{0.8}U_{0.2})O_2$ with 20 wt% Ln_2O_3 show an increase as compared to those of the parent compounds. Authors related the higher values of coefficient of expansion to the partial substitution of U^{4+} or Th^{4+} with Ln^{3+} resulting in weakening the interatomic bonding in the solid solution matrix. Grover et al. [88] found that coefficient of linear thermal expansion of (Th, Ce, Zr)O_2 is higher than ThO_2 and increases with increase of cerium and zirconium content in (Th, Ce, Zr)O_2. Dilatometric measurement on ThO_2–10.09 % UO_2 and ThO_2–20.02 mol% UO_2 by Springer et al. [79] and XRD determination on ThO_2–50.05 % UO_2 by Kempter and Elliot [56] have been reported. Variation in expansion with composition in any case is reported to be quite small.

4.4 ThO_2–PuO_2

ThO_2 and PuO_2 form a continuous series of solid solutions over the entire range of composition. At the Pu-rich end, mixed oxide may be heterogeneous if prepared under reducing conditions, as a result of the formation of Pu_2O_3. The lattice parameter of $(Th_{1-y}Pu_y)O_2$ decreases linearly from pure ThO_2 to pure PuO_2 [26]. Lattice parameters of (Th, Pu)O_2 with various PuO_2 contents are given in Table 11. Assuming ideal solid solution behavior at high temperatures for ThO_2 and PuO_2, it would be expected that this linear decrease in lattice parameter would also happen at elevated temperatures.

The available data on ThO_2–PuO_2 are scanty. One way to overcome this problem is to use CeO_2 in place of PuO_2 as they both have quite similar physico-chemical properties viz., ionic radii in octahedral and cubic coordination, melting points, standard enthalpy of formation and specific heat etc. Thus, the plutonium chemistry can be well simulated using CeO_2 in place of highly active PuO_2. Mathews et al. [89] have recently measured bulk thermal expansion of (Th, Ce)O_2 system. Bulk and lattice thermal expansion studies on $(Th_{1-y}Ce_y)O_2$ ($y = 0.0$, 0.04, 0.08 and 1.0) were carried out by dilatometry and high temperature XRD from room temperature to 1,123 and to 1,473 K, respectively. The average linear thermal expansion coefficients of ThO_2, $Th_{0.96}Ce_{0.04}O_2$, $Th_{0.92}Ce_{0.08}O_2$, and CeO_2 were found to be 9.04×10^{-6}, 9.35×10^{-6}, 9.49×10^{-6}, and 11.58×10^{-6} K^{-1}, respectively, between 293 and 1,123 K. Some data on $(Th_{1-y}Pu_y)O_2$ generated at BARC was reviewed in IAEA-TECDOC [40]. Thermal expansion curve for $(Th_{1-y}Pu_y)O_2$ for $y = 0.02$, 0.04, 0.06, 0.10 are shown in Fig. 10.

The thermal expansion of the solid solutions $(Th_{1-y}Pu_y)O_2$ could be reasonably approximated at various temperatures by taking linear interpolated expansion data of ThO_2 and PuO_2 as per their weight fraction. IAEA-TECDOC [40] used "interpolation method," using the recommended equation by Touloukian [61] for ThO_2 and the following equation for pure PuO_2 as recommended by MATPRO. The equation for linear strain calculations is as given below:

$$\varepsilon = K_1 \cdot T - K_2 + K_3 \cdot \exp(-E_D/k_B T), \tag{32}$$

Table 11 Lattice parameter of ThO_2–PuO_2 solid solution at 298 K [26]

PuO_2 (mol%)	Lattice parameter (pm)
0	559.6 ± 0.1
15	556.8 ± 0.4
26	554.62 ± 0.04
36.9	552.6 ± 0.1
46.7	550.2 ± 0.1
63.5	546.93 ± 0.04
82.5	542.8 ± 0.1
100	539.60 ± 0.03

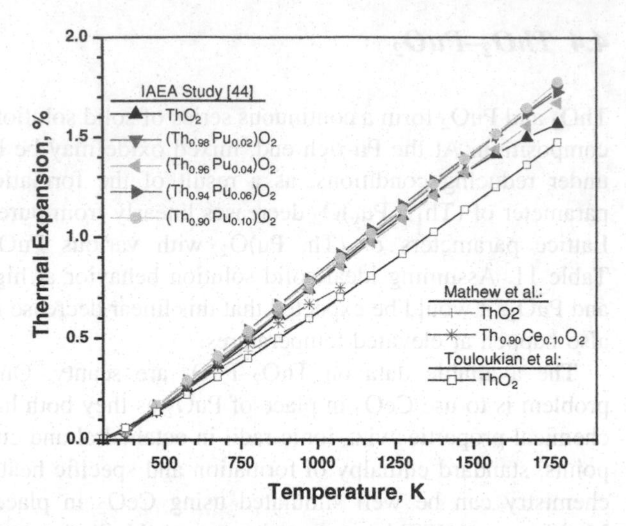

Fig. 10 Thermal expansion curves for $(Th_{1-y}Pu_y)O_2$ for $y = 0.02, 0.04, 0.06$ and 0.10 [40]. (permission from IAEA)

where, ε is the linear strain which is taken as zero at 300 K, T is the temperature (K), k_B is Boltzman's constant (1.38×10^{-23} J/K), and E_D, K_1, K_2, and K_3 are constants having values 7×10^{-20} (J), 9×10^{-6} (K^{-1}), 2.7×10^{-3} (unit less), and 7×10^{-2} (unit less), respectively.

Percentage linear thermal expansion for $(Th_{1-y}Pu_y)O_2$ $(0 < y < 1)$ obtained by linear interpolation of the data of ThO_2 and data for PuO_2 and can be expressed as (in the temperature range of 300–1,773 K) [40].

$(Th_{1-y}Pu_y)O_2$ where $0 < y < 1$:

$$(\Delta L/L_0) \times 100 = -0.179 - 0.049 \cdot y + \left(5.079 \times 10^{-4} + 2.251 \times 10^{-4} \cdot y\right) \cdot T + \left(3.732 \times 10^{-7} - 2.506 \times 10^{-7} \cdot y\right) \cdot T^2 + \left(-7.594 \times 10^{-11} + 12.454 \times 10^{-11} \cdot y\right) \cdot T^3 \tag{33}$$

As part of thorium-based fuel development program for fast breeder reactors, the thermophysical properties of mixed thorium–plutonium oxide pellets of both thorium- and plutonium-rich compositions were evaluated in India [58]. The plutonium-rich mixed oxide pellets contained 70–80 % PuO_2 which could be considered as candidate fuel for small LMFBR core like the operating fast breeder test reactor (FBTR). The thorium-rich compositions contained 20–30 % PuO_2 which could be considered as alternative fuel for large LMFBRs like the forthcoming prototype fast breeder reactor (PFBR-500). The mixed oxide pellets were prepared by "powder-pellet" route involving mechanical mixing of ThO_2 and PuO_2 powders followed by cold pelletization and high temperature sintering. Small amount of Nb_2O_5 (0.25 wt%) or CaO (0.5 wt%) powder were used as "sintering aid" and admixed with the powder during co-milling. The coefficient of thermal expansion of mixed $(Th_{0.3}Pu_{0.7})O_2$, $(Th_{0.7}Pu_{0.3})O_2$, and $(Th_{0.8}Pu_{0.2})O_2$

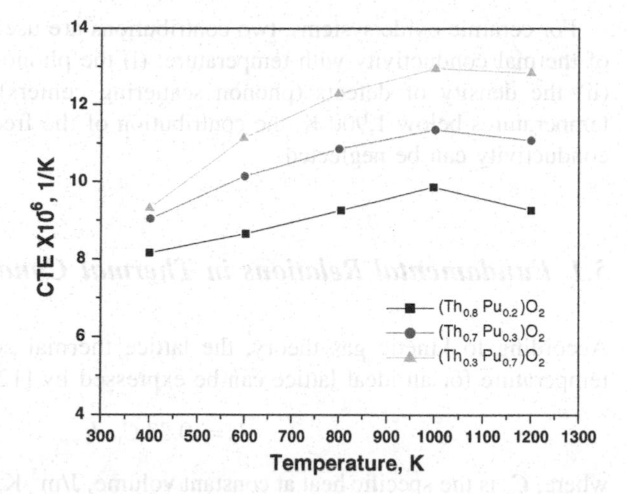

Fig. 11 Thermal expansion curves for high Pu bearing $(Th_{1-y}Pu_y)O_2$ samples [58]. (permission from IAEA)

were evaluated by a high-temperature dilatometer and is summarized in Fig. 11. XRD pattern of ThO_2 and the pellets containing lower amounts of PuO_2 (30 % PuO_2) sintered in either Ar or Ar-8 % H_2 showed only single-phase isostructural with fluorite phase. But ThO_2–PuO_2 pellets with higher plutonium content such as ThO_2–50 % PuO_2 and ThO_2–75 % PuO_2 pellets sintered in either Ar or Ar-8 % H_2 showed the presence of two phases. In addition to the phase that is isostructural with PuO_2 (fluorite), another phase which is isostructural with bcc α-Pu_2O_3 has been observed [90]. Hence, no conclusion could be drawn from the above results.

5 Thermal Conductivity

Among the various thermal properties, thermal conductivity is the most useful property for the nuclear scientist. It is the ability of the material to transfer heat from a region of high temperature to a region of low temperature. In normal conditions, thermal conductivity and linear power determine the peak fuel operating temperature. Under the accident conditions, the thermal conductivity of the fuel determines the maximum permissible linear rating, χ_{max}, if central melting is to be avoided [10, 38]. The thermal conductivity, λ, allows the determination of centre temperature of fuel, T_c, when the surface temperature T_s, is known by using the conductivity integral,

$$\chi = 4\pi \int_{T_s}^{T_c} \lambda dT \tag{34}$$

where χ is the linear rating.

For ceramic oxide systems, two contributions are used to describe the behavior of thermal conductivity with temperature: (i) the phonon–phonon interaction and (ii) the density of defects (phonon scattering centers) in the lattice [91]. For temperatures below 1,900 K, the contribution of the free electrons to the thermal conductivity can be neglected.

5.1 Fundamental Relations in Thermal Conductivity

According to kinetic gas theory, the lattice thermal conductivity above Debye temperature for an ideal lattice can be expressed by [12]

$$\lambda_l = 0.33\, C_v v L \quad (35)$$

where, C_v is the specific heat at constant volume, J/m^3·K, v is the velocity of sound in solid, m/s, and L is the mean free path of scattered waves (the phonon wavelength), m.

Above the Debye temperature, thermal conductivity of electric insulators decreases with increasing temperatures. Since atomic vibrational frequency increases with temperature, an increase in wave scattering is anticipated which are shown to be due to phonon interaction. Thermal energy is transferred by the Umklapp process, in which two phonons interact to form a third [12]. According to this theory, lattice thermal conductivity is inversely proportional to absolute temperature and this becomes a minimum when phonon wavelength becomes less than the mean distance between the scattering centers. For crystalline solid, the minimum distance between the scattering centers is the interatomic distance which is the lattice parameter, a_o. Therefore, the above equation becomes [12]:

$$(\lambda_l)_{\min} = 0.33\, C_v v a_o \quad (36)$$

In solids, phonon–phonon scattering is due to the anharmonic components of crystal vibrations. Lattice anharmonicity increases with the mass difference between anions and cations in the ionic material and is greatest in UO_2 or PuO_2 [38, 91]. As a result, the thermal conductivity of the oxides of the actinide metals is considerably lower than that of most other crystalline oxides. The kinetic theory of gases shows that the collision mean free path is given by the reciprocal of the product of the collision cross-section σ_P and the density of scattering points (n_P):

$$L = (1/\sigma_P n_P) \quad (37)$$

The deviation from stoichiometry and the presence of foreign atoms or porosity result in lower values of λ in actinide oxides. Further, it can be shown that the phonon mean free path should vary as *1/T*. In general, phonon–phonon scattering and phonon–impurity scattering are the dominant mechanisms of the thermal conductivity in ceramics. Klemens [92] has proposed a heat conduction model in materials where the phonon–phonon (Umklapp) scattering and the phonon–impurity scattering occur simultaneously. Theoretically, the phonon component of the thermal conductivity λ may be written as:

$$\lambda = (A + \mathrm{BT})^{-1} \tag{38}$$

where A and B are constants and T is the absolute temperature.

Thermal resistivity (R), which is the reciprocal of thermal conductivity (λ), of the above oxides can be described by the following equation:

$$R = 1/\lambda = A + \mathrm{BT} \tag{39}$$

The first term, A, in Eq. (39) represents the defect thermal resistivity. This results from the phonon interactions with lattice imperfections, impurities, isotopic, or other mass differences as well as bulk defects such as grain boundaries in the sample. The influence of substituted impurities on the thermal conductivity is described by the increase of the parameter A. The second term in Eq. (39), namely BT, represents the intrinsic lattice thermal resistivity caused by phonon–phonon scattering [1, 2]. As the temperature increases, this term becomes predominant. The parameter B remains nearly constant by substitution. The constants A and B can be obtained from the least squares fitting of the experimental data.

The thermal conductivity of nuclear ceramics is strongly influenced by the stoichiometry. Deviations from stoichiometry produce point defects, most likely oxygen vacancies or metal interstitials in hypostoichiometric compounds and oxygen interstitials or metal vacancies in hyperstoichiometric compounds. Introduction of point defects into the oxygen ion sublattice or substitution of Th for U on the cation sublattice provides additional centers from which phonon scattering occurs. It is reported that there is a drastic change in the uranium vacancy concentration on varying O/M ratio around the stoichiometric composition. Many reports are available on the effect of stoichiometry on the thermal conductivity of UO_2 and (U, Pu)O_2 samples [1]. Thermal conductivities decrease as their hyperstoichiometry, x, increases. At low temperatures, thermal conductivity of mixed oxide can be described by a modified equation of (38) as [93, 94]:

$$\lambda = 1/[A(x,q) + B(x,q)T] \tag{40}$$

where, x and q denote the extent of nonstoichiometry and the Pu/Th content in the UO_2 lattice, respectively. The limited amount of experimental information available suggests that the coefficient A depends primarily on the O/M ratio and only very weakly on the plutonium content. The coefficient A may be written as

$$A = A_0 + \Delta A(x) \tag{41}$$

where, A_0 is very nearly equal to the A value of pure UO_2. The perturbation ΔA arises from interactions of point defects with lattice. The magnitude of ΔA is proportional to the defect atom fraction and to a measure of the cross section of the defect for phonon scattering. The latter is proportional to the square of the difference between the atomic radius of the defect (r_1) and that of the host atom (r). The mass difference between the impurity atom and the host atom may also influence A, but this contribution is not significant in mixed oxide fuel materials. A can also be represented by the following equation as

$$A = [(\pi^2 V\theta)/(3\text{hv}^2)] \sum_i \Gamma_i, \tag{42}$$

where V, θ, h, and v denote the average atomic volume, Debye temperature, Planck's constant, and phonon velocity, respectively. The term $\Sigma_i \Gamma_i$ is the sum of the cross-sections of all the phonon-defect scattering centers. The analysis of the lattice defect thermal resistivity and the evaluation of phonon scattering by the various defect scattering centers in pure and mixed actinide oxides have been carried out by several authors [91–94]. Accordingly, A of Eq. (42) can be given as

$$A = C(\Gamma_u + \Gamma_o), \tag{43}$$

where $C = (\pi^2 V\theta)/(3\text{hv}^2)$. Γ_u is the scattering cross-section arising from U substitution and Γ_o is that from all other native defects present in the sample. The scattering cross-section Γ_u can be expressed in terms of the mass and size difference of the substituted atom over that of the host [95]:

$$\Gamma_u = x(1-x) \times [(\Delta M/M)^2 + E(\Delta r/r)^2], \tag{44}$$

where, x is atomic fraction of substituted U in place of Th, ΔM and Δr are the mass and radius difference between U and Pu/Th atom, respectively, M and r are average mass and radius of the substituted atom, and E is an adjustable parameter which represents the magnitude of lattice strain. From the above, it is clear that scattering cross-section depends upon the mass difference between Th and U atoms, size difference between Th and U atoms, charge of U ion and microstructure.

Thermal transport by electrical charge carriers can also contribute to thermal conduction at high temperatures. The ratio between thermal and electrical conductivities of metals can be expressed in terms of the ratio:

$$L_c = \lambda/\sigma T = \pi^2 k^2/3e^2 = 2.45 \times 10^{-8}\, W\Omega/K^2, \tag{45}$$

which may be called the Wiedemann–Franz ratio or the Lorenz constant. In the above equation, σ is the electrical conductivity, e is the electronic charge, and k is Boltzmann's constant. Thermal conductivity of a solid can be measured by two methods:

1. By determining the stationary heat flow through the specimen, which gives λ directly,
2. By determining the variation of the temperature at a fixed plane, that is a specimen surface, due to an induced nonstationary heat flow which gives the thermal diffusivity, α.

Since the second method is more versatile and requires smaller specimen, it has become a standard method for determining λ for $T > 600$ K. For lower temperatures, the first method is more suited.

For the thermal diffusivity measurement, the sintered pellet was sliced into discs of about 10 mm diameter and 2 mm thickness using a low speed cut-off wheel. A pulse of laser was projected on to the front surface of the pellet and the temperature rise on the rear side of the pellet was recorded as a transient signal by using an infrared detector. The thermal diffusivity (α_t) was calculated from the following relationship:

$$\alpha_t = WL^2/\pi t_{1/2} \tag{46}$$

where $t_{1/2}$ is the time required in seconds to reach half of the maximum temperature rise at the rear surface of the sample and L is the sample thickness in millimeter. W is a dimensionless parameter which is a function of the relative heat loss from the sample during the measurement. The data have to be corrected for radiation heat losses by the method of Clark and Taylor [96].

Unlike UO_2 or PuO_2, ThO_2 is a semitransparent material to wavelengths of the infrared region. For a laser flash experiment, all the energy of the laser pulse is not absorbed on the front face of the sample, but also in volume. Also, the temperature measurement on the rear face is skewed as the pyrometer may receive radiation produced not only at the sample surface, but also in volume. These difficulties are overcome if the faces of the samples are given a coating by graphite. A coating of graphite on both faces was used in order to make sure that the energy of the laser is absorbed on the front face and to improve the temperature recording on the rear face.

Heat transport through materials is described by two properties: thermal conductivity, λ (under steady state conditions) and thermal diffusivity, α_t (under transient conditions). These two properties are related by the expression:

$$\lambda(T) = \alpha_t(T) \cdot \rho(T) \cdot C_p(T), \tag{47}$$

Where, ρ the density of the material and C_p its specific heat at constant pressure. The specific heat of mixed oxide like $(Th_{1-y}Pu_y)O_2$ solid solutions was calculated from the literature values of specific heats of pure ThO_2 and PuO_2 and subsequently using Neumann–Kopp's rule. The following equations were used to calculate C_p of $(Th_{1-y}Pu_y)O_2$:

$$C_p(Th_{1-y}Pu_y)O_2 = (1-y) \cdot C_p(ThO_2) + y \cdot C_p(PuO_2), \tag{48}$$

where y is the weight fraction of PuO_2.

Effect of Porosity on Thermal Conductivity

Attempts to evaluate the decrease in thermal conductivity due to porosity (P) have been made by Eucken in as early as 1932. There are many relations in the literature describing the effect of porosity on thermal conductivity. Some of them are listed below [1, 64, 97–106]:

1. Loeb	$\lambda_M = (1 - P)\,\lambda_{TD}$	(i)
2. Modfied Loeb	$\lambda_M = (1 - \alpha P)\,\lambda_{TD}$ where $2 < \alpha < 5$	(ii)
3. Kampf and Karsten	$\lambda_M = \left(1 - P^{2/3}\right)\lambda_{TD}$	(iii)
4. Biancharia	$\lambda_M = [(1 - P)/(1 - (\beta - 1)P)]\,\lambda_{TD}$ $\beta = 1.5$ for spherical pores	(iv)
5. Maxwell-Eucken	$\lambda_M = [(1 - P)/(1 + \beta P)]\,\lambda_{TD}$	(v)
6. Brand and Neuer	$\lambda_M = (1 - \sigma P)\,\lambda_{TD}$ where $\sigma = 2.6 - 0.5\,(T + 273)/1000$	(vi)
7. Schultz	$\lambda_M = (1 - P)^{\gamma}\,\lambda_{TD}$	(vii)

(λ_M and λ_{TD} are the thermal conductivities, respectively, in presence and absence of porosity P, $0 < P < 1$).

Schultz [106] has theoretically shown that, for spherical pores distributed randomly, γ of Eq. (vii) has a value of 1.5. However, in reality the above coefficients for fuel pellets are larger ($\gamma > 1.5$), due to the porosity being neither spherical nor uniformly distributed [1]. IAEA [40] has recommended the value of $\alpha = 2.5 \pm 1.5$ for the modified Loeb equation for $0 < P < 0.1$. Inoue, Abe, and Sato [107] experimentally showed that $\gamma = 2.4$ for $0.044 < P < 0.470$ and reported that $\beta = 2$ (Eq. (v)). The IAEA recommendation ($\alpha = 2.5 \pm 1.5$) is in agreement with other experimenters [1, 40]. Among the above, Eq. (i) under predicts the data and Eq. (iv) accounts for the shape of the pores.

5.2 Thermal Conductivity of ThO_2

The thermal conductivity of ThO_2 up to 1,800 K is reasonably well established (Table 12). Most of the data were derived from thermal diffusivity measurements. Peterson and Curtis [26] compiled data on thermal conductivity of ThO_2 to about 2,000 K. Bakker et al. [38] systematically evaluated the data of various authors. They analyzed the data of Pears [102], Rodriguez et al. [82], McEwan and Stoute [103], Belle et al. [104], Peterson et al. [105], Faucher et al. [108], Kingery et al. [109], McElroy et al. [110], ARF [111], Weilbacher [112] and DeBoskey [113].

Assessing the A and B parameters has the advantage that data sets that were determined in different temperature ranges can easily be compared and that data sets with extremely large or small A and B parameters can be rejected. On this basis, Bakker et al. [38] rejected many data and accepted only that data which shows a small variation between the A and B parameters. Hence data of Murabayashi [114], McElroy et al. [110], Koenig [115] and Springer et al. [57] are only used in their assessment. The A and B parameters were averaged, which yielded $A = 4.20 \times 10^{-4}$ mKW^{-1} and $B = 2.25 \times 10^{-4}$ mW^{-1} and these values can be used as the recommended values for 95 % dense ThO_2 in the temperature

Table 12 Thermal conductivity measurement for ThO_2

Author	Year	Remarks
Kingery et al. [109]	1954	373–1,273 K
ARF [111]	1957	93.3 % TD: 527–824 K
Peterson et al. [105]	1966	373–1,473 K
Belle et al. [104]	1967	393 K
Springer et al. [57]	1968	573–2,173 K
McElroy et al. [110]	1968	92.7 % TD: 200–1,400 K
McEwan and Stoute [103]	1969	95.0 % TD: 333 K
Murabayashi et al. [114]	1970	293 K
Faucher et al. [108]	1970	91.6 % TD: 1,900–2,600 K
Touloukian [61]	1970	Review paper
Weilbacher [112]	1972	97 % TD: 400–2,550 K
Rodriguez et al. [82]	1981	773–1,773 K
Srirama Murti and Mathews [124]	1991	92.0 % TD: 573–1,573 K
Bakker et al. [38]	1997	Review paper
Pillai and Raj [93]	2000	300–1,200 K
Jain et al. [126]	2006	373–1,773
Kutty et al. [84]	2008	298–1,500 K
Lu et al. [145]	2012	Modeling

range between 300 and 1,800 K. Hence, thermal conductivity of pure ThO_2 can be expressed as:

$$\lambda_{ThO_2}(W/mK) = \left(4.20 \times 10^{-4} + 2.25 \times 10^{-4}\,T\right)^{-1} \tag{49}$$

Belle and Berman [12] reported the following equation for the thermal conductivity of 100 % dense ThO_2 in the temperature range 298–2,950 K,

$$\lambda_{ThO_2}(W/mK) = \left(0.0213 + 1.597 \times 10^{-4}\,T\right)^{-1} \tag{50}$$

To evaluate the thermal conductivity beyond 2,950 K, Belle and Berman first obtained an expression for thermal diffusivity up to 2,950 K as

$$\alpha_{ThO_2}(m^2/s) = (-34191.1 + 561.28\,T)^{-1} \tag{51}$$

Assuming there is no discontinuity, Belle and Berman [12] extrapolated thermal diffusivity data values from 2,950 to 3,400 K. Their results along with others are shown in Fig. 12. The only high temperature data available is that of Weilbacher [112] which was fitted by a dashed line. The fitted data of Cozzo et al. [95] and Kutty et al. [84] represent the lowest and highest values in the low temperature range.

Figure 13 shows the calculated value of thermal conductivity for fully dense ThO_2 from ambient to 3,400 K. There is a sudden increase in conductivity at 2,950 K discontinuity, from 2.03 to 3.05 W/m·K, which represents the change in the heat capacity occurring at that point. The lowest thermal conductivity in the

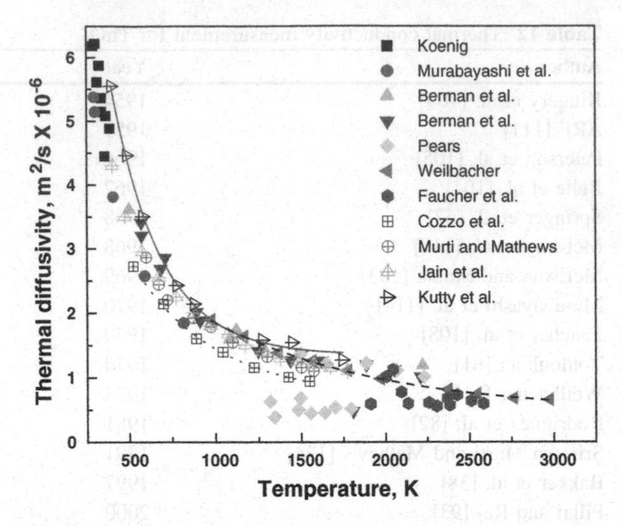

Fig. 12 Thermal diffusivity of ThO_2 as a function of temperature. Data of various authors are plotted together. Dotted line are fitted data of Cozzo et al. [95], dashed line that of Weilbacher [112] and solid line that of Kutty et al. [84]

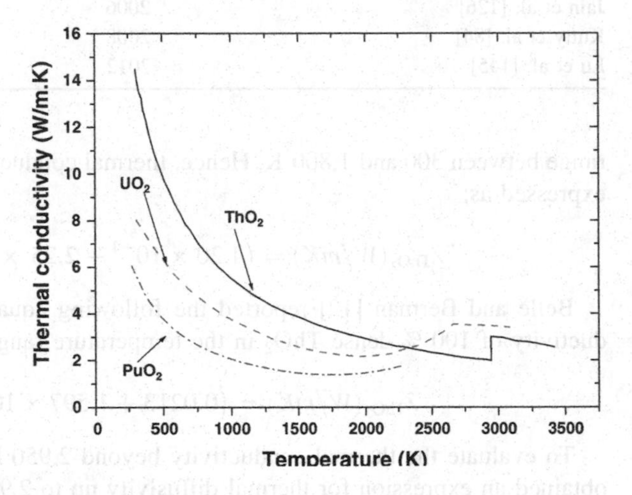

Fig. 13 Thermal conductivity of ThO_2, UO_2 and PuO_2 corrected for porosity and extrapolated to higher temperatures [6, 12, 40, 50]

entire temperature range was 2.03 W/m·K at 2,950 K. Belle and Berman [12] estimated minimum values in conductivity using Eq. (36). They assumed that:

1. Phonon velocity can be approximated to $(E/\rho)^{0.5}$,
2. Equation dealing temperature variation in E can be extrapolated to higher temperatures, and
3. Minimum value of phonon mean free path can be approximated to lattice parameter.

They calculated the minimum in thermal conductivity for ThO_2 at 2,950 K as 2.07 W/m·K which is very near to the value (2.03 W/m·K) shown in Fig 13. Thermal conductivity data of UO_2 [50] and PuO_2 [6, 40] are also shown in the same figure. The lowest value for UO_2 is 2.19 W/m·K at 1,970 K. The upswing in

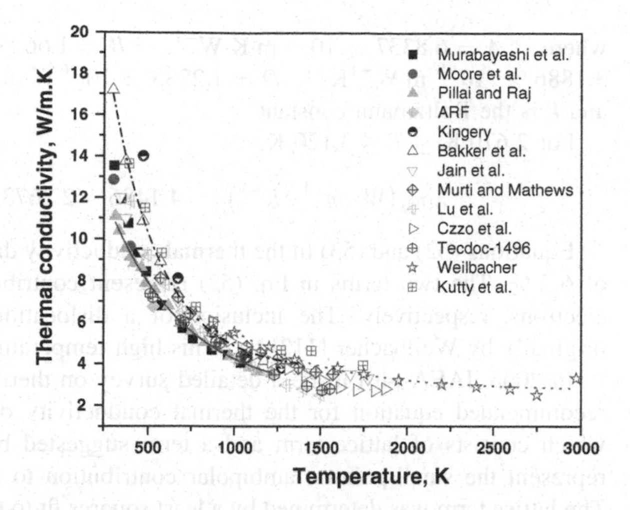

Fig. 14 Thermal conductivity of ThO_2 reported by various authors are corrected for 100 % density and plotted together. Solid line are fitted data of Pillai et al. [93], dotted line that of Weilbacher [112] and dashed line that of Bakker et al. [38]

thermal conductivity in UO_2 at $\sim$1,970 K can be explained in terms of the electronic contribution. On the other hand, increase in thermal conductivity beyond 2,950 K in ThO_2 is not result of electronic contribution, but is associated with increase in heat capacity. Thermal conductivity data of ThO_2 reported by various authors are shown in Fig. 14.

5.3 Thermal Conductivity of UO_2

There are many publications numbering over hundreds dealing with thermal conductivity of UO_2. Washington [116], Brandt and Neuer [117], and Fink et al. [50] made appraisals of the conductivity data found in the open literature. Brandt and Neur [117] presented a mean correlation curve of thermal conductivity versus temperature for UO_2 by using data from number of sources. Their equation had three terms: the first two terms are for phonon and electronic conductions, respectively. The third term stood for the decrease in thermal conductivity resulting from dislocations created at higher temperatures. Fink et al. [50] used a different model to fit the voluminous data on UO_2. They showed the evidence of a phase transition for UO_2 at 2,670 K from the enthalpy measurements and suggested a similar transition with temperature for thermal conductivity. Fink et al. [50] suggested a relation conforming with the enthalpy and heat capacity equations. Their relation is given below:

$$\lambda_{\mathrm{UO_2}}(W.m-1.K-1) = \left(A + \mathrm{BT} + \mathrm{CT}^2\right)^{-1} + \mathrm{DT}e^{-E/\mathrm{kT}}, (298 \le T \le 2670\ K) \tag{52}$$

where $A = 6.8337 \times 10^{-2}$ m·K·W^{-1}, $B = 1.6693 \times 10^{-4}$ m·W^{-1}, $C = 3.1886 \times 10^{-8}$ m·W^{-1}K^{-1}, $D = 1.2783 \times 10^{-1}$ W·m^{-1} K^{-2}, $E = 1.1608$ eV, and k is the Boltzmann constant.

For 2,670 K $\leq T \leq$ 3,120 K,

$$\lambda_{UO_2}\left(W \cdot m^{-1} \cdot K^{-1}\right) = 4.1486 - 2.2673 \times 10^{-4}\, T \quad (53)$$

Equations (52) and (53) fit the thermal conductivity data within an error margin of 6.2 %. The two terms in Eq. (52) represent contributions from phonons and electrons, respectively. The inclusion of a dislocation term as recommended originally by Weilbacher [112] to fit his high temperature data was not justified.

In 2006, IAEA [40] made a detailed survey on thermal conductivity data and recommended equation for the thermal conductivity of 95 % dense solid UO_2 which consists of lattice term and a term suggested by Ronchi et al. [118] to represent the small-polaron ambipolar contribution to the thermal conductivity. The lattice term was determined by a least squares fit to the lattice contributions to the thermal conductivity obtained by Ronchi et al. [118], Hobson et al. [119], Bates [120], Conway et al. [121] and Godfrey et al. [122]. The recommended equation for thermal conductivity of solid 95 % dense UO_2 is:

$$\lambda_{UO_2} = \left[100/\left(7.5408 + 17.692t + 3.6142\,t^2\right)\right] + \left(6400/t^{2.5}\right)\exp(-16.35/t) \quad (54)$$

where, t is T/1,000, T is in K, and λ is the thermal conductivity in W·m^{-1} K^{-1}. Thermal conductivity values for 100 % dense UO_2 or for a different density may be calculated using the porosity relation derived by Brandt and Neurer [117], which is:

$$\lambda_0 = \lambda_p/(1 - \sigma p), \quad (55)$$

where, $\sigma = 2.6\text{–}0.5t$. Here, t is T/1,000 where T is in K, p is the porosity fraction, λ_p is the thermal conductivity of UO_2 with porosity p, and λ_0 is the thermal conductivity of fully dense UO_2.

Uncertainties in thermal conductivity values for 298–2,000 K are 10 %. From 2,000 to 3,120 K, the uncertainty increased to 20 % because of the large discrepancies between measurements by different investigators [40]. Typical thermal conductivity of 95 % dense UO_2 as a function of temperature is given in Fig. 15.

The lattice term has traditionally been determined by fitting the low temperature thermal conductivity data because the lattice contribution dominates the thermal conductivity at low temperatures. Figure 16 shows the total thermal conductivity, the lattice contribution, and the ambipolar contribution as a function of temperature that have been calculated from the equation of Ronchi et al. [118], which is given below:

$$\lambda_{UO_2} = [100/(6.548 + 23.533t)] + \left(6,400/t^{2.5}\right)\exp(-16.35/t), \quad (56)$$

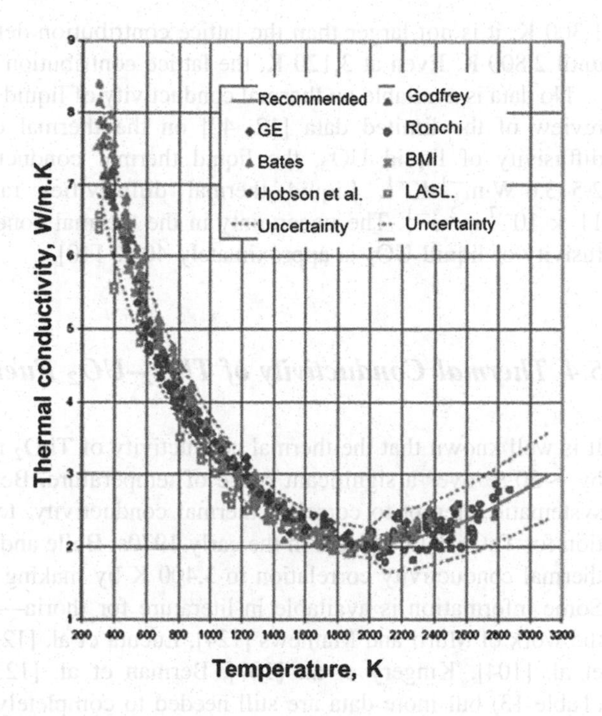

Fig. 15 Thermal conductivity data of 95 % dense UO_2 [40]. (permission from Elsevier)

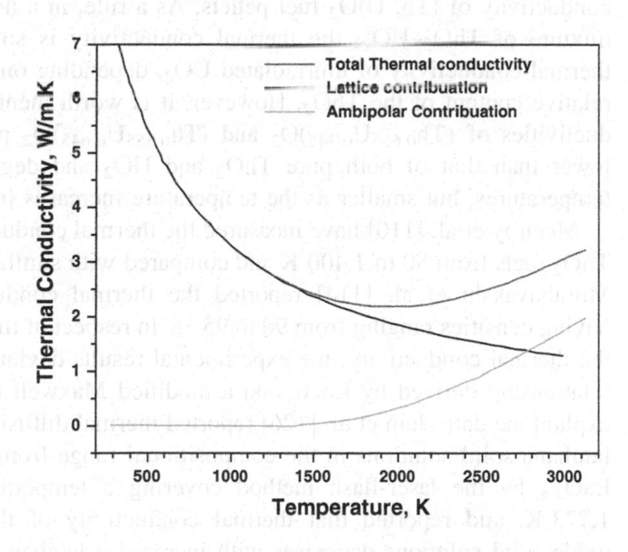

Fig. 16 Thermal conductivity of UO_2 showing lattice and electronic contributions [40]. (permission from IAEA)

where t is $T/1{,}000$, T is in K, and λ is the thermal conductivity for 95 % dense UO_2 in $W{\cdot}m^{-1}{\cdot}K^{-1}$. Below 1,300 K, the ambipolar term is insignificant and the total thermal conductivity equals the lattice contribution. Although the ambipolar term begins to have a significant contribution to the total thermal conductivity above

1,300 K, it is not larger than the lattice contribution determined by Ronchi et al. until 2,800 K. Even at 3,120 K, the lattice contribution is still significant.

No data is available on thermal conductivity of liquid ThO_2. Based on an initial review of the limited data [12, 40] on the thermal conductivity and thermal diffusivity of liquid UO_2, the liquid thermal conductivity is in the range of 2.5–3.6 $W{\cdot}m^{-1} K^{-1}$. Liquid thermal diffusivities range from 6×10^{-7} to 11×10^{-7} $m^2 s^{-1}$. The uncertainty in the thermal conductivity and thermal diffusivity of liquid UO_2 is approximately 40 % [40].

5.4 Thermal Conductivity of ThO_2–UO_2 Fuel

It is well known that the thermal conductivity of ThO_2 is higher than that of UO_2 by ~50 % over a significant range of temperature. Berman et al. [123] made a systematic attempt to correlate thermal conductivity, temperature, and composition for ThO_2–UO_2 system in the early 1970s. Belle and Berman [12] updated the thermal conductivity correlation to 3,400 K by making use of the enthalpy data. Some information is available in literature for thoria—urania mixtures are from the work of Murti and Mathews [124], Lucuta et al. [125], Pillai et al. [93], Belle et al. [104], Kingery et al. [109], Berman et al. [123], IAEA-TECDOCs etc. (Table 13) but more data are still needed to completely characterize the thermal conductivity of (Th, U)O_2 fuel pellets. As a rule, in a homogeneous unirradiated mixture of ThO_2–UO_2, the thermal conductivity is somewhat higher than the thermal conductivity of unirradiated UO_2, depending on the temperature and the relative content of the ThO_2. However, it is worth mentioning that thermal conductivities of $(Th_{0.655}U_{0.345})O_2$ and $(Th_{0.355}U_{0.645})O_2$ pellets were found to be lower than that of both pure ThO_2 and UO_2 and degradation is large at low temperatures, but smaller as the temperature increases [67].

Mcelroy et al. [110] have measured the thermal conductivity of sol–gel-derived ThO_2 fuels from 80 to 1,400 K and compared with similar measurements on UO_2. Murabayakshi et al. [114] reported the thermal conductivity of ThO_2 pellets having densities ranging from 90 to 95 %. In respect of the porosity dependence of the thermal conductivity, the experimental results deviated significantly from the relationship derived by Loeb, and a modified Maxwell model was introduced to explain the data. Jain et al. [126] reported thermal diffusivity of a range of thoria–lanthana solid solutions in the compositional range from pure thoria to 10 mol% $LaO_{1.5}$ by the laser-flash method covering a temperature range from 373 to 1,773 K, and reported that thermal conductivity of thorium oxide–lanthanum oxide solid solutions decreases with increasing lanthanum content and temperature. Ronchi et al. [118] measured thermal conductivity of $(Th_{0.88}U_{0.12})O_2$ in the temperature range of 573–1,573 K. Ferro et al. [127] evaluated diffusivity of $(Th_{0.94}U_{0.06})O_2$ and $(Th_{0.90}U_{0.10})O_2$ from 650–2,700 K. Lemehov et al. [4] presented a model for the lattice thermal conductivity of pure and mixed oxides based on the Klemens-Callaways approach for the dielectric heat conductance modeling

Table 13 Thermal conductivity measurements for $(Th_{1-y}U_y)O_2$

Authors	Year	Temperature range, *K*	Composition, % UO_2	Remarks
Kingery [109]	1959	373–1,070	0, 10, 26, 31, 100	
DeBoskey [113]	1962	570–1,100	0, 8, 10	
Harbinson et al. [155]	1966	1,073–2,073	10, 100	
Moore et al. [156]	1967	293–423	4.7, 6.1, 6.3	
Belle et al. [104]	1967	393	0, 10, 20, 30, 50, 90, 100	
Springer et al. [57]	1968	573–2,173	3, 5, 7, 10, 13, 20, 25, 30, 100	
Ferro et al. [127]	1968	873–1,673	1, 4, 10	
McElroy et al. [110]	1969	303–393 K		
MacEwan et al. [103]	1969	333	0, 1, 3	
Jacob [128]	1969	573–2,123 K	3.1, 7.7, 10.0	
Murabayashi [114]	1970	293–1,073	1, 3, 5, 10	
Berman et al. [123]	1972	573–2,273	0, 2, 5.10, 20	
Ferro et al. [127]	1972	923–2,973	6, 10	
KWU [76]	1979	370–1,663 K	5	
Young [5]	1979	ThO_2	0	Modeling
Rodriguez et al. [82]	1981	773–1,773	0, 20, 100	
Bask et al. [130]	1989	800–2,100 K	2	
Murti et al. [124]	1991	573–1,573 K		ThO_2–$LaO_{1.5}$ solid solutions
Konings et al. [39]	1995	273–2,200 K	<30 %	Review paper
Bakker et al. [38]	1997	273–1,073 K	<20 %	Review paper
Pillai et al. [93]	2000	300–1,200 K	2	
INEEL [67]	2002	293–1,673 K	65, 35	
Ronchi et al. [24]	2003	573–1,573	12	
Jain et al. [126]	2006	373–1,773 K	ThO_2–$LaO_{1.5}$ (<10 mol%)	ThO_2-lanthana
IAEA –TECDOC [40]	2006	873–1,873 K	2, 4, 6, 10	
Kutty et al. [84]	2008	298–1,500	4, 10, 20	

and on some correlations between thermoelastic properties of solids. The thermal conductivity of ThO_2 and $Th_{0.98}U_{0.02}O_2$ was measured from 300 to 1,200 K by Pillai and Raj [93] and they showed that the decrease in thermal conductivity of $Th_{0.98}U_{0.02}O_2$ over that of ThO_2 is due to the enhanced phonon–lattice strain interaction in the oxide. Murti and Mathews [124] measured thermal conductivity on thorium–lanthanum mixed oxide solid solutions covering a temperature range from 573 to 1,573 K and a compositional range from 0 to 30 mol% $LaO_{1.5}$ and reported that thermal conductivity of the solid solutions were found to decrease with increase in lanthanum oxide content or temperature. Kutty et al. [84] measured thermal conductivity of ThO_2, ThO_2–4 % UO_2, ThO_2–10 % UO_2 and ThO_2–20 % UO_2 made by coated agglomerate pelletization (CAP) process and reported that thermal conductivity decreased with UO_2 content. A study carried out by

INEEL [67] shows that ThO_2 has a higher thermal conductivity than UO_2, but (Th, U)O_2 containing 65 or 35 wt% ThO_2 has similar in thermal conductivity of UO_2.

An assessment of thermal conductivity data of both irradiated and unirradiated ThO_2 and $Th_{1-y}U_yO_2$ solid solutions has been made by Berman et al. [123]. They analyzed data of Springer et al. [57], Jacobs [128], Matolich and Storhok [129] and Belle et al. [12]. Berman et al. [123] suggested complex behavior of the parameters A and B of Eq. (39) on variation of the uranium content which is inconsistent with theory and data on other ThO_2 or UO_2 compounds containing substitutions. The assessment by Bakker et al. [38] used only those data sets that contain pure ThO_2, which show a systematic decrease of the thermal conductivity on increasing UO_2 content (for UO_2 concentrations up to 20 %). Since good agreement exists between the variation of the A and B parameter on substitution as determined by Murabayashi [114] and the variation of A and B of comparable compounds as well as that predicted by theory, these parameters are used to obtain a recommended thermal conductivity for $Th_{1-y}U_yO_2$. The uranium concentration dependence of the thus obtained A and B parameters were fitted to obtain an equation that is valid for uranium concentration up to 10 % and a theoretical density of 95 %:

$$A = 4.195 \times 10^{-4} + 1.112 \cdot y - 4.499 \cdot y^2, \tag{57}$$

$$B = 2.248 \times 10^{-4} - 9.170 \times 10^{-4} \cdot y + 4.164 \times 10^{-3} \cdot y^2 \tag{58}$$

The recommended equation for $(Th_{1-y}U_y)\ O_2$ containing up to 10 % UO_2 is:

$$\lambda_{(Th-U)O_2} = \Big[4.195 \cdot 10^{-4} + 1.112 \cdot y - 4.499 \cdot y^2 + \left(2.248 \cdot 10^{-4} - 9.170 \cdot 10^{-4} \cdot y + 4.164 \cdot 10^{-3} \cdot y^2\right) \cdot T\Big]^{-1} \tag{59}$$

The above equation is valid in the temperature range 300–1,173 K. Figure 17 shows thermal conductivity of ThO_2–UO_2 for various UO_2 contents.

An elaborative study has been reported in IAEA-TECDOC [40] on ThO_2 containing 4, 6, 10, and 20 % of UO_2. The following are the recommended equations for the thermal conductivity (λ) as a function of temperature (T/K) which is valid from 873 to 1,873 K:

$$\lambda[Th_{0.96}U_{0.04}]O_2 = 1/(-0.04505 + 2.6241 \cdot 10^{-4} \cdot T) \tag{60}$$

$$\lambda[Th_{0.80}U_{0.20}]O_2 = 1/(0.02771 + 2.4695 \cdot 10^{-4} \cdot T) \tag{61}$$

Subsequently, best-fit equation for thermal conductivity of $(Th_{1-y}U_y)O_2$ of 95 % theoretical density as a function of composition (y in wt%) and temperature (T/K) has been derived, which is valid through 873–1,873 K.

$$\lambda(y,\ T) = 1/[-0.0464 + 0.0034 \cdot y + (2.5185 \cdot 10^{-4} + 1.0733 \cdot 10^{-7} \cdot y) \cdot T] \tag{62}$$

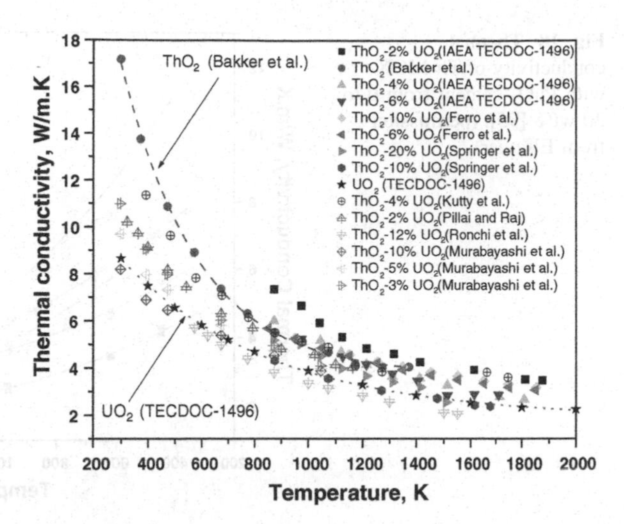

Fig. 17 Thermal conductivity of ThO_2-UO_2 fuels for various UO_2 contents reported in the literature. All the data used are corrected for 100 % density. Dashed line are fitted data of Bakker et al. [38] for pure ThO_2 and dotted line that of IAEA-TECDOC-1496 [40] for pure UO_2. (permission from IAEA)

5.5 Thermal Conductivity of ThO_2–PuO_2 Fuel

Although thoria-based fuels have been studied extensively in the past, namely in the 1970s, to our knowledge very little open literature is available for (Th, Pu)O_2 [40, 130]. Only a few measurements of thermal conductivity have been made for ThO_2–PuO_2 fuel. Since CeO_2 and PuO_2 have similar thermodynamic and crystallographic properties [131], Murbayashi [114] tried to simulate the thermal conductivity as a function of temperature and CeO_2 up to 10 wt% using Laser flash method. Jeffs [132, 133] determined the integral thermal conductivity of irradiated $(Th_{1-y}Pu_y)O_2$ containing 1.10, 1.75, and 2.72 wt% of PuO_2 using a steady state method. The thermal conductivity of a mixture of ThO_2 and 4 wt% PuO_2 was also measured by Basak et al. [130] using the laser flash technique for the temperature range of 950–1,800 K. Recently, Cozzo et al. [95] reported that at 500 K the thermal diffusivity of the Th-MOX can be down to 50 % of that of its pure oxide components ThO_2 and PuO_2. The presence of the two different oxides inside the Th-MOX lattice, generate a high amount of phonon scattering centers. When temperature increases, the plutonium concentration affects the thermal diffusivity of the fuel to a lesser extent, because the phonon–phonon scattering mechanism increases with temperature and becomes predominant when compared to the lattice strains due to the presence of either Th or Pu atoms in the lattice [95]. However, the thermal conductivity of pure PuO_2 was found to be higher than that of ThO_2 for all temperatures covered by their study. This is somewhat surprising and contradicts the understanding that ThO_2 always have a higher thermal conductivity than the other actinide oxides.

In Fig. 18, the thermal conductivity of Th-MOX with PuO_2 content varying from 0 to 30 wt% are shown. At low temperature, the thermal conductivity of the Th-MOX with a PuO_2 content from 0 to 30 wt% decreases with an increase of the

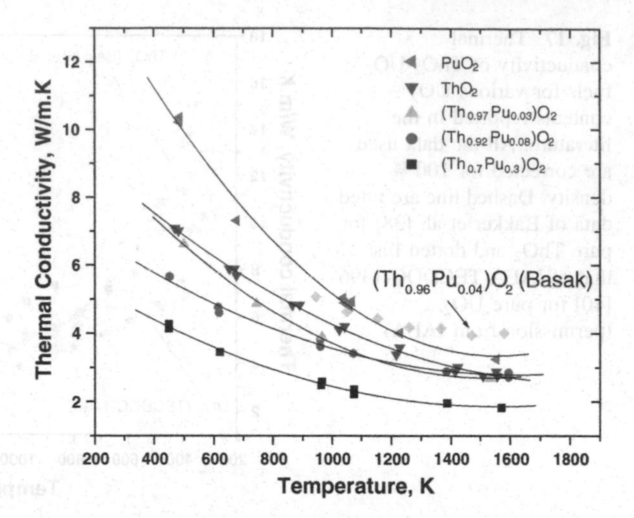

Fig. 18 Thermal conductivity of Th-MOX with PuO_2 content from 0 to 30 wt% [95]. (permission from Elsevier)

PuO_2 content. At higher temperature (above 1,000 K), the thermal conductivity of Th-MOX with a PuO_2 content from 0 to 8 wt% is almost independent from the concentration of plutonium. The conductivity of Th-MOX with 30 wt% PuO_2 at high temperatures is much lower [95].

The thermal conductivity λ, of $(Th_{1-y}Pu_y)O_2$ as a function of temperature and PuO_2 content is reported by IAEA study [40]. Figure 19 shows a systematic decrease of thermal conductivity with increasing PuO_2 content and temperature. The data are comparable with those obtained by Murabayashi [114] on simulated fuel samples of the composition ranging from 0 to 10 wt% CeO_2. The best-fit equation for the thermal conductivity, λ [W/m·K], of $(Th_{1-y}Pu_y)O_2$ as a function of composition, y [wt%], and temperature, T [K], was derived for the temperature range from 873 to 1,873 K [40].

$$\lambda(y, T) = 1/[-0.08388 + 1.7378 \cdot y + (2.62524 \cdot 10^{-4} + 1.7405 \cdot 10^{-4} \cdot y) \cdot T] \tag{63}$$

In order to introduce the influence of the plutonium content on parameter A, one can rely on the simplified theory of Abeles [95]. The parameter A has a second-order dependence on both the relative mass and radius differences as per the above theory. A polynomial equation of the second degree was chosen to define $A_{(PuO2)}$:

$$A_{(PuO_2)} = A_0 + A_1 \cdot [PuO_2] + A_2 \cdot [PuO_2]^2, \tag{64}$$

($[PuO_2]$ = Concentration of PuO_2 in wt%).

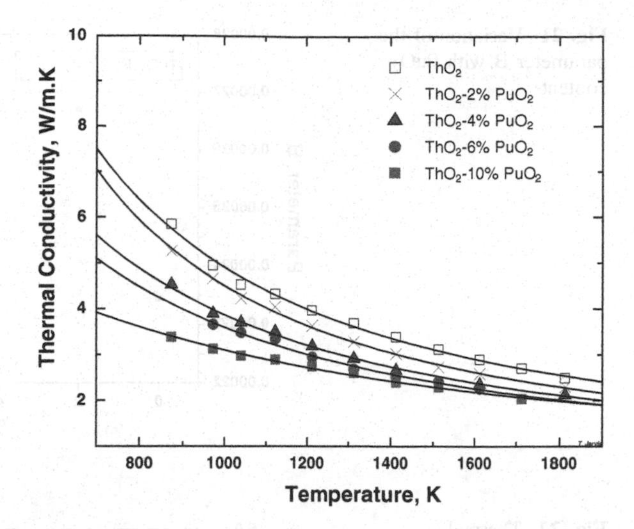

Fig. 19 Systematic decrease of thermal conductivity with increasing PuO_2 content and temperature for ThO_2–PuO_2 system [40]. (permission from IAEA)

The values of the parameters are [95]:

$$A_0 = 6.071 \times 10^{-3}\, mKW^{-1},$$
$$A_1 = 5.72 \times 10^{-1}\, mKW^{-1},$$
$$A_2 = -5.937 \times 10^{-1}\, mKW^{-1}.$$
$$B = 2.4 \times 10^{-4}\, mW^{-1}.$$

Figures 20 and 21 show the variation *A* and *B* parameters with PuO_2 content for ThO_2–PuO_2 system. The parameter *A* increases with increase in PuO_2 while the variation of *B* with PuO_2 content was found to be random.

The experimental thermal conductivity data of high Pu bearing hypostoichiometric and stoichiometric mixed thorium–plutonium oxide of compositions,

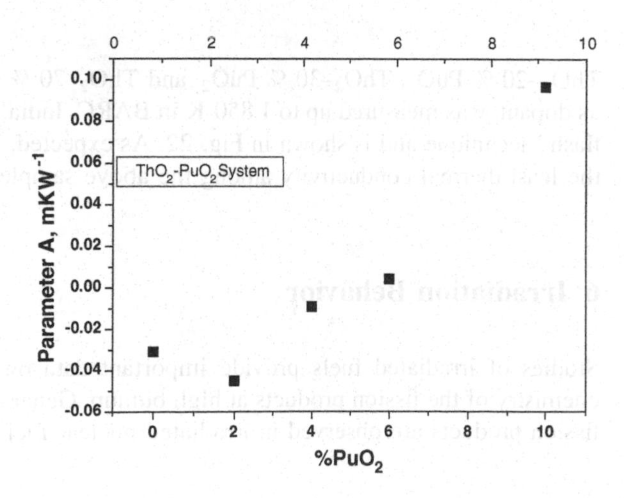

Fig. 20 Variation of parameter A with PuO_2 content

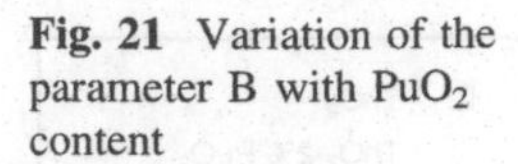

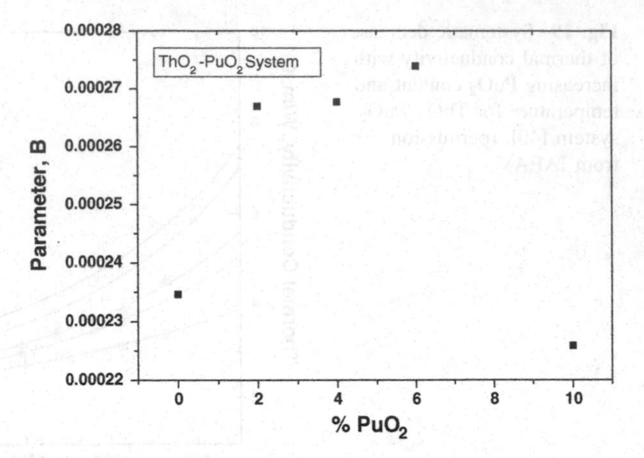

Fig. 21 Variation of the parameter B with PuO_2 content

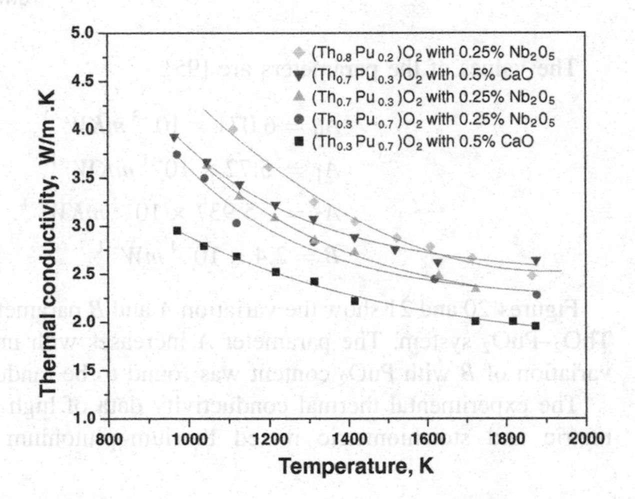

Fig. 22 Thermal conductivity data of ThO_2–20 % PuO_2, ThO_2–30 % PuO_2, and ThO_2–70 % PuO_2 with CaO or Nb_2O_5 as dopant [58]. (permission from IAEA)

ThO_2–20 % PuO_2, ThO_2–30 % PuO_2, and ThO_2–70 % PuO_2 with CaO or Nb_2O_5 as dopant, was measured up to 1,850 K in BARC, India, by employing the "Laser-flash" technique and is shown in Fig. 22. As expected, ThO_2–70 % PuO_2 showed the least thermal conductivity among the above sample.

6 Irradiation Behavior

Studies of irradiated fuels provide important data on the thermodynamics and chemistry of the fission products at high bumup. Generally, four distinct groups of fission products are observed in irradiated nuclear fuel [85, 125]:

(a) oxides dissolved in the matrix: Sr, Zr, Nb, Y, La, Ce, Pr, Nd, Pm, Sm;
(b) metallic precipitates: Mo, Tc, Ru, Rh, Pd, Ag, Cd, In, Sb, Te;
(c) oxide precipitates: Ba, Zr, Nb, Mo, (Rb, Cs, Te);
(d) gases and other volatile elements, e.g.: Kr, Xe, Br, I.

6.1 Studies on ThO_2

In India, ThO_2 has been extensively used in PHWRs for neutron flux flattening of the initial core during start-up. During the last few years, ThO_2 bundles have been utilized for this purpose in seven units of PHWR 220 including the two units each at Kakrapar Atomic Power Station (KAPS 1&2), Kaiga Atomic Power Station (KGS 1&2), Rajasthan Atomic Power Station (RAPS 3&4), and in RAPS 2 after mass coolant channel replacement. So far, some 232 thoria bundles have been successfully irradiated in the operating PHWR up to a maximum power of 408 kW and burnup of 13,000 MWd/Te HM without any failure [134]. Details of irradiation of ThO_2 bundles in India are shown in Table 14. In-pile irradiation of Zircaloy-clad (Th, Pu)O_2 fuel pins have been successfully carried out in the pressurized water loop of CIRUS research reactor. A six-pin cluster of free-standing Zircaloy–2 cladded ThO_2–4 % PuO_2 was successfully irradiated up to a burnup of 18,400 MWd/Te. Subsequently, two additional six pin clusters of collapsible Zircaloy–2 cladded pins containing high density ThO_2 and ThO_2–6.75 % PuO_2 were successfully irradiated up to a burnup of 10,300 MWd/Te without failure. The peak pin-power rating was 40 kW/m [135]. Details of ThO_2–PuO_2 pin irradiation are shown in Table 15.

Although several isolated irradiation experiments on ThO_2 and ThO_2–UO_2 materials have been conducted, the bulk of the irradiation data generated has come US from four goal-oriented programs [8, 12, 39, 136]:

1. Boiling water reactor (BWR) Program, which culminated in the irradiations performed in the BORAX-IV and Elk River reactors.
2. Thorium Utilization Program, which included the irradiation of vibratory compacted and pelletized fuels.

Table 14 Irradiation of ThO_2 bundles in Indian PHWRs

Reactor	Number of bundles
Madras-1	4
Kakrapar-I	35
Kakrapar-II	35
Rajasthan-II	18
Rajasthan-III	35
Rajasthan-IV	35
Kaiga-I	35
Kaiga-II	35

Table 15 Details of the irradiation of ThO_2–PuO_2 fuels

Fuel	Cladding type	Burnup (GWD/t)	Linear rating (kW/m)
ThO_2–4 % PuO_2	Free-standing	18.5	40
ThO_2–6.75 % PuO_2	Collapsible	10.2	42

3. Babcock and Wilcox developmental work on pressurized water reactor (PWR) fuels.
4. Light water breeder reactor (LWBR) Program at Bettis Atomic Laboratory which resulted in the core loading of the Shippingport Reactor.

The majority of experiments were performed on fuels with less than 10 wt% UO_2. From the above experiments and other experiments carried out in India, Russia, Canada, Japan, and Korea the following conclusions can be drawn about thermophysical properties.

6.2 Melting Point

The incorporation of transmuted elements into a solid phase is expected to change the melting point significantly. Because of the high melting points of urania and thoria, it would normally be expected that the introduction of other oxides would cause a decrease in the melting point [85]. No direct measurements were available on the effects of irradiation on melting points on ThO_2 or ThO_2–UO_2 solid solutions, but there are some data on UO_2 and simulated mixed oxide fuel. Christensen et al. [137] have measured the melting point of UO_2 and measured the effect of irradiation on melting point up to 13 fission units. They reported a nearly linear decrease of melting point of irradiated UO_2 with irradiation at a rate of 9 K per fission unit.[1] Konno et al. [138] studied melting temperature of simulated high burnup mixed oxide fuels for fast reactor corresponding to different burnup and found that melting temperature decreases with increase in fuel burnup. Yamamoto et al. [139] and Hirosawa et al. [140] also obtained similar results for FBR MOX fuel. Belle and Berman [12] derived that an irradiation of 1 fission unit (10^{26} fissions/m^3) should depress the melting point of thoria–urania fuel by about 5 K. This value differs from the 9 K per fission unit for urania may be due to a difference in the oxygen content or to the accumulation of about 0.3 mol% plutonia in the urania fuel [85].

[1] One fission unit = 4072 MWd/MTM (megawatt days per metric ton of heavy metal content) [85].

6.3 Thermal Conductivity

In general, the thermal conductivity is expected to decrease as a material is irradiated. This decrease is attributed to radiation damage to the lattice, and pellet cracking. The thermal conductivity measurements of irradiated and unirradiated materials can be approximately equal when the lattice damage is annealed during irradiation at high temperatures or thermal cycling as would occur during out-of-pile thermal conductivity measurements.

Effect of Irradiation on Conductivity

The thermal conductivity of irradiated nuclear fuel is affected by [12],

i. solid fission products—dissolved, and precipitated,
ii. pores and fission gas bubbles,
iii. deviation from stoichiometry,
iv. radiation damage, and
v. circumferential cracks.

The expression for the parametric dependence of irradiated fuel thermal conductivity, λ can be provided in a form of contributing factors for each individual effect [141–143]:

$$\lambda = k_I(\beta)\,x\,k_2(P)\,x\,k_3(x)\,x\,k_4(r)\,x\lambda_0(T)\,(W/mK), \tag{65}$$

where, $k_I(\beta)$ is the burnup (β) dependence factor, $k_2(P)$ is the porosity/bubbles (p) contribution, $k_3(x)$ describes the effect of *O/M* ratio (x-deviation from stoichiometry), $k_4(r)$ refers to the radiation damage, and λ_0 is the analytical expression for thermal conductivity of unirradiated fuel. Such an analytical expression can be easily adapted into fuel codes. The solid fission products formed during irradiation (dissolved and precipitated) affect the fuel thermal conductivity by changing the lattice contribution. Attempts have been made theoretically or experimentally to evaluate the effect of the fission products on thermal conductivity. It is shown that the dissolved fission products lower the thermal conductivity, whereas the precipitated fission products increase the thermal conductivity [5].

Radiation damage from neutrons, α-decay and fission increases the number of lattice defects and consequently reduces the thermal conductivity of the fuel. Early work at Chalk River showed that reactor radiation damage, for short irradiations, and below 1,000 K, results in a maximum reduction of about 25 % [8]. The decrease was very rapid at the beginning of the irradiation (10^{19}–10^{21} n/m^2, hence order of minutes to hours) and no further effect was found above about 10^{23} n/m^2. A limiting value of 3.5 W/m·K was reported for longer irradiations (up to 2.8×10^{24} n/m^2, corresponding to about a month of irradiation). Similar results were also obtained on irradiated ThO_2–1.3 wt% UO_2 in out-of-pile measurements [12].

Microstructural changes, lattice damage, fission depletion effects, and concomitant effect of pellet cracking leads to lowering thermal conductivity in irradiated fuels. The thermal resistivity relation developed by Belle and Burman et al. [12] for oxide fuels can be modified by taking into the effects of irradiation by adding two terms to the resistivity of unirradiated material as:

$$R_{irr} = R_a + C/T + DF/T, \quad (66)$$

where, C and D are constants equal to 32.4 and 11.1, respectively. R_{irr} and R_a are in (m·K)/W, F is in fission units, and T is in K. The term C/T term accounts for the increase in thermal resistivity during the early stage of irradiation due to the lattice defects concentration caused by neutron and fission. The term DF/T account for the decrease in thermal conductivity, due to the accumulation of transmuted elements. These fission product atoms become more mobile with increasing temperature and tend to agglomerate and precipitate. A large number of reports on thermal conductivity of irradiated UO_2 fuel are available. It shows that below 773 K, thermal conductivity is reduced as a result of induced lattice defects. Between 773 K to melting point of UO_2, no substantial change in thermal conductivity is observed.

Jacobs [128] compared the unirradiated and in-pile-irradiated thermal conductivities of ThO_2–10 % UO_2. The data were fitted to equation and the thermal conductivity of the unirradiated specimen (λ_{un}) varied with temperature, T (°C), as shown in Equation:

$$\lambda_{un} = (8.4703 + 0.02551\, T)^{-1} \quad (67)$$

The thermal conductivity of the irradiated specimen (λ_{irr}) varied with temperature, T as

$$\lambda_{irr} = (10.5181 + 0.02003\, T)^{-1} \quad (68)$$

Jacobs [128] concluded that there is no statistical difference at the 0.99 confidence level between the thermal conductivity of unirradiated ThO_2–10 wt% UO_2 and ThO_2–10 wt% UO_2 irradiated to 0.26×10^{18} fissions/cm^3. MacEwan and Stoute [103] reported that thermal conductivity of ThO_2–1.3 mol% UO_2 at 333 K before and after irradiation and found that irradiation at low temperature decreased the thermal conductivity. Irradiation exposure varied from 3.1×10^{-6} to 4.7×10^{-4} fission units. Increasing the irradiation temperature to 600–629 K reduced the extent of decrease in thermal conductivity which may be due to the annealing of damage during irradiation. Thermal conductivity of ThO_2 and ThO_2–UO_2 compositions was measured by Jacobs [128] during the reactor irradiation and reported no statistically significant difference between in reactor and unirradiated values for ThO_2 and ThO_2–9.8 mol% UO_2 below 1,273 K. Matolich and Storhok [129] determined post-irradiation thermal conductivity of irradiated ThO_2–3 % UO_2, ThO_2–9.8 % UO_2, and ThO_2–14.8 % UO_2 (composition in mol%). They could not find any significant differences between irradiated and unirradiated measurements for ThO_2–3 % UO_2 which they concluded that due to the annealing of the lattice damages at high temperature irradiation at 1,173 K. For ThO_2–9.8 % UO_2, anomalous results were obtained but the results were uncertain. The measured post-irradiation conductivity of ThO_2–14.8 % UO_2 was lower. After annealing between 573 and 1,473 K, irradiated thermal conductivities were about 55 % below the unirradiated values. But, the materials had internal cracks and annealing effect was improper. The above results point to the fact that no quantitative conclusions can be made regarding the lattice damage and fission depletion effects on the thermal conductivity of ThO_2 and ThO_2–UO_2.

The differences in properties will cause ThO_2–UO_2 fuel rods to behave somewhat differently than UO_2 fuel rods during both normal operation and design basis accident conditions. During normal operation, ThO_2–UO_2 fuel will operate with somewhat lower fuel temperatures and less fission gas release than UO_2 fuel at corresponding powers and burnups. Several important conclusions can be drawn from this analysis. The mixed ThO_2–UO_2 fuel, using a mixture of 70 wt% ThO_2 and 30 wt% UO_2, where the uranium is initially enriched to 19.5 wt% U-235, appears to have sufficient reactivity to be used for extended burnup cycles to 72 GWD/t in LWRs. Likewise, a mixture containing 35 wt% UO_2, with the same enrichment and 65 wt% ThO_2, appears suitable for extended cycles approaching 90 GWD/t of initial heavy metal. The in situ breeding of U-233 maintains a more uniform reactivity during the course of irradiation and reduces the need for burnable poisons [67].

6.4 Fission Gas Release Features

Fission gas release (FGR) is primarily dependent on fuel temperature and fuel burnup [58, 144]. Figure 23 show a plot of FGR versus element power. It compares fission gas behavior for both ThO_2 and UO_2. Below a linear rating of ~40 kW/m, fuel microstructure plays a minimal role in FGR due to the low fuel temperature. The Fig. 23 compares the performance between UO_2 and granular ThO_2 although non-granular thorium demonstrates superior performance. A granular ThO_2 results in elevated central temperatures (despite the higher thermal conductivity of the ThO_2), which subsequently causes increased fission gas release which is comparable to that of similarly operated UO_2 [3]. The granular fuel often contains networks of tunnels that assist fission gas transport to the free void volume. The ratio of the granular FGR to non-granular FGR above 40 kW/m is approximately 2–4. Therefore, high quality non-granular thorium fuel exhibits significantly less fission gas release (two to four times less), even at higher power ratings and burnups.

6.5 Dimensional Stability

Very little irradiation-induced swelling occurs in ThO_2–UO_2 fuels up to 4 at.% burnup. In general, the volume change is less than 1 % for each at.% burnup. This relationship holds up to 10 at.%.

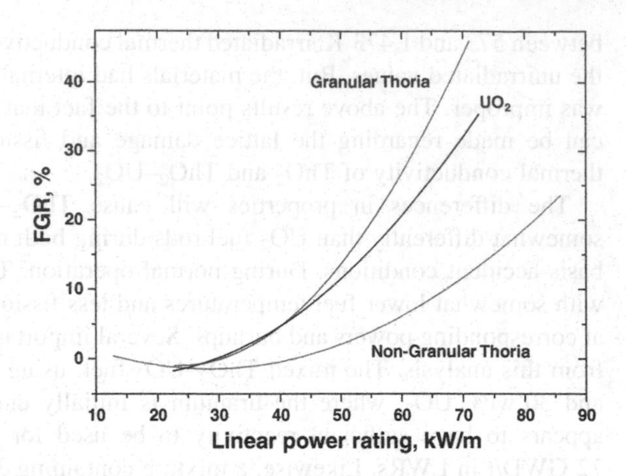

Fig. 23 Fission gas release from UO_2 and ThO_2 fuel [144]. (permission from IAEA)

6.6 Structural Stability

Post-irradiation examination (PIE) studies carried out on ThO_2 fuel bundles of KAPS irradiated up to 11,725 MWD/T of burnup showed excellent behavior of ThO_2 fuel during reactor operation. Gamma scanning was carried out on irradiated ThO_2 and UO_2 fuel pins of different burn-up to generate information on axial burnup distribution and power distribution profile in the fuel bundle using the ^{137}Cs as fission monitor. The relative power profile of UO_2 and ThO_2 fuel bundle showed significant difference. The inner ring fuel elements in ThO_2 fuel bundle showed higher power than the outer fuel elements in the bundle showing effect of epithermal neutrons. Figure 24 shows β–γ autoradiographs of the outer elements of UO_2 and ThO_2 having similar burn-up. Figure 24 shows that presence of Cs in cracks and also it can be noticed that number of cracks in ThO_2 is not as extensive like UO_2. The uniform presence of Cs in ThO_2 indicates that thermal gradients in ThO_2 are not large enough to cause migration of fission products [134].

6.7 ThO_2–PuO_2

PIE was carried out on experimental MOX (UO_2–4 % PuO_2) fuel elements of AC-4 cluster, that contained fuel elements pellets produced by different fabrication routes and varying pellet clad gaps and filler gas composition, after test irradiation in pressurized water loop (PWL) of CIRUS. Fuel pins with controlled porosity pellets and pellets with central hole showed very low fission gas release even at 110 W/cm^2 heat flux. Fuel pin containing low temperature sintered pellets showed abnormal gas release [134].

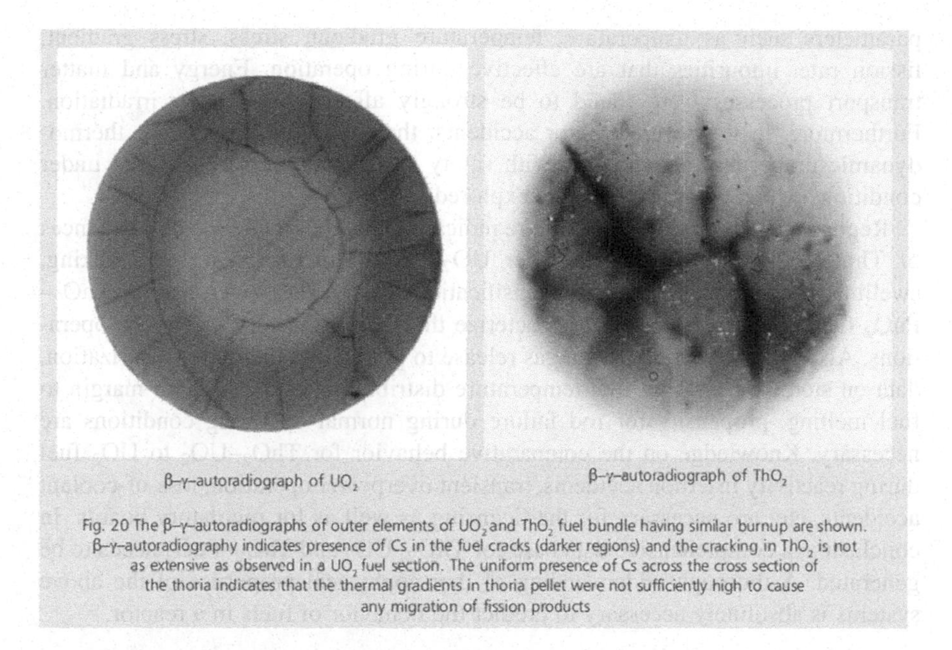

β–γ–autoradiograph of UO2
β–γ–autoradiograph of ThO2
Fig. 20 The β–γ–autoradiographs of outer elements of UO2 and ThO2 fuel bundle having similar burnup are shown. β–γ–autoradiography indicates presence of Cs in the fuel cracks (darker regions) and the cracking in ThO2 is not as extensive as observed in a UO2 fuel section. The uniform presence of Cs across the cross section of the thoria indicates that the thermal gradients in thoria pellet were not sufficiently high to cause any migration of fission products

Fig. 24 β–γ autoradiographs of the outer elements of UO_2 and ThO_2 having similar burnup [134]

7 Conclusions

Improvement of the nuclear fuel exploitation has been one of the main objectives of reactor technology during the last decades. In recent years, considerable progress has been made in the understanding of fuel thermal performance. Accurate knowledge of the thermal properties of the fuel material is needed for assessment of reactor behavior under transient conditions. Assurances are needed that thermophysical properties available are sufficiently accurate and require further verification, documentation, peer review of existing data etc. Few selective measurements are also needed to obtain new data at conditions for which data are currently lacking or highly inadequate. There is also a need for international database and collaborative research on out-of-pile and in-pile property evaluation and irradiation testing especially on Th-based fuels.

Thermophysical data of ThO_2-based fuel, especially of ThO_2–PuO_2 system, are limited during irradiation and must be known prior to their use in existing or advanced reactors. Also high temperature properties of Th-based fuels at temperatures >2,000 K are very sparse and fresh attempts should be made to measure these data so that they can be effectively used by the fuel designers. A high confidence level on the fuel performance can only be reached from a good interpretation of the irradiation data followed by post-irradiation examinations. A prerequisite for this is to have data on out-of-pile properties such as thermal conductivity or thermal expansion that allows to understand the influence of parameters such as temperature, temperature gradient, stress, stress gradient, fission rate, impurities that are effective during operation. Energy and matter transport processes were found to be strongly affected by reactor irradiation. Furthermore, in the severe reactor accidents, the fuel high temperature thermodynamic properties must comply with safety requirements to be satisfied under conditions which have been not yet explored.

Reports are available in the literature indicating that the irradiation performances of ThO_2–UO_2 fuel are superior over UO_2 fuels. However, data on cracking, swelling, microstructural changes, densification etc. for ThO_2–UO_2 and for ThO_2–PuO_2 fuels are needed to fully characterize these fuels during steady state operations. Also information on fission gas release to predict the internal pressurization, data on stored energy and fuel temperature distribution, knowledge on margin to fuel melting, propensity for rod failure during normal operating conditions are necessary. Knowledge on the comparative behavior for ThO_2–UO_2 to UO_2 fuel during reactivity insertion accidents, transient overpower operation, loss-of-coolant accidents, etc. are necessary for the licensing as well as for regulatory boards. In conclusion, a comprehensive data base for ThO_2–UO_2 and ThO_2–PuO_2 needs to be generated. A thorough understanding of thermophysical properties of the above systems is absolutely necessary to predict the behavior of fuels in a reactor.

Appendix 1

The coefficient of linear thermal expansion is defined as the differential change in length per unit change in temperature [32]. The instant temperature coefficient of the expansion is thus expressed as:

$$\alpha_l(T) = (1/L)\,(\delta L/\delta T)_P \tag{A1}$$

where, L is the length of the substance at temperature, T. The instant volume expansion coefficient, similarly defined, has the well-known connection to $\alpha_l(T)$ as $\alpha_V(T) \equiv (1/V)\ (\delta V/\delta T)_P = 3\alpha_l(T)$. The corresponding mean coefficient over a temperature interval (T_0 and T) is given by

$$\alpha_l = (1/L_0).(L - L_0)/(T - T_0) \equiv (1/\Delta T)(\Delta L/L_0), \tag{A2}$$

where, $\Delta T = T - T_0$ and $\Delta L = L - L_0$. Equation (A2) is more useful because most data compilations are given in terms of the dilation, $\Delta L/L_0$. For materials with isotropic thermal expansion property, the $\Delta L/L_0$ data can be directly correlated to the volume expansion and density lowering. This is so as the dilation factor (L/L_0) involved in Eq. (A2) can be used to express the corresponding volume ratio V/V_0 and equivalently, the density ratio (ρ_0/ρ) of the material as $V/V_0 \equiv \rho_0/\rho = (L/L_0)^3$, where $L/L_0 = 1 + \alpha_l\,\Delta T$. Like α_l one defines α_V, the mean volume expansion coefficient as $\alpha_V = (1/\Delta T)\,(\Delta V/V_0)$, which results in $V/V_0 = 1 + \alpha_V\,\Delta T$. Use of this result in the volume, or, density ratios leads to the correlation between the two coefficients as $1 + \alpha_V \, \Delta T = (1 + \alpha_l \, \Delta T)^3$, which in other words leads to,

$$\alpha_v = 3\alpha_l + 3\alpha_1^2 \Delta T + \alpha_1^3 (\Delta T)^2 \tag{A.3}$$

The result reproduces the well-known correlation of instantaneous values of linear and volume expansion coefficients, $\alpha_V(T) = 3\alpha_l(T)$. The additional higher order terms involved in the correlation between the two mean coefficients generally making small contribution one can use the approximated form, $\alpha_v \approx 3\alpha_l$, for all practical purpose. The density ratio having the form,

$$\rho_0/\rho = (L/L_0)^3 \equiv (1 + \alpha_l \, \Delta T)^3, \tag{A.4}$$

can be approximated to $\rho_0/\rho \approx (1 + 3\alpha_1 \, \Delta T)$, or equivalently, to $\rho_0/\rho \approx (1 + 3\Delta L/L_0)$. Thus, the fractional density change $(\rho_0 - \rho)/\rho_0 \equiv \Delta\rho/\rho_0$, is written as

$$\Delta\rho/\rho_0 \approx 3(\Delta L/L_0) \equiv 3\alpha_l \Delta T \tag{A.5}$$

The original relation $\rho_0/\rho = (L/L_0)^3$ can be approximated as

$$(L - L_0)/L_0 \equiv (\Delta L/L_0) \approx (\rho_0/\rho)^{0.33} - 1 \tag{A.6}$$

for expressing thermal dilations from tabulated values of density at different temperatures.

References

1. Olander DR (1976) Fundamental aspects of nuclear reactor fuel elements. TID-26711-P1, Energy Research and Development Administration, Oak Ridge
2. Matzke H (1986) Science of advanced LMFBR fuels: a monograph on solid state physics, chemistry and technology of carbides, nitrides and carbonitrides of uranium and plutonium. North Holland, Amsterdam
3. Olander DR (2009) Nuclear fuels—present and future. J Nucl Mater 389:1–22
4. Lemehov SV, Sobolev V, Van Uffelen P (2003) Modelling thermal conductivity and self-irradiation effects in mixed oxide fuels. J Nucl Mater 320:66–76
5. Young RA (1979) Model for the electronic contribution to the thermal and transport properties of ThO_2, UO_2, and PuO_2 in the solid and liquid phases. J Nucl Mater 87:283–296
6. Popov SG, Carbajo JJ, Ivanov VK, Yoder GL (2000) Thermophysical properties of MOX and UO_2 fuels including the effects of irradiation. Report ORNL/TM-2000/351, Oak Ridge National Laboratory, Oak Ridge
7. Kutty TRG, Ganguly C, Sastry DH (1995) Thermal expansion of Al-U and Al-U-Zr alloys. J Nucl Mater 226:197–205
8. Herring JS, MacDonald PE (1999) Characteristics of mixed thorium—uranium dioxide high burnup fuel. ANS annual meeting, June 6–10, Idaho National Engineering and Environmental Laboratory Report INEEL/CON-99-00141
9. Herring JS, MacDonald PE, Weaver KD, Kullberg C (2001) Low cost, proliferation resistant, uranium–thorium dioxide fuels for light water reactors. Nucl Eng Des 203:65–85

10. Oggianu SM, Kazimi MS (2000) A review of properties of advanced nuclear fuels. MIT report MIT-NFC-TR-021. Nucl Eng Dept MIT
11. Kazimi MS, Czerwinski KR, Driscoll MJ, Hejzlar PJ, Meyer JE (1999) On the use of thorium in light water reactors. MIT report MIT-NFC-TR-016. Nucl Eng Dept MIT
12. Belle J, Berman RM (1984) Thorium dioxide: properties and nuclear applications. Naval Reactors Office, United State Department of Energy, Government Printing Office, Washington
13. Olson GL, McCardell RK, Illum DB (1999) Fuel summary report: shippingport light water breeder reactor. Idaho National Engineering and Environmental Laboratory Report INEEL/EXT-98-00799 Rev.1
14. Lindemann FA (1910) The calculation of molecular vibration frequencies. Physik Z 11:609–612
15. Haire RG (1995) Comparison of the chemical and physical properties of f-element metals and oxides: their dependence on electronic properties. J Alloys Compd 223:185–196
16. Brooks MSS, Johansson B, Skriver HL (1984) Handbook on the physics and chemistry of the actinides, vol I. North-Holland, New York, pp 153–270
17. Yun Y, Oppeneer PM, Kim H, Park K (2009) Defect energetics and Xe diffusion in UO_2 and ThO_2. Acta Mater 57:1655–1659
18. Ruff O, Ebert F, Woitinek H (1929) Contribution to the ceramics of very refractory materials III: the system ZrO_2-ThO_2. Z Anorg Allgem Chem 180:252–256
19. Wartenberg H, Reusch HJ (1932) Schmelzdiagramme höchstfeuerfester oxyde. VI Berichtigung Z Anorg Allgem Chem 208:380–381
20. Geach GA, Harper ME (1953) Arc melting of non-metallic materials and a redetermination of the beryllia-thoria system. Metallurgia 47:269
21. Lambertson WA, Mueller MH, Gunzel FH (1953) Uranium oxide phase equilibrium systems: IV, UO_2–ThO_2. J Am Ceram Soc 36:397–399
22. Benz R (1969) Thorium-thorium dioxide phase equilibria. J Nucl Mater 29:43–49
23. Chikalle TD, McNeilly CE, Bates JL, Rasmussen JJ (1972) Collogues international C.N.R.S. No. 205, Etude Des transformations cristallines a haute temperature, Odeillo, 27–30 September, 1971
24. Ronchi C, Hiernaut JP (1996) Experimental measurement of pre-melting and melting of thorium dioxide. J Alloys Compd 240:179–185
25. Rand MH (1975) Thorium: physico-chemical properties of its compounds and alloys. Atom Energy Rev 5:35
26. Peterson S, Curtis CE (1970) Thorium ceramics data manual volume I—oxide. Report ORNL-4503 Oak Ridge Nation Laboratory, Oak Ridge
27. Christensen JA (1962) Irradiation effects on uranium dioxide melting. Report HW-69234, Hanford Atomic Product Division, Richland
28. Hohorst JK (ed) (1990) SCDAP/RELAP5/MOD2 code manual, vol 4: MATPRO a library for material properties for light water reactor accident analysis. NUREG/CR-5273
29. Brassfield HC, White JF, Sjodahl L, Bittel JT (1968) Recommended property and reactor kinetics data for use in evaluating a light water coolant reactor loss-of-coolant incident involving Zircaloy-4 or 304-SS-Clad UO_2. Report GEMP-482, General Electric Co Cincinnati, Ohio
30. Lyon WL, Baily WE (1967) The solid-liquid phase diagram for the UO_2-PuO_2 system. J Nucl Mater 22:332–339
31. Popov SG, Carbajo JJ, Ivanov VK, Yoder GL (2000) Thermophysical properties of MOX and UO_2 fuels including its effects of irradiation. ORNL/TM-2000/351, Oak Ridge National Laboratory, Oak Ridge
32. Rand MH, Ackermann RJ, Gronvold F, Oetting FL, Pattoret A (1978) Rev Int Des Hautes Temp et des Refract 15:355–365
33. Harding JH, Martin DG and Potter PE (1989) Thermophysical and thermochemical properties of fast reactor materials. Commission of European Communities Report EUR 12402

34. Latta RE, Fryxell RE et al (1970) Determination of solidus-liquidus temperatures in the UO_{2+x} system ($-0.50 < x < 0.20$). J Nucl Mater 35:195–210
35. Christensen JA (1964) Irradiation effects on uranium dioxide melting. United States Report HW-69234, Hanford Atomic Product Division, Richland
36. Ronchi C, Hiernaut JP, Selfslag R, Hyland GJ (1993) Laboratory measurement of the heat capacity of urania up to 8000 K: I. Experiment. Nucl Sci Eng 113:1–19
37. Christensen JA (1963) UO_2-ThO_2 phase studies. General Electric Report HW-76559, Richland, Washington
38. Bakker K, Cordfunke EHP, Konings RJM, Schram RPC (1997) Critical evaluation of the thermal properties of ThO_2 and $Th_{1-y}U_yO_2$ and a survey of the literature data on $Th_{1-y}Pu_yO_2$. J Nucl Mater 250:1–12
39. Konings RJM, Blankenvoorde PJAM, Cordfunke EHP, Bakker K (1995) Evaluation of thorium based nuclear fuels: chemical aspects. ECN-R-95-007, Netherlands Energy Research Foundation, ECN
40. IAEA-TECDOC 1496 (2006) Thermophysical properties database of materials for light water reactors and heavy water reactors. International Atomic Energy Agency, Vienna
41. Freshley MD, Mattys HM (1962) Ceramics research and development operation quarterly report October–December 1962. Report HW-76300. Hanford Power Products Division, Richland, Washington
42. Freshley MD, Mattys HM (1963) Properties of sintered ThO_2–PuO_2. Report HW-76302. Hanford Power Products Division, Richland, Washington
43. Gmelin Handbook der Anorgnischen Chemie (1978) Thorium, Suppl, vols C1 and C2 8th edn. Springer, Berlin
44. Mulford RNR, Ellinger FH (1958) ThO_2–PuO_2 and CeO_2–PuO_2 solid solutions. J Phys Chem 62:1466–1467
45. Dawson JK (1952) Magnetochemistry of the heaviest elements. Part VI. PuO_2–ThO_2 and PuF_4–ThF_4 solid solutions. J Chem Soc 1882–1886
46. Gronvold F (1955) High temperature X-Ray study of uranium oxides in UO_2–U_3O_8 region. J Inorg Nucl Chem 1:357–370
47. Martin DG (1988) The thermal expansion of solid UO_2 and (U, Pu) mixed oxides: a review and recommendations. J Nucl Mater 152:94–101
48. Harding JH, Martin DG, Potter PE (1989) Thermophysical and thermochemical properties of fast reactor materials. Commission of European Communities Report EUR 12402
49. Hutchings MT (1987) High-temperature studies of UO_2 and ThO_2 using neutron scattering techniques. J Chem Soc Faraday Trans II 83:1083–1103
50. Fink JK, Chasanov MG, Leibowitz L (1981) Thermophysical properties of uranium dioxide. J Nucl Mater 102:17–25
51. Benedict M, Pigford T, Levi HW (1981) Nuclear chemical engineering. McGraw-Hill
52. Cohen I, Berman RM (1966) A metallographic and X-ray study of the limits of oxygen solubility in the UO_2–ThO_2 system. J Nucl Mater 18:77–107
53. Olsen CS (1979) Fuel thermal expansion (FTHEXP). In: Hagraman DR, Reymann GA (ed) MATPRO- version 11: a handbook of materials properties for use in the analysis of light water reactor fuel rod behavior. US Nuclear Regulatory Commission Report NUREG/ CR-0497
54. Momin AC, Venkateswarlu KS (1982) Thermophysical properties of ThO_2, UO_2 and mixed (Th,U)O_2. BARC report BARC-1171, Bhabha Atomic Research Centre, Trombay
55. Momin AC, Karkhanvala MD (1978) Temperature dependence of Gruneisen parameter and lattice vibrational frequencies of UO_2 and ThO_2 in the range 298-2300 K. High Temp Sci 10:45
56. Kempter CP, Elliott RO (1959) Thermal expansion of UN, UO_2, UO_2–ThO_2 and ThO_2. J Chem Phys 30:1524–1526
57. Springer JR, Eldridge EA, Goodyear MU, Wright TR, Lagerdrast JF (1967) Fabrication, characterization and thermal-property measurements of ThO_2–UO_2 fuel materials. Report No. BMI-X-10210, Battelle Memorial Institute Columbus

58. IAEA-TECDOC-1450 (2005) Thorium fuel cycle-Potential benefits and challenges. International Atomic Energy Agency, Vienna
59. Anthonysamy S, Panirselvam G, Vasudeva Rao PR (2000) In: Proceedings of the 12th national symposium on thermal analysis THERMANS 2000, Gorakhpur, p 29
60. Momin AC (1991) High temperature X-ray diffractometric studies on the lattice thermal expansion behaviour of UO_2, ThO_2 and $(U_{0.2}Th_{0.8})O_2$ doped with fission product oxides. J Nucl Mater 185:308–310
61. Touloukian YS, Kirby RK, Taylor RE, Lee LTYR (1970) Thermal expansion: nonmetallic solids. IFI/Plenum, New York
62. Breitung W, Reil KO (1990) The density and compressibility of liquid (U, Pu)-mixed oxide. Nucl Sci Eng 105:205–217
63. Dronting WD (1981) Thermal expansion of molten uranium dioxide. In: Proceedings of the 8th symposium on thermophysical properties. National Bureau of Standard, Gaithersburg, Maryland
64. Fink JK (2000) Thermophysical properties of uranium dioxide. J Nucl Mater 279:1–8
65. Christensen JA (1963) Thermal expansion and change in volume of uranium dioxide on melting. J Am Ceram Soc 46:607–608
66. Porter PE (2009) Over forty years of thermodynamics of nuclear materials. J Nucl Mater 489:29–44
67. MacDonald PE (2002) Advanced proliferation resistant, lower cost, uranium-thorium dioxide fuels for light water reactors. Idaho National Engineering and Environmental Laboratory Report INEEL/EXT-02-01411, Idaho
68. Shuller LC, Ewing RC, Becker U (2011) Thermodynamic properties of $Th_xU_{1-x}O_2$ $(0 < x < 1)$ based on quantum–mechanical calculations and Monte-Carlo simulations. J Nucl Mater 412:13–21
69. Yamashita T, Nitani N, Tsuji T, Inagaki H (1997) Thermal expansions of NpO_2 and some other actinide dioxides. J Nucl Mater 245:72–78
70. Goodman GL (1992) The electronic structure of actinide dioxides. J Alloys Compd 181:33–48
71. Taylor D (1984) Thermal expansion data: II. Binary oxides with the fluorite and rutile structures, MO_2, and the antifluorite structure, M_2O. Br Ceram Trans J 83:32–37
72. Hoch M, Momin AC (1969) High temperature thermal expansion of ThO_2 and UO_2. High Temp High Press 1:401–407
73. Ohnysty B, Rose FK (1964) Thermal expansion measurements on thoria and hafnia to 4500 °F. J Am Ceram Soc 47:398–400
74. Aronson S, Clayton JC (1960) Thermodynamic properties of non-stoichiometric Urania–Thoria solid solutions. J Chem Phys 32:749–754
75. Hirata K, Moriya K, Waseda Y (1977) High temperature thermal expansion of ThO_2, MgO and Y_2O_3 by X-ray diffraction. J Mater Sci 12:838–839
76. KFA, Nuclebras, KWU, NUKEM (1988) Program of research and development on the Thorium utilization in PWR's, final report (1979–1988). KFA, Jülich
77. Tyagi AK, Mathews MD (2000) Thermal expansion of ThO_2-2 wt% UO_2 by HT-XRD. J Nucl Mater 278:123–125
78. Anthonysamy S, Panneerselvam G, Bera Santanu, Narasimhan SV, Vasudeva Rao PR (2000) Studies on thermal expansion and XPS of thoria-urania solid solutions. J Nucl Mater 281:15–21
79. Springer JR, Eldrige EA, Goodyear MU, Wright TR, Langedrost JF (1967) Fabrication, characterization and thermal property measurements of ThO_2-UO_2 fuel materials. Battelle Memorial Institute Report BMI-X-10210, Columbus
80. Turner DN, Smith PD (1967) Linear thermal expansion of thoria, urania-thoria and their dispersion in beryllia in the range of 20–1000 °C together with the improved data of beryllia. Australian Atomic Energy Commission report AAEC E183, Lucas Heights
81. Lynch ED, Beals RJ (1962) Argonne national laboratory, annual report for 1962. ANL-6677, p 101

82. Rodriguez P, Sundaram CV (1981) Nuclear and materials aspects of the thorium fuel cycle. J Nucl Mater 100:227–249
83. Powers RM, Shapiro H (1959) Quarterly technical progress report. Sylvania Corning Nuclear Corporation, SCNC-301
84. Kutty TRG, Kulkarni RV, Sengupta P, Khan KB, Sengupta AK, Panakkal JP, Kamath HS (2008) Development of CAP process for fabrication of ThO_2-UO_2 fuels Part II: characterization and property evaluation. J Nucl Mater 373:309–318
85. Bhagat RK, Krishnan K, Kutty TRG, Kumar Arun, Kamath HS, Banerjee S (2012) Thermal expansion of simulated thoria–urania fuel by high temperature XRD. J Nucl Mater 422:152–157
86. Subramanian CGS, Panneerselvam G, Syamala KV, Antony MP (2009) Synthesis, characterization and thermal expansion studies on ThO_2–$SmO_{1.5}$ solid solutions. Ceram Int 35:2185–2190
87. Mathews MD, Ambekar BR, Tyagi AK (2006) Lattice thermal expansion studies of $Th_{1-x}Nd_xO_{2-x/2}$ solid solutions. Ceram Int 32:609–612
88. Grover V, Tyagi AK (2005) Lattice thermal expansion studies on single-phase compositions in CeO_2–ThO_2–ZrO_2 system. Ceram Int 31:769–772
89. Mathews MD, Ambekar BR, Tyagi AK (2001) Bulk thermal expansion studies of $Th_{1-x}Ce_xO_2$ in the complete solid solution range. J Nucl Mater 288:83–85
90. Kutty TRG, Khan KB, Hegde PV, Pandey VD, Sengupta AK, Majumdar S, Kamath HS (2002) Microstructure of ThO2–PuO2 pellets with varying PuO2 content. In: Ganguly C, Jayaraj PN (ed) Proceedings of CQCNF. Hyderabad, India
91. Fink JK, Petri MC (1997) Thermophysical properties of uranium dioxide. Argonne National Laboratory Report, ANL/RE-97/2
92. Klemens PG (1985) Theory of thermal conductivity of nonstoichiometric oxide and carbides. High Temp High Press 17:41–54
93. Pillai CGS, Raj P (2000) Thermal conductivity of ThO_2 and $Th_{0.98}U_{0.02}O_2$. J Nucl Mater 277:116–119
94. Degueldre C, Arima T, Lee YW (2003) Thermal conductivity of zirconia based inert matrix fuel: use and abuse of the formal models for testing new experimental data. J Nucl Mater 319:6–14
95. Cozzo C, Staicu D, Somers J, Fernandez A, Konings RJM (2011) Thermal diffusivity and conductivity of thorium–plutonium mixed oxides. J Nucl Mater 416:135–141
96. Clark LM, Taylor RE (1975) Radiation loss in the flash method for thermal diffusivity. J Appl Phys 46(2):714–719
97. Loeb AL (1954) Thermal conductivity VIII- a theory of thermal conductivity of porous materials. J Am Ceram Soc 37:96–99
98. Macewan JR, Stoute RL, Notley MF (1967) Effect of porosity on the thermal conductivity of UO_2. J Nucl Mater 24:109–112
99. Biancheria A (1966) The effect of porosity on thermal conductivity of ceramic bodies. Trans Am Nucl Soc 9:15
100. Ondracek G, Schultz B (1973) The porosity dependence of the thermal conductivity for nuclear fuels. J Nucl Mater 46:253–258
101. Vancraeynest JC, Stora JP (1970) Effect de la Porositesur la Variation de Conductubilite Thermique du Bioxyded'Uranium en Fonction de la Temperature. J Nucl Mater 37:153–158
102. Pears CD (1963) The thermal properties of twenty six solid materials to 5000 °F or their destruction temperatures. Report ASD-TR-62-765
103. MacEwan JR, Stoute RL (1969) Annealing of irradiation induced thermal conductivity changes in ThO_2-1.3 wt% UO_2. J Am Ceram Soc 52:160–165
104. Belle J, Berman RM, Bourgeois WF, Cohen I, Daniel RC (1967) Thermal conductivity of bulk oxide fuels. WAPD-TM-586, Bettis Atomic Power Laboratory, West Mifflin
105. Peterson S, Adams RE, Douglas DA (1966) Panel on utilization of thorium in power reactors. International Atomic Energy Agency Report STI/DOC/10/52, Vienna

106. Schultz B (1981) Thermal conductivity of porous and highly porous materials. High Temp High Press 37:649
107. Inoue M, Abe K, Sato I (2000) A method for determining an effective porosity correction factor for thermal conductivity in fast reactor uranium-plutonium oxide fuel pellets. J Nucl Mater 281:117–128
108. Faucher M, Cabannes F, Anthony AM, Piriou B, Simonato J (1970) Measurement of thermal diffusivity and total emissivity of solids between 1500 K and melting point. Rev Int Hautes Temp Refract 7:290–297
109. Kingery WD, Francl J, Cobble RL, Vasilos T (1954) Thermal conductivity data for several pure oxide materials corrected to zero porosity. J Am Ceram Soc 37:107–110
110. McElroy DL, Moore JP, Spindler PH (1968) Status and progress report for thorium fuel cycle development for 1967 and 1968. Oak Ridge National Laboratory Report ORNL-4429, pp 121–132
111. Armour Research Foundation (1957) ARF-Project No 6-025, final report
112. Weilbacher JC (1972) Measurement of the thermal diffusivity of mixed uranium plutonium oxides. High Temp High Press 4:431–438
113. DeBoskey WR (1962) Irradiation testing of thoria-urania fuel for the Indian point reactor. In: Proceedings of the Thorium fuel cycle symposium, TID-7650, Gatlinburg, TN, pp 630–642
114. Murabayashi M (1970) Thermal conductivity of ceramic solid solutions. J Nucl Sci Technol 7:559–563
115. Koenig JH (1953) Progress report no. 2 from March 1 to June 1, 1953. Rutgers Univ NJ Ceram. Research Sta AD-13154
116. Washington ABG (1973) Preferred values for the thermal conductivity of sintered ceramic fuel for fast reactor use. United Kingdom Atomic Energy Authority TRG-Report -2236
117. Brandt R, Neuer G (1976) Thermal conductivity and thermal radiation properties of UO_2. J Non-Equilib Thermodyn 1:3–23
118. Ronchi C, Sheindlin M, Musella M, Hyland GJ (1999) Thermal conductivity of uranium dioxide up to 2900 K from simultaneous measurement of the heat capacity and thermal diffusivity. J Appl Phys 85:776–789
119. Hobson IC, Taylor R, Ainscough JB (1974) Effect of porosity and stoichiometry on the thermal conductivity of uranium dioxide. J Phys D Appl Phys 7:1003–1015
120. Bates JL (1970) High-temperature thermal conductivity of round robin uranium dioxide. Battelle Memorial Institute Pacific Northwest Laboratories Report BNWL-1431
121. Conway JB, Feith AD (1969) An interim report on a round robin experimental program to measure the thermal conductivity of stoichiometric uranium dioxide. General Electric Report GEMP-715
122. Gofrey TG, Fulkerson W, Kollie TG, Moore JP, Mcelroy DL (1964) The thermal conductivity of uranium dioxide and armco iron by an improved radial heat flow technique. Oak Ridge national Laboratory Report ORNL-3556
123. Berman RM, Tully TS, Belle J, Goldberg I (1972) The thermal conductivity of polycrystalline thoria and thoria-urania solid solution. LMWR Development Program WAPD-TM-908
124. Murti PS, Mathews CK (1991) Thermal diffusivity and thermal conductivity studies on thorium-lanthanum mixed oxide solid solutions. J Phys D 24:2202–2209
125. Lucuta PG, Matzke H, Hastings IJ (1996) A Pragmatic approach to modeling thermal conductivity of irradiated UO_2 fuel: review and recommendations. J Nucl Mater 232:166–180
126. Jain D, Pillai CGS, Rao BS, Kulkarni RV, Ramdasan E, Sahoo KC (2006) Thermal diffusivity and thermal conductivity of thoria lanthana solid solutions up to 10 mol% $LaO_{1.5}$. J Nucl Mater 353:35–41
127. Ferro C, Patimo C, Piconi C (1972) Thermal diffusivity of mixed $(Th_{1-x}U_x)$oxides and some materials to be used as a reference in the range 650–2700 K. J Nucl Mater 43:273–276

128. Jacobs DC (1969) The in-pile thermal conductivity of selected ThO_2–UO_2 pellets at low depletions. Report WAPD-TM-758, Bettis Atomic Power Laboratory, West Mifflin
129. Matolich J, Storhok VW (1970) Thermal diffusivity measurements of irradiated oxide fuels. Battelle Memorial Institute Report BMI-RX-10274, Columbus
130. Basak U, Sengupta AK, Ganguly C (1989) Hot hardness and thermal conductivity of ThO_2–PuO_2 and ThO_2–UO_2 sintered pellets. J Mater Sci Lett 8:449–450
131. Altas Y, Tel H (2001) Structural and thermal investigations on cerium oxalate and derived oxide powders for the preparation of (Th, Ce)O2 pellets. J Nucl Mater 298:316–320
132. Jeffs AT, Boucher RR, Norlock LR (1967) Fabrication of UO_2–PuO_2 and ThO_2–PuO_2 experimental fuel. Report AECL-2675, Atomic Energy Canada Limited
133. Jeffs AT (1969) Thermal conductivity of ThO_2–PuO_2 under irradiation. AECL-3294, Atomic Energy of Canada Ltd, Chalk River
134. Anantharaman S, Ramadasan E, Singh JL, Dubey JS, Mishra Prerna (2010) Post Irradiation examination of thermal reactor fuels. BARC Newsletter 315:76–91
135. BARC Highlights (2006) Nuclear fuel cycle, post irradiation examination on fuel. Bhabha Atomic Research Centre, Trombay, p 39
136. Akabori M, Shiba K (1986) Dimensional changes in irradiated (Th, U)O_2. J Nucl Sci Technol 23:594–601
137. Christensen JA, Allio RJ, Biancheria A (1964) Melting point of irradiated uranium oxide. Trans Am Nucl Soc 7:390–391
138. Konno K, Hirosawa T (1999) Melting temperature of simulated high-burnup mixed oxide fuels for fast reactors. J Nucl Sci Technol 36:596–604
139. Yamamoto K, Hirosawa T, Yoshikawa K, Morozumi K, Nomura S (1993) Melting temperature and thermal conductivity of irradiated mixed oxide fuel. J Nucl Mater 204:85–92
140. Hirosawa T, Sato I (2011) Burnup dependence of melting temperature of FBR mixed oxide fuels irradiated to high burnup. J Nucl Mater 418:207–214
141. Bailly Henri, Menessier Denise, Prunier Claude (1999) The nuclear fuel of pressurized water reactors and fast neutron reactors—design and behaviour. Lavoisier Publishing Inc. Paris, France, p 138
142. Hart PE, Griffin CW, Hsieh KA, Matthews RB, White GD (1979) ThO_2 based fuels-their properties, methods of fabrication and irradiation performance: a critical assessment of the state of technology and recommendations for future work. Pacific Northwest Laboratory Report PNL-3069, Richland, Washington
143. Lucuta PG, Verrall RA, Matzke H, Palmer BJ (1991) Microstructural features of SIMFUEL—simulated high burn-up UO_2 based nuclear fuel. J Nucl Mater 178:48–60
144. IAEA Nuclear Energy Series No NF-T-2.4 (2012) Role of thorium to supplement fuel cycles of future nuclear energy systems. IAEA, Vienna
145. Lu Y, Yang Y, Zhang P (2012) Thermodynamic properties and structural stability of thorium dioxide. J Phys Condens Matter 24:225801
146. Aitken EA, Evans SK (1968) A thermodynamic data program involving urania at high temperatures. General Electric Report GEAP-5672
147. Adamson MG, Aitken EA, Caputi RW (1985) Experimental and thermodynamic evaluation of the melting behavior of irradiated oxide fuels. J Nucl Mater 130:349–365
148. Hagrman DT (1995) MATPRO - A library of materials properties for Light Water Reactor accident Analysis. SCDAP/RELAP5/MOD 3.1/Code manual, Vol. 4. MATPRO, USNRC Report NUREG/CR-6150 (EGGG-2720)
149. Komatsu J, Tachibana T, Konashi K (1988) The melting temperature of irradiated oxide fuel. J Nucl Mater 154:38–44
150. Marples JAC (1976) Plutonium 1975 and other actinides. In: Blank H, Lindner R (ed) North-Holland, Amsterdam, p 353
151. Fahey JA, Turcotte RP, Chikalla TD (1974) Thermal expansion of actinide oxides. Inorg Nucl Chem Lett 10:459–465
152. Zachariasen WH (1948) Crystal radii of heavy elements. Phys Rev 73:1104–1105

153. Baldock PJ, Spindler WE, Baker TW (1966) The X-ray thermal expansion of near—stoichiometric UO_2. J Nucl Mater 18:305–313
154. Katz JJ, Morse LR, Seaborg GT (1986) The chemistry of the actinide elements. Katz JJ, Morse LR, Seaborg GT (eds) Chapman and Hall, New York, vol 2, Ch 14, p 1121
155. Harbinson EN, Walker RJ (1966) Thermal conductance of ThO_2–10%UO_2 under simulated operating conditions. Trans Am Nucl Soc 9:26–27
156. Moore JP, Graves RS, Kollie TG, McElroy DL (1967) Thermal conductivity measurements on solids between 20 and 150 °C using a comparative longitudinal apparatus: results on M_gO, BeO, ThO_2, $Th_xU_{1-x}0_{2+y}$, and Al-UO_2 cermets. Oak Ridge National Laboratory Report ORNL-4121, Oak Ridge

Phase Diagrams and Thermodynamic Properties of Thoria, Thoria–Urania, and Thoria–Plutonia

R. Agarwal and S. C. Parida

Abstract To understand the in-pile behavior of a fuel it is very important to have full understanding of its thermodynamic properties and phase diagram of the system. Phase diagram gives the information of the possible phases of the system that will be present under different temperatures and changing compositional conditions existing during fuel burnup. The phase diagram defines the transition temperatures of stable phases and their oxygen solubility limits. Phase diagrams of Th–O, Th–U–O systems, and solidus–liquidus of UO_2–ThO_2 and PuO_2–ThO_2 are given in the present chapter. In case of oxide fuels, understanding of oxygen solubility in dioxide phase is very important. This is especially true for thoria-based fuel because of very limited non-stoichiometry in pure thoria; thus, the non-stoichiometry of this fuel is mainly due to added actinide (U/Pu). This in turn affects the oxygen potential of the fuel, which can put a severe constraint during reactor operation, because high oxygen potential of the fuel can oxidize the clad and make it brittle. The oxygen potential of $(Th,U)O_{2+x}$ and $(Th,Pu)O_{2-x}$ systems are discussed here as a function of composition and temperature. Heat capacity is another important thermodynamic parameter that gives information about the heat contained in the reactor at any given time. It also controls the heat transferred from the fuel center to the surface, as thermal conductivity is defined as a product of heat capacity, thermal diffusivity, and density ($Cp.\kappa.\rho$). Heat capacities of pure oxides and mixed oxides are compared with each other to show the trend. Heat capacity equations as a function of temperature are given for different compositions of $(Th,U)O_2$. Performance of the fuel at very high reactor operation temperatures is related to the partial pressures of actinide-bearing species. These pressures control the actinide redistribution due to the temperature gradient in the

R. Agarwal (✉) · S. C. Parida
Product Development Division, Bhabha Atomic Research Centre Trombay, Mumbai 400085, India
e-mail: arenu@barc.gov.in

S. C. Parida
e-mail: sureshp@barc.gov.in

D. Das and S. R. Bharadwaj (eds.), *Thoria-based Nuclear Fuels*,
Green Energy and Technology, DOI: 10.1007/978-1-4471-5589-8_3,

reactor and can also result in change in oxygen to metal ratio due to incongruent evaporation. It is also found that the fuels with high actinide vapor pressures show more restructuring, though fuel restructuring is not a simple phenomenon and has many controlling parameters. In this chapter, partial pressures of actinide bearing species over $(Th,U)O_2$ and ThO_{2-x} are discussed as a function of temperature and composition.

1 Phase Diagrams

1.1 Solid State Solubilities

ThO_2 has a very stable closed packed fluorite structure; therefore, Gibbs energy of formation of thorium oxide is very low. As thorium can exist only in +4 oxidation state, hence, its structure is almost defect free (free of O^{2-} ion vacancies and interstitials) at low temperatures. It shows non-stoichiometry at very high temperatures. That explains a steep decrease in oxygen potential of thoria even for very small reduction in O/M. A sharp increase in 'Th' bearing species is observed in the vapor phase with reduction in O/M, mainly due to the increase in partial pressure of ThO(g). Ionic radii of Th(IV) (1.05 Å), U(IV) (0.997 Å), and Pu(IV) (0.962 Å) are very similar, with difference between the ionic radii of 'Th' and 'U' being less than that between 'Th' and 'Pu'. The dioxides of these actinides are fluorite-structured FCC with similar lattice parameters, 5.5975, 5.4704, and 5.396 Å, respectively. As they are members of f-block elements with loosely bound outer shell electrons resulting in similar electron densities and electronegativities, Hume-Rothery rules aptly apply to their solid solutions. Thus ThO_2, UO_2, and PuO_2 make homogeneous, substitutional solid solutions over the whole composition range. The lattice of ThO_2 contracts with addition of either urania or plutonia (Fig. 1). As per most of the reported literature, lattice parameters of their solid solutions follow Vegards law with a slight deviation up to 2 at.% ThO_2 in case of UO_2–ThO_2 and up to 5 at.% PuO_2 in case of PuO_2–ThO_2. EXAF analysis of these solid solutions shows that the Th–O distance decreases linearly from 2.424 (ThO_2) to 2.406 for $(Th_{0.09}U_{0.91})O_2$ and to 2.407 for $Th_{0.34}Pu_{0.66}O_2$, which is understandable because the ionic radii of Pu(IV) is lower than that of U(IV), which in turn is lower than Th(IV). Hence, addition of 'Pu' leads to sharper lattice shrinkage compared to addition of 'U' [1]. In $(Th_{1-y}U_y)O_2$ system, $R_{Th\text{-}O}$ and $R_{U\text{-}O}$ can be calculated using 2.424–0.019y (Å) and 2.394–0.026y (Å) relations, respectively. In $(Th_{1-y}Pu_y)O_2$ system, $R_{Th\text{-}O}$ and $R_{Pu\text{-}O}$ can be calculated using 2.424–0.025y (Å) and 2.391–0.055y (Å) relations, respectively. As seen from these linear relations, the lowest slope (−0.055) and thus the sharpest decrease is seen for Pu–O distance with increase in 'Pu' content. But, the change in actinide-oxygen distance does not change much with composition, whereas, considerable change in M–M distance is seen with change in composition, $R_{M\text{-}M}$ (Å) = 3.959

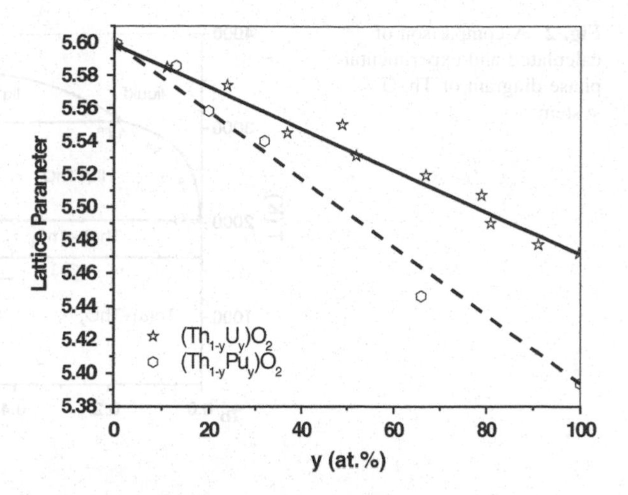

Fig. 1 Change in lattice parameter of $(Th,U)O_2$ and $(Th,Pu)O_2$ at 298 K with change in thorium content [1]

$-$ 0.09y in $(Th_{1-y}U_y)O_2$ and $R_{M\text{-}M}$ (Å) = 3.959 $-$ 0.145y in $(Th_{1-y}Pu_y)O_2$. For incineration of 'Pu', thoria–plutonia fuel is more attractive than urania–plutonia fuel as the former does not breed 'Pu'. Thoria is easy to handle and store compared to urania, because the presence of only one oxidation state, Th(IV), does not allow thoria to get oxidized. Whereas, urania can get oxidized to form U_3O_8 because uranium can achieve oxidation states higher than 'IV'. Therefore, thoria–urania can get oxidized due to the presence of uranium in it and the degree of oxidation depends on the fraction of 'U'. Higher the 'U' content, higher can be the value of 'x' in $(Th,U)O_{2+x}$.

Th–O binary phase diagram has been investigated by many workers [2–9]. Some of these have reported the presence of ThO and ThO_2 solid oxides. However, latest work by Benz [2] and Eyring [8] showed the presence of only ThO_2. Eyring [8] has indicated that ThO may become stable in the presence of considerable 'C' or 'N' impurities. Thoria exists as a stoichiometric compound till ~1300 K and shows hypostoichiometry at higher temperatures (Fig. 2). An investigation of non-stoichiometry of thoria indicated that it is oxygen deficient at oxygen partial pressure of 1 atm and T = 1,300–2,200 K. Oxygen deficiency, 'x' in ThO_{2-x}, can be expressed as $\log(x) = -1.870 - 3400/T - \log(P)/6$, at temperature T (K) and pressure P (atm). The stoichiometric thoria is white in color; however, the 'Th' vacancies (p-type) generated above 1,300 K, in high-oxygen partial pressure, result in deep red color. On the other hand, oxygen vacancies created (n-type) on heating thoria in vacuum result in pale yellow color.

Thoria–urania solid solution exhibits considerable range of hyperstoichiometry (Fig. 3). As this non-stoichiometry is stabilized by the oxidation state of uranium, the range of hyperstoichiometry shows very strong dependence on Th/(Th+U) ratio of the solution. This hyperstoichiometry is reasonably large for $(Th,U)O_{2+x}$ system with U/(Th + U) $\geq$ 0.5. The XRD analysis of this system has shown the presence of a compound, $(Th_{0.5}U_{0.5})_4O_9$ with structure very similar to that of U_4O_9. Some authors have indicated that instead of the presence of a new

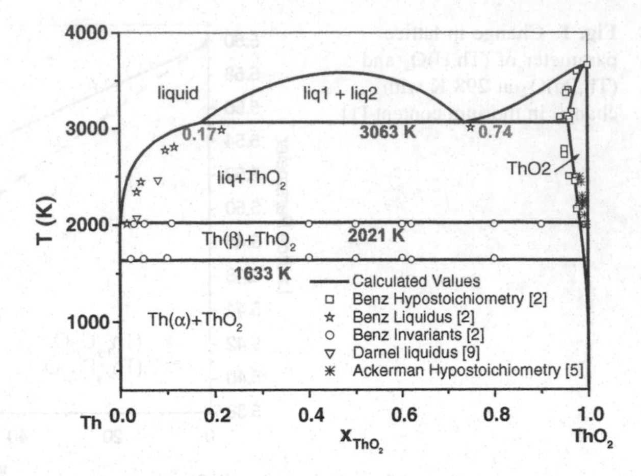

Fig. 2 A comparison of calculated and experimental phase diagram of Th–O system

compound, it may be just a large solubility of 'Th' in UO_{2+x} and U_3O_8 lattices. Rachev et al. [10] reported a compound $(U_{0.5}Th_{0.5})_4O_9$, similar to U_4O_9, but, Paul [11] and Paul and Keller [12] indicated that it is not a new compound. According to them it is U_4O_9 with superlattice lines, with maximum 50 % substitutional solubility of ThO_2 in U_4O_9. Braun et al. [13] reported a complete solid solution in the composition range from pure ThO_2 to $(U_{0.5}Th_{0.5})_4O_9$ using spectroscopic methods. Similar compound is not reported for thoria–plutonia system, but that may be due to lack of sufficient experiments to establish such a compound. At sufficiently high temperatures, thoria–urania is known to show some hypostoichiometry too. However, there is not much reported data of that phase boundary either. Reasonable amount of thoria is reported to dissolve in higher oxides of uranium, U_4O_9 [13] and U_3O_8 [14], especially at high temperatures (>1523 K). It is found that all the U(IV) in U_4O_9 ($U^{IV}{}_2O_4 + U^{V}{}_2O_5$) can be replaced by Th(IV) to form $(Th_{0.5}U_{0.5})_4O_9$ ($Th^{IV}{}_2O_4 + U^{V}{}_2O_5$). Anderson et al. [15] found that a MOX with more than 50 % 'U' can contain 16–17 % extra 'O' in its lattice while retaining its fluorite structure. But the oxygen tolerance in the lattice decreases with increase in 'Th' content when 'U' content is lower than 50 %. The presence of +5 oxidation state and not +6 oxidation state of 'U' in $(Th,U)O_{2+x}$ was confirmed by Anderson et al. [15] on the basis of crystallographic data, which indicated that the unit cell contracted as the uranium valency increases from +4 to +5 because ionic radii of U^{5+} is lower than U^{4+}, but further oxidation to UO_2^{+2} results in lattice expansion.

Paul and Keller [12] carried out phase diagram analysis of UO_2–ThO_2–UO_3 system from 873 to 1,823 K. They found that 'FCC+Orthorhombic' phase field shrinks with increase in temperature. At 873 K, the thoria-rich corner of the phase field is at Th/(Th,U) ≈ 0.48 which reduces to ≈0.39 at 1,823 K (Fig. 3). Investigations carried out by Cohen and Berman [16], at 1,473 K, clearly show the existence of biphasic system FCC+Orthorhombic for systems with U/(U + Th) > 0.9 and O/(U + Th) > 2.25 (corresponding to M_4O_9 composition). A hypothesis similar to that of substitution of U^{4+} by Th^{4+} in U_4O_9 may be applied

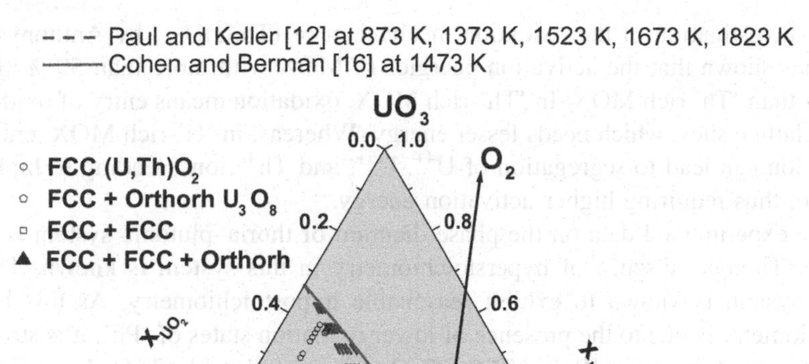

Fig. 3 Pseudo ternary phase diagram of UO_2–UO_3–ThO_2, showing non-stoichiometry of FCC phase at different temperatures (at 873 K, 1373 K, 1523 K, 1673 K, 1823 K [12] and at 1473 K [16])

for U_3O_8 to understand solubility of thoria in U_3O_8. U_3O_8 has uranium in +4, +5, and +6 oxidation states in the ratio of 1:2:3, respectively. Even if all the U^{+4} get replaced by Th^{+4}, the solubility of thorium in this oxide cannot exceed ~16 at.%. However, replacing all the U^{+4} (0.89 Å) by Th^{+4} (0.94 Å) in a lattice that contains much smaller ions of U^{+5} (76 Å) and U^{+6} (73 Å) will destabilize the lattice; therefore, the solubility of thoria in this compound should be much lower than 16 at.%. The orthorhombic U_3O_8 ($6.72 \times 11.96 \times 4.15$ Å^3) is a more compact lattice than U_4O_9 (a (FCC) = 5.447 Å); thus, it may not be able to accommodate bigger Th^{4+} ions [17]. Lynch [14] heated U_3O_8 and ThO_2 mixtures in different environment, to different temperatures and indicated that the solubility of cubic uranium hyperdioxide in thoria results in its stabilization to such a degree that U_3O_8 gets destabilized. Their experiments showed that when heating pure U_3O_8 in air, the O/U ratio changes from 2.667 to 2.65 at 1,073 K and to 2.59 at 1,523 K. Whereas, O/U of a mixture of ThO_2+U_3O_8 (70.4 at.% $UO_{2.667}$) reduced to 2.43 at 1,523 K. Heating this mixture in air showed the presence of both phases at 1,073 K. Heating to 1,273 K showed the beginning of formation of cubic solid solution of $(Th,U)O_{2+x}$. At 1,473 K, no unreacted ThO_2 was observed and the system had become $(Th,U)O_{2+x}$ + U_3O_8. Above 1,523 K, only $(Th,U)O_{2+x}$ was present and U_3O_8 had disappeared. These experiments show that U_3O_8 has negligible solubility for ThO_2 and it gets destabilized with the addition of ThO_2, whereas, U_4O_9 and UO_{2+x} phases which dissolve reasonable amount of ThO_2, become stable at the expense of U_3O_8.

An investigation of the oxidation mechanism of $(Th,U)O_{2+x}$ by Anthonysamy [18] has shown that the activation energies of MOX with more than 50 % 'U' is higher than 'Th' rich MOX. In 'Th'-rich MOX, oxidation means entry of oxide ion in the lattice sites, which needs lesser energy. Whereas, in 'U'-rich MOX, entry of oxide ion can lead to segregation of U^{4+}, U^{5+}, and Th^{4+} ions, leading to biphasic system, thus requiring higher activation energy.

The experimental data on the phase diagram of thoria–plutonia system is very scarce. Though, absence of hyperstoichiometry in this system is known, ThO_2–PuO_2 system is known to exhibit reasonable hypostoichiometry. As this hypostoichiometry is due to the presence of lower oxidation states of 'Pu', it is strongly linked to the fraction of 'Pu' in ThO_2–PuO_2 solid solution. As ThO_2 has a fluorite structure similar to that formed by most of the actinides, therefore, the minor actinides formed during reactor operation will dissolve in fuel matrix and no separate phase should appear.

1.2 Liquidus–Solidus

Liquidus–solidus of UO_2–ThO_2 system is investigated by many workers [19–22]. As shown in Fig. 4, the experimental data clearly indicates that the solid solution of ThO_2–UO_2 does not follow ideal solution model. The presence of a minimum in liquidus–solidus at ~ 5 at.% ThO_2 indicates an asymmetric composition deviation from ideal solution model.

Thoria–urania is a homogeneous solid solution over the whole composition range, but they do not form an ideal solution. As seen from the solidus–liquidus curves given in Fig. 4, there is a dip in liquidus near UO_2 at ~ 5 mol % ThO_2 and a deviation to higher temperatures on 'ThO_2' end. If liquid solution is assumed to be an ideal solution, then UO_2-rich solid solution shows a positive deviation,

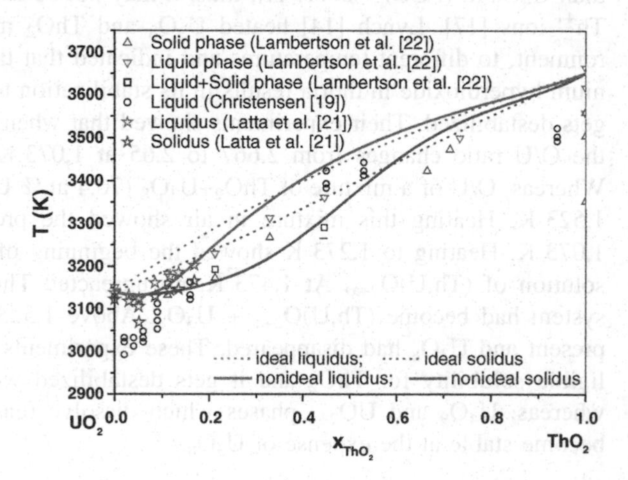

Fig. 4 A comparison of experimental and calculated solidus–liquidus of ThO_2–UO_2 system, with O/(U + Th) $\approx$ 2.0

whereas, ThO_2-rich solid solution shows negative deviation. The nonideal liquidus–solidus lines in Fig. 4 were drawn using nonideal solution interaction parameters for UO_2–ThO_2 solid solution. The excess Gibbs energy of ThO_2–UO_2 solid solution was calculated using the relation: $G^{ex} = L^1 x_{ThO2} x_{UO2} (x_{ThO2} - x_{UO2}) + L^2 x_{ThO2} x_{UO2} (x_{ThO2} - x_{UO2})^2$, where, interaction parameters, $L^1_{ThO2-UO2}$ and $L^2_{ThO_2-UO_2}$ were taken as -10 and $+13$ kJ/mol, respectively. An asymmetric interaction parameter, $L^1 x_{ThO_2} x_{UO_2} (x_{ThO_2} - x_{UO_2})$, with coefficient value -10 kJ/mol means a negative deviation from ideality for ThO_2 rich solid solutions, but a positive deviation from ideality for UO_2 rich solid solution. In liquidus–solidus, this deviation is reflected by decrease in solidus–liquidus temperature for UO_2-rich alloys and increase in these transition temperatures for ThO_2-rich alloys. The liquid solution is assumed to be an ideal solution, therefore, a negative deviation from ideality for ThO_2-rich solid alloys means that ThO_2-rich alloys become more stable compared to their hypothetical ideal solution. This results in increase in the temperature of their transition from solid to liquid phase. Thus, negative deviation from ideality results in stabilization of that phase over a wider temperature range as compared to its ideal solution. Similarly, a positive deviation from ideality results in destabilization of that phase to narrower temperature range compared to its ideal solution behavior. This is reflected in lowering of liquidus–solidus temperatures in UO_2 rich solutions. A comparison of melting points of pure UO_2 and ThO_2 given by Christensen [19] and Lambertson et al. [22] with presently accepted values shows a deviation of 100 K. This should also be reflected in their experimental liquidus–solidus temperatures of ThO_2–UO_2. The more recent work by Latta et al. [21], though confined to UO_2-rich system, is more careful investigation of solidus–liquidus. To prevent change of composition of the system during measurement, they first carefully brought the system to stoichiometric composition and then sealed the sample during transition temperature measurements. Whereas, previous authors had carried out the measurements with samples exposed to the furnace environment. This could have resulted in deviation from stoichiometric composition, depending upon the reducing or oxidizing conditions in the furnace. In either conditions, the melting point is expected to reduce because near stoichiometric UO_2 has the highest melting point. Due to very high melting point of ThO_2, the readings in thoria-rich region are not reliable. The error could be introduced due to temperature measurement or change in composition due to vapor loss of U/UO_2. The experimental solidus data of ThO_2–PuO_2 system also indicates a positive deviation from ideal solution behavior for PuO_2 rich solid solution. A comparison of experimental data of Freshly and Mattys [23] with the liquidus and solidus lines of this system calculated by assuming ideal liquid and solid solution (Fig. 5) shows asymmetric deviation from ideality in PuO_2 rich. Therefore, liquidus–solidus were redrawn using a nonideal interaction parameter for PuO_2–ThO_2 solid solution. The excess Gibbs energy function for the solid solution used for the calculations of liquidus–solidus curves drawn in the figure was, $G^{ex} = L^1 x_{PuO2} x_{ThO2} (x_{PuO2} - x_{ThO2})$. Where, the asymmetric interaction parameter $L^1_{PuO_2-ThO_2}$ was taken as +25 kJ/mol. Similar to the explanation given in

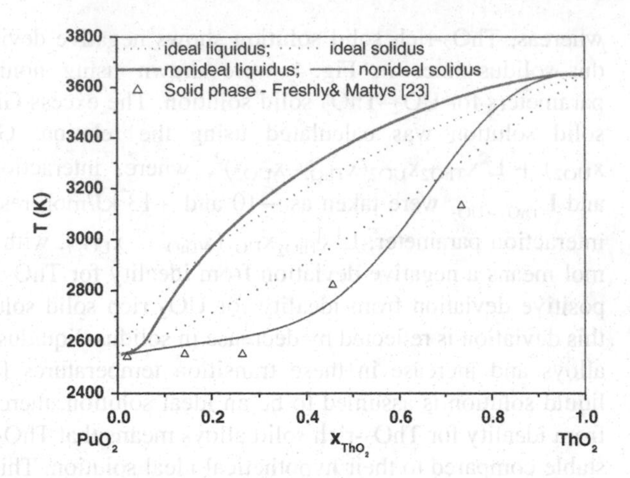

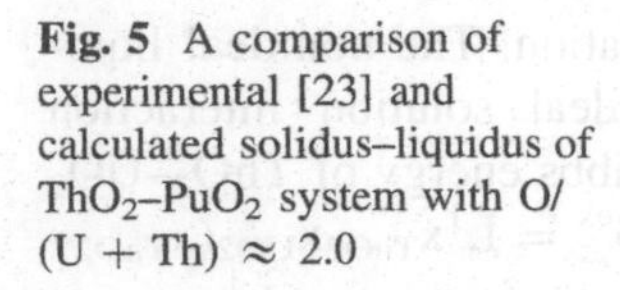
Fig. 5 A comparison of experimental [23] and calculated solidus–liquidus of ThO_2–PuO_2 system with O/(U + Th) ≈ 2.0

UO_2–ThO_2 system, this positive asymmetric interaction parameter results in positive deviation from ideality for PuO_2-rich solid solutions and negative deviation from ideality for ThO_2-rich solid solutions. This is seen by the decrease in solidus–liquidus for PuO_2 corner and increase in solidus–liquidus for ThO_2 corner.

No experimental data is available for phase diagram or thermodynamic properties of PuO_2–ThO_2–UO_2 system; however, based on the properties of binary oxides, it can be safely assumed that all three stoichiometric dioxides will completely dissolve in each other over the whole composition range. Reeve [24] suggested that the presence of PuO_2 and ThO_2 may stabilize UO_2 to such a degree that even in the presence of excess oxygen, it will stay in UO_2 lattice instead of getting oxidized to U_3O_8.

2 Specific Heats of Thoria-Based Fuels

Thermophysical properties of nuclear fuels are very important to predict in-pile performance of nuclear fuel, selection of a fuel and cladding material for a new reactor design concept and to generate computer codes for prediction of fuel performance under normal/accidental conditions [25]. Specific heat is one of the basic thermodynamic properties of any material and is also a very important design input parameter for analyzing the behavior of the nuclear fuels in the accidental or off-normal reactor operative conditions. Moreover, specific heat is an important input parameter to the calculation of thermal conductivity, which is one of the most important thermo-physical properties of the fuel. During reactor accident situation due to loss of coolant accident (LOCA) or breached cladding scenario, the specific heat of the fuel becomes the basis of heat transfer calculation for the core cooling. Thoria-based fuel systems have recently drawn considerable attention due to the possibility of use in advanced reactors with higher safety criteria. Hence, it is important to have an update of the thermophysical properties of thoria-based fuels for reactor applications which is addressed in this part of the article.

For a complete analysis of thermophysical data of fuel materials, it is necessary to have specific heat data starting from near 0 K to temperatures as high as >3,000 K, where core melting due to accident occurs. The low temperature (~0–298.15 K) specific heat data gives the accurate value of standard entropy at 298.15 K $\left(S^{\circ}_{298.15\text{K}}\right)$, as well as the fundamental parameters like Debye temperature (θ_D), electronic coefficient of specific heat (γ), the magnetic exchange parameters (J) etc. These fundamental data are extensively used to model thermophysical properties by coupling to high temperature specific heat data through various theoretical models. However, there is no universal technique which can measure the specific heat of these type solid materials in the entire temperature range starting from near 0 K to high temperatures such as >3,000 K with reasonable accuracy. There is no universal theoretical model which can predict the specific heats of these types of solids in the entire temperature range. Hence, in usual practice, a combination of experimental data in the amenable temperature range with semi-empirical/abinitio methods can yield specific heat data in the entire temperature range with good accuracy. Hence, this article considers both the experimental and theoretical models used so far to derive the specific heats of thoria-based materials. For general readership, it is essential to discuss the fundamental theories of specific heats of solids as well as the general experimental methods used to measure this property. Excellent approach to the theories of specific heats of solids can be found out in many text books on thermodynamics and solid state physics [26–29].

2.1 Theoretical Models for Specific Heats of Solids

The total heat capacity of a real solid is the sum of the heat capacities of all excited modes in the system.

$$C = C_{\text{phonons}} + C_{\text{electrons}} + C_{\text{magnons}} + C_{\text{twolevel}} + \cdots\cdots \quad (1)$$

where the subscripts denote the contributions arising from different modes of excitations. It has been observed that for most of the metals and other electrical conductors, the heat capacity contribution due to lattice vibration (C_{phonons}) is the dominant term at high temperatures whereas at low temperatures; the contribution from electrons ($C_{\text{electrons}}$) becomes appreciable. Similarly, for diamagnetic materials, the magnetic contribution (C_{magnons}) is negligibly small. However, for substances having paramagnetic ions/atoms, the contribution from magnons is substantial. For a diamagnetic insulator, the internal energy and hence the heat capacity depend only on the vibrational contributions. If the atomic nucleus has a magnetic moment, it may have a set of energy levels in an effective field arising from orbital and conduction electrons. Therefore, this interaction will give rise to nuclear level splitting. Moreover, if the nucleus has a quadrupole moment, its interaction with the field gradients produced by neighboring atoms will cause small level splitting. The change in population of these energy levels will contribute to specific heat, which is generally described as $C_{\text{two level}}$ for a two-level splitting system. Such effects are observed in antiferromagnets and more prominently in many ferromagnetic rare earths. Because of the smallness of the nuclear moments compared to the electronic moments, the nuclear Schottky effects occur in the region of 10^{-2} K, whereas the electronic Schottky effects are observed within 1–10 K. Nuclear Schottky effects may arise even in diamagnetic materials. Most of the oxides studied here show magnetic order–disorder transitions. The present study aims at resolving the lattice, electronic, and magnetic contributions to the total isobaric heat capacity of these oxides. To carry out such an exercise, it is essential to discuss the theory of heat capacity of solids.

The vibrational heat capacity of solids (C_{phonon}) is usually calculated using Debye's continuum approximation. The Debye function is given as:

$$\frac{C_V}{3R} = \frac{3}{x^3}\int_0^x \frac{x^4 e^x}{(e^x-1)^2}\mathrm{d}x = f(\text{Debye}) \tag{2}$$

where, $x = \frac{h\upsilon}{k_B T}$. The Debye function is a universal function for which table is available. Debye's expression for (C_{phonon}) is given as:

$$C_{\text{phonon}} = 9Nk_B\left(\frac{T}{\theta_D}\right)^3 \int_0^{x_D} \frac{x^4 e^x}{(e^x-1)^2}\mathrm{d}x \tag{3}$$

where, θ_D is Debye temperature and $x_D = \frac{h\upsilon_D}{k_B T} = \frac{\theta_D}{T}$. In the low temperature limit, $T \ll \theta_{\text{D}}$, it can be shown that:

$$C_V = \frac{9Nk_B}{x_D^3}\left(\frac{4}{15}\pi^4\right) = \frac{12}{5}Nk_B\pi^4\left(\frac{T}{\theta_D}\right)^3 \tag{4}$$

which shows that $C_V \alpha T^3$, in agreement with the experiment (T^3 law). In the high temperature limit, $T \gg \theta_{\text{D}}$ and series expansion gives:

$$C_V = 3Nk_B\left[1 - \frac{1}{20}\left(\frac{\theta_D}{T}\right)^2 + \frac{1}{560}\left(\frac{\theta_D}{T}\right)^4\right] \tag{5}$$

In the limit $\theta_D/T \to 0$, $C_V = 3Nk_B = 3R$, which is the classical limit of Dulong and Petit.

The electronic contribution to heat capacity (C_{electron}) is derived by *free electron Fermi Gas Model*. The corresponding expression for C_{electron} is given as:

$$C_{\text{electron}} = \frac{1}{3}\pi^2 Nk_B^2\left(\frac{T}{T_F}\right) = \gamma T \tag{6}$$

At room temperature, the electronic contribution to the heat capacity is insignificant compared to the lattice contribution. However, the situation is quite different at low temperatures. C_{electron} decreases linearly with T, whereas C_{phonon} decreases as T^3. Therefore, at some temperature, the two terms become equal; at still lower temperatures, C_{electron} is larger than C_{phonon}. In general, at very low temperatures (Liquid He-temperature), both terms are comparable and the observed specific heat is of the form:

$$C = C_{\text{phonon}} + C_{\text{electron}} = aT^3 + \gamma T \tag{7}$$

Therefore, a plot of C/T against T^2 should be a straight line with the intercept equal to γ and slope equal to a.

The magnetic contribution to heat capacity (C_{magnon}) is derived by *Weiss Molecular Field Theory* (WMF). For ferro- and ferrimagnets, the expression for C_{magnon} is given by:

$$C_{\text{magnetic}} = c_f N k_B \left(\frac{k_B T}{2JS}\right)^{3/2} = 0.113 k_B \left(\frac{k_B T}{D}\right)^{3/2} \tag{8}$$

where J is the exchange constant and S is the total spin quantum number. Therefore, at low temperatures, the ferromagnetic contribution to specific heats is proportional to $T^{3/2}$.

For the antiferromagnetic case, the heat capacity relation at low temperatures becomes:

$$C_{\text{magnetic}} = c_f N k_B \left(\frac{k_B T}{2J'S}\right)^{3} \tag{9}$$

Therefore, at low temperatures, the antiferromagnetic contribution to specific heats is proportional to T^3. The temperature dependence is the same as for the lattice contribution in the Debye model at low temperatures. This makes an experimental separation of spin wave and lattice specific heats almost impossible in metals and very difficult in nonmetallic antiferromagnets.

An equation for the Schottky heat capacity can be developed from the methods of statistical mechanics (Boltzmann statistics). The electronic specific heat due to Schottky effect is given by:

$$\Delta C_{\text{Schottky}} = Z^{-2} \cdot k_B^{-1} \cdot T^{-2} \left\{ Z \cdot \sum_{i=0}^{n} g_i \varepsilon_i^2 \exp\left(-\frac{\varepsilon_i}{k_B T}\right) - \left[\sum_{i=0}^{n} g_i \varepsilon_i \exp\left(-\frac{\varepsilon_i}{k_B T}\right)\right]^2 \right\}$$
$$= k_B^{-1} \cdot T^{-2} \left[\overline{E^2} - \overline{E}^2\right] \tag{10}$$

where Z is the partition function. Therefore, the electronic Schottky specific heat shows T^{-2} dependence. The difference between the mean square energy and the square of the mean energy is called fluctuation in energy. In its expanded form, this equation is especially useful for the evaluation of the electronic Schottky heat capacity of multilevel systems.

The maximum heat capacity for a simple two-level system will be given approximately by $(g_1/g_0)(0.87)$. For a more complex system of levels, the shape and the maximum heat capacity will depend upon the spacing and degeneracies of several levels.

Now the total low temperature constant volume heat capacity can be expressed as:

$$\left.\begin{aligned} C_V &= \frac{12}{5}Nk_B\pi^4\left(\frac{T}{\theta_D}\right)^3 + \frac{1}{3}\pi^2 Nk_B^2\left(\frac{T}{T_F}\right) + 0.113k_B\left(\frac{k_BT}{D}\right)^{3/2}_{\text{ferro/ferri}} + k_B^{-1}\cdot T^{-2}\left[\overline{E^2} - \overline{E}^2\right] \\ C_V &= \frac{12}{5}Nk_B\pi^4\left(\frac{T}{\theta_D}\right)^3 + \frac{1}{3}\pi^2 Nk_B^2\left(\frac{T}{T_F}\right) + c_f Nk_B\left(\frac{k_BT}{2J'S}\right)^3_{\text{antiferro}} + k_B^{-1}\cdot T^{-2}\left[\overline{E^2} - \overline{E}^2\right] \end{aligned}\right\} \tag{11}$$

The low temperature experimental heat capacity is thus fitted into different equations depending on the nature of the solid. For a Debye solid having only lattice contribution, the low temperature heat capacity is generally plotted as C_V against T^3 which yields a straight line passing through origin. The Debye temperature is then calculated from the slope of the line.

For a diamagnetic solid with electrical conduction, the low temperature heat capacity is expressed by the first two terms of Eq. (11) as:

$$C_V = aT^3 + \gamma T \tag{12}$$

Thus in such a case, the plot of C_v/T against T^2 yields a straight line. The slope and intercept represent a and γ, respectively. The value of a is used to calculate the Debye temperature (θ_D) where as the value of γ is used to calculate the Fermi temperature (T_F).

For a ferro- or ferrimagnetic solid with electrical insulator, the low temperature heat capacity is represented as:

$$C_V = aT^3 + bT^{3/2} \tag{13}$$

Thus, a plot of $C_v/T^{3/2}$ against $T^{3/2}$ will give rise to a straight line.

If the Schottky heat capacity is present, then the heat capacity will show a broad hump. The high temperature tail of this hump is extrapolated to the lower temperature side in order to separate $\Delta C_{\text{Schottky}}$ from the total heat capacity. Then the value of $(C_V - \Delta C_{\text{Schottky}})$ is modeled with Eq. (13) in order to calculate the constants a and b.

2.2 Calculation of θ_D as a Function of Temperature from Heat Capacity Data

It can be approximated that Debye's T^3 law is valid at very low temperatures (up to T $\approx$ $\theta_D/20$). So, the value of θ_D calculated from the low temperature heat capacity data using the T^3 law is called θ_D at 0 K. At higher temperatures ($T > \theta_D/20$), the

Debye temperature varies with temperature. In general, the electronic, magnetic, and Schottky contribution is separated from the total heat capacity to get the lattice heat capacity. The values of θ_D as a function of temperature are calculated in the following way.

1. For a solid oxide of general formula $A_xB_yO_z$, the lattice heat capacity is first divided by the stoichiometric number $(x + y + z)$. Thus, we get $C_{phonon}/(x + y + z)$.
2. The values thus obtained in step (1) are again divided by $3R$ in order to get the Debye function: $\frac{C_V}{3R} = \frac{3}{x^3}\int_0^x \frac{x^4e^x}{(e^x-1)^2}dx = f(\text{Debye})$. The tabulated values of $f(\text{Debye})$ are used to find out the values of x at each individual temperature. The values of x are then used to calculate θ_D at each temperature.

In the high temperature regime, many oxides show substantial contribution to specific heat due to vacancies, ionic defects, and electronic disorder as pointed out by Hoch [30]. Similarly, the anharmonic contribution to specific heat dominates at high temperatures in which case Debye model is not adequate to account for the specific heats of solids. In that case, the specific heat temperature can be approximated by the formula:

$$C_v = 3Rf(\text{Debye}) + dT^3 \tag{14}$$

where, d represents the anharmonic lattice contribution. Besides these basic models, there are a number of simulation techniques which are used to derive specific heat as a function of temperature. In order to use these simulation techniques, it is essential to know some of the fundamental crystallographic and physicochemical parameters of these oxides which are tabulated below (Table 1).

The following experimental techniques are usually employed to experimentally measure the specific heats.

(1) The low temperature specific heats (<298.15 K) are usually determined by adiabatic calorimetry or AC calorimetry. The accuracies of these methods are usually within ±0.5 to 1 %.
(2) Specific heats in the temperature range of 298.15–800 K are usually measured by either power compensated or heat flux differential scanning calorimeters. The accuracies are generally within ±1 to 2 %.

Table 1 Crystallographic and physicochemical properties of ThO_2, UO_2, PuO_2, and their solid solutions at STP [31–34]

Material	Crystallographic structure/space group	Lattice constants (Å)	Theoretical density (g/cc)	Melting points (K)	Enthalpy of fusion (kJ/mol)
ThO_2	FCC fluorite type	5.5968	10.002	3651 ± 17	90
UO_2	FCC fluorite type	5.4704	10.956	3120 ± 30	78
PuO_2	FCC fluorite type	5.3960	11.458	2701 ± 35	67

(3) Special differential scanning calorimetric sensors in the heat flux mode are used to measure specific heats in the temperature range of 298.15–1,500 K. These types of sensors give accuracies within ± 1 to 2 % in the low temperature range and $\sim\pm 5$ to10 % in the high temperature range.
(4) Above 1,500 K, the specific heat can be measured using laser flash method. However, the precision and accuracies of these measurements are uncertain. The general practice at higher temperatures is to measure the enthalpy increments (H_T–$H_{298.15\ K}$) by drop calorimetry and derive the heat capacity from the enthalpy data by differentiation with respect to temperature. This method can give heat capacity data up to very high temperature ($\sim$3,000 K) depending on the sensor used to measure the heat flux. The accuracies of this method in the high temperature regime is within ± 5 to 10 %.

2.3 Specific Heat of ThO_2

Utilizing the above experimental and theoretical methods, the specific heats of ThO_2 and solid solutions of ThO_2 with UO_2 have been determined by many researchers [25, 32–49]. A summary of the literature information about the compounds investigated, methods of measurement/analysis and the temperature range of investigation are given in Table 2. The literature data shows considerable scatter which can be attributed to different methods of preparation of samples which ultimately lead to samples of different stoichiometry, density, and impurity contents. It is difficult to comprehend the source of discrepancy of specific data for these samples unless the above parameters are known. Hence, it is imperative for fuel designer to measure the specific heat of the samples prepared by their own methods.

Table 3 shows the available data on specific heats in analytical forms. The low temperature specific heat data of ThO_2 obtained by Osborne and Westrum [35] are considered to be most accurate and used by many researchers to derive the standard entropy and enthalpy of ThO_2 at 298.15 K which are the two fundamental parameters required for any thermodynamic calculations. Hoch [30] has carried out modeling of high temperature specific heats of ThO_2 using the enthalpy data of Fischer et al. [39] and found out the anharmonic contribution to specific heats of ThO_2. Young [41] has carried out a detailed modeling of specific heats of ThO_2, UO_2, and PuO_2 using band structure calculations. The conclusion arrived from Young's study is that unlike UO_2, there is no evidence that there is anomalous contribution to specific heat of ThO_2 up to 3000 K. Sobolev [32] and Sobolev and Lemehov [33] have simplified model of the carried out modeling of thermal properties and specific heats of actinide oxides based on a phonon spectrum, on the statistical thermodynamics, and on the generalized Klemens model for thermal conductivity. They have derived some useful relationships bounding the specific heat capacity, the thermal expansion coefficient, the bulk modulus, and the thermal

Table 2 Summary of literature data on specific heats of ThO_2 and $(Th_{1-y}U_y)O_2$ solid solutions

Authors	Compounds studied	Property measured/ analyzed	Method of measurement/ analysis	Temperature range
Banerjee et al. [25]	$(Th_{1-y}U_y)O_2$ with $y = 0, 0.961, 0.902, 0.804, 0.506, 0.204$	Heat capacity	Differential scanning calorimetry	$(298.15 \le T/K \le 1650)$ by heat flux DSC
Hoch [30]	ThO_2	Heat capacity	Thermodynamic modeling	2,500–3,500 K
Sobolev [32]	ThO_2	Heat capacity and equation of states	Thermodynamic modeling	0–2,000 K
Sobolev and Lemehov [33]	ThO_2	Heat capacity	Thermodynamic modeling	0–2,000 K
Bakker et al. [34]	ThO_2, $(Th_{1-y}U_y)O_2$	Enthalpy and heat capacity	Assessment of literature data	298–3,500 K
Osborne and Westrum [35]	ThO_2	Heat capacity	Adiabatic calorimetry	10–305 K
Southard [36]	ThO_2	Drop calorimetry	Drop calorimetry	298–1,790 K
Takahashi and Murabayashi [37]	ThO_2 ThO_2-UO_2	Heat capacity	Laser flash method	80–1,000 K
Hoch and Johnston [38]	ThO_2	Enthalpy increment and heat capacity	Drop calorimetry	1,273–2,773 K
Fischer et al. [39, 40]	ThO_2	Enthalpy and heat capacity	Drop calorimetry	298–3400 K
Young [41]	ThO_2	Heat capacity	Assessment of data and modeling	1,500–3,500 K
Dash et al. [42]	ThO_2	Heat capacity	Thermodynamic modeling	0–2,000 K
Krishnan and Nagarajan [43]	ThO_2	Heat capacity	Temperature modulated DSC	400–800 K
Agarwal et al. [44]	ThO_2, $(Th_{1-y}U_y)O_2$ with $y = 0.0196, 0.039, 0.059, 0.098$	Enthalpy increment	Drop calorimetry	375–991 K
Kurina et al. [45]	$(U,Th)O_2$	Heat capacity	Calorimetry	573–1473 K

(continued)

Table 2 (continued)

Authors	Compounds studied	Property measured/analyzed	Method of measurement/analysis	Temperature range
Ralph [46]	ThO_2, $(Th_{1-y}U_y)O_2$ with $y = O.3, 0.15$ and 0.08	Heat capacity	Thermodynamic modeling and data analysis	298 to melting point
Fishcer et al. [40]	$(Th_{1-y}U_y)O_2$ with $y = 0.30, 0.15, 0.08$	Enthalpy increment	Induction heated drop calorimetry	2,300–3,400 K
Shuller et al. [47]	$Th_{1-y}U_yO_2$ $(0 < y < 1)$	Enthalpy and Gibbs energy of mixing	Monte-Carlo simulation	300—000 K
Kandan et al. [48]	$(Th_{1-y}U_y)O_2$ with $y = 0.1, 0.5, 0.9$	Enthalpy increment	Drop calorimetry	479–1,805 K
Anthonysamy et al. [49]	$(Th_{1-y}U_y)O_2$ with $y = 0.1, 0.5$ and 0.9	Enthalpy increment	Drop calorimetry	473–973 K

Table 3 Compilation of specific heat data of ThO_2, UO_2, and $(Th_{1-y}U_y)O_2$ solid solutions

Authors (Year) [References]	C_p $(ThO_2)/J.K^{-1}.mol^{-1}$	C_p $(Th_{1-y},U_y)O_2/J.K^{-1}.mol^{-1}$	Remarks
Banerjee et al. [25]	$68.7 + 0.01045 \cdot (T/K) - 1074291 \cdot (T/K)^{-2}$	$61.3 + 0.01584 \cdot (T/K) - 464670 \cdot (T/K)^{-2}$ $(y = 0.039)$	$(298.15 \leq T/K \leq 1650)$
		$60.9 + 0.01816 \cdot (T/K) - 460881 \cdot (T/K)^{-2}$ $(y = 0.098)$	By heat flux DSC
		$61.5 + 0.01879 \cdot (T/K) - 499317 \cdot (T/K)^{-2}$ $(y = 0.196)$	
		$61.7 + 0.02145 \cdot (T/K) - 413410 \cdot (T/K)^{-2}$ $(y = 0.494)$	
		$63.6 + 0.0204 \cdot (T/K) - 491434 \cdot (T/K)^{-2}$ $(y = 0.796)$	
Dash et al. [42]	$70.94 + 0.00772 \cdot (T/K) - 1.0968 \cdot 10^6 \cdot (T/K)^{-2}$	$72.17 + 0.00668 \cdot (T/K) - 1.1234 \times 10^6 \cdot (T/K)^{-2}$ $(y = 0.0196)$	$(304 \leq T/K \leq 1690)$ by combined fit of data obtained from heat flux DSC and drop calorimetry
		$70.51 + 0.00653 \cdot (T/K) - 1.1750 \cdot 10^6 \cdot (T/K)^{-2}$ $(y = 0.0392)$	
		$71.40 + 0.00812 \cdot (T/K) - 1.1139 \cdot 10^6 \cdot (T/K)^{-2}$ $(y = 0.098)$	
		$72.392 + 0.00807 \cdot (T/K) - 1.1308 \cdot 10^6 \cdot (T/K)^{-2}$ $(y = 0.1964)$	
Agarwal et al. [44]	$71.6726 + 8.2532 \times 10^{-3} \cdot (T/K) - 1116094 \cdot (T/K)^{-2}$	$76.8601 + 3.598 \cdot 10^{-4} \cdot (T/K) - 1888448 \cdot (T/K)^{-2}$ $(y = 0.0196)$	$(298.15 \leq T/K \leq 1000)$ by drop calorimetry
		$67.8243 + 8.798 \cdot 10^{-3} \cdot (T/K) - 684523 \cdot (T/K)^{-2}$ $(y = 0.039)$	
		$78.8135 - 6.083 \cdot 10^{-3} \cdot (T/K) - 1897724 \cdot (T/K)^{-2}$ $(y = 0.059)$	
		$78.3578 + 4.657 \cdot 10^{-4} \times (T/K) - 1774856 \cdot (T/K)^{-2}$ $(y = 0.098)$	

(continued)

Table 3 (continued)

Authors (Year) [References]	C_p (ThO_2)/J.K^{-1}.mol^{-1}	C_p (Th_{1-y},U_y)O_2/J.K^{-1}.mol^{-1}	Remarks
Bakker et al. [34]	$55.962 + 0.05126 \cdot (T/K) - 3.6802 \cdot 10^{-5} \cdot (T/K)^2 + 9.2245 \cdot 10^{-9} \cdot (T/K)^3 - 5.7403 \cdot 10^5 \cdot (T/K)^{-2}$		$298.15 \leq T/K \leq 3500$ by assessment of literature data
Young [41]	$K1 \cdot (\theta/T)2 \cdot \exp(\theta/T) \cdot (\exp(\theta/T) - 1)^{-2} + K2T$ ($K1 = 69.29$ J.K^{-1}.mol^{-1}; $K2 = 9.34 \cdot 10^{-3}$ J.K^{-2}.mol^{-1}; $\theta = 398$ K)		By compilation and fitting of available experimental data
Southard [36]	$68.761 + 9.806 \cdot 10^{-3} (T/K) - 8.878 \cdot 105 \cdot (T/K)^{-2}$		$298 \leq T/K \leq 1790$ by drop calorimetry
Kandan et al. [48]		$71.63 + 10.65 \cdot 10^{-3} \cdot (T/K) - 13.56 \cdot 10^5 \cdot (T/K)^{-2}$ ($y = 0.10$) $73.47 + 11.48 \cdot 10^{-3} \cdot (T/K) - 14.23 \cdot 10^5 \cdot (T/K)^{-2}$ ($y = 0.50$) $73.63 + 14.79 \cdot 10^{-3} \cdot (T/K) - 12.89 \cdot 10^5 \cdot (T/K)^{-2}$ ($y = 0.90$)	$298 \leq T/K \leq 800$ by drop calorimetry
Kurina et al. [45]		$64.564 + 1.183 \cdot 10^{-2} \cdot (T/K) - 1311934.8 \cdot (T/K)^{-2}$ ($y = 0.2458$) $64.126 + 8.01 \cdot 10^{-3} \cdot (T/K) - 466255.8 \cdot (T/K)^{-2}$ ($y = 0.2458$) (corrected for 100 % TD)	
Anthonysamy et al. [49]		$71.923 + 17.3295 \cdot 10^{-4} \cdot (T/K) - 107.20 \times 10^4 \cdot (T/K)^{-2}$ ($y = 0.1$) $74.249 + 17.557 \cdot 10^{-4} \cdot (T/K) - 121.54 \cdot 10^4 \cdot (T/K)^{-2}$ ($y = 0.5$) $72.344 + 33.727 \cdot 10^{-4} \cdot (T/K) - 115.177 \cdot 10^4 \cdot (T/K)^{-2}$ ($y = 0.9$)	

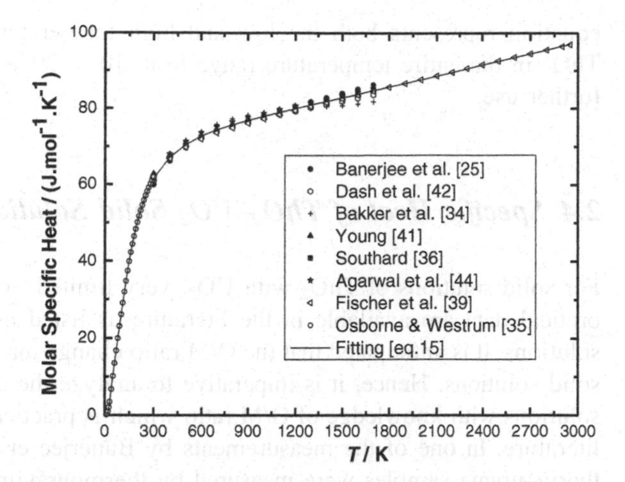

Fig. 6 Comparison of specific heats of ThO_2

conductivity of dioxides in a quasi-harmonic approximation in the temperature range from 0 to 2000 K. The developed models were first verified with UO_2 and then applied for prediction of the isobaric specific heat, the isobaric thermal expansion coefficient, and the thermal conductivity of ThO_2. The results obtained are in very good agreement with the experimental data. A comparison of specific heats of ThO_2 available in the literature is made in Fig. 6.

It is useful to generate a single analytical expression for the specific heat of ThO_2 in the entire temperature range of 0–3,000 K by least squares analysis of all the literature reported data. In order to calculate the smoothed heat capacity and thermodynamic functions, Woodfield et al. [50] have modified the method of King and King [51] by a six-term fitting equation based on Debye and Einstein functions as given below.

$$C^{\mathrm{o}}_{p,\mathrm{m}} = 3R \cdot \{m \cdot D(\Theta_{\mathrm{D}}/T) + n \cdot E(\Theta_{\mathrm{E}}/T)\} + A_1 \cdot (T/\mathrm{K}) + A_2 \cdot (T/\mathrm{K})^2 \quad (15)$$

where, $D(\Theta_D/T)$ and $E(\Theta_E/T)$ are Debye and Einstein functions, respectively; m, n, Θ_D, Θ_E, A_1 and A_2 are adjustable parameters; and $(m + n)$ should be close to the number of atoms in the molecule. The linear term in Eq. (15) takes into account the high temperature heat capacity. The heat capacity data shown in Fig. 6 are smoothened using Eq. (15). The fitting parameters are listed in Table 4. Since this

Table 4 Coefficients obtained from a fit of the experimental heat capacity with Eq. (15)

Parameters	Coefficients
m	2.178 ± 0.021
N	0.696 ± 0.021
Θ_D (K)	684 ± 6
Θ_E (K)	153 ± 4
A_1 (J K^{-2} mol^{-1})	$7.086 \times 10^{-3} \pm 2.511 \times 10^{-4}$
A_2 (J K^{-3} mol^{-1})	$5.341 \times 10^{-7} \pm 6.854 \times 10^{-8}$

equation represents both the low and high temperature molar specific heats of ThO_2 in the entire temperature range from 10 to 2950 K, it is recommended for further use.

2.4 Specific Heat of ThO_2–UO_2 Solid Solutions

For solid solutions of ThO_2 with UO_2, very limited experimental as well as theoretical data are available in the literature as listed in Table 3. For these solid solutions, it is anticipated that the O/M ratio change may affect the specific heat of solid solutions. Hence, it is imperative to analyze the experimental data on solid solutions with knowledge of O/M ratio which is practically lacking in most of the literature. In one of the measurements by Banerjee et al. [25], the O/M ratio of thoria–urania samples were measured by thermogravimetry and was found to be 2.0 before measurement. The carrier gas used for specific heat measurements by Banerjee et al. [25] was high purity argon (oxygen content <5 ppm, moisture content <5 ppm) which was further purified for oxygen and moisture by passing it through several purification traps and finally through hot uranium getter. Hence, it was anticipated that the O/M ratio of samples investigated during heat capacity measurements do not change appreciably. Further, to confirm the stoichiometry, they have analyzed a representative sample having higher UO_2 content and found out no detectable change in O/M ratio which led to their assumption that the O/M ratio for all other solid solutions remain unaffected during specific heat measurement.

A comparison of specific heats of these solid solutions is made by Banerjee et al. [25] which is shown in Fig. 7. It is evident from this figure that the specific heats of ThO_2–x wt% UO_2 solid solutions show a gradual increase with increase in UO_2 content. Further, it is apparent from Fig. 7, that the specific heats of ThO_2–4 % UO_2 is lower than that of pure ThO_2 at lower temperatures which is in conformity with another observation by Dash et al. [42]. However, at high

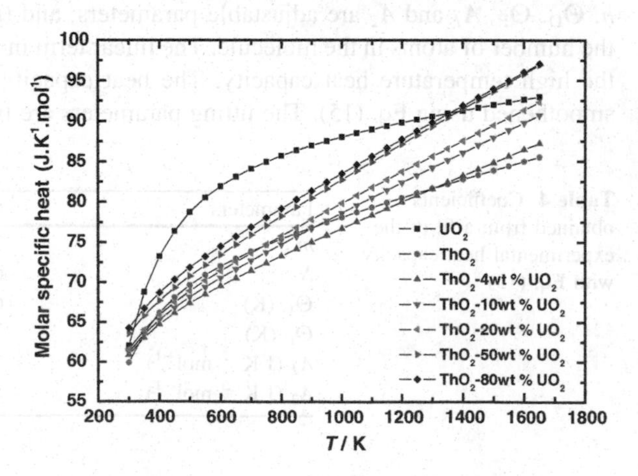

Fig. 7 Comparison of molar specific heats of UO_2, ThO_2, and ThO_{2-x} wt % UO_2 solid solutions by Banerjee et al. ([25] Copyright Elsevier)

temperatures, the specific heats of ThO_2–50 % UO_2 and ThO_2–80 % UO_2 become higher than that of pure UO_2. Further, Banerjee et al. [25] have observed that considering experimental uncertainties and other factors such as sample inhomogeneity, impurities etc., for lower concentration of UO_2 (i.e., 4 wt% and less), the specific heats of ThO_2–UO_2 solid solutions do not show ideal solution behavior in the entire temperature range from 300 to 1,800 K whereas for higher UO_2 content, the solid solutions can be approximated as close to ideal solution behavior.

In summary, the specific heat data on solid solutions of ThO_2–PuO_2 are not available in the literature and hence there is wide scope in both experimentation and modeling the specific heats of these solutions. Similarly, the specific heats of ThO_2–UO_2 solid solutions have been investigated to limited temperature range and hence high temperature measurements are necessary. Moreover, the specific heat data on simulated thoria fuel (SIMFUEL) with appropriate additions of fission products is extremely important to access the real situation in nuclear reactor.

3 Oxygen Potential of Thoria-Based Oxide Fuels

Oxygen potential of oxides in nuclear fuel is the most important thermodynamic parameter that affects almost all aspects of fuel behavior. The most significant contribution of this parameter is to affect the clad corrosion. Oxygen potential also affects the chemical state of the fission products formed during burnup. For example, at low oxygen potentials the corrosive fission product species, iodine reacts with other fission products and form stable compounds, CsI, ZrI_4 etc.; thus, it is held in the fuel matrix. However, on increase in oxygen potential of the fuel during burnup, these metallic fission products form more stable oxides, e.g., Cs_2MoO_4, Cs_2ZrO_4, Cs_2UO_4 etc. hence, the bound iodine gets released from the fuel and goes in the gaseous phase. This iodine can thus easily migrate to the clad through interlinked porosity or grain surface diffusion process. In case of zirconium-based cladding, zircaloy, iodine can result in stress corrosion cracking, which is detrimental to the fuel pin behavior. Oxygen potential also controls the composition of actinide bearing species in vapor phase. A prediction of the chemical states of fuel and fission product elements during burnup is related to temperature, oxygen potential, and atomic composition of the fuel. Reactor physicist can compute atomic composition of the fuel for a defined reactor design and starting fuel composition. Based on these features, approximate radial and axial temperature gradient can also be computed. Due to change in chemical composition and restructuring of the fuel, its temperature gradient keeps on changing with burnup. All these parameters are intricately related to each other. Chemical composition depends on the temperature, whereas, chemical composition also controls the temperature by affecting thermal conductivity and material transport, thus restructuring. But if an average temperature gradient is taken at a defined burnup, then chemical composition of the fuel can be estimated by the knowledge of its oxygen potential.

Oxygen potential is determined by various methods, e.g., emf, gas equilibrium coupled to thermo-gravimetric instrument, isopiestic method etc. [52–59]. All these techniques are for reasonably high temperature measurements, where kinetics is sufficiently fast for the solid phase equilibrium. However, EMF method cannot be used for temperatures above 1473 K, as the condition of nearly unit transport number for oxide ion is not fulfilled by most of the known solid-state electrolytes at very high temperatures. In case of mixed oxides $(Th,U)O_{2+x}$ and $(Th,Pu)O_{2-x}$, the measurement becomes very erroneous near stoichiometric composition as the change in oxygen potential is very steep near $x \cong 0.0$. In thermogravimetric method, the mixed oxide is brought in equilibrium with a gas mixture of CO/CO_2 or H_2O/H_2 with known gas ratio. The ratio of the gas mixture controls the oxygen potential over the system, thus the system either looses or gains weight to be in equilibrium with the cover gas. For this, the starting weight is preferably of stoichiometric composition, thus making a direct relation between oxygen potential and composition. In case of $(Th,O)_{2+x}$, the lowest stoichiometry possible is 2.0, thus the weight will increase with increase in 'x' or O/(Th + U). On the other hand, in $(Th,Pu)O_{2-x}$, the highest stoichiometry is 2.00, so the weight of the oxide will decrease with increase in 'x' or O/(Th + Pu). This method can be used for the systems with both hyper and hypostoichiometry. At very high temperatures, the starting O/M is not known and most of the mixed oxides have both hypo and hyperstoichiometry at high temperatures. In such system, one needs to monitor oxygen potential as a function of weight change. A steep change in oxygen potential near stoichiometric composition helps in identifying the composition where O/M is 2.0. Then back calculations of the values are used to find a relation between oxygen potential and O/M.

There is reasonable amount of experimental oxygen potential data on $(Th,U)O_{2+x}$ system for different temperatures and compositions. Bakker et al. [34] have compiled thermodynamic and thermophysical data of thoria, thoria–urania and thoria–plutonia. The oxygen potential of $(Th_{1-y}U_y)O_{2+x}$ increases with increase in temperature and increase in 'x'. Though oxygen potential of thoria is lower than that of urania, but in thoria–urania, oxygen potential increases with increase in thorium content. This is due to increase in uranium valency in the solid solution with increase in 'Th' content to accommodate the same amount of 'O'. Different approaches were used for computing oxygen potential of thoria–urania. Schram [60] used defect model and thermochemical model to assess oxygen potential of $(Th,U)O_{2+x}$. Some others reported works were based on polynomial fits of oxygen potential data as a function of temperature and composition. However, due to complicated relationship of these parameters with oxygen potential, different equations had to be used for different composition regions. Agarwal [61] used different oxygen potential models earlier applied to $(Pu,U)O_{2+x}$ system and compared the calculated values with experimental oxygen potential data. The hypothesis of these models is that the change in oxygen potentials of both $(Th,U)O_{2+x}$ and $(U,Pu)O_{2+x}$ with change in 'x' value is primarily due to change in oxidation state of uranium (U^{4+},U^{5+}), as both 'Pu' and 'Th' are in '4+' oxidation states. Though this assumption may be correct, but the models which

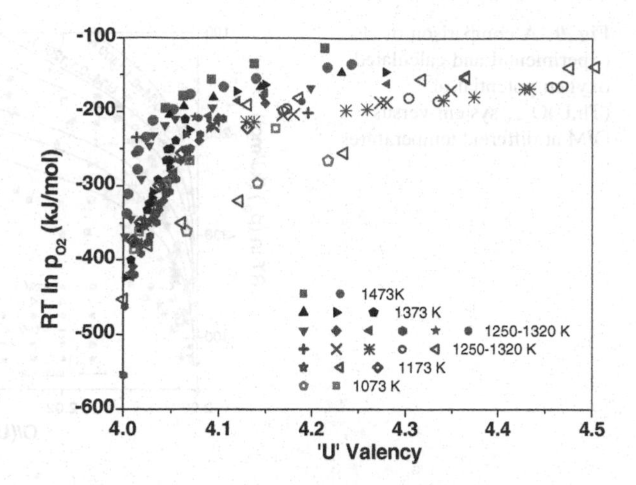

Fig. 8 A plot of oxygen potential of $(Th,U)O_{2+x}$ system versus 'U' valency at different temperatures. Same color code is used for the data at same temperatures

gave reasonably good agreement with experimental data for urania–plutonia system were not sufficiently suitable for urania–thoria system. To understand the effect of 'U' valency on oxygen potential, it is important to understand the relation between composition and uranium valency. The valency of 'U' in mixed oxide $(Th_{1-y}U_y)O_{2+x}$ is related to both 'x' and 'y' by the relation: U-valency = $4+2x/y$. If oxygen potential is merely a function of uranium valency, then a plot of oxygen potential versus uranium valency should show a single curve for a selected temperature, independent of 'x' (O/M) and 'y' (Th/M) values. This hypothesis assumes that stoichiometric ThO_2 does not affect the oxygen potential of the system and all the changes in oxygen potential are generated by the urania component. This assumption also means that the interaction between UO_2 and $UO_{2.5}$ alone affects the oxygen potential of the system. A plot of experimental oxygen potential of this system with uranium valency (Fig. 8), clearly shows a scatter that is not merely temperature dependent but also depends upon the composition, 'x' and 'y' values of $(Th_{1-y}U_y)O_{2+x}$. In this plot, data at different temperatures are shown with different color symbols. If the above hypothesis was acceptable, then all the same color symbols should be falling on the same curve, independent of their compositions. But, as seen in Fig. 8, some of high temperature data, red symbols corresponding to reading at 1,473 K are mixing with black symbols for oxygen potential data at 1,373 K. Similar mixing is seen between low temperature data too. Moreover, the measurements at higher O/M, corresponding to higher 'U' valency are diverging from each other. As seen in the scatter of same color symbols with increase in 'U' valency. This is possible only when oxygen potential does not depend only on valency state of uranium but also on nonideal interaction parameters between ThO_2–UO_2 and ThO_2–$UO_{2.5}$. Therefore, Agarwal calculated interaction parameters of ThO_2–UO_2–$UO_{2.5}$ using CALPHAD method (Fig. 9). She has also calculated interaction parameters of ThO_2–PuO_2–Pu_2O_3 system to estimate oxygen potential of $(Th,Pu)O_{2-x}$ system and compared these values with

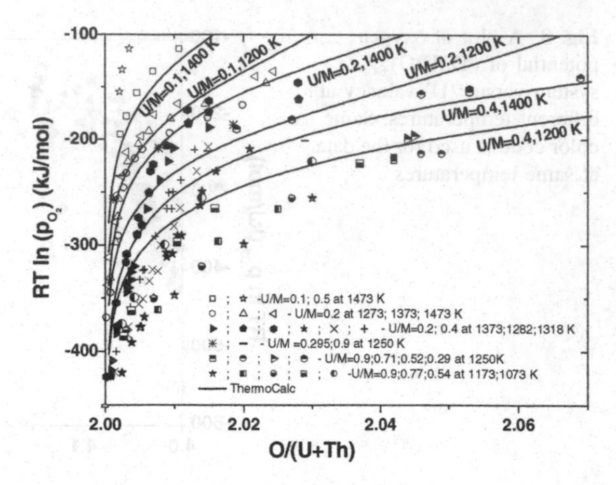

Fig. 9 A comparison of experimental and calculated oxygen potential of $(Th,U)O_{2+x}$ system versus O/M at different temperatures

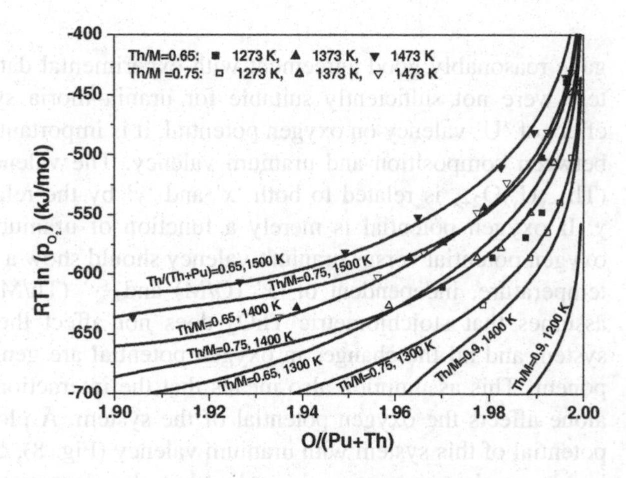

Fig. 10 A comparison of experimental data of Woodley and Adamson [59] with calculated oxygen potential of $(Th,Pu)O_{2+x}$ system versus O/M at different temperatures

experimental results (Fig. 10). The only reported data on oxygen potential of ThO_2–PuO_2 is the one determined in Hanford Engineering Development Laboratory [62], measured by thermogravimetry. The data is available only for Pu/(Th + Pu) = 0.25 and 0.35, at 1,273, 1,373, and 1,473 K, up to hypostoichiometry region of its stability. As seen from the figure, oxygen potential of $(Th,Pu)O_{2-x}$ increases with decrease in 'x' value. Unlike in the case of $(Th,U)O_{2+x}$, where oxygen potential increases with decrease in 'U' content, in $(Th,Pu)O_{2-x}$ system, oxygen potential decreases with decrease in 'Pu' content. Table 5 lists all the experimental oxygen potential data available in the literature for thoria, thoria–urania, and thoria–plutonia.

Using statistical thermodynamics of point defects, Kosuge [63] derived the following relation between intrinsic defect fraction 'C', non-stoichiometry 'x' and p_{O2}.

Table 5 A list of all the available experimental oxygen potential data for ThO_{2-x}, ($(Th,U)O_{2+x}$, and $(Th,Pu)O_{2-x}$ systems

Composition	Temperature (K)	Technique	References
ThO_{2-x} O/M = 1.95 to 2.0	2,400, 2,500, 2,600, and 2,655	H_2/H_2O gas equilibria and TG	Ackermann and Tetenbaum [58]
U/M = 0.2	1,273	TG and EMF	Ugajin et al. [52]
O/M = 2.0002–2.0199	1,373		
O/M = 2.0007–2.0232	1,473		
O/M = 2.0004–2.0243			
U/M = 0.29,	1,150–1,350	EMF and TG	Aronson and Clayton [53]
O/M = 2.019–2.073			
U/M = 0.52,			
O/M = 2.044–2.152			
U/M = 0.71,			
O/M = 2.046–2.154			
U/M = 0.9,			
O/M = 2.042–2.157			
U/M = 0.03, 0.063, 0.244	1,003–1,203	TGA/Volumetric method	Anderson et al. [15]
U/M = 0.0053–0.0597	1,123	PM	Roberts et al. [54]
U/M = 0.295,	1,250	EMF	Tanaka et al. [55]
O/M = 2.002–2.0289			
U/M = 0.195,			
O/M = 2.0012–2.0185			
U/M = 0.2,	1,273	TG and EMF	Ugajin [56]
O/M = 2.0054–2.0199	1,373		
O/M = 2.0048–2.0232	1,473		
O/M = 2.0004–2.0243	1,473		
U/M = 0.1,			
O/M = 2.0022–2.0107			
U/M = 0.05,			
O/M = 2.0021–2.0035			

(continued)

Table 5 (continued)

Composition	Temperature (K)	Technique	References
U/M = 0.2,	1,282	TGA with H_2/CO_2 gas equilibria	Matsui and Naito [57]
O/M = 2.0008–2.015	1,318		
U/M = 0.4,	1,373		
O/M = 2.0001–2.021			
U/M = 0.2,			
O/M = 2.0004–2.028			
U/M = 0.4,			
O/M = 2.0013–2.013			
U/M = 0.2,			
O/M = 2.00024–2.028			
U/M = 0.4,			
O/M = 2.0016–2.013			
U/M = 0.54, 0.77, 0.9	1,073 and 1,173	H_2, H_2/Ar, H_2/CO_2 gas equilibrium with TG	Anthonysamy et al. [18]
O/M = 2.001–2.01			
Pu/M = 0.25	1,273, 1373 and 1,473	TG and gas equilibrium	Woodley and Adamson [59]
O/M = 1.94–2.00			

$$\frac{p_{O_2}(x)}{p_{O_2}(0)} = 1 + \frac{x^2 + x\sqrt{(x^2+4C^2)}}{2C^2};\ Where\ C = A\exp\left(-\frac{E}{2\kappa T}\right), \quad (16)$$

where, $p_{O2}(x)$ is the partial pressure of oxygen over non-stoichiometric oxide $(Th,U)O_{2+x}$ or $(Th,Pu)O_{2-x}$ and $p_{O2}(0)$ is oxygen partial pressure over stoichiometric oxide. '*C*' is intrinsic defect fraction and '*E*' is formation energy of intrinsic defects that is related to band-gap energy. '*A*' is a constant for selected oxide and 'κ' is Boltzmann constant. From the knowledge of change in oxygen potential with '*x*', one can find the value of formation energy of intrinsic defects.

In non-stoichiometric oxides, defects are generally charged species. In most of the oxides, the relative dielectric constant is small, therefore, Coulomb interactions play an important role in Gibbs energies of formation of defects in non-stoichiometric oxides. The most important thermodynamic parameter that gets affected by the formation of these charged point defects is partial pressure of oxygen and oxygen potential, $\mu_{O2} = RT \ln p_{O2}$. The oxygen potential is madeup of partial enthalpy and partial entropy of oxygen. These values show a maximum near the stoichiometric composition. $(Th,U)O_{2+x}$ has more negative partial molar enthalpy and entropy of oxygen than UO_{2+x}. This is due to considerably low partial enthalpy and entropy of oxygen in ThO_2. Low partial enthalpy or entropy values of oxygen in an oxide means it has strong oxygen holding capability; thus, its oxygen releasing capability or oxygen potential will be low. The change in shape of partial enthalpy and partial entropy curve as a function of composition, gives information on the type of defect in the lattice. For example, a clear maximum was seen in the plot of $\mathbf{S}_{O_2}$ and $\mathbf{H}_{O_2}$ versus O/M, in urania-rich oxides. It becomes less sharp and the O/M position of the maximum increases with increase in 'Th' content. This may be due to increase in oxygen vacancies and increase in vibrational partition function of oxygen ions. The relation between defect structure and oxygen partial pressure of MO_{2+x} is estimated from the value of '*n*' in $\boldsymbol{x} \propto \boldsymbol{p}_{O_2}^{1/n}$, which can be calculated from the slope of log *x* versus log p_{O2}. The value of '*n*' has direct link to hyperstoichiometry of these oxides and was found to be almost independent of temperature. The value of '*n*' was found to be almost same for a range of hyperstoichiometry. Spectroscopic structural analysis of these crystals have found a relation between the values of '*n*' and the structure of defect clusters created due to hyperstoichiometry. In UO_{2+x} and $(Th,U)O_{2+x}$, hyperstoichiometry is generated by creating oxygen vacancies ($\mathbf{V}_O$) and addition of oxygen ions in two different types of interstitial sites ($\mathbf{O}_i^a$ and $\mathbf{O}_i^b$). Depending upon the number of such defects, the cluster will have a negative charge, '*m*', which is balanced by generation of '*m*' number of holes ($\mathbf{h}^o$). For near stoichiometric compositions, the cluster is almost neutral, thus '*m*' is zero. The value of '*m*' increases with increase in hyperstoichiometry, so does the value of '*n*', though the cluster type remains same, $(2O_i^aO_i^b2V_O)^m$, as shown in Table 6. At high hyperstoichiometry ($x > 0.006$ for UO_{2+x} and $x > 0.008$ for $(Th,U)O_{2+x}$), the cluster type changes and it has lesser vacancies, $\{2(O_i^aO_i^b2V_O)\}^m$. Hence, the values of '*m*' and '*n*' suddenly decrease

Table 6 The relation between defect structure and values of '*m*' and '*n*' for different hyperstoichiometry

System	Defect cluster type	'*x*' values	*m*	*n*
UO_{2+x}	$(2O_i^aO_i^b2V_O)^m$	$x < 0.003$	0	2
	$(2O_i^aO_i^b2V_O)^m$	$0.003 < x < 0.006$	5	12
	$\{2(O_i^aO_i^bV_O)\}^m$	$x > 0.006$	1	2
$(Th,U)O_{2+x}$	$(2O_i^aO_i^b2V_O)^m$	$x < 0.001$	0	2
	$(2O_i^aO_i^b2V_O)^m$	$0.001 < x < 0.003$	1	4
	$(2O_i^aO_i^b2V_O)^m$	$0.003 < x < 0.008$	5	12
	$\{2(O_i^aO_i^bV_O)\}^m$	$x > 0.008$	3	4

above this '*x*'. The formation of these two types of complexes can be shown by the following reactions:

$$2V_i^a + V_i^b + 2O_0 + 1/2O_2(g) = \{2O_i^aO_i^b2V_O\}^m + mh^o \quad (17)$$

$$2V_i^a + 2V_i^b + 2O_0 + O_2(g) = \{2(O_i^aO_i^b2V_O)\}^m + mh^o \quad (18)$$

This has been confirmed by neutron diffraction analysis of defect structure of UO_{2+x}, by Willis [64]. Another method of assessing the defect structure is by relating the electrical conductivity (σ) of oxides with oxygen partial pressure. The relation between these two parameters is same as that between '*x*' and p_{O2}, $\sigma \propto p_{O_2}^{1/n}$. A comparison of experimentally determined electrical conductivity values of UO_{2+x} with its oxygen partial pressure has indicated that value of 'n' varies between 2 and 12, indicating a defect cluster of $O_i^aO_i^bV_O$, whereas, for compositions very close to stoichiometry, conductivity was found to be almost independent of p_{O2}, indicating the presence of mostly neutral defects. Lee [65] have measured electrical conductivity of $(Th,U)O_{2+x}$ and they also established that near stoichiometric region, where partial pressures of oxygen are low, electrical conductivity is almost independent of oxygen partial pressure; thus, the defects are mainly neutral clusters of $2O_i^aO_i^bV_O$. Whereas, for higher non-stoichiometry, a relationship with $n = 4$ is used to explain the results.

4 Partial Pressure of Actinide Species

The partial pressures of actinide bearing species play a very important role in fuel performance during burnup. The actinide with higher partial pressure will migrate to cooler temperature zone, resulting in actinide redistribution. The lower partial pressures of 'Th' bearing species compared to 'U' bearing species will result in preferential migration of 'U' to cooler surface zone, resulting in 'Th' enrichment in the hot central zone of the fuel. The higher partial pressure of actinide species with O/M > 2, e.g., UO_3, will result in depletion of O/M in the central zone. On the other hand, higher partial pressure of actinide species with O/M < 2, e.g., UO, ThO, U, Th, will result in increase in O/M in the central zone.

Ackermann and Tetenbaum [58] measured total pressures of 'Th' bearing gaseous species over ThO_{2-x}, by gas equilibrium method, using H_2/H_2O carrier gas and measuring the weight loss. They found an interesting linear relationship between logarithm of hypostoichiometry of congruently vaporizing composition and inverse of temperature. Similar relation is also given for lower phase boundary of hypostoichiometric ThO_{2-x} and temperature.

Lower Phase Boundary of ThO_{2-x}: $\log(x) = 0.506\ (\pm\ 0.166) - 4934\ (\pm\ 372)/T$

Congruently Vaporizing Compositions of ThO_{2-x}: $\log(x) = 2.019\ (\pm\ 0.493) - 11980\ (\pm\ 1280)/T$

The partial pressure of ThO(g) over Th(l)–ThO_2 system in the temperature range 2,000–2,500 K is given as: $\log p_{ThO(g)} = 7.58 - 28630/T$.

Yamawaki et al. [66] have given enthalpy of sublimation of UO_2(g) from $(Th,U)O_2$ as 564 kJ/mol and found it to be independent of actinide composition. This is lower than sublimation value of UO_2(g) over UO_2(s), 592 kJ/mol [67, 68]. Aitken et al. [69] measured total pressure of actinide bearing species over thoria–urania single phase system using transpiration technique in the temperature range 1473–1873 K by passing dry air as carrier gas over $(Th_{1-y}U_y)O_{2+x}$(s) with y = 0.063, 0.2, 0.25, 0.5. Alexander et al. [70] measured partial pressure of UO_2(g) over the stoichiometric $(Th_{1-y}U_y)O_2$(s) solid solutions with $y = 0.08$ and 0.2 in the temperature range 2,373–2,773 K by passing Ar–H_2 as a carrier gas. Yamawaki et al. [66] measured partial pressures of actinide species using mass spectrometer, in the temperature range 2,298–2,465 K for y = 0.1, 0.2, 0.4, 1. It is clear from Fig. 11 that partial pressure of 'U' bearing species is higher than that of 'Th' bearing species.

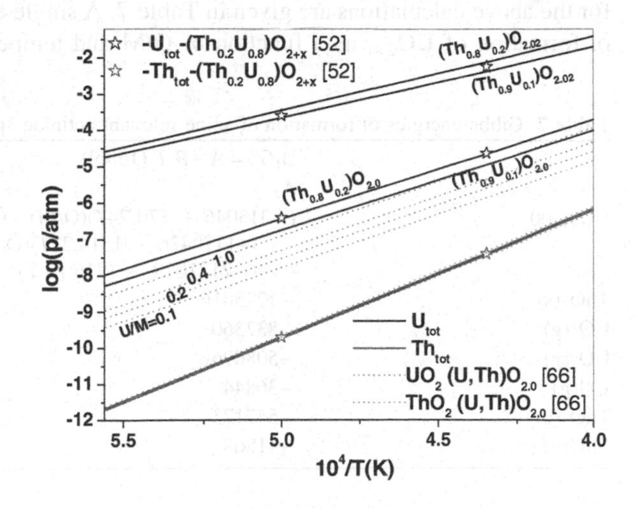

Fig. 11 Total partial pressures of 'U' and 'Th' bearing species versus temperature inverse for given compositions of $(Th,U)O_{2+x}$ compared with total pressures given by Ugajin et al. [52] and partial pressures of UO_2(g) and ThO_2(g) given by Yamawaki et al. [66]

This figure compares total partial pressures of 'U' bearing species and 'Th' bearing species as a function of temperature inverse, at different U/M and O/M compositions, where, M = Th + U. Some experimentally obtained data are also plotted in the figure for comparison. This shows clearly that partial pressure of 'U' bearing species is more composition dependent than 'Th' bearing species. Moreover, change in O/M has greater influence on partial pressure than change in metallic composition. The total partial pressures of 'Th' bearing species is almost equal to the partial pressure of $ThO_2(g)$ and is almost independent of composition. On the other hand, total partial pressures of 'U' bearing species is reasonably higher than that of $UO_2(g)$, indicating significant contribution of $UO_3(g)$. At this stage, it is also required to point out that the effect of temperature on partial pressures of 'U' and 'Th' bearing species is almost similar, as indicated by parallel slopes of these plots. The total partial pressure of 'Th' bearing species over ThO_{2-x}, as given by Ackermann and Tetenbaum [71], is higher than that of $ThO_2(g)$, over $(Th,U)O_{2+x}$ system, indicating significant contribution from ThO(g) in hypostoichiometric thoria. The pressures of actinide species, UO_3, UO_2, UO, ThO_2, and ThO over $(U,Th)O_{2+x}$, were calculated using Gibbs energies of formation of UO_{2+x}, ThO_2, all the gaseous species, partial Gibbs energy of formation of oxygen and activity coefficients of actinide oxides. The following relationships were used:

$$RT\ln p_{UO_3(g)} = \Delta_f G^o{}_{UO_{2+x}(s)} + RT\ln a_{UO_{2+x}(s)} + 0.5\,RT\ln po_2 - \Delta_f G^o{}_{UO_3(g)}$$

$$RT\ln p_{UO_2(g)} = \Delta_f G^o{}_{UO_3(g)} + RT\ln p_{UO_3(g)} - 0.5\,RT\ln po_2 - \Delta_f G^o{}_{UO_2(g)}$$

$$RT\ln p_{UO(g)} = \Delta_f G^o{}_{UO_2(g)} + RT\ln p_{UO_2(g)} - 0.5\,RT\ln po_2 - \Delta_f G^o{}_{UO(g)}$$

$$RT\ln p_{ThO_2(g)} = \Delta_f G^o{}_{ThO_2(s)} + RT\ln x_{ThO_2} - \Delta_f G^o{}_{ThO_2(g)}$$

$$RT\ln p_{ThO(g)} = \Delta_f G^o{}_{ThO_2(s)} - 0.5\,RT\ln po_2 - \Delta_f G^o{}_{ThO(g)}.$$

The standard Gibbs energies of formation of the solid and gaseous species used for the above calculations are given in Table 7. A single equation for Gibbs energy of formation of UO_{2+x} as a function of O/M and temperature, given in Table 7,

Table 7 Gibbs energies of formation of some relevant actinide species

	$\Delta_f G^o = A + B\,T$ (J/mol)	
	A	*B*
$UO_{2+x}(s)$	$(-313046 + 370.74T)(O/M)^2 +$ $(+1170426 - 1514.27T)(O/M) +$ $(-2171018 + 1716.17\,T)$	
$ThO_2(s)$	−1225016	182.80
$UO_3(g)$	−837360	81.22
$UO_2(g)$	−508696	22.82
UO(g)	−36844	−43.12
$ThO_2(g)$	−547173	27.88
ThO(g)	−41868	−51.16

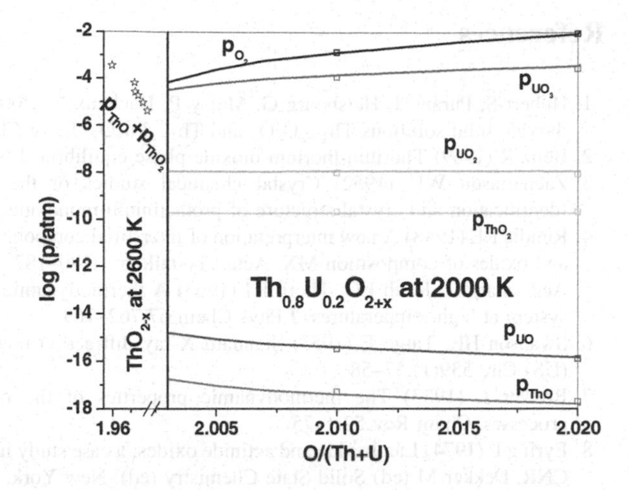

Fig. 12 A comparison of partial pressures of different actinide species as a function of O/M for $(Th_{0.8}U_{0.2})O_{2+x}$ at 2,000 K [52] and ThO_{2-x} [71]

was calculated using temperature-dependent Gibbs energy of formation equations given for UO_2, $UO_{2.01}$, $UO_{2.05}$, and $UO_{2.1}$ by Ugajin et al. [52]. The hyperstoichiometric thoria–urania solution was assumed to be an ideal solution of UO_{2+x} and ThO_2 and 'x' of UO_{2+x} were calculated assuming thorium to be in 'IV' oxidation state. The po_2 over UO_{2+x} were calculated using the method given by Blackburn [72]. As can be seen from the Fig. 11, the partial pressure of 'U' bearing species is many orders of magnitude higher than that of 'Th' bearing species and this difference increases with increase in O/M of the system. It is also seen that partial pressure of 'U' bearing species is more sensitive to composition than 'Th' bearing species. The reason behind this observation becomes clearer from Fig. 12. This figure compares partial pressure of different species and it is seen that partial pressure of UO_3 is higher than that of UO(g) and ThO(g). With increase in O/M, partial pressure of UO_3(g) becomes increasingly higher than that of UO_2(g) and ThO_2(g) and there is a corresponding decrease in the partial pressures of monoxide gaseous species. Hence, even at O/M = 2.00, partial pressures of 'U' bearing species is higher than that of 'Th' bearing species, but with increase in O/M, this difference increases further. This means that at high temperature not only 'U' will start moving to cooler zones, but it will also move oxygen along with it. However, due to lesser mass transport and restructuring in thoria-based mixed oxides compared to urania-based mixed, thoria fuel should not see much redistribution. Though some of the post-irradiation results and simulation analysis have indicated that in thoria–urania annular pellets, urania moves to the rim of the annular hole or to the outer edge of the columnar grain region, as reported by Padden et al. [73].

References

1. Hubert S, Purans J, Heisbourg G, Moisy P, Dacheux N (2006) Local structure of actinide dioxide solid solutions $Th_{1-x}U_xO_2$ and $Th_{1-x}Pu_xO_2$. Inorg Chem 45(10):3887–3894
2. Benz R (1969) Thorium-thorium dioxide phase equilibria. J Nucl Mater 29:43–49
3. Zachariason WH (1952) Crystal chemical studies of the 5f-series of elements. XVI. Identification and crystal structure of protactinium monoxide. Acta Crystallogr 5:9–11
4. Rundle RE (1948) A new interpretation of interstitial compounds—metallic carbides, nitrides and oxides of composition MX. Acta Crystallogr 1:180–187
5. Ackerman RJ, Rauh EG, Thorn RJ (1963) A thermodynamic study of the thorium–oxygen system at high temperatures. J Phys Chem 67:762–769
6. Swanson HE, Tatge E (1953) Standard X-ray diffraction powder patterns. Natl Bur Stand (US) Circ 539(1):57–58
7. Brewer L (1953) The thermodynamic properties of the oxides and their vaporization processes. Chem Rev 52:1–75
8. Eyring L (1974) Lanthanide and actinide oxides: a case study in solid state chemistry. In: Rao CNR, Dekker M (ed) Solid State Chemistry (ed). New York, pp 565–634
9. Darnell AJ, McCollum WA (1960). Report NAA-SR-6498. Off Tech Sers, Dept comm, Washington 25 DC
10. Rachev VV (1968) A new compound $Th_2U_2O_9$. Izv Akad Nuak SSSR Neorg Mater 4(3):475–476. Inorg Mater (English Trans) 4:408–409
11. Paul R (1970) Phase equilibria in the system, ThO_2–UO_2, CeO_2–UO_{2+x} and NpO_2–UO_2. KFK 1297 ORNL/tr-2483
12. Paul R, Keller C (1971) Phasengleichgewichte in den Systemen UO_2–$UO_{2.67}$–ThO_2 und UO_2–NpO_2. J Nucl Mater 41:133–142
13. Braun R, Kemmler-Sack S, Rudorff W (1970). Spektroskopische Untersuchungen im System $(U_{0.5}Th_{0.5})O_{2.25}$–$ThO_2$ und $UO_{2.25}$–ThO_2. Z Naturfuresch 25(b):424–426
14. Lynch ED (1965) Studies of stoichiometric and hyperstoichiometric solid-solutions in the thoria–urania system. AEC Research and Development Report, ANL-6894
15. Anderson JS, Edigington DN, Roberts LEJ, Wait E (1954) The oxides of uranium. Part IV. The system UO_2–ThO_2–O. J Chem Soc 3324–3330
16. Cohen T, Berman RM (1966) A metallographic and X-ray study of the limits of oxygen solubility in the UO_2–ThO_2 system. J Nucl Mater 18:77–107
17. Rousseau G, Desgranges L, Charlot F, Millot N, Niepce JC, Pijolat M, Valdiviesoc F, Baldinozzi G, Berar JF (2006) A detailed study of UO_2 to U_3O_8 oxidation phases and the associated rate-limiting steps. J Nucl Mater 355:10–20
18. Anthonysamy S, Nagarajan K, Vasudeva Rao PR (1997) Studies of oxygen potentials of $(U_yTh_{1-y})O_{2+x}$ solid solutions. J Nucl Mater 247:273–276
19. Christensen JA (1962) UO_2–ThO_2 phase studies. Ceramics Research and development operation quarterly report HW-76559, pp 11.5–11.6
20. Slowinski E, Elliott N (1952) Lattice constants and magnetic susceptibilities of solid solutions of uranium and thorium dioxide. Acta Crystallogr 5(6):768–770
21. Latta RE, Duderstadt EC, Fryxell RE (1970) Solidus and liquidus temperatures in the UO_2–ThO_2 system. J Nucl Mater 35:347–349
22. Lambertson WA, Mueller MH, Gunzel Jr FH (1953) Uranium oxide phase equilibrium systems: IV, UO_2–ThO_2. J Am Ceram Soc 36(12):397–399
23. Freshly MD, Mattys HM (1962) Properties of sintered ThO_2–PuO_2. Ceramics research and development operation quarterly report HW-76556, pp 11.6–11.9
24. Reeve KD (1966) The stability of fissile-fertile oxide solid solutions in air. AAEC/TM-352

25. Banerjee J, Parida SC, Kutty TRG, Kumar A, Banerjee S (2012) Specific heats of thoria–urania solid solutions. J Nucl Mater 427:69
26. Gopal ESR (1966) Specific heats at low temperatures. Plenum, New York
27. Ashcroft NW, Mermin ND (1976) Solid state physics, Holt, Rinehart and Winston. New York. ISBN 0-03-083993-9
28. Kittel C (2005) Introduction to solid state physics, 8th edn. Wiley, New York
29. Bevan Ott J, Boerio-Goates Juliana (2000) Chemical thermodynamics: principles and applications. Elsevier Ltd
30. Hoch M (1985) High-temperature specific heat of UO_2, ThO_2, and Al_2O_3. J Nucl Mater 130:94
31. Adamson MG, Aitken EA, Caputi RW (1985) Experimental and thermodynamic evaluation of the melting behavior of irradiated oxide fuels. J Nucl Mater 130:349
32. Sobolev V (2005) Modeling thermal properties of actinide dioxide fuels. J Nucl Mater 344:198
33. Sobolev V, Lemehov S (2006) Modelling heat capacity, thermal expansion, and thermal conductivity of dioxide components of inert matrix fuel. J Nucl Mater 352:300
34. Bakker K, Cordfunke EHP, Konings RJM, Schram RPC (1997) Critical evaluation of the thermal properties of ThO_2 and $Th_{1-y}U_yO_2$ and a survey of the literature data on $Th_{1-y}Pu_yO_2$. J Nucl Mater 250:1
35. Osborne DW, Westrum EF Jr (1953) The heat capacity of thorium dioxide from 10 to 305 K. The heat capacity anomalies in uranium dioxide and neptunium. J Chem Phys 21:1884
36. Southard JC (1941) A modified calorimeter for high temperatures. The heat content of silica, wollastonite and thorium dioxide above 25°. J Am Chem Soc 63:3142
37. Takahashi Y, Murabayashi M (1975) Measurement of thermal properties of nuclear materials by laser flash method. J Nucl Sc Tech 12:133
38. Hoch M, Johnston HL (1961) The heat capacity of aluminium oxide from 1000 to 2000° and of thorium dioxide from 1000 to 2500. J Phys Chem 65:1184
39. Fischer DF, Fink JK, Leibowitz L (1981) Enthalpy of thorium dioxide to 3400 K. J Nucl Mater 102:220–222
40. Fischer DF, Fink JK, Leibowitz L, Belle J (1983) Enthalpies of thoria–urania from 2300 to 3400 K. J Nucl Mater 118:342
41. Young RA (1979) Model for the electronic contribution to the thermal and transport properties of ThO_2, UO_2, and PuO_2 in the solid and liquid phases. J Nucl Mater 87:283–296
42. Dash S, Parida SC, Singh Z, Sen BK, Venugopal V (2009) Thermodynamic investigations of ThO_2–UO_2 solid solutions. J Nucl Mater 393:267–281
43. Venkata Krishnan R, Nagarajan K (2010) Evaluation of heat capacity measurements by temperature-modulated differential scanning calorimetry. J Therm Anal Calorim 102:1135–1140
44. Agarwal R, Prasad R, Venugopal V (2003) Enthalpy increments and heat capacities of ThO_2 and $(Th_yU_{(1-y)})O_2$. J Nucl Mater 322:98–110
45. Kurina IS, Gudkov LS, Rumyantsev VN (2002) Investigation of ThO_2 and (U, Th)O_2. At Energ 92:461
46. Ralph J (1987) Specific heats of UO_2, ThO_2, PuO_2 and the mixed oxides $(Th_xU_{1-x})O_2$ and $(Pu_{0.2}U_{0.8})O_{1.97}$ by enthalpy data analysis. J Chem Soc Faraday Trans 2(83):1253–1262
47. Shuller LC, Ewing RC, Becker U (2011) Thermodynamic properties of $Th_xU_{1-x}O_2$ $(0 < x < 1)$ based on quantum–mechanical calculations and Monte-Carlo simulations. J Nucl Mater 412:13–21
48. Kandan R, Babu R, Manikandan P, Venkata Krishnan R, Nagarajan K (2009) Calorimetric measurements on (U, Th)O_2 solid solutions. J Nucl Mater 384:231–235
49. Anthonysamy S, Joseph J, Vasudea Rao PR (2000) Calorimetric studies on urania–thoria solid solutions. J Alloys Compd 299:112–117

50. Woodfield BF, Boerio-Goates J, Shapiro JL, Putnam RL, Navrotsky A (1999) Molar heat capacity and thermodynamic functions of zirconolite $CaZrTi_2O_7$. J Chem Thermodyn 31:245–253
51. King KK, King EG (1961) Contribution to the data on theoretical metallurgy. Part XIV entropies of the elements and inorganic compounds. Bulletin 592 (U.S. Bureau of Mines)
52. Ugajin M, Shiratori T, Shiba K (1983) Thermodynamic properties of $Th_{0.80}U_{0.20}O_{2+x}$ solid solution. J Nucl Mater 116:172–177
53. Aronson S, Clayton JC (1960) Thermodynamic properties of Nonstoichiometric urania–thoria solid solutions. J Chem Phys 32(3):749–754
54. Roberts LEJ, Russell LE, Adwick AG, Walter AJ, Rand MH (1958) The actinide OXIDES (F, S). In: Proceedings of the second United Nations international conference on the peaceful uses of atomic energy, Geneva, 1–13 September, vol 28, Paper no. P/26, p 215
55. Tanaka S, Kimura E, Yamaguchi A, Moriyama J (1972) Thermodynamic properties of Nonstoichiometric UO_2–ThO_2 solid-solutions by electromotive force measurements. J Japan Inst Metals 36:633–637
56. Ugajin M (1982) Oxygen potentials of (Th, U)O_{2+x} solid solution. J Nucl Mater 110:140–146
57. Matsui T, Naito K (1985) Oxygen potentials of UO_{2+x} and $(Th_{1-y}U_y)O_{2+x}$. J Nucl Mater 132:212–221
58. Ackermann R, Tetenbaum M (1980) High-temperature thermodynamic properties of the thorium–oxygen system. High Temp Sc 13:91–105
59. Woodley RE, Adamson MG (1979) Thermodynamic and chemical properties of thoria–urania and thoria–plutonia solid-solution fast breeder reactor fuels. Thermodyn Nucl Mater 1:333–355, IAEA, IAEA-SM-236/62, Vienna (1980)
60. Schram RPC (2005) Analysis of the oxygen potential of $Th_{1-y}U_yO_{2+x}$. J Nucl Mater 344:223–229
61. Agarwal R (2012) Oxygen potential of $(Th_1\text{-}yUy)O_{2+x}$ and $(Th_1\text{-}yPuy)O_{2-x}$ systems, NuMat 2012: the nuclear materials conference, 22–25 October 2012, Osaka, Japan, O32
62. Swanson Gerald C (1975) Oxygen Potential of Uranium-Plutonium Oxide as determined by Controlled-Atmosphere Thermogravimetry. Dissertation, LA-6083-T University of California, Hanford Engineering Development Laboratory, Richland
63. Kosuge K (1994) Chemistry of non-stoichiometric compounds. Oxford University Press, New York, pp 21–26
64. Willis BTM (1963) Positions of the oxygen atoms in $UO_{2.13}$. Nature 197:755–756
65. Lee HM (1973) Electrical conductivity of UO_2–ThO_2 solid solutions. J Nucl Mater 48:107–117
66. Yamawaki M, Nagasaki T, Kanno M (1985) Vaporization of thoria–urania solid solution. J Nucl Mater 130:207–216
67. Ackermann RJ and Rauh EG (1980) Thermodynamics of nuclear materials, 1979 (Proc. Symp. Julich, 1979) IAEA Vienna, vol l, p 11
68. Pattoret A, Drowart J and Smoes S (1968) Thermodynamics of nuclear materials, 1967 (Proc. Symp. Vienna 1967) IAEA Vienna, vol 1, p 613
69. Aitken EA, Edwards JA, Joseph RA (1966) Thermodynamic study of solid solutions of uranium oxide. I. Uranium oxide–thorium oxide. J Phys Chem 70:1084–1090
70. Alexander CA, Ogden JS, Cunningham GW (1967) Thermal stability of zirconia and thoria based fuels. Report BMI-1789
71. Ackermann RJ, Tentenbaum M (1979) High temperature thermodynamic properties of the thorium–oxygen system. In: International colloquium on materials for high temperature energy, Toronto, Canada, Contract W-31-109-Eng-38, Argonne National Laboratory, U.S. Department of Energy

72. Blackburn PE (1973) Oxygen partial pressures over fast breeder reactor fuel, (I) a model for $UO2\pm x$. J Nucl Mater 46:244
73. Padden TR, Burtn R, Campbell CE (1967) Electron probe microanalysis of irradiated materials using a small sample approach. Report WAPD-TM-644

Thermochemistry of Thoria-based Fuel and Fission Products Interactions

S. R. Bharadwaj, R. Mishra, M. Basu and D. Das

Abstract Thermochemical studies on fuel and fission products have significantly contributed to our understanding of the chemical behavior of nuclear fuels during normal operation and in possible accident situations. These studies help in modeling of the operating fuel materials and can predict the fuel performance. An understanding of the vaporization behavior of fuels and fission products is essential for estimating the redistribution of various elements in the steep temperature gradients that exist across the fuel pellets. The buildup of fission product elements influence the change of thermodynamic potentials of oxygen, carbon, or nitrogen in oxide, carbide, or nitride fuels. The change in oxygen potential in an oxide fuel element affects the chemical constitution of the fuel. The oxygen potential of the fuel at different oxygen to metal ratio is the key parameter in understanding the oxidation behavior and the resultant chemistry of the fission products inside the oxide fuels. The thoria-based fuels with low urania content exhibits a very fast growth of the potential at the onset of hyperstoichiometry, while that of pure urania exhibits a better buffering action against the increase in the potential. In order to understand the oxygen redistribution among fission products and clad, knowledge of the oxygen potentials of their metal/metal oxide systems and chemical transport property of the oxide fuel for oxygen should be known. The thermodynamic properties of the possible chemical states of alkali and alkaline earth fission products and their relative stabilities as thorates, zirconates, iodides, tellurides, uranates, molybdites, molybdates, etc. also influence the distribution of oxygen among the fission products. A review of

S. R. Bharadwaj (✉) · R. Mishra · M. Basu · D. Das
Chemistry Division, Bhabha Atomic Research Centre, Trombay, Mumbai 400085, India
e-mail: shyamala@barc.gov.in

R. Mishra
e-mail: mishrar@barc.gov.in

M. Basu
e-mail: deepa@barc.gov.in

D. Das
e-mail: dasd1951@gmail.com

D. Das and S. R. Bharadwaj (eds.), *Thoria-based Nuclear Fuels*,
Green Energy and Technology, DOI: 10.1007/978-1-4471-5589-8_4,

thermochemical properties of possible binary and ternary oxides resulting from the fuel-fission products' interactions which influence the fuel performance will be presented considering the oxygen transport and other kinetic aspects.

1 Introduction

The fuel chemistry of thoria is not expected to be the same as that experienced with urania in the conventional reactors, though the same set of fission products with similar yields are formed. Both urania and thoria form fluorite type MO_2 lattice, have similar crystal radii of the cations (Th^{4+} and U^{4+}) and accommodate the fission products in the fluorite crystal lattice. The difference in chemistry originates from the unique tetra valency of thorium (Th^{4+}) and its compounds as against flexible valency of four to six for uranium. Urania can be oxidized to hyper-stoichiometric composition UO_{2+x}, and by this it can buffer oxygen released during fission to a large extent. Oxygen is released during the fission of actinide metal M in MO_2 matrix (M = U, Pu), because most of the fission products such as alkali metals, alkaline earth metals, rare earths, and noble metals bind less oxygen, as compared to Th, U, or Pu. The chemical affinity of Th for oxygen is lower in comparison with U, and therefore the buffering action will be weaker in thoria-based fuels. For the same reason, the oxygen transport in the fuel is predominantly by self-diffusion unlike the case in urania where oxygen transport is faster through the chemical affinity driven diffusion process. These subtle features can lead to faster growth of oxygen pressure in thoria-based fuels and alter the thermo-chemistry of fission products.

The oxygen released during fission gets redistributed among reactive fission products, fuel, and clad. The states of oxidized fission products inside urania matrix are fairly well-established for a typically high burn-up fuel of pressurized water reactor [1].

Similar to urania, the fluorite lattice of thoria can accommodate many of the fission products in oxides, metallic, and gaseous phases. The alkaline earth thorates $MThO_3$ can be chemical components in the gray (perovskite) phase. Additionally, there can be tetragonal/cubic phases due to alkali thorates viz. M_2ThO_3.

Considering, however, the stated subtle features of thoria, one needs to address a number of specific issues such as how fast and to what extent the oxygen pressure inside the thoria-based fuels grows with burn up, whether the oxygen pressure growth could be buffered by the oxidation of reactive fission products and clad, or, whether there could be oxygen transport impediment in thoria rich matrix to result in much higher oxygen potential than in urania. If the oxygen potential is high then there are other issues pertaining to increase in free iodine concentration and formation of oxygen rich bulky phases. These later issues become important when Mo oxidation to buffer oxygen potential is slow [2]. In this chapter, an attempt is made to address these issues. For this, a detailed balance of oxygen released during fission, its consumption by the fission products, fuel, and clad according to their oxidation hierarchies to different chemical states has been worked out. This analysis requires thermodynamic as well as kinetic data for the oxidation of fission products, oxygen transport properties inside the fuel matrix and also across the fuel-clad gap and oxidation kinetics of the zircaloy clad.

2 Chemical States of Fission Products in Thoria-Based Fuel Matrix

The chemistry of the fission products is mainly decided by the matrix within which they are produced. The fuel chemistry with thoria is not expected to be the same as that experienced with urania of the conventional reactors, though the same set of fission products with similar yields are formed and settle down inside the fluorite lattices of MO_2 ($M = Th^{4+}, U^{4+}$), with similar crystal radii of the cations. The yields of the fission products of different fuels are given in Table 1. It can be seen in the table that there is general similarity in the fission products yields for the two cases with the exceptions that the thoria-based fuel results in comparatively more gaseous and less metallic products.

The difference in chemistry for thoria and urania fuels originates from the rigid nature of the four valency of thorium in its compounds as against the flexible valency (four to six) of uranium, which is known to acquire the higher valencies in compounds with alkali and alkaline earth fission products. With increase in oxygen partial pressure, urania undergoes oxidation from the stoichiometric UO_2 to the hyper-stoichiometric composition UO_{2+x}, whereas this aspect is absent in thoria. The valency rigidity of thorium will result in a higher oxygen pressure growth in the thoria rich fuels during their burn up. For the same reason the oxygen transport in thoria rich matrix is expected to be predominantly by self-diffusion unlike the case in urania where the oxygen makes faster transport through the chemical affinity driven diffusion process [3].

Table 1 Integrated values of fission products yields from reactor fuels at their respective burn ups [2]

FP (atom%)	^{235}U PWR (45 GWD T^{-1})	MOX (7.8 % PuO_2) PWR (45 GWD T^{-1})	^{233}U AHWR (20 GWD T^{-1})	^{239}Pu FBR (100 GWD T^{-1})
Xe + Kr	12.8	12.8	16.2	12.6
Pd + Ru + Tc + Rh	16	23.8	7.1	22.9
Mo	11.7	11	10.3	10.7
Zr	13.3	9.5	16.7	9.8
Y + RE	25.3	23.1	26.6	23.4
Ba + Sr	7.2	5.3	9.4	5.3
Cs + Rb	11	10.8	9.8	11
Metalloids + Halogens	1.8	2.2	2.8	2.4
Ag + Sn + Cd	0.7	1.4	0.3	1.2

Fission of tetravalent cation in MO_2 ($M = {}^{235}U$, ${}^{233}U$ or ${}^{239}Pu$) generates significantly zero to lower valent fission products, which according to their valencies and hierarchies of oxidation consume oxygen released in the fission. Simultaneous to the consumption, a part of the fission released oxygen undergoes transport under the steep concentration gradient generated by the oxidation of fuel as well as the clad. The oxygen redistribution takes place under a steep thermal as well as concentration gradient. For thoria-based fuel in AHWR configuration, Table 2 lists the stoichiometric oxides of the different fission products indicating the oxygen consumption according to their fission yields. The tabulated species indicate that the fission product cerium can form tri and tetravalent oxides, whereas molybdenum can form its tetravalent oxide and can also remain in alloyed form. For the various chemical states, the oxygen consumption by the fission products differs and therefore the oxygen balance sheet need not be as straightforward as indicated in Table 2. The fission product oxides generally remain as dissolved component inside the fuel or get precipitated in the form of multicomponent phases in which U and Mo can even be in higher oxidation states than four.

As in the case of urania and plutonia, thoria can be one of the chemical components in the precipitated phases. Several binary oxides with thoria such as alkali and alkaline earth thorates, thorium molybdate and tellurite are known in the literature [4–8]. In order to understand the chemistry of fission products in thoria, one should take note of the chemistry of the conventionally used fuels.

The chemical states of fission products inside urania matrix are fairly well-established and summarized in Table 3, for a typical high burn-up PWR fuel.

As in case of urania, the fluorite lattice of thoria can accommodate many of the fission products. The alkaline earth-based thorates $MThO_3$ can be chemical

Table 2 Oxygen consumption by the fission products of AHWR fuel, $(Th,{}^{233}U)O_2$

Fission products (${}^{233}U = 2FP + 2O$)	Yield per 50 fissions	Oxide type/other possibilities	Oxygen consumption
Y + RE	26.6	M_2O_3	39.9
Ba + Sr	9.4	MO	9.4
Zr	16.7	MO_2	33.4
Ce part in RE	7.4	CeO_2 from Ce_2O_3	3.7
Therefore, oxygen uptake by the reactive fps (Y, RE, Ba, Sr, Zr) = 86.4			
Cs + Rb	9.8	M_2O	< 4.9
		M_2UO_4 from $M_2O + UO_2$ (Some Cs as CsI, Cs_2Te)	< 4.9
Therefore, oxygen uptake by the reactive FPs + alkali metals <96.2			
Mo	10.3	>1.9 out of 10.3 Mo as MoO_2 (O-uptake by clad neglected)	> 3.8
Pd + Ru + Tc + Rh	7.1	Alloy Phase formation with remaining Mo (<8.4)	Nil
Xe + Kr	16.2	Gas	
Metalloid + halogen	3 (1.8 + 1.2)	CsX, Cs_2Te	

Approximately, 4 % of the fission generated oxygen needs to be taken up by Mo. ~18.4 % of the generated Mo can get oxidized and form Mo/MoO_2 buffer.

Table 3 Chemical states of fission products in typical oxide fuels

Type of fission products	Chemical states
Non-volatiles	
Rare earths (Nd, La, Ce, etc.) and transition metals (Y, Zr, Nb, etc.)	Dissolved state in fuel matrix (MO_2)
Alkaline earths (Sr, Ba) and Zr, U, Pu, Mo, Ce	Perovskite, MMO_3 (gray) phase
Noble metals (Pd, Ru, Rh, Tc) and Mo	Alloy phases (white inclusions)
Gases and volatiles	
Inert gases (Xe, Kr)	Dispersed microbubbles in fuel fractional release at fuel—clad gap and at plenum
Alkali metals (M = Cs, Rb)	MI, M_2Te at low-moderate burn up conversion of MI, M_2Te to uranate/molybdate at high burn up
Cd, Te, Sb	In alloy phases
Te, I, Br	As alkali metal compounds

component in the gray (perovskite) phase. Additionally, there can be tetragonal/cubic phases due to alkali thorates of composition M_2ThO_3. The status of noble metals, gases, and volatile components of the fission products can remain the same as in urania. Considering, however, the subtle features of thoria that it does not have any buffering action for oxygen and has high impediment in oxygen transport because of the insignificant presence of chemical affinity driven diffusion, one can predict the possibility of faster growth of oxygen pressure in thoria-based fuels as compared to the case in urania.

Thus, there are several issues that need to be addressed for understanding the chemical state of fission products in thoria-based fuels. Some of these are listed below:

1. The extent to which the oxygen pressure grows inside thoria-based fuels as a function of burn up.
2. Interaction of oxygen with reactive fission products.
3. Oxygen transport in thoria rich matrix.
4. Thermochemistry of fission products under the higher oxygen pressures in thoria-based fuels.
5. Possibility of release of free iodine.
6. Possibility of formation of oxygen rich bulky phases that may lead to in-clad incompatibility problem.
7. Role of fission product Mo in buffering the oxygen pressure in thoria-based fuels.

In order to address the above mentioned issues, one requires information on the thermochemistry of fission products, the oxidation kinetics of clad and Mo alloyed with other fission products, and the transport kinetics of oxygen and the corrosive fission products iodine, tellurium, etc. This information also helps analyzing oxygen pressure profile in the fuel and the cumulative release of the volatiles in the fuel-clad gap.

A burn up of 50–100 GWD T^{-1} will be common in the thoria-based fuel considering its once through cycle strategy. During irradiation, the fuel bears steep thermal profile (typically of 1000–800 K in thermal reactors and 2000–1000 K in fast reactors) from its center to periphery, and the profile becomes occasionally steeper in power ramp situation. With burn up it is known that the thermo-mechanical properties of the cladded fuel pin undergoes gradual degradation primarily resulting from the swelling of the fuel due to the accumulation of fission products inside the fuel matrix. The clad undergoes corrosion by the reactive volatile fission products. At a steep thermal gradient radially across the fuel rod, the fission gases xenon, krypton, and the volatile corrosive fission products, iodine, tellurium, etc. undergo transport toward the pin periphery and get released into the fuel-clad gap (typically 50–100 microns in size).

The major fraction of the gases remains dispersed inside the fuel matrix as microbubbles. Generally, all the non-volatiles, the volatiles to a large extent (99 %) and a major fraction (90–95 %) of the gaseous fission products remain within the matirx [1] (Fig. 1).

The dispersed bubbles as well as the undissolved solid phases formed by other fission products impart internal stress to the matrix, under which the fuel pin swells, develops radial microcracks and voids and as a result, the thermal

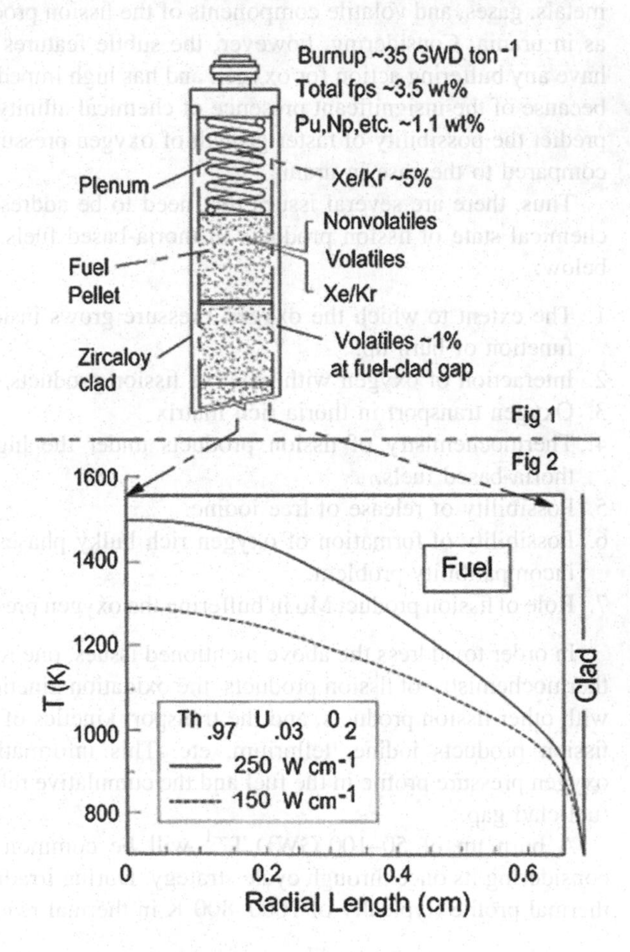

Fig. 1 General view of fission product distribution in moderate burn-up fuel ([2], Copyright Elsevier)

conductivity of the fuel matrix deteriorates. The gap conductivity also deteriorates progressively due to the released gas that dilutes the helium column filling the gap at several tens of bars pressures. The volatile fission products like I and Te, though produced in small amounts, can cause corrosion of the clad material [1, 9, 10]. Chemical interaction of I and Te with the clad results in degradation of its mechanical property and, in extreme situation, leads to its stress corrosion cracking.

Expansion of the fuel pellets due to high internal temperatures, cracking due to thermal and mechanical stresses and irradiation induced swelling may lead to contact of the fuel with the cladding. Fuel pellet–cladding interaction (FCI) at the stressful contacts may lead to cladding failure and subsequent release of fission products into the reactor coolant. The severity of PCI is more under power ramping conditions.

For the fuel-clad integrity evaluation, the thermodynamic and transport behaviors of the fission products and fission released oxygen are analyzed for their redistributions spatially inside a closed system and chemically among the different phases. Thermal expansion, thermal conductivity, and heat capacity of the fuel and clad are important properties that decide the detrimental effects on the fuel pins. These properties with respect to thoria-based fuels have been discussed in detail in chapters "Thermophysical Properties of ThoriaBased Fuels" and "Phase Diagrams and ThermodynamicProperties of Thoria, Thoria–Urania,and Thoria–Plutonia" of this book. The thermodynamic and kinetic analyses are also tied up with steady as well as transient thermal profiles in the fuel pin. The information such as the fission product phases and their molar volumes relative to that of the fuel, the extent of retention of gaseous fission products in the fuel matrix are required for the understanding and evaluation of the stresses. Similarly, the oxygen partial pressure (p_{O2}) or the oxygen potential ($\Delta G_{O2} = RT\ln(p_{O2}/\text{bar})$ of the fuel at different oxygen to metal ratios (O/M) is the key parameter in understanding the oxidation behavior and the resultant chemistry of the fission products inside oxide fuels $M_{1-y}M'_yO_{2+x}$ (O/M $= 2 + x$).

The pressure profile must be known as a function of temperature (T), composition (y), and O/M ratio in order to elucidate the chemical states of fission products. This topic has been dealt with in detail in chapter "Phase Diagrams and Thermodynamic Properties of Thoria, Thoria–Urania,and Thoria–Plutonia" of this book.

Using the parameters of the thermochemical approach due to Schram [11], the oxygen potentials of $Th_{1-y}U_yO_{2+x}$ fuels of urania contents (y) of 0.03, 0.06, 0.1 and 0.2 are compared at different O/M ratios (See Fig. 2).

In order to understand the oxygen redistribution among the fission products and clad for oxidations, the oxygen potentials of their metal/metal oxide systems are plotted as a function of temperature in Fig. 3a. The oxygen potentials of the conventional urania fuel are also included in the figure for comparison. The variation of the oxygen potential with O/M ratio for the advanced heavy water reactor fuel with a typical urania content of 3 mol% ($Th_{0.97}U_{0.03}O_{2+x}$) is similarly evaluated and included in Fig. 3b. The corresponding variation for pure urania fuel is also given in the figure.

Because of the low urania content, the AHWR fuel is seen exhibiting very fast growth of the potential at the onset of the oxygen hyper-stoichometry, where pure

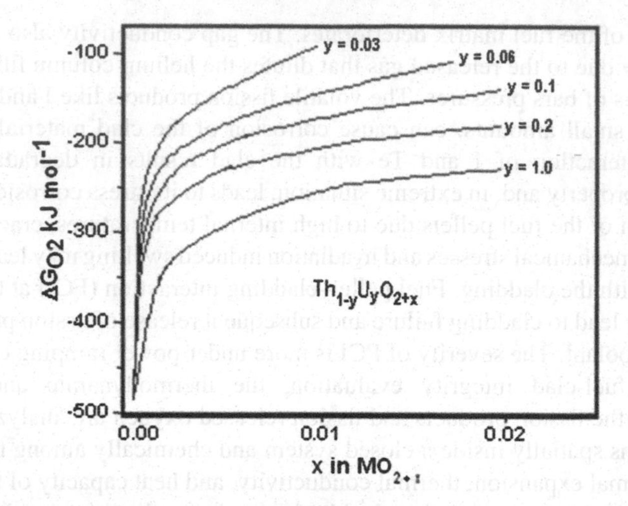

Fig. 2 Oxygen potential of $Th_{1-y}U_yO_{2+x}$ solid solution and of pure urania at 1250 K

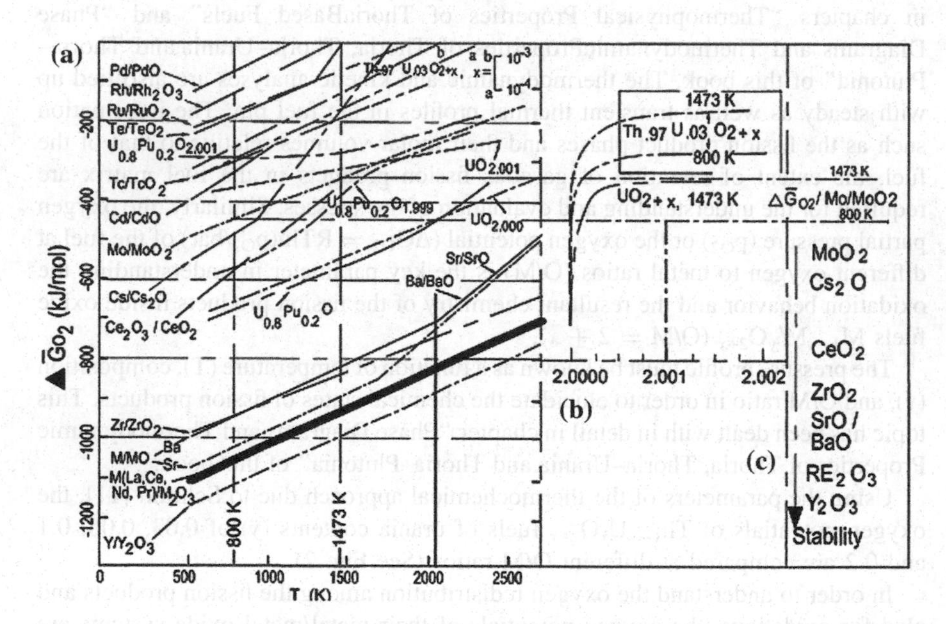

Fig. 3 **a** Oxygen potentials of fission product oxide systems. **b** Oxygen potentials in $(Th,U)O_{2+x}$ and UO_{2+x} at 1473 K. **c** Oxidation hierarchies with increased stabilities of FPs oxides ([2], Copyright Elsevier)

urania exhibits better buffering action against the growth. For example, at 1,473 K, thoria-3 mol% urania would attain O/M = 2.0008 at an oxygen potential of about −127 kJ/mol, that is at 3.1×10^{-5} bar O_2 pressure. At the same O/M, the oxygen potential in pure urania case is −400 kJ/mol (i.e., 6.5×10^{-15} bar O_2).

Oxidation hierarchy of the fission products and clad is primarily governed by the oxygen potentials of the two-phase mixtures of the elements and their respective oxides. Lower the oxygen potential or oxygen pressure (p_{O2}) of the mixture, higher is the hierarchical preference.

It is evident from the figure, that at first yttrium and rare earths (RE) are oxidized to their trivalent oxides and this will be followed by the oxidation of alkaline earths (Sr, Ba) and zirconium to di- and tetravalent oxides, respectively. The Zr in Zircaloy clad will be oxidized approximately at the Zr/ZrO_2 potential. The result of hierarchial oxidation is given in Fig. 3c.

The extent of oxidation of the other fission products(Cs, Mo, etc.), less oxidizing than Zr cannot be deduced without considering clad oxidation. Oxygen uptake by clad depends on the oxygen transport to clad and its oxidation kinetics. We can discuss this aspect later in this chapter.

Although Cs/Cs_2O potential is way above that of the fuel (Fig. 3a), oxidations of Cs and fuel start almost parallel because of formation of highly stable oxide-based compounds like alkali uranates, molybdates, zirconates, and thorates. The thoria rich fuel such as $Th_{0.97}U_{0.03}O_{2+x}$ with an idealized O/M ratio of 2 can, however, share an insignificant part of the fission released oxygen before the oxygen potential rises to that of Mo/MoO_2 system. The concourse of oxygen potential rise of such a fuel at different temperatures is indicated in Fig. 3b. The fuel with typically fabricated O/M ratio of 2.0001 has higher oxygen potential than in Mo/MoO_2 (Fig. 3a), and therefore, it will not be oxidized before molybdenum. On the contrary, in urania fuel, the oxygen uptake continues up to at least an order of magnitude higher O/M ratio ($\sim$2.001 at 1473 K, Fig. 3b) before Mo oxidation can start. In UO_2 fuel, therefore, the MoO_2(s) formation starts at much higher burn up as compared to the case of thoria-based fuel. The extent of Mo oxidation, however, depends on the detailed balance between oxygen generation in fission and consumption by the more reactive fission products and transport to the clad for oxidation. The analysis needs thermodynamic as well as kinetic aspects of attaining the local oxygen potentials across the fuel pin.

Taking note of the fission product yields in thoria-based fuel such as AHWR fuel (Table 1) and the oxidation hierarchies of the fission products (Fig. 3c), one can infer that about 82.7 atom% of oxygen atom released in fission is consumed by the yttrium and rare earth (RE), alkaline earth and zirconium in forming their respective binary oxides, RE_2O_3, SrO/BaO, and ZrO_2 (see Table 2).

The kinetics of oxygen transport to clad now decides whether oxidations of the other fission products with higher oxygen potential than that of the clad components Zr in zircaloy or Cr in steel clad takes place.

If the oxygen transport is extremely slow in thoria-based fuels as compared to that in urania, then taking note of the reported analysis of oxygen consumption by the fission products in urania [12], we can work out distribution of the remaining 17.3 atom% of the fission released oxygen: 3.7 % will be used in the oxidation of cerous oxide (Ce_2O_3), and at the most another 9.8 % by the alkali metal fission products in forming their oxide ternaries such as uranates, zirconates, and thorates. This is schematically shown in Fig. 4, where we can see that about 4 % of the fission

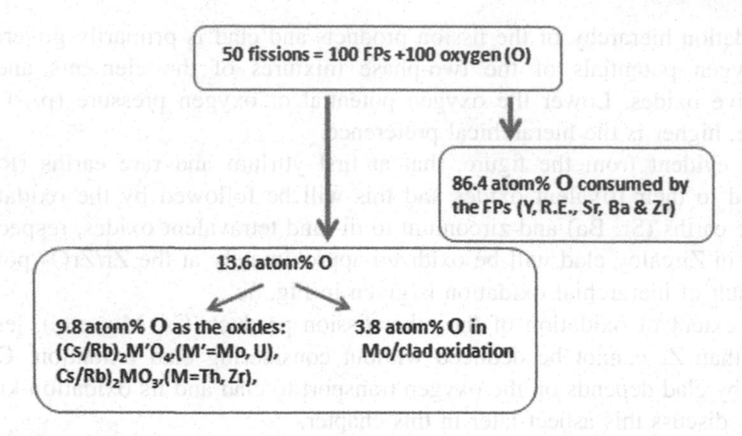

Fig. 4 Tentative oxygen uptake by FPs and clad

released oxygen is leftover and is available for oxidation of the hierarchically next oxidizable fission product and parallely, the clad material via oxygen transport process. It is to be noted that the oxygen consumption by the alkali metals forming their molybdates as $Mo + 2O_2 + 2Cs = Cs_2MoO_4$ is more in comparison to the uranate formation from urania. As will be seen later, the oxygen partitioning in ternary compounds depends on temperature besides the oxygen potential inside the fuel.

The oxygen buildup leads to oxidation of Mo in the fission products to form compounds according to their stabilities at the prevailing oxygen potential. However, this conjectured distribution of oxygen among the fission products is to be founded on thermodynamic properties of the possible chemical states of the alkali and alkaline earth fission products and their relative stabilities as thorates and zirconates, iodides, tellurides and tellurites, uranates, molybdites, and molybdates. The fraction of Cs/Rb engaged in chemical states other than uranates or molybdates ($M_2M'O_4$, M = Cs/Rb, M′ = U/Mo), i.e., thorates or zirconates ($M_2M'\ O_3$, M = Cs/Rb, M′ = Th/Zr) consume lesser oxygen per alkali atom than in alkali uranate/molybdate.

The formation of alkali thorates and molybdates can be represented by :

$$ThO_2(s) + 2M\,(s) + 1/2\,O_2(g) = M_2ThO_3\,(s)$$
$$Mo\,(s) + 2M\,(s) + 2O_2(g) = M_2MoO_4\,(s)$$

According to the thermodynamic preference of formations of mentioned states, the conjectured oxygen balance will be altered. For the same reason, the thermodynamic stabilities of the alkaline earth (A) zirconates, thorates, cerates, uranites, molybdites of general formula, $AM'O_3$ (M′ = Zr, Th, Ce, U, Mo) are to be compared with their uranates and molybdates ($AM'O_4$), as the formation of the later states consume more oxygen in comparison.

The knowledge of the possible chemical states of the fission products and their thermochemistry are essential for drawing proper balance sheet of the fission released oxygen.

2.1 Chemistry of the Fission Products and Their Thermodynamic Properties

Similar to the case of conventional urania/plutonia-based fuel, the fission products produced in thoria matrix distribute themselves in different phases: (i) gas or vapor, (ii) metallic, and (iii) oxide phases. The chemical states of the fission products depend on temperature as well as partial pressures of oxygen and also on the partial pressure of volatile species such as iodine, tellurium, and cesium, involved in the chemical bindings with other fission products. The thoria matrix dissolves a major chunk of the fission products as in the case of urania. The most reactive fission products, yttrium and rare earths, alkaline earths, zirconium, and nobium dissolve as their oxide components with characteristic solubilities in the fluorite phase of thoria. The solubility limits, however, vary significantly among these oxides. Ugajin et al. [13] from their study on thoria-based SIMFUEL at 1873 K have reported that about 16 mol% of the fission product oxides, ZrO_2, CeO_2 and rare earth sesquioxides, remain in the dissolved state in the fuel matrix. Zirconium has a high fission yield ($\sim$17 %), but only limited solubility ($\sim$2.6 atom% at 1400 K and $\sim$12 atom% at 2000 K) in thoria [14], though augmentation in the solubility is reported in UO_2 like fluorite lattice in the presence of rare earths [15]. The undissolved part of Zr predominantly forms the zirconate (pervoskite) phase [13, 16] with the alkaline earth fission products. Yttrium and rare earth elements accounting about 27 atom% of the fission products have significant solubilities (60–100 mol% of $MO_{1.5}$) in the thoria matrix [17, 18].

The tetravalent rare earth cerium forms continuous range of solid solution with thoria. The partitioning of Ce in tri and tetravalent states dissolved in throia depends on the oxygen partial pressure. The higher valent state occurs at oxygen potential well above that of zirconium oxidation (Fig. 3). The inclusion of alkaline earth cerates ($MCeO_3$, M = Sr. Ba) in the perovskite phase is governed by the relative stability of CeO_2 component in the fluorite and the perovskite phases. The thermodynamic properties of perovskite phases [19–29] are included in Table 4. Ugajin et al. [13] have observed the presence of Ce in the perovskite phase in thoria-based SIMFUEL.

Data regarding the dissolution of Sr or Ba in thoria are scarce. Fava et al. [30] have reported 13 mol% solubility of SrO in thoria at 2273 K from lattice parameter measurements. In comparison, urania dissolves almost the same concentration of SrO at much lower temperature (12 mol% SrO at 1773 K) [31].

O'Hare et al. in their studies of thermochemistry of alkaline earths in nuclear fuels have observed a solubility of 3–5 mol% SrO at 2073 K [32]. Subasri et al. [33]

Table 4 Lattice type, density and thermodynamic stability of $AM'O_3$ perovskites

$AM'O_3$ perovskites (lattice type and theoretical density in g $cm^{-3)}$	$\Delta_f G_T^0$ KJ mol^{-1}($AM'O_3$) (800 ≤ T ≤ 1500 K)	$\Delta_m G^0$ for the reaction, AO(s) + BO_2(s) = ABO_3(s)
$BaUO_3$ Pseudo cubic; 7.59 [19]	−1687.2 + 0.2705 T [19]	−42.2 −0.008 T
$BaPuO_3$ Cubic; 8.62 [19] Orthorhombic;7.02 [19]	−1656.7 + 0.2924 T [19]	−55.5 −0.005 T
$BaThO_3$ Cubic;7.63 [5]	−1802 + .279 *T* [This study]	−24.6 −0.005 T
$SrThO_3$ Orthorhombic; 7.07 [5]	−1825 + 0.29 *T* [This study]	−10.1 + 0.006 T
$BaMoO_3$ Cubic; 7.08 [19]	−1228.8 + 0.2615 *T* [19]	−91.8 −0.014 T
$SrMoO_3$ Cubic; 6.127 JCPDS 24-1224	−1269.5 + 0.2702 *T* 800 ≤ T ≤ 1500 K	−95.0 −0.006 T
$BaZrO_3$ Cubic; 6.23 [19]	−1786.8 + 0.2954 T [19]	−129.0 + 0.002 T
$SrZrO_3$ Orthorhombic-I; 5.46 (ambient) Orthorhombic-II; 5.37(1033 K) Tetragonal; 5.30 (1173 K) Cubic; 5.26 (1443 K) [19]	−1765.8 + 0.3039 T [19]	−70.5 + 0.010 T
$BaCeO_3$ Tetragonal; 6.363 JCPDS 35-1318	−1702.9 + 0.3066 T [28, 29]	−50.2 −0.012 T
$SrCeO_3$ Monoclinic; 5.765. JCPDS 47-1689	−1681 + 0.30 T [This study]	9.1 −0.0196 T

have reported 1 mol% solubility of SrO at 1573 K from lattice parameter and SrO activity studies.

The high temperature solubility of Ba in thoria is limited to 0.5 mol% [34] and this is similar to that reported in urania (0.6 at% at 2273 K) [35]. In the reported study [13] of the thoria-based SIMFUEL (at 1873 K), it is shown that out of the added 0.29 mol% BaO about 40 % remain in the dissolved state inside thoria while the remaining 60 % has been taken to be in the ABO_3 perovskite phases. The noted BaO dissolution of 0.116 mol% in the SIMFUEL is significantly lower than 0.5 mol% reported with virgin thoria. However, the noted dissolution unlike the case of pure thoria is decided by BaO activity in the $BaZrO_3$ based perovskite phase present in the SIMFUEL. Cesium oxide solubility in thoria will be negligibly small similar to the reported result in urania [35]. Recently, theoretical estimates of solubilities of the fission products oxides in urania matrix have been reported [36].

3 Thermochemistry of Possible Ternary Fuel-Fission Products—Oxides

The alkaline earth oxides form perovskite phases $SrThO_3$ and $BaThO_3$ with thoria. The thorates have 30–40 % lower densities as compared to thoria (see Table 4). Thermodynamic stabilities of these two compounds have been studied by several techniques, viz., Knudsen effusion [20, 23] and transpiration [21, 22] with reactive gas for $BaThO_3$(s), and Galvanic emf and manometry for $SrThO_3$(s) [24].

The types of heterogeneous equilibrium used in the different studies and the measured stability data of the two thorates are given in Table 5. The standard free energy of formation of $BaThO_3$ obtained by the two vapor pressure techniques agrees within the experimental uncertainties.

The standard enthalpy of formation of the compound from the effusion studies is 1802 kJ mol^{-1} at the mean temperature of the study (1955 K), and corroborates with the estimated value of 1800 kJ mol^{-1} considering the empirical rule [37] using Goldschmidt's tolerance factor (t) as noted in the ABO_3 type perovskite formation. Considering the close agreement of the results of effusion and transpiration studies [20–22], the stability equation is recalculated in the normal working temperatures of the fuel (800–1500 K) and is given by

$$\Delta_f G^0\,(BaThO_3(s)) = -1802 + 0.279T\ (\pm 15\,kJ/mol)$$

The extrapolation of the thermodynamic property to lower temperature is carried out with estimated heat capacity of the thorate. As for $SrThO_3$, the results obtained from the effusion study corroborates to the very small stability relative to the constituent oxides, SrO, and ThO_2. Such result is expected from the low enthalpy and entropy of formation of the perovskite, the enthalpy being empirically derivable

Table 5 Summary of thermodynamic studies on thorate and cerate in this laboratory

System	Method	Reaction studied	Measured properties
$BaThO_3$ (s)	Knudsen Effusion [20]	$BaThO_3(s) = BaO(g) + ThO_2(s)$ $1770 \le T \le 2140$ K	Δ_fG^0 $(BaThO_3(s)) = -1802 + 0.28\,T$ (± 15 kJ mol^{-1})
	Transpira- tions [21, 22]	$BaThO_3(s) + H_2O(g) = Ba(OH)_2(g) + ThO2(s)$, $1548 \le T \le 1683$ K $BaO(s) + H2O(g) = Ba(OH)_2(g)$, $1346 \le T \le 1451$ K	Δ_fG^0 $(BaThO_3(s)) = -1791.8 + .27\,T$ ($\pm$ 10 kJ mol^{-1}) Δ_fG^0 $(Ba(OH)_2$ (g)) = -667.3 + 0.12 T (± 5 kJ mol^{-1})
$SrThO_3$ (s)	Knudsen Effusion [23]	(i) $5SrThO_3(s) + W(s) = 3Sr(g) + Sr_2WO_5(s) + 5ThO_2(s)$, $1670 \le T \le 2040$ K (ii)$4SrThO_3(s) + W(s) = 3Sr(g) + SrWO_4(l) + 4ThO2(s)$, $2135 \le T \le 2420$ K	Δ_fG^0 $(SrThO_3(s)) = -1954 + 0.37\,T$ ($\pm$ 10 kJ mol^{-1}) Δ_fG^0 $(SrThO_3(s)) = -1960 + 0.37\,T$ (± 15 kJ mol^{-1})
$SrCeO_3$	Tensimtery [27]	$SrCO_3(s) + CeO_2(s) = SrCeO_3(s) + CO_2(g)$. $1113 \le T \le 1184$ K	Δ_fG^0 $(SrCeO_3(s)) = -1680.5 + 0.31\,T$ (± 7 kJ mol^{-1})
$Cs_2ThO_3(s)$	Knudsen Effusion [38]	$Cs_2ThO_3(s) = Cs_2O(g) + ThO_2(s)$, $1100 \le T \le 1250$ K	Δ_fG^0 $(Cs_2ThO_3(s)) = -1780 + 0.44\,T$ (± 20 kJ mol^{-1})
$Cs_2ZrO_3(s)$	-do-[39]	$Cs_2ZrO_3(s) = Cs_2O(g) + ZrO_2(s)$, $1140 \le T \le 1275$ K	$\Delta_fG^0$$(Cs_2ZrO_3(s)) = -1672 + 0.44\,T$ (± 18 kJ mol^{-1})
$Rb_2ThO_3(s)$	-do-[40]	$Rb_2ThO_3(s) = 1/2O_2(g) + 2Rb(g) + ThO2(s)$, $1058 \le T \le 1232$ K	Δ_fG^0 $(Rb_2ThO_3(s)) = -1822 + .45T$ (± 8 kJ mol^{-1})

from the consideration of the tolerance factor. Out of the two results of the effusion experiment that is based on the reaction, (i) (see Table 5):

$$5SrThO_3(s) + W(s) = 3\,Sr(g) + Sr_2WO_5(s) + 5\,ThO_2(s)$$
$$(1670 \leq T/K \leq 2040)$$

The Gibbs energy of formation $\Delta_m G^0$ of $SrThO_3(s)$ from constituent oxides in their standard states is derived as [23], $\Delta_m G^0(SrThO_3(s))$ (kJ/mol) $= -4.9 + 0.003$ T at the mean temperature of 1855 K. The other result based on the reaction (ii) (see Table 5), that is,

$$4\,SrThO(s) + W(s) = 3\,Sr(g) + SrWO_4(liq) + 4\,ThO_2(s)\ (2135 \leq T/K \leq 2420),$$

involves uncertainties from the activity assignment in the $SrWO_4$ liquid phase and is not considered.

The result of reaction (i) is used to extrapolate the stability in the working temperature of the fuel, (800–1500 K) and is given by

$$\Delta_f G^0(SrThO_3(s)) = -1825 + 0.297\ (\pm 10\,kJ/mol).$$

Necessary modification for the change of standard state of Sr component below the boiling point of 1685.5 K [41] is included in the extrapolated result.

It may be noted that the reported results of tensimetric [24] and galvanic studies [24] are mutually inconsistent as indicated below :

Tensimetric results [24]:

$$SrCO_3(s) + ThO_2(s) = SrThO_3(s) + CO_2(g)$$
$$\Delta_f G^0(SrThO_3(s)) = -1865 + 0.309\,T\ (\pm 14\,kJ/mol)\ (1075 \leq T/K \leq 1195).$$

Galvanic study results [24],

Cell—I:

$$O_2/SrO, SrF_2//CaF_2//SrThO_3, ThO_2, SrF_2/O_2$$
$$\Delta_f G^o(SrThO_3(s)) = -1829 + 0.273\,T\ (\pm 10\,kJ/mol)\ (978 \leq T/K \leq 1154).$$

Cell—II:

$$O_2/SrThO_3, ThO_2, SrF_2//CaF_2//CaO, CaF_2/O_2$$
$$\Delta_f G^o(SrThO_3(s)) = -1853 + 0.287\,T\ (\pm 20\,kJ/mol)\ (1008 \leq T/K \leq 1168).$$

The result of reaction in Cell-I closely agrees with the recommended stability of $SrThO_3$. As compared to the alkaline earth thorates, their zirconates are more stable. Thermodynamically, the thorates can undergo substitution reaction with the fission product Zr, forming zirconates. For example,

$$BaThO_3\,(s) + ZrO_2\,(s) = BaZrO_3 + ThO_2\,(s)$$
$$\Delta G^0 = -90.6 + 0.0056\,T\,kJ/mol.$$

The reported study on thoria-based SIMFUEL at 1873 K showed the absence of the thorate and also the $BaUO_3$ component in the identified perovskite phase (Ba,Sr), (Zr,Ce) O_3. Qualitatively, the preferential formation of the zirconate can be understood by comparing the BaO activity in the equilibrium reactions, BaO + MO_2(s) (M = Zr, Th, U) = $BaMO_3$(s), which is given by $RTln(a_{BaO}) = \Delta G^0$; the one with lowest value of the free energy can have the preference. Considering that the ΔG^0 values (Table 4) are essentially governed by the reaction enthalpies for the cases, the zirconate is the most preferred component. The $BaThO_3$ component thus can at the most exist as minor chemical component dissolved in the perovskite phase. Details will be elaborated later in the text.

As for the $BaUO_3$ component in the perovskite, though the a_{BaO} value in the $BaUO_3$(s) + UO_2(s) phase mixtures can be evaluated to be about three time lower compared to that in the biphasic mixture, $BaThO_3$(s) + ThO_2(s), the low concentration of UO_2 in the SIMFUEL of the base matrix composition, $(Th_{0.933}U_{0.067})O_2$ fails to get stabilized as $BaUO_3$. As against the case of SIMFUEL, the reported FBR and HTR fuels [35, 42] have significantly higher UO_2 activities to promote the formation of $BaUO_3$ component, the presence of which is noted in the cubic perovskite phase (Ba,Sr,Cs) (U,Pu,Zr,RE,Mo)O_3. The $BaUO_3$–$BaZrO_3$ psendobinary is reported to have continuous range of solid solutions at 1973 K [43]. Similar to $BaUO_3$ perovskite, $BaPuO_3$ and $BaCeO_3$ have about -50 kJ mol^{-1} enthalpic stabilities with respect to their constituent oxides. The stabilities included in Table 4 are derived from the calorimetric data [19, 28, 29, 44]. For $BaPuO_3$ the required thermal function for the derivation was estimated considering the available data for $BaUO_3$. Stability data of $SrPuO_3$ though available from the reported mass spectrometer study [45] could not be considered as there is internal inconsistency in the second and third law evaluations of sublimation enthalpy of the compound and the derived standard enthalpy of formation significantly differs from the systematic trend in the enthalpy values reported for the ABO_3 perovskites of the alkaline earths and the actinides [46]. The alkaline earth cerates and plutonites are seen in the perovkite phase of FBR and HTR fuels [35, 42]. In the SIMFUEL study [13], the observed presence of cerate in the zirconate phase, (Ba, Sr) (Zr, Ce) O_3, wherein the $BaUO_3$ component of similar stability is absent, could be explained considering comparable molar volumes of barium zirconate and cerate, and higher molar volume of uranite (Table 4).

The stability of $SrCeO_3$ obtained from tensimetric study [27] is in good agreement with that derived room solution calorimeter study [28, 29]. The recommended stability as included in Table 4 is the mean value taken from two types of studies. Among the two molybdites, $SrMoO_3$ and $BaMoO_3$, the former has molar volume closer to $BaZrO_3$ and is expected to be the main Mo bearing component in the zirconate phase. The stability functions of molybdites included in Table 4 are based on the reported bomb calorimetric data for the formation enthalpies of the compounds and their thermal property data [25, 26]. For $SrMoO_3$ the standard entropy value is estimated considering its constituent oxides while for $BaMoO_3$, the reported entropy is used in deriving their stabilities. The oxidation of $BaMO_3$ (M = Mo, U) components of the perovskite phase to the oxygen rich $BaMO_4$ phases are primarily governed by the oxygen potentials of the pseudo-binaries, $BaMoO_4$–$BaMoO_3$ and $BaUO_4$–$BaUO_3$; for the phase formation to take place these potentials are to be lower than the prevailing oxygen potential in the fuel. Considering the reported thermodynamic data of the ternary oxides included in (Tables 4, 7), the potentials are expressed for the temperature range $800 \leq T \leq 1500$ K as,

$$\Delta G_{O2} = -653.4 + 0.251\,T\,(BaMoO_4 / BaMoO_3),$$
$$\text{and}\ \Delta G_{O2} = -603.2 + 0.1784\,T\,(BaUO_4 / BaUO_3).$$

Comparison of these potentials with $\Delta\bar{G}_{O2}$ values of Mo/MoO_2 system, viz. $\Delta\bar{G}_{O2} = -580.11 + 0.1698\ T$), suggests that the perovskite components can undergo oxidation to the oxygen rich phases when the fission product Mo undergoes oxidation to $MoO_2(s)$. The BaO activity in the perovskite phase will be the additional factor in the oxidation feasibility as $BaO + MoO_2\,(s) + 1/2O_2(g) = BaMoO_4(s)$.

There is hardly any data on the solubility of alkali metal fission products in thoria. The reported EPMA result on thoria-based SIMFUEL [13] heat treated at 1973 K, showed complete loss of added Cs. The low solubility of Cs in urania (1.2 atom% at 1273 K [35]), suggests similar behavior in thoria matrix. Cs as well as Rb form their thorates phases [38, 40, 46] Thermodynamic stability data of the two alkali metal thorates measured in this laboratory by Knudsen effusion technique are included in Table 6. In the absence of heat capacity data of the thorates, the stability derived by the second law analysis of the vaporization result is expressed for the normal working temperatures (800–1500 K) of the fuel. For Rb_2ThO_3, the reported stability by tensimetric study of the reaction, Rb_2CO_3 (s) + ThO_2 (s) = Rb_2ThO_3 + CO_2 (s) [46] is in agreement with the effusion result [40]. The thermodynamic properties of Cs_2ZrO_3 are reported from vaporization [39] and calorimetric studies [47]. These components volatilize incongruently as Cs and Rb

Table 6 Thermodynamic data for alkali metal thorates, zirconates and tellurites

M_2 (Th/Zr/Te)O_3 (lattice type and theoretical density in g cm^{-3})	$\Delta_f G^0$, kJ mol^{-1} $1000 \leq T \leq 1500$ K
Cs_2ThO_3 Cubic; 6.349 [4])	$-1780 + 0.44T$ ($\pm$20 kJ mol^{-1}) [38]
Rb_2ThO_3 Cubic 6.069 [4]).	$-1822 + .45T$ ($\pm$8 kJ mol^{-1}) [40]
Cs_2ZrO_3 Orthorhombic; 5.2 [19]).	$-1672 + 0.44T$ ($\pm$ 18 kJ mol^{-1}) [39]
Rb_2ZrO_3 Orthorhombic; 2.25 [49].	$-1661 + 0.3093$ T [50]
Cs_2TeO_3 Tetragonal; 4.606 [51]	$-998.2 + 0.309$ T (952–1085 K) estimated

bearing oxides and elemental vapors in monomer and dimer forms [48]. The reported phase $Rb_2ZrO_3(s)$ [49] and its stability estimate [50] are included in Table 6. The table also includes the reported phase [51], and stability data for $Cs_2TeO_3(s)$ considering the available thermodynamic information [52]. Under the prevailing oxygen potential inside the fuel, the predominant species will be gaseous elements and the normal oxides of stoichiometry M_2O. The alkali thorates as compared to alkali molybdates and uranates [19, 53] are less preferred above certain oxygen potential. Thus, considering the inter-conversion reaction, Cs_2ThO_3 (s) + MO_2 (s) + ½ O_2 (s) = Cs_2MO_4 (s) + ThO_2 (s) (M = Mo, U) one expresses the oxygen potential for the case, M = Mo, as $\Delta\bar{G}_{O2}$ (kJ/mol) = −1020 + 0.183 T, and for M = U as $\Delta\bar{G}_{O2}$(kJ/mol) = −824 + 0.255 T. In the evaluation, the urania activity in $Th_{0.97}U_{0.03}O_2$ solid solution is taken to be 0.03 and MoO_2 is considered to be in phase pure form. The required stability data for the Cs-molybdates and uranates, and other higher Mo and U compounds of the fission products are summarized in Table 7.

The oxygen potential of thoria-based fuels remaining above −500 kJ/mol, (see Fig. 2), the above-mentioned oxidation reactions will be driven forward displacing thoria from the thorates. The same conclusion holds good when one considers metallic Mo (a_{Mo} ~ 0.3 in the alloy phase) instead of MoO_2 as reactant, or, the compound $Cs_2U_4O_{12}$ instead of Cs_2UO_4 as the product phase. Like the case of alkali thorates, alkali zirconates undergo similar conversion to molybdates and uranates. The oxygen potential governing the conversion for Cs_2ZrO_3 to Cs_2MoO_4 is given by $\Delta\bar{G}_{O2}$ (kJ/mol) = −989 + 0.194 T, and therefore, the conversion takes place at potentials much below that of Mo/MoO_2 system ($\Delta\bar{G}_{O2}$ = −580.11 + 0.1698 T). The Cs_2ZrO_3 stability data derived from the vaporization study [39] poorly agrees with the one obtainable from the reported calorimetric measurement of the compound by Cordfunke et al. [47]. The feasibility of formation of alkali metal zirconates is limited to the zircaloy surface and its neighborhood, where the oxygen potential remains below the limit of conversion of the zirconates to molybdates and uranates.

Table 7 Thermodynamic stability of uranates/molybdates of alkali and alkaline earths

A_2BO_4/ABO_4 type uranates/molybdates, of alkali and alkaline earths (lattice type and density in g cm^{-3} at ambient)	Δ_fG^0, kJ mol^{-1} 1000 ≤ T≤1500 K
Cs_2UO_4 Tetragonal; 6.60 [53]	−1926 + 0.4108T, T < 952 K; −2062 + 0.5532 T, T > 952 K [53]
Cs_2MoO_4 Orthorhombic; 4.38 [19] Cs_2MoO_4(liq)	−1514 + 0.3605 T, T < 952 K; [53] −1649.6 + 0.5209 T, T > 952 K [53] −1611.3 + 0.4897 T, T > 1229.5 K [53]
$SrMoO_4$ Tetragonal; 4.701 [54]	−1540.6 + 0.3576 T (estimated from calorimetric data [54])
$BaMoO_4$ Tetragonal; 5.01 [19]	−-1555.5 + 0.387 T (estimated from calorimetric data [54])
$SrUO_4$ Tetragonal; 7.888 [53]	−1980.2 + 0.34 T [53]
$BaUO_4$ Orthorhombic; 7.56 [53]	−1988.8 + 0.3597 T [53]

Though the Cs_2MO_4 (M = U, Mo) phases are chosen in the above discussion, there are number of their Cs based compounds with different stoichiometry in the Cs–M–O (M = U, Mo) ternary equilibria. The compounds with oxygen potential lower or in the neighborhood of that of Mo/MoO_2 are considered and their properties [53] are tabulated in Table 8. As can be seen in this Table, the Cs–M–O equilibria falling in the oxygen rich side of the ternary sections of the respective cases, Cs_2UO_4 (s)–$Cs_2U_4O_{12}$ (s) – UO_2 (s) and Cs_2MoO_4 (s) – MoO_2 (s) – Mo(s), have oxygen potentials higher than that of Mo/MoO_2 system.

These oxygen rich equilibria are generally uncalled for in the fuel if the fission product, Mo can truly buffer the potential without encountering kinetic hindrance in its oxidation. Thus, it can be seen that the chemistry of fission products in thoria fuel rests heavily on the molybdenum oxidation kinetics.

A small fraction of the alkali metals will always form volatile species as halides MX (X = I/Br), tellurides M_2Te, hydroxides MOH, molybdates M_2MoO_4 and tellurites (M_2TeO_3), and undergo transport to the cooler regions of the pin. Their concentrations will be governed by the thermodynamic stabilities of the volatile species and the concentrations of reactants leading to their formations. The possible ways of partitioning of Cs, I, and Te in the fuel pin are given in Tables 9, 10 and 12. The relevant thermodynamic data for Cs mono and dimeric species, Cs-iodide, hydroxide, and telluride vapors used in the partitioning analysis are based on the reported information [19, 55]. As for Cs_2MoO_4 (g), the reported results of thermodynamic studies on the equilibrium between Cs_2MoO_4 (s) and Cs_2MoO_4 (g) [56–58] were considered. The vapor pressures from transpiration study [56] carried out above the melting point of the compound (1229.5 K) is in agreement with those from mass spectrometric studies [57, 58]. The mass spectrometric studies by Johnson et al. [57] and Yamawaki et al. [58] on the solid molybdate have resulted in consistent data on the vaporization of the molybdate. The thermodynamic data of Cs_2MoO_4 (g) included in Table 10 is based on mean values from the reported vaporization studies on Cs_2MoO_4 (s). The data for Cs_2TeO_3 (g) in the same table is based on the reported mass spectrometric results [59].

Table 8 Oxygen and cesium potentials in Cs–M–O systems (M = U, Mo)

Ternary equilibrium	$\Delta\bar{G}_{O2}$, and also$\Delta\bar{G}_{Cs}$(within bracket) [53]
Cs(l)-$Cs_2UO_{3.56}$(s)-UO_2(s) $550 \leq T \leq 952$ K	−837.2 + 0.2101 T (−71.98 + 0.0752 T)
Cs_2UO_4-$Cs_2UO_{3.56}$(s)-UO_2(s), $550 \leq T \leq 1300$ K	−846.7 + 0.2933 T (−65.46 + 0.0408 T)
Cs_2UO_4-$Cs_2U_4O_{12}$(s)-UO_2(s), $550 \leq T \leq 1350$ K	−394.7 + 0.1292 T (−291.4 + 0.1229 T)
Other ternary equilibria sans UO_2 involving Cs compounds with high valent U	Higher oxygen potentials, and lower Cs potentials
Cs(l)–$Cs_6Mo_2O_9$(s)–Mo(s) $300 \leq T \leq 952$ K	−767.0 + 0.1812 T (−71.98 + 0.0752 T)
Cs_2MoO_4–$Cs_6Mo_2O_9$(s)–Mo(s), $300 \leq T \leq 1150$ K	−695.7 + 0.1549 T (−129.1 + 0.1055 T)
Cs_2MoO_4–MoO_2(s)–Mo(s) $300 \leq T \leq 1150$ K	−580.1 + 0.1698 T (−246.1 + 0.0931 T)
Cs_2MoO_4-$Cs_2Mo_2O_7$-MoO_2(s), $300 \leq T \leq 1150$ K	−419.6 + 0.1204 T (−325.9 + 0.117 T)
Five other ternary equilibria involving polymeric molybdates with MoO_2(s)	Higher oxygen potentials, and lower Cs potentials

Table 9 Thermodynamic properties of cesium iodide and its interaction with MoO_2, UO_2 and Zr components in the fuel

Reactions	ΔG^{O} (kJ mol^{-1})	p_I estimates
$Cs(l) + I(g) = CsI(s)$ $CsI(s) = CsI(g)$	$-458.7 + 0.1479$ T $190.3 - 0.1403$ T $500 \leq T \leq 899$ K	$p_I = 7.5\times10^{-18}$ p_{CsI} at 800 K \ $p_{CsI} = 0.81$ Pa at 800 K ($p_{CsI} = 10^{-5}$ Pa at 573 K)
$Cs(l) + I(g) = CsI(l)$ $CsI(l) = CsI(g)$	$-481.9 + 0.1737$ T $147.9 - 0.0935$ T $899 \leq T \leq 1500$ K	$p_I = 3.7 \times 10^{-9}$ p_{CsI} at 1500 K $p_{CsI} = 54867$ Pa at 1500 K
$2CsI(s) + MoO_2(s) + O_2 = 2I(g) + Cs_2MoO_4(s)$	$-18.4 - 0.1018$T ($500 \leq T \leq 899$ K) $\ln(p_I/Pa) = 17.6 + 1106/T + \Delta\bar{G}_{O2}/2RT$	$P_I = 3.6\times10^{-7}$ Pa on the fuel surface at 800 K and $\Delta\bar{G}_{O2} = -450$ kJ mol^{-1} (Mo/MoO_2)
$2CsI(s) + UO_2$, ($a_{UO2} = 0.03$) $+ O_2 = 2I(g) + Cs_2UO_4(s)$	$75.4 - 0.0532$ T ($500 \leq T \leq 899$ K) $\ln(p_I/Pa) = 11.82 - 4534/T + \Delta\bar{G}_{O2}/2RT$	$P_I = 1.0\times10^{-12}$ Pa on the fuel surface at 800 K and $\Delta\bar{G}_{O2} = -450$ kJ mol^{-1} (Mo/MoO_2)
$2CsI(s) + Zr(s) + 3/2O_2 = 2I(g) + Cs_2ZrO_3(s)$	$-754.6 + .144$ T ($500 \leq T \leq 899$ K) $\ln(p_I/Pa) = 2.87 + 45381/T + 3\,\Delta\bar{G}_{O2}/4RT$	$P_I = 2\times10^{-32}$ Pa on the clad surface (573 K) and $\Delta\bar{G}_{O2} = -985$ kJ mol^{-1} (Zr/ZrO_2)
CsI radiolysis [63] $CsI + Xe^* = Cs + I + Xe^*$	p_I(radiolysis) $= 3.1\times10^{-5}$ Pa at $p_{csI} = 1\times10^{-5}$Pa Dose $= 4 \times 10^{19}$ fission m^{-3}s^{-1}	p_I (radiolysis) $= 3 \times 10^7$ times p_I (thermochemical; $CsI\text{-}UO_2\text{–}Cs_2UO_4$ at 800 K).

Table 10 Cs partitioning in vapor phase (Cs potential fixed by the equilibria, $Cs_2UO_4 = UO_2(s) + 2Cs(g) + O_2$ and $MoO_2(s) = Mo(s) + O_2$: $\Delta\bar{G}_{Cs} = -199 + 0.103T$ KJ mol^{-1}

Vapor species	$\Delta_f G_T^0$ KJ mol^{-1}	Cs partitioning as other volatile species
Cs	71.98-0.0752T $500 \le T \le 952$ K	
Cs_2	98.28-0.0819T $500 \le T \le 952$ K	$\ln(p_{Cs2}/p_{Cs}) = -18405/T + 4.10$, ($2Cs = Cs_2$) Negligible partitioning as $Cs_2(g)$
Cs_2O	$-152.9 - 0.0223$ $500 \le T \le 952$ K	$\ln(p_{Cs2O}/p_{Cs}) = -24718/T + 7.15$, ($Cs + 1/2O_2(g) = Cs_2O$). Negligible partitioning as $Cs_2O(g)$
CsOH	$-269.0 - 0.0121T$ $500 \le T \le 952$ K	$\ln(p_{CsOH}/\sqrt{p_{H2}}) = 17.58 - 11647/T$, ($Cs + 1/2O_2 + 1/2H_2 = CsOH$). Almost all H_2 as CsOH(g)
Cs_2MoO_4	$-1250 + 0.24$ T, $T \le 1229.5$ K $-1385 + 0.40T$, $T \ge 1229.5$ K	$\ln(p^0_{Cs2MoO4}/p_{Cs}) = 5.08 + 3139/T$. Almost all Cs(g) as $Cs_2MoO_4(g)$
Cs_2TeO_3	$-755 + 0.31$ T, $T \le 952$ K (estimated from mass spectrometric data)	$\ln(p_{Cs2TeO3}/p_{Cs}) = -25 + 3470/T + \ln a_{Te}$ (Negligible $Cs_2TeO_3(g)$)

Cumulatively, the halide/telluride amount in the fuel pin is limited by the total halogen and tellurium contents, which is less than 3 atom percent for the thoria-based fuel (See Table 2). The thermodynamic properties of CsI in its solid and liquid forms (Table 9) suggest that iodine should remain mostly as CsI in its vapor and condensed phases. However, the generally observed faster transport of iodine than cesium [60] indicates that all iodine is not gettered by cesium in the dynamic situation prevailing during irradiation. It is known that radiolysis of CsI is a major factor for deviation from the thermodynamically obtainable concentration of iodine [61, 62] considering reaction like,

$$2CsI\,(g) + MO_2\,(s) + O_2\,(g) = Cs_2MO_4\,(s) + 2I\,(g) \quad [M = U, Mo]$$

Iodine pressure generated by radiolysis can be orders of magnitude higher [63] at the fuel-clad gap than expected by thermochemical evaluation. Typical results are included in Table 9. Higher oxygen potential facilitating the thermochemical generation of iodine can, however, assume significance in thoria-based fuel, where the buffering for oxygen is poor. The thermodynamic data of the other volatile species of Cs given in Table 10, suggest that Cs partitioning into dimeric Cs_2 (g), Cs_2O (g) or Cs_2TeO_3 (g) forms are insignificant as compared to that in Cs_2MoO_4(g) and CsOH(g) forms. Hydrogen and its isotope content inside the fuel resulting from fission, diffusional ingress from the coolant and from the initial moisture impurity remain mostly as alkali hydroxide vapor. The partitioning of cesium as Cs_2Te (g) is governed by the stability of Te bearing alloys and compounds at the local temperature and oxygen partial pressure. At oxygen potentials well below that of Mo/MoO_2 system, tellurium can remain distributed in vapor state within the dispersed gas bubbles, and chemically fixed as alkali metal tellurides, M_2Te [19] (Cs_2Te has melting point of 1083 K), and as dissolved components in the metallic fission products, Tc, Ru, Rh, Mo, and Pd with each of which it forms quite stable intermetallics [19, 64]. Thermodynamic stabilities of some of the intermetallics in Ru-Te, Rh-Te binaries that were measured in this laboratory [65, 66] are included in Table 11. The table also includes the reported calorimetric data of the two intermetallic compounds in Pd–Te binary.

Due to the general volatility, Te inside the fuel matrix undergoes transport toward the cooler zone and gets deposited on the inner surface of the clad. Like in

Table 11 The thermodynamic data of Ru, Rh and Pd tellurides

Intermetallic compound	Method/thermodynamic properties
$RhTe_{0.9}$ (s)	Knudsen effusion, $\Delta_fG^0(1026 \leq T \leq 1092) = -74.7 + 0.015\ T$ (± 3 kJ mol^{-1}) [65]
Rh_3Te_2 (s)	Knudsen effusion, $\Delta_fG^0(1151 \leq T \leq 1234) = -176.9 + 0.039\ T$ ($\pm$ 7 kJ mol^{-1}) [65]
$RuTe_2$(s)	Knudsen effusion, Δ_fG^0 $(831 \leq T \leq 1148) = -161.7 + 0.1\ T$ ($\pm$ 7 kJ mol^{-1}) [66]
PdTe(s)	Solution calorimetry, $\Delta_fH^0(298.15\ K) = -51.93$ (± 0.49 kJ mol^{-1}) [67]
$PdTe_2$(s)	Solution calorimetry, $\Delta_fH^0(298.15\ K) = -75.75$ (± 0.68 kJ mol^{-1} [67]

the case of iodine, the free tellurium results from Cs_2Te by the preferential formation of cesium molybdates and uranates at higher oxygen potentials. It is seen in Table 12 that the molybdate formation is favored, when the oxygen potential is exceeding the value of Mo/MoO_2 system. At lower oxygen potentials, considering that Cs vapor pressure is fixed by the Cs_2UO_4 (s)–UO_2 (s) phase equilibrium, Cs_2Te formation is preferred to free Te or to metal telluride forms. Tellurium concentration in the alloy will also be low. Thermodynamic data of Cs and Te bearing species presented in Table 12 show that Cs_2Te(s) formation is preferred as compared to $ZrTe_2$ (s) on zircaloy surface. Table 12 also gives the oxygen potential for the tellurium oxidation to form tellurite phase. The potential being significantly higher than that of Zr/ZrO_2 system, the tellurite Cs_2TeO_3 (s) cannot form on zircaloy surface. Generally, the Te oxidation is not feasible under normal operating condition of reactor. Cs_2MoO_4 vapor, the thermodynamic property of which is given in Table 10 is produced by molybdenum oxidation in the presence of cesium vapor as,

$$[Mo]\ alloy + 2O_2\ (g) + 2Cs\ (g) = Cs_2MoO_4\ (g)$$

At high oxygen potential such as that expected in the thoria-based fuels, Mo will make early partitioning from the alloy phase of the noble metals to the ternary oxide phases. MoO_2 solubility in the fuel being low (about 0.2 mol% in unstructured peripheral part of high burn up urania–plutonia [68]), most of the partitioned Mo will be in the form of the alkali and alkaline earth molybdites and molybdates (see Tables 4, 7). In the reported study with thoria-based SIMFUEL [13], it was noted that sample annealed at 1773 K under the oxygen potential -122 kJ.mol^{-1}, which is well above that of Mo/MoO_2 (-273 kJ/mol), showed Mo partitioning from the alloy phase to form (Sr, Ba)MoO_4(s). At lower oxygen potential ($\Delta\bar{G}_{O2} = \sim -310$ kJ mol^{-1} this study showed that Mo remains in the alloy phase. This study further showed that Ba thus consumed out of the $BaZrO_3$ based perovskite resulted in the precipitation of the mixed pyrochlore phase Nd_2 (Zr, Ce)$_2O_7$. The pertinent reaction is $2BaZrO_3 + 2NdO_{1.5} + 2MoO_2 + O_2 = 2BaMoO_4 + Nd_2Zr_2O_7$.

At intermediate oxygen potential one expects Mo partitioning as alkaline earth molybdite ($MMoO_3$) components into the perovskite phase, such as the one with general composition (Ba, Sr, Cs) (U, Pu, Zr, RE, Ce, Mo)O_3 (gray phase) that has been observed in urania-based FBR, and HTR fuels [35, 42]. Mo and Ru are the main components of the alloy phases accounting about 17 atom% yield of the total fission products in thoria-based fuel. Table 13 gives the metallic fission products yields calculated using the ORIGEN code [69].

Along with Mo and Tc, the noble metal fission products, Ru, Rh, and Pd are known to form fine particulate of alloy precipitates (while inclusions) in the fuel matrix. At an oxygen potential well below that of Mo/MoO_2, most of Mo are seen in the alloy phases generally consisting of the Ru-based hexagonal (ε) phase and a second phase, the nature of which varies according to the fuel temperature. The phase equilibrium studies of Mo-Ru-Pd ternary and Mo–Ru–Rh–Pd quaternary system reported by different investigations [35, 70–74], are in agreement with the

Table 12 Thermodynamic data to explain the formation of various Cs bearing species

Reactions	ΔG^0(kJ mol^{-1})	Evaluated results
$Cs(l) + Te(s) = Cs_2Te(s)$	$-367.8 + 0.0542\ T$ $300 \le T \le 722$ K	On clad at 573 K, $Cs_2Te(s)$ formation is possible when $a_{Te} \sim 1$ and a_{Cs} is fixed by UO_2–Cs_2UO_4/Mo–MoO_2 at the fuel surface, at T ~ 800 K
$Cs(l) + Te(l) = Cs_2Te(s)$	$-384.7 + 0.0782\ T$ $722 \le T \le 1080$ K	
$Cs_2Te(s) = Cs_2Te(g)$	$132.2 - 0.0521\ T$ $\Delta_f G^0(Cs_2Te(g)) = 252.5 + 0.0261T$ kJ mol^{-1} $722 \le T \le 1080$ K	
$ZrTe_2 + Cs(l) = Cs_2Te(s) + Zr(s)$	$-71.74 + 0.0361\ T$ $300 \le T \le 722$ K	Hence, no $ZrTe_2$ on Zircaloy surface
$Cs_2Te(s) + 3/2O_2 = Cs_2TeO_3(s)$	$-613 + 0.2269T$, $300 \le T \le 722$ K $\Delta\bar{G}_{O2}$=-306.5 + 0.1134T	The evaluated $\Delta\bar{G}_{O2}$is higher than in Mo/MoO_2. Hence, no $Cs_2TeO_3(s)$ formation
$2Cs(g) + Te = Cs_2Te(g)$	$-396.5 + 0.1765T$ $(\ln(p_{Cs2Te}/a_{Te}) = -(720+\Delta\bar{G}_{O2})/RT$ $+42.1$, p_{Cs} controlled by UO_2–Cs_2UO_4/Mo–MoO_2)	As per the equilibrium reaction $p_{Cs2Te} = 3x10^4\ a_{Te}$ at 1500 K, $p_{Cs2Te} = 1.3a_{Te}$ at 800 K
$Cs_2Te(s) + MoO_2(s) + O_2 = Te(l) + Cs_2MoO_4(s)$	$-686.7 + 0.2762T$; when $T < 1000$ K and $\Delta\bar{G}_{O2}$(Reaction) $<\Delta\bar{G}_{O2}$(Mo/MoO_2)	$Cs_2Te(s)$ to $Cs_2MoO_4(s)$ conversion possible at lower temperatures
$Cs_2Te(s) + UO_2(a_{UO2} = 0.03) + O_2 = Te(l) + Cs_2UO_4(s)$	$-588.3 + 0.3269\ T$; $\Delta\bar{G}_{O2}$(Reaction) $< \Delta\bar{G}_{O2}$(Mo/MoO_2) attainable below 50 K only	No conversion of $Cs_2Te(s)$ to $Cs_2UO_4(s)$

Table 13 Mo and noble metal yields in thoria-based fuel

Metallic fps	Yield (atom%)	Total (atom%)
Mo	10.25	17.36
Ru	3.53	
Tc	2.35	
Rh	0.65	
Pd	0.58	

observed nature of the metallic inclusions. From the reported study of the quaternary system, one expects the second phase to be Mo-based bcc (β) phase below 2173 K and the Mo_5Ru_3 intermetallic based tetragonal (σ) phase above 2173 K [73]. The post irradiation studies on the conventional oxide fuels show the presence of the two alloy phases [71]. Mo and Ru contents in the hexagonal phase are, respectively, found to be 38–40 wt% and 28–35 wt% in FBR and 24–25 wt% and 44–52 wt% in LWR [35]. The phase diagram information generated at 1500 K by Fukuzawa et al. [74] shows the shrinking of the ε phase field at lower temperature such as in LWR fuel.

At high burn up where much of the Mo component oxidizes and partitions out into the oxide phases, the Mo-rich β or σ can disappear. In FBR pin, the central temperature being quite high, Pd component volatilizes out and concentrates in the cooler region as the Pd-based fcc (α) phase [35]. At the high oxygen potential, Mo and Ru components also volatilize out to some extent as their oxides, e.g., MoO_3 (g) and RuO_3 (g).

With a view to knowing the buffering capacity of Mo for fission released oxygen in the oxide fuel, Mo activity in the alloy phase has been assessed [70–74]. Yamawaki [75] et al. have measured the Mo activity in the hexagonal phase in the temperature range of 1200–1300 K by using galvanic cell technique. Naito et al. [76] and Matsui et al. [77] have measured Pd vapor pressure together with the use of regular solution model due to Kaufman and Berntein [78] and have reported the Mo activity in the $Mo_xRu_yPd_z$ alloys ($x = 0.20 - 0.45$, $y = 0.15 - 0.60$, $z = 0.20 - 0.40$). The evaluated Mo activities in the hexagonal phase at different compositions are plotted as a function of temperature in Fig. 5. It is to be noted that the oxygen potential of $[Mo]_{alloy}/MoO_2$ will be higher by$-RT \ln(a_{Mo})$, a_{Mo} being the Mo activity in the alloy phase.

3.1 Ternary Oxide Formations of Fission Products at Higher Burn ups

The stabilities of the different ternary oxides of the fission products described above are now evaluated to see their feasibility of formations with increase of the oxygen potential inside the fuel. The feasibility can be represented in the oxygen potential diagram (Fig. 6) of the relevant pseudo-binaries and ternaries involving

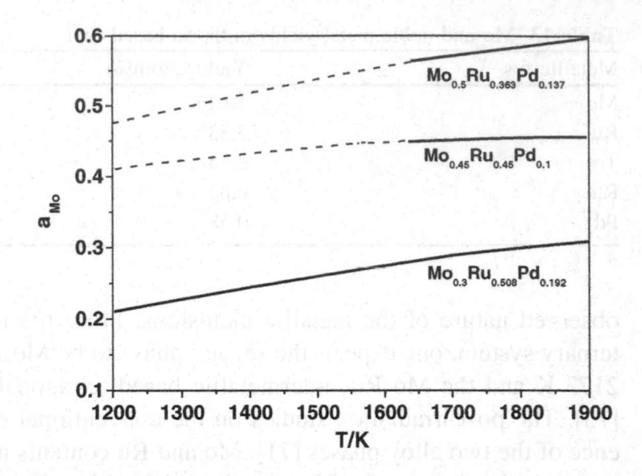

Fig. 5 Mo activity in hexagonal phase of Mo–Ru–Pd ternary

the oxides. At any temperature, the respective ternary oxides will be formed as the potential inside the fuel crosses the characteristic values represented in Fig. 6. The notional states of the different fission products considered generally pertain to those that can be formed below the oxygen potential of $[Mo]_{alloy}/MoO_2$.

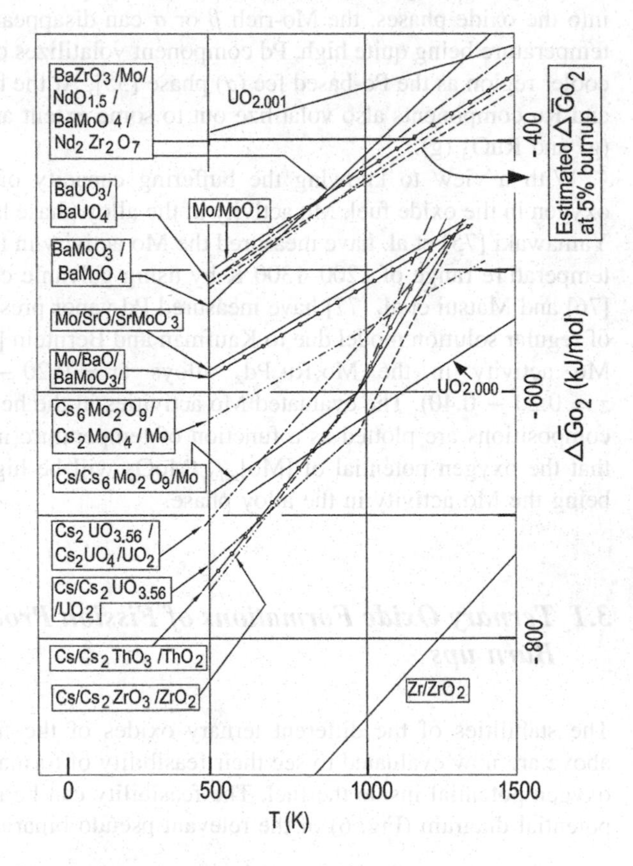

Fig. 6 Oxygen potentials of relevant pseudo-binaries, ternaries, etc. at different temperatures (all phases taken to be pure) ([2], Copyright Elsevier)

From Fig. 6, it is evident that the oxygen potential to rise up to the value of Mo/MoO_2, all Cs should be in the form of ternary oxides $Cs_6Mo_2O_9$ and Cs_2MoO_4 that are in equilibrium with Mo. With the constant availability of the fission generated oxygen most of the Cs will be as Cs_2MoO_4, consuming there by two oxygen atom per Cs. Also, a major fraction of the alkaline earth oxides, can be fixed as molybdites and uranites (ABO_3, A = Ba/Sr and B = Mo/U) consuming thereby three oxygen atoms per Ba or Sr in case of molybdite and one oxygen atom per Ba or Sr in case of uranite formations. Fig. 6 suggests that at the higher oxygen potential there can be even molybdates and uranates (ABO_4) formations particularly toward the cooler part of the fuel. However, the Zr yield being nearly double of the alkaline earth fission products and the stability of alkaline earth zirconates being higher than molybdites, the Mo and U based ABO_3 and ABO_4 type formations depends on the Zr redistribution into other phases like rare earth pyrochlore. The Zr-based pyrochlore formation is evidenced in the SIMFUEL study [13] at high oxygen potential.

In view of the above-mentioned facts, it is quite possible that the remaining 3.8 % of the fission released oxygen (Fig. 4) can lead to the formation of alkaline earth ternary oxides in thoria-based fuels, had it not been transported out to clad. As against this, pure urania fuel distinguishes itself by efficient buffering and redistributing the left over oxygen. At 1473 K for example, urania postpones Mo oxidation until O/U reaches to about 2.001. Thus, with 150 W cm^{-1} power rating, if the whole of unused 3.8 % oxygen is available for the fuel oxidation, it takes about 2 years (15 GWD T^{-1} burn up) to attain the O/U value. In actual situation this happens at much higher burn up, as a major fraction of the available oxygen flux is transported out of the fuel for clad oxidation.

In thoria-based fuels, whether the remaining 3.8 % oxygen could be consumed as the molybdite/uranite and molybdate/uranate phases of the alkaline earths or not is decided by the kinetic aspects. For the apportioning of the leftover oxygen between the clad zircaloy and Mo, one considers the kinetic aspects involved in the Zr and Mo oxidation. Besides the oxidation processes, the kinetic aspects include the oxygen transport inside the fuel and across the fuel-clad gap. These aspects are considered below.

3.2 Redistribution Kinetics of Fission Released Oxygen

One of the important factors for the redistribution of fission released oxygen among the fission products and clad is the oxygen transport inside the fuel. The transport in thoria-based fuel is principally due to self-diffusion contribution $D_O(T)$. The reported $D_O(T)$ values in pure thoria and urania matrices are quite comparable [3, 79], as given below:

$$\log(D_o/m^2s^{-1}) = -14362/T - 3.35(\text{in thoria})$$
$$\text{and } \log(D_o/m^2s^{-1}) = -13086/T - 4.58(\text{in urania}).$$

Oxygen transport properties of urania are well-studied and it has been shown that the self-diffusion contribution is several orders of magnitude lower than the one from chemical diffusion $D_{O,c}$ (T), which is given by Bayogln et al. [80] as

$$\log(D_{o,c}/m^2s^{-1}) = -4680/T - 5.74.$$

For thoria-based matrix, there is very little information on the chemical diffusion property of oxygen. Matsui et al. [81] have reported the oxygen chemical diffusion property in $Th_{1-y}U_yO_{2+x}$ ($y = 0.2$ and 0.4) with activation energies quite similar in magnitude to that of pure urania. Their data reflect progressive decrease of the transport coefficient with lowering of the urania content in the ternary solution as shown in Fig. 7. The sharp fall in transport is expected from the consideration of the thermodynamic factor (F) in the oxygen diffusion in MO_{2+x} fluorite lattice and is expressed as $D_{O,c} = D_O F$, with $F = (d \ln a_O/d \ln x_O)$, a_O being the activity and x_O the mole fraction of interstitial oxygen. $F = (1 + d \ln \gamma_o / d \ln x_o)$, γ_O is the activity coefficient.

As the cationic diffusion is comparatively negligible, the variable x_O can be expressed as $x/(2 + x)$, or dx_O as $2\,dx/(2 + x)^2$ so that,

$$F \approx (x/2RT)d(\Delta\bar{G}_{O2})/dx$$

Considering the thermochemical approach [11] the F factor nonlinearly changes with the urania content (y) because of the limiting behavior of the slope, $d(\Delta\bar{G}_{O2})/dx = s$(say), at $x \to 0$ and $y \to 0$, namely, $(ds/dy) = 1/y$. The oxygen transport study in the ternary solution with much lower urania content carried out in this laboratory [82] conforms to the nonlinear decline in the transport. The thermogravimetry based study on the highly sintered composition $Th_{0.98}U_{0.02}O_{2+x}$

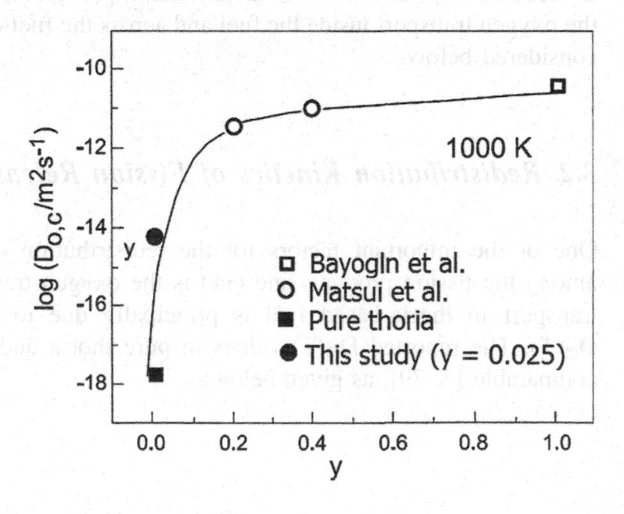

Fig. 7 Trend of oxygen transport coefficient with urania content in $Th_{1-y}U_yO_{2+x}$ at x = 0 ([2], Copyright Elsevier)

doped with the fission products (Sr, Ba, Y, Nd, La, Ce, Zr, Mo, Ru) representing 20 GWD T^{-1} burn up, showed that the diffusion coefficient of oxygen is orders of magnitude lower than the chemical diffusion coefficient reported for pure urania [80] as expressed above. The experimental data can be represented at the limit of $x = 0$, as

$$\log D_{O,c}\left(\pm 0.25\,m^2s^{-1}\right) = -5383/T - 8.82.$$

Figure 7 that incorporates the data of this study [82] along with those of Bayogln et al. [80]. Matsui et al. [81] shows that there is nonlinear drop in the oxygen transport with decreasing values of the urania content in $Th_{1-y}U_yO_2$ and in the limit of $y = 0$, the extrapolation meets the reported data of self-diffusion of oxygen in thoria [79].

The thoria-based fuels with low urania content will have significant impediment in the oxygen transport as compared to the urania case. The oxygen profile evolution in the two fuel matrices can be made considering the redistribution of a given percentage of the left over oxygen out of the fission process in the respective cases. As a typical example one may consider the redistribution of the left over oxygen after the oxidations of reactive fission products up to the alkali metals as schematically shown in Fig. 4. The 3.8 atom percent of the fission generated oxygen constantly available from 12.5 mm dia fuel pellet at 150 W cm^{-1} power rating will have the oxygen flux of 4.5×10^{-8} g.atom d^{-1} cm^{-1}, which is equivalent to the flux density of 4.6×10^{-13} g.atom.$cm^{-3}s^{-1}$.

3.3 Oxygen Transport in Fuel Matrix

The leftover part of fission generated oxygen in the oxide matrix forms a constant volume flux B ($B = 4.6 \times 10^{-13}$ g.atom $cm^{-3}s^{-1}$) that undergoes diffusion transport to the fuel boundary, where its concentration is usually fixed by the low oxygen potential of the clad. Zircaloy clad potential is set at Zr/ZrO_2 while stainless steel clad potential at [Cr] alloy/Cr_2O_3. The transport is to be accounted under the prevailing radial concentration and temperature gradients. The ponderomotive force for the transport derives from the entropy rise for the oxygen transfer from the higher to lower concentration and also for the associated heat transport in the temperature gradient. On the approximation of radial diffusion of species from the pellet, the oxygen flux J (r) across the cylindrical surface of radius r and unit height under the steady state of chemical and thermal potential gradients is given by

$$J(r) \equiv \pi r^2 B = 2\pi r C_O\left[-D_{O,c}\left(\partial \ln x_O/\partial r\right) - D_O\left(Q/RT^2\right)\left(\partial T/\partial r\right)\right]$$

C_O is the concentration of interstitial oxygen, (C_O/x_O) being the total oxygen concentration in MO_{2+x}. The fluorite lattice of $Th_{1-y}U_yO_{2+x}$ has the self and chemical diffusion transport characteristics D_O and $D_{O,c}$, and R is the gas constant. The net heat transport Q differs from the heat energy Q_O taken from the higher temperature (T + δT) end of the infinitesimal section, because a part of Q_O is expended as work input in increasing the standard chemical potential of the transported oxygen. For the work input of $\delta W = \mu°_O(T) - \mu°_O(T + \delta T) \equiv S°_O$ δT, the minimum heat expense is the reversible value, T (δW/δT) = Q (say); $S°_O$ being the standard partial molar entropy of oxygen in the ternary solid solution at x (r). The net heat flow Q is thus $Q_O - Q$, $Q \geq T(\delta W/\delta T) \equiv TS°_O$. In steady-state transport, Q_O essentially representing the infinitesimal quantity $N^0_{Th}\, C^0_{p,O}\, \delta T$ ($C^0_{p,O}$, the standard partial molar heat capacity of oxygen), Q assumes negative values. The negative Q has been reported [83] from the measured oxygen redistribution in hyper and hypo stoichiometric urania-plutonia fuel under steady state. The measured Q at different uranium valency (V_U) under hyperstoichiometry in empirically fitted as $Q = -8.3 \times 10^{33} \exp(-17\, V_U)$. For the thoria-based fuel, as there is no reported data on the oxygen redistribution, the steady-state oxygen profile in the $Th_{0.98}U_{0.02}O_{2+x}$ composition-based SIMFUEL is evaluated considering that Q is approximately equal to T $S°_O$. With the steady flux, $B = 4.6 \times 10^{-13}$ g.atom $cm^{-3}s^{-1}$, and $Q \approx -TS°_O$, the measured coefficient $D_{O,c}$ in $Th_{0.98}U_{0.02}O_{2+x}$ SIMFUEL, and the molar mean value of D_O's of the two fluorite lattices MO_2 (M = Th and U), the diffusion equation of the fuel is solved. Considering the reported self-diffusion coefficients of ThO_2 and UO_2 [3, 79], the mean value used is $\log D_O\, (m^2s^{-1}) = -14336/T - 3.37$ In the evaluation, $S°_O$ value has been obtained from the reported oxygen potential behavior of the fuel in the thermochemical approach of Schram [11]. Considering the temperature dependence of the oxygen potential in the fuel [11], the standard molar entropy value of about 12 J K^{-1} $g.atom^{-1}$ of oxygen can be assigned at the mean temperature (1250 K). This mean value is assigned to $S°_O$.

The oxygen profile in $Th_{0.98}U_{0.02}O_{2+x}$ SIMFUEL evaluated at the steady state of temperature as well as the concentration radially across the cylindrical matrix of diameter 12.5 mm is included in Fig. 8. The profile is evaluated without considering oxygen uptake by Mo to form its oxide. For the sake of comparison, the

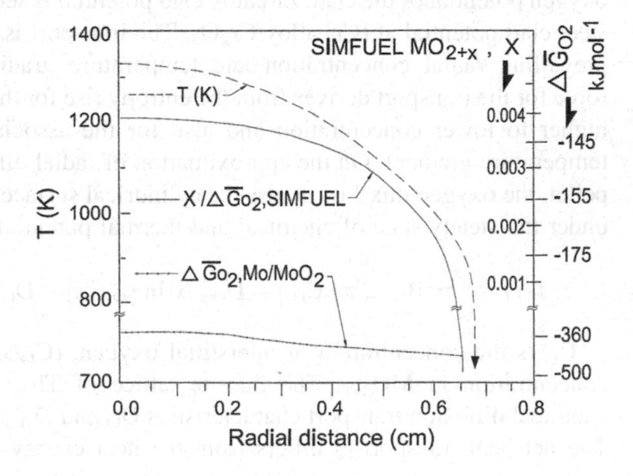

Fig. 8 Steady-state radial-oxygen-profile over thoria-2 mol% urania with FPs at 20 GWD T^{-1} in absence of Mo oxidation(Mo/MoO_2 potential according to the thermal profile included for comparison) ([2], Copyright Elsevier)

figure also includes the oxygen profile in pure urania matrix under the same temperature gradient as that in the thoria-based fuel. The orders of magnitude difference in the oxygen concentration in the two matrices are evident. The difference is more to do with the comparatively high oxygen diffusion in urania than the approximate use of the same temperature gradient as in thoria; the gradient in reality should have been steeper in urania because of its lower thermal conductivity. The steady-state oxygen profile is attained involving a time constant $\tau = (D_{O,c}a_n^2)^{-1}$($\alpha_n$ is the geometric factor). Due to the dissimilar values of oxygen transport coefficients in thoria and urania-based fuels, the indicated profile attains much slowly in the former marix as compared to that in urania under a given geometry. Considering a simpler situation with an average diffusivity over the thermal profile of the respective case in the cylindrical geometry of diameter 2a (=12.5 mm), the geometric factor α_n is the nth root of the zeroth order Bessel function Jo (aα) = 0, and the concentration evolution Co (r,t)/Co (r,∞) is expressed as [84],

$$\left[1 - 8\sum e^{(-t/\tau)}\left(J_0(r\alpha)/\alpha_n^2\right)\left(J_1(a\alpha)/(a^3 - r^2a)\right)\right]$$

With a flat temperature distribution of 1273 K, the time constants for the two fuel matrices are effectively 1.5 y for thoria and less than an hour for urania. Under the slow redistribution of the constantly accumulating left over (3.8 %) of the fission released oxygen the O/M ratio in the thoria-based fuel can be quite high and this can lead to oxygen potential well above that of Mo/MoO_2, which under the thermal profile varies within −355 to −423 kJ mol^{-1} (Fig. 8). This suggests that Mo thus far seen present as molybdites and alloy phases will undergo oxidations to molybdates of the alkaline earth fission products. The uranite if present as chemical component in the perovskite will also be oxidized to the uranates. The thermodynamic result readily follows from the oxygen potentials representation given in Fig. 8. If UO_2 concentration in thoria is increased, the oxygen profile shown in Fig. 8 will be lowered due to better oxygen transport with increasing $D_{O,c}$ values. Referring to the trend of $D_{O,c}$ with UO_2 concentration given in Fig. 7, one finds that the oxygen profile will be close to that of pure UO_2 as the concentration is increased to 20 mol% and higher. As in pure urania, the thoria-based fuels with high urania content can effectively transport out the residual 3.8 % of fission released oxygen and keep the oxygen potential below that of Mo/MoO_2. High burn-up matrix also can increase the oxygen transport to clad and thus control the local rise of oxygen potential.

For fuels with low urania content, whether the oxygen potential will be effectively buffered by the Mo presence, rests upon the kinetic aspect of attaining the oxidized states of Mo compounds. If the Mo oxidation is not fast enough, the tendency of the oxygen potential growing beyond Mo (alloy)/MoO_2 is apparent in Fig. 8. As such there is little information on the kinetics.

3.4 Molybdenum Oxidation Kinetics

Although the Mo oxidation kinetics and mechanism have been extensively studied on the metal surface under various temperatures and oxygen partial pressures [85–87], there is hardly any such study in fine granular forms of metal or its alloys as are produced in the fission products. The oxidation rate is bound to be quite different in fine grains of the metals and its alloys. In the absence of exact experimental data, one can consider the reported result of the metal oxidation to work out limiting rate. For this one takes the possible scenario that the alloying kinetics is slower than the oxidation considering faster diffusion of oxygen than the metallic fission products, and that oxygen reacts more easily with finely divided particles. Gulbransen et al. [85] have reported the oxidation rate of molybdenum surface as

$$dN_{Mo}/dt = A\exp\left(-E/RT\right),$$

The oxygen consumption rate according to the reported result is as high as about 91×10^{-4} g.atom $cm^{-2}d^{-1}$ even at the lowest temperature of 800 K prevalent at the fuel surface. This value being orders of magnitude higher than the net oxygen availability of 1.1×10^{-8} g atom $cm^{-2}d^{-1}$ out of the fission process, one may conclude that finely divided state of Mo can easily buffer against the possible surpassing of oxygen potential over Mo/MoO_2 value. For the oxidation of alloyed Mo, or the Mo in ABO_3 perovskite phase, the conclusion may not be so straightforward.

The extent of Mo oxidation is nevertheless subjective as a certain amount of oxygen will be transported out reaching clad for its oxidation. In this context, it is also necessary to know the transport property of the thin gap (typically 50 microns) as well as the oxidation kinetics of clad like zircaloy as included below.

3.5 Zircaloy Oxidation Kinetics

The oxide layer (S) growth kinetics in zircaloy has been reported [1] and initially, it follows cubic law until attaining a critical thickness, when there is transition to the linear kinetics. Among the available empirical fittings describing the kinetics, those used by COCHISE CEA [1] is given below to understand the flux limit for quantitative oxygen uptake by the clad.

(i)

$$dS^3/dt = k\exp\left(-Q/RT\right)$$

$$k = 11.4 \times 10^{10}\ \mu m^3 d^{-1},\quad Q = 142.8\ \mathrm{kJ/mol}$$

(ii)

$$dS/dt = k' \exp(-Q'/RT)$$

$$k' = 4 \times 10^{11}\ \mu m\, d^{-1},\ Q' = 152.9\, kJ/mol.$$

The reported source [1] also describes the time period to attain the critical thickness as,

$$t_{(critical)}\ (\text{in hour}) = 8.857 \times 10^{10} \exp[1830/RT - 0.035\, T]$$

At the clad temperature of ~600 K, the critical layer thickness and its attainment time according to the kinetic equation are 1.7 μm and 13 days, respectively. With the growth rate of 5.59×10^{-6} μm/d at the critical ZrO_2 layer formation, the oxygen consumption is 3.85×10^{-6} g.atom $d^{-1}cm^{-1}$ which is about two orders of magnitude higher than the oxygen flux of 4.5×10^{-8} g.atom $d^{-1}cm^{-1}$ out of the 3.8 a/o fission released oxygen at 150 W cm^{-1} power rating. If all of the 3.8 % oxygen flux were available, the 1.7 μm oxide layer (equivalent to 5×10^{-5} g.atom oxygen) is attainable within 1160 days (~27.8 GWD T^{-1}). Subsequent linear growth rate of 0.83×10^{-6} μm/d (5.70×10^{-7} g.atom $d^{-1}cm^{-1}$) is again an order of magnitude higher than the available oxygen flux.

Because of the higher rate kinetics of zircaloy oxidation, as compared to the available oxygen flux rate, the oxygen pressure at the fuel boundary will be as low as that in Zr/ZrO_2 system. This conclusion is valid when the flux transport across the fuel-clad gap does not encounter impediment. The factors deciding the transport across the fuel-clad gap is given below.

3.6 Oxygen Transport Kinetics Across Fuel-Clad Gap

The oxygen flux from pin to reach the clad, has to undergo transport across a thin gap (typically 50 micron) filled with helium gas at high pressure (~25 bar). The transport coefficient D_g of oxygen gas, according to the kinetic theory, is of the order of $u_{av}/(3\pi N\sigma^2)$, where u_{av} is average speed of O_2 molecules at the gap temperature of ~ 800 K, N is total gas density, and σ the mean collision diameter of $He + O_2$ gas mixture. Dg value works out to be 0.054 cm^2s^{-1} (5.4×10^{-6} ms^{-1}). With this value of the kinetic coefficient one thus addresses to the question whether the oxygen flux load of 4.5×10^{-8} g.atom $d^{-1}cm^{-1}$ of the pin during the steady power generation of 150 W cm^{-1} can be transported radially outwards across the thin annular gap of 50 micron. For this transport to occur through gas diffusion, there is need of oxygen concentration of 6×10^{-12} g mol cm^{-3}, or p_{O2} pressure of 3.5×10^{-10} bar on the fuel surface. This requirement cannot be met by the low oxygen pressure prevailing on the surface, which at best attains the buffering pressure of Mo/MoO_2 system ($P_{O2} \sim 1 \times 10^{-28}$ bar at 800 K). The change of flow regime from diffusion to turbulence obviously is not helpful for the required transport to take place. The oxygen transport across the gap in fact can, however, take place more efficiently by augmentation of oxygen bearing gaseous species on the fuel surface via the involvement of various chemical equilibria with reactive gaseous/vapor species present therein, such as hydrogen, Cs(g), etc. These species establish chemical vapor transport of oxygen. The reactive species generate their oxidized products such as H_2O(g), Cs_2O(g), and CsOH(g) in sufficiently high concentration meeting the required transport rate of the fission released oxygen [88]. Transport cycle of the reactive species set up between the fuel and the clad is exemplified below:

One may consider whether the chemical transport of oxygen via H_2/H_2O gas mixture can cope up with the oxygen flux of the steady state. It is a well-established fact that the fabricated fuel contains hydrogen and moisture impurities together to the extent of 0.1 ppm. With 0.1 ppm hydrogen impurity, the H_2/H_2O equilibrium set up at the pin surface containing, say, $p_{O2} \sim 1 \times 10^{-28}$ bar at 800 K would result in the p_{H2O}/p_{H2} value of about 5.1, that is, $p_{H2O} \sim 2 \times 10^{-8}$ bar, which is orders of magnitude higher than the required pressure of 3.5×10^{-10} bar of the oxygen bearing species on the fuel surface. The reaction product H_2O (g) then undergoes diffusional transport to the clad and is quantitatively consumed in oxidizing the clad as $2H_2O$ (g) + Zr (zircaloy) = $2H_2$ (g) + ZrO_2 (g) ($pH_2/pH_2O \sim 4.3 \times 10^{17}$ at 800 K [41]). The product H_2 (g) then returns and gets reused for the chemical transport of oxygen from the fuel matrix. Considering that the chemical transport of oxygen from MO_{2+x} phase to H_2O (g) is fast enough to take up the oxygen flux load from fission, the stated kinetic path effectively helps the O/M ratio of the peripheral region of pin at or below the oxygen potential of Mo/MoO_2.

Thus the chemical transport of oxygen across the gap can be as fast as the clad oxidation. The diffusion inside the matrix is slowest of all the kinetics of transporting out oxygen. The Mo oxidation process inside the matrix thus plays decisive role in controlling the oxygen potential inside the fuel.

4 Evaluation of High Burn-Up Fuel Chemistry in Thoria-Based Fuel Matrix

The considerations of the thermophysical, thermodynamic and transport properties have shown that the thoria-based fuels have the merit of less thermal dilation, excellent thermal transport properties, good mechanical strength, chemical inertness, and also less thermal release of the gaseous and volatile fission products [89]. At the same time, some demerits are also seen in thoria-based matrix, viz., its poor chemical transport of oxygen and worse buffering of oxygen potential, particularly for the low urania content fuels as discussed already. This results in subtle difference in the redistribution of the fission released oxygen among the fission products and clad in the fuel. Molybdenum oxidation kinetics plays decisive role on the potential buffering all through the burning process. The chemical states of the oxygen rich phases of the fission products that can be formed in the neighborhood of the $[Mo]_{alloy}/MoO_2$ potential are relevant in the integrity analysis of the thoria-based matrix.

Assessment of the chemical states of the oxidized Mo considers the effect of thermodynamic activities of Mo and other components participating in the oxidation equilibria. The equilibrium with the lowest oxygen potential establishes first before others, unless kinetic hindrance delays equilibrium being established. The comparative representation for the potential of pseudo binary, ternary, etc. equilibria involving pure phases of the fission products given in Fig. 6 indicates that in the proximity of Mo oxidation, several oxygen rich phases have their formation possibilities. Taking into account the activities of Mo and its relevant oxidized states in their respective dissolved phases, one compares the oxygen potentials of the following stated equilibria:

$$(i)\ 2[Mo]_{alloy} + 2O_2\ (g) + 2[AO] = 2\,[AMoO_3],$$
$$(ii)\ 2[Mo]_{alloy} + 3O_2\ (g) + 2[AO] = 2\,[AMoO_4],$$
$$(iii)\ 2[AZrO_3] + R_2O_3 + 2[Mo] + 3O_2\ (g) = [R_2Zr_2O_7]2\,[AMoO_4],$$

where, $AMoO_3$ and $AMoO_4$ respectively represent perovskite and scheelite-type phase components with A as divalent Ba/Sr, and $R_2Zr_2O_7$ is the pyrochlore phase component of the rare earths (R = La, Nd, etc.). Along with these equilibria, the potential of UO_2 oxidation is also considered as (iv) 2 $[UO_2]$ + O_2 (g) + 2 [AO] = 2 $[AUO_4]$. For the evaluation of the oxygen potential deciding the order of participation of the different oxidative equilibria in the oxygen buffering process in the fuel, one has to consider the Mo and MoO_2 activities in the oxidative equilibrium involving the alloyed Mo as $[Mo]_{alloy} + O_2 = [MoO_2]$. For the a_{Mo} value one considers the Mo content in the alloy, which generally varies from 20 to 40 % [72] depending on the fuel composition, irradiation pattern, and burn up. For example, in AHWR case at 50 GWD T^{-1}, the Mo content works out to be about 34 atom% if one considers the total Mo produced and the part consumed at a lower oxygen potential as $(Cs, Rb)_2\,MoO_4$ (Fig. 6) and also as MoO_2 dissolved (<0.2 mol%) in the oxide fuel. Recalling the experimental results of Yamawaki et al. [75], the hexagonal alloy phase with 34 atom% Mo content can have a_{Mo} values with 0.25– 0.3 in the AHWR fuel at temperatures of 800–1500 K. In the present analysis, a value of 0.3 for a_{Mo} is used all through. The a_{MoO2} values on the other hand, can be estimated from the stipulation that 3.8 % leftover of the fission released oxygen (Table 2) is incessantly getting fixed either as $BaMoO_3$ and $SrMoO_3$ components in the perovskite phase or as MoO_2 phase itself. Molybdites formation depends on BaO and SrO activities as well as stabilities of AMO_3 (A = Sr, Ba) components. The relative values of activities of the two oxides decide the MoO_2 partitioning in the equilibrium, $[BaMoO_3]$ + [SrO] = [BaO] + $[SrMoO_3]$, that has nominal change in the standard free energy, $\Delta G^0_{reaction} = -3.2 + 0.0087 kJmol^{-1}$. The estimation of SrO and BaO activities, the key parameters deciding the oxidized state of Mo are elaborated below.

4.1 Estimation of SrO Activity

From consideration of higher stability of zirconate than cerate of Sr, the reported presence of the alkaline earth in the perovskite phase (Ba, Sr) (Zr, Ce) O_3 [13] can be taken to be mainly as the $SrZrO_3$ component. Whatever, little Ce present in the phase is essentially to be taken as $BaCeO_3$ component that has better stability than $SrCeO_3$ (Table 4). Again from the relative stabilities of the two zirconate components, ZrO_2 partitioning as $SrZrO_3$ is less than as $BaZrO_3$. The higher solubility of SrO as compared to BaO in thoria also facilitates the partitioning behavior. On this basis one can initiate the analysis assuming the $SrZrO_3$ content in the perovskite phase to be below 50 mol% and calculate the SrO activity limit from the equilibrium property of the reaction,

$$[ZrO_2] + [SrO] = [SrZrO_3]$$

Thus one substitutes, with $SrZrO_3$ as an ideal component of the perovskite phase, the activity limit $a_{SrZrO3} < 0.5$ in the equilibrium relation, $RT\ln(a_{SrO}) = \Delta G^o_{reaction} + RT\ln(a_{SrZrO3}/a_{ZrO2})$, obtaining thereby an upper limit of SrO activity. $\Delta G^0_{reaction}$ is the standard free energy change for the reaction. The ZrO_2 activity in the relation is obtainable from the concentration of dissolved part of ZrO_2 in thoria at a burn up, B (at%), considering its solution property in thoria. Until saturation in the fuel matrix, the ZrO_2 concentration is approximately expressed as, $X_{ZrO2} = (Y_{Zr} - Y_{Ba})\ B/100$, where Y_{Zr} ($\sim$0.33) and Y_{Ba} ($\sim$0.064) are the average yields of Zr and Ba per fission. Thus, $X_{ZrO2} \sim 0.0027$ B. Here it is assumed that the Ba equivalent of Zr is practically used up in the zirconate formation.

The reported evaluation of ZrO_2 solubility in thoria [14] as expressed by $\ln(X^O_{ZrO2}) = 1.67 - 7510/T$, provides the value of the regular solution parameter (45.2 kJ mol^{-1}), which can be used to estimate the oxide activity as

$$RT\ln(a_{ZrO_2}/X_{ZrO_2}) = 45.2\,(1 - X_{ZrO_2})^2$$

From the above-mentioned considerations, the upper limit of the SrO potential is thus expressed as $RT\ln a_{SrO} < -115.7 + (0.053 - 0.0083\ln B)T$ kJ mol^{-1}. For the reported study at 1873 K on SIMFUEL (burn up of 21.5 atom%; Nd yield, $X_{Nd} \sim 0.042$) [13], the upper limit works out to be -64 kJ mol^{-1}. Noting this activity limit and their observation that $NdO_{1.5}$ in the fuel phase did not form the pyrochlore via the reaction, $2SrZrO_3 + 2NdO_{1.5} = Nd_2Zr_2O_7 + 2SrO$, one expresses the lower limit of SrO potential as,

$$RT\ln a_{SrO} > \left[-\Delta G^0\ \text{reaction} + 2RT\ln(a_{SrZrO_3}\,a_{NdO_{1.5}})\right]/2$$

With an initial value of a_{SrZrO3} as 0.5 and with the assumption of ideal mixing of $NdO_{1.5}$ in thoria, the limiting potential at 1873 K works out to be -75 kJ mol^{-1}. The selection of median value of the potential as -69.5 kJ mol^{-1} amounts to the correction of $SrZrO_3$ activity from the previously chosen value of 0.5–0.35.

The iterative use of the changed values of a_{SrZrO3} and the median potential in the thermodynamic criterion of pyrochlore's absence in the reported result leads to the convergence of the $SrZrO_3$ activity to 0.25. In ideally mixed state of the two zirconates, this correction will be registered in the temperature coefficient of the median potential as,

$$RT \ln a_{SrO} = -115.7 + (0.047 - 0.0083 \ln B)T$$

In the above analysis, the dissolved part of ZrO_2 in the fuel will be modified due to the ZrO_2 partitioning into $SrZrO_3$, though the extent of modification is small. For example, with the ZrO_2 partitioning as 25 mol% $SrZrO_3$ and 75 mol% $BaZrO_3$, the X_{ZrO2} would reduce to 0.0024B from the previously accounted value of 0.0027B. It is to be noted that the assigned concentrations of $BaZrO_3$ and $SrZrO_3$ can deviate due to partial partitioning of Ba and Sr into $ACeO_3$ and $AMoO_3$ (A = Ba, Sr) components in the perovskite phase.

To get the MoO_2 and ZrO_2 partitioning, respectively, into the molybdites and zirconates of the two alkaline earths, the knowledge of BaO activity is also required as elaborated below.

4.2 Estimation of BaO Activity

BaO being common to all perovskite components, $BaMO_3$ (M = Zr, Ce, U, Th), its activity can be assessed from the following equilibria, with MO_2 as the dissolved component in the fuel phase:

$$[MO_2] + [BaO] = [BaMO_3]$$

One considers $BaZrO_3$ as the predominant ideal component in the perovskite phase with its concentration lying within the two limits, 50 mol and 75 mol%, as arrived above. Accordingly, the lower and upper limits of BaO potentials (in kJ mol^{-1}) are expressed as,

$$-174 + (0.048 - 0.0083 \ln B)T < RT \ln a_{BaO} < -174 + (0.045 - 0.0083 \ln B)T$$

The median of the two limits is

$$-174 + (0.0465 - 0.0083 \ln B)T \ (\text{kJ mol}^{-1})$$

The temperature dependent BaO potential obtained from the mentioned two limits is represented in Fig. 9. Besides these two closely spaced upper and lower limits of a_{BaO}, the figure includes several other thermochemically assessed upper limits all of which are well above the median of the said two limits. These upper limits are assessed from the consideration of presence or absence of other AMO_3 components in the perovskite phase of the SIMFUEL reported [13] and also from

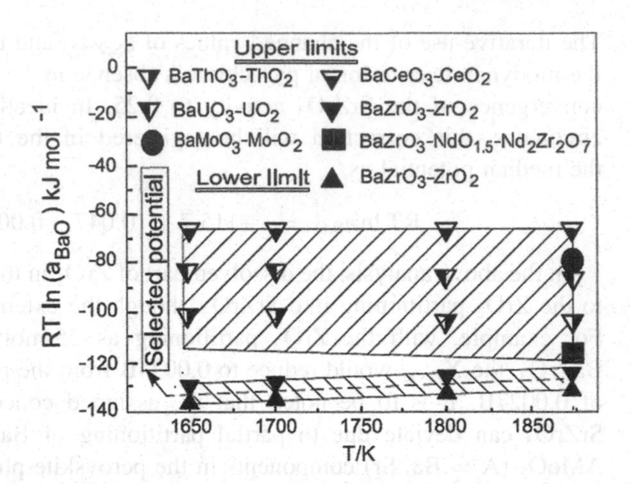

Fig. 9 BaO potential ([2], Copyright Elsevier) obtained from different thermochemical considerations

the absence of rare earth pyrochlores in the result. For completion sake a brief account of the assessments is included below.

The reported SIMFUEL ($Th_{0.933}U_{0.067}O_2$ matrix doped with the rare earths Y, Zr, Sr, Ba, Mo, Ru, and Rh for 21.5 % FIMA [13]) was annealed at 1873 K under an oxygen potential of -309.6 kJ mol^{-1}, which is well below that of Mo–MoO_2 [41]. The study revealed that the perovskite phase has the composition, $(Ba_{1-x}\ Sr_x)$ (Ze, Ce) O_3 and that the Th, U, and Mo contents in it remained below the EPMA resolution limit (about 1 atom percent). The thoria and urania activities in the SIMFUEL remaining, respectively, at 0.933 and 0.067, establish equilibria with the perovskite components $BaThO_3$ and $BaUO_3$ as, [BaO] + [AcO_2] = [$BaAcO_3$] (Ac = Th, U). Considering the EPMA resolution limit of the two perovslite components, the BaO activity is obtainable from the relation, $RT \ln (a_{BaO}) = \Delta G^o_{reaction} + RT \ln (a_{BaAcO3}/a_{AcO2})$.

In $BaUO_3$ case, one can approximate $BaUO_3$ activity to its concentration from the observed complete miscibility of $BaUO_3$ in $BaZrO_3$ with the lattice parameters [90] nearly fitting to the Vegard's law in the dilute region, [$BaUO_3$] $\leq$ 20 mol%. The limit, $a_{BaUO3} < 0.01$ (EPMA limit), results in an upper bound of BaO activity in the fuel at 1873 K. One can make similar assumption for $BaThO_3$ component also. The temperature trend of this upper bound of activity could be assessed down to 1650 K where the thoria matrix attains saturation with 5.7 mol% ZrO_2 content at 21 % FIMA of the reported SIMFUEL[13]. In the assessment, a_{BaAcO3} value is considered to evolve according to the equilibrium, [$BaAcO_3$] + [ZrO_2] = [$BaZrO_3$] + [AcO_2], from the noted EPMA limit of $a_{BaAcO3} < 0.01$ at 1873 K and for the given values of a_{ZrO2}, a_{BaZrO3} and a_{AcO2}.

BaO activity can also be obtained from $BaCeO_3$–CeO_2 equilibrium by considering a upper bound of the cerate concentration in the perovskite phase as 10 mol% [13]. BaO partitioning as cerate is expected to be low because of its comparatively low solubility with respect to zirconate. The required information of $BaCeO_3$ activity is obtainable from the reported tensimetric data (640-780 K) of cerate-carbonate equilibrium [91] in the $BaZrO_3$–$BaCeO_3$ solution. The regular solution parameter (-40.16 kJ mol^{-1}) obtained thereof can be used for the high temperature estimation of the activity values from the relation, RT ln (a_{BaO}) = $\Delta G^o_{reaction}$ + RT ln (a_{BaCeO3}/a_{CeO2}). For the CeO_2 concentration in the fuel, the upper limit is set by the inventory of the oxide as CeO_2 is completely miscible in thoria [17, 18].

Besides the above-mentioned estimates, other limiting values of the BaO activity follow from the observations that Mo is absent in the perovskite phase of the SIMFUEL [13], and that there is no formation of pyrochlone phase $(La, Nd)_2 (Zr, Ce)_2 O_7$. Application of these observations in the thermodynamic analysis of the reactions, [Mo] + O_2 (g) + [BaO] = [$BaMoO_3$] and R_2O_3 (R = La, Nd) + 2 [$BaZrO_3$] = $R_2Zr_2O_7$ + 2 [BaO], lead to the respective limits of BaO activity. The former one gives an upper limit, while the later one under assumption of ideal solution of $NdO_{1.5}$ in thoria gives a lower limit. There is, however, indication of negative deviation from ideality. The reported pseudo binary of ThO_2–Nd_2O_3 shows that the solid solution shares phase field with a stoichiometric ternary [17].

The upper limits obtained from the different consideration are seen to fall over a large stretch of the potential values (Fig. 9). In view of their large uncertainties, the two limits obtained out of $BaZrO_3$–ZrO_2 system defining the lower shaded region of the potential are only considered in defining the median potential (the dotted line in Fig. 9) as

$$RT \ln aBaO = -174 + (0.0465 - 0.0083 \ln B)T \, kJ \, mol^{-1}$$

4.3 MoO_2 partitioning

Considering the derived activities of SrO and BaO at a given burn up (B atom%) in thoria-based fuel and the standard free energy change of MoO_2 partitioning in the reaction, [$BaMoO_3$] + [SrO] = [BaO] + [$SrMoO_3$], ($\Delta G^0_{reaction} = -3.2 + 0.0087 kJmol^{-1}$), one expresses the partitioning behavior as

$$RT \ln (a_{BaMoO3} / a_{SrMoO3}) = -61.5 + 0.0082T \, kJ \, mol^{-1},$$

which is independent of the burn up.

The result is indicative of MoO_2 partitioning predominantly as $SrMoO_3$ and it is essentially due to significantly low BaO potential out of large stability of $BaZrO_3$. One can examine the thermodynamic state of the fuel considering that all of the 3.8 % leftover of the fission released oxygen is fixed as Sr molybdite. The 3.8 % leftover from 50 fissions is equivalent to 0.076 oxygen atom per fission. If all of the Ba yield ($y_{Ba} = 0.064$ per fission) is fixed as $BaZrO_3$ while the 3.8 % remain of oxygen is fixed as $SrMoO_3$, then the ratio of molybdite to zirconate is 0.038/0.064 = 0.594. Recalling that the value X_{SrZrO3} is about 0.25, and that there is some $BaCeO_3$ component in the perovskite phase, one also writes $X_{BaZrO3} + X_{SrMoO3} < 0.75$. In this result, the substitution of $X_{SrMoO3} = 0.594\ X_{BaZrO3}$ and the inverse form, $X_{BaZrO3} = 1.684 X_{SrMoO3}$ respectively yield the limiting values, $X_{BaZrO3} < 0.47$, and $X_{SrMoO3} < 0.28$. The molybdite solubility in the zirconate phase would decide whether such a high concentration of $SrMoO_3$ can be accommodated in the perovskite or whether the molybdite would separate out. As there is no reported data of $SrMoO_3$ solubility in the zirconate, one refers to the reported 46 mol% solubility of $BaMoO_3$ in $BaZrO_3$ at 1973 K in order to get some estimate about the saturation limit (X^s).

The regular solution parameter ($\alpha \sim 43.68$ kJ mol^{-1}) obtainable from the $BaMoO_3$ solubility (X^s) in $BaZrO_3$ at 1973 K leads to the extrapolated limits of X^s at lower temperatures (T) using the Van't Hoff relation, $d\ln X^s/d(1/T) = -H_{excess}/R$, $H_{excess} = \alpha(1 - X^s)^2$. Such exercise leads to the values of saturation solubility of $BaMoO_3$ in $BaZrO_3$ phase at different temperatures, namely, 30 mol% at 1486 K, 20 mol% at 1325 K, 10 % at 1155 K, 5 mol% at 1025 K, etc. Having dissimilarity in the cationic size, the $SrMoO_3$ solubility in $BaZrO_3$ is expected to be poorer in comparison. Therefore, the zirconate phase would partially accommodate $SrMoO_3$ component, the rest of which separates out. The fraction separated is progressively more with decreasing temperature along the pin radius.

Similar consideration for the standard free energy change of ZrO_2 partitioning in the reaction, $[SrZrO_3] + [BaO] = [SrO] + [BaZrO_3]$, results in the relation, $RT \ln (a_{SrZrO3}/a_{BaZrO3}) = -0.2 - 0.0075\,T$ (kJmol^{-1}). With all the uncertainties involved in the potential estimations, the result is in order, viz. the $SrZrO_3$ activity is lower than that of $BaZrO_3$.

4.4 Oxygen Potential Estimate

The Mo oxidation in thoria rich fuel is expected to begin from the initial burn-up stage. It is quite possible that molybdite component would separate out of the zirconate-based perovskite phase before reaching the temperature dependent steady concentrations discussed earlier. For such situation the oxygen potential can be estimated by considering the equilibrium $[Mo] + O_2\,(g) + [SrO] = [SrMoO_3]$. Assuming the Mo activity as 0.3, and making use of the estimated SrO potential (Section), one expresses the oxygen potential for the nearly pure phase of $SrMoO_3$ as

$$\Delta G_{O2} = -559.4 + (0.117 + 0.0083 \ln B)\,T\,\left(\text{kJ mol}^{-1}\right).$$

The burn up dependence arises from the SrO potential decrease due to ZrO_2 activity augmentation with burn up. At 5 atom% burn up, the oxygen potential is

$$\Delta G_{O2} = -559.4 + 0.13\,T\,\left(\text{kJ mol}^{-1}\right),$$

and this results in marginally lower values of the potential than those of Mo/MoO_2 in the working temperature region of the fuel.

4.5 Molybdite to Molybdate Conversion Possibilities

One can examine whether the phase equilibria involving oxygen rich components (Fig. 6) other than the molybdite can have significance in the oxygen buffering at any burn up. The feasibility of strontium molybdate formation by the oxidation of $SrMoO_3$ is obtainable from the free energy change in the reaction,

$$2[SrMoO_3] + O_2\,(g) = 2[SrMoO_4\,(s)]$$

As compared to the arrived oxygen potential of the fuel, the potential of the molybdite/molybdate system being higher by 17.2 + (0.058 − 0.0083 ln B) T (kJ mol^{-1}), the oxidation will not take place at all at the practical burn up. The other possible situation that the molybdite can disproportionate forming the molybdate and Mo as,

$$3[SrMoO_3] + [ZrO_2] = [Mo] + 2\,SrMoO_4\,(s) + [SrZrO_3],$$

will be feasible if $\Delta G^0_{reaction} + RT\ln[(a_{Mo}.a_{SrZrO3})/(a^3_{SrMoO3}a_{ZrO2}) < 0$. If $SrMoO_3$ (s) phase separately exists due to low saturation solubility (X^s) then one gets

$$2137\,K < T < 104.3/(0.0488 - 0.0083\,\ln B)$$

Thus, the disproportionation resulting in molybdate can occur. The reaction being associated with redistribution of one-third of the SrO components to the dissolved part of ZrO_2 in the fuel, much of its kinetics depends on the interphase diffusion of SrO. Through the disproportionation, the oxygen uptake by Mo alloy will ultimately result in the formation of strontium molybdate phase.

With the oxygen potential one can examine whether there can be BaO partitioning from zirconate phase to result in pyrochlore and molybdates formation as,

$$2\,[BaZrO_3] + 2\,[NdO_{1.5}] + 2\,[Mo] + 3O_2\,(g) = Nd_2Zr_2O_7\,(s) + 2\,BaMoO_4\,(s)$$

Considering the Nd concentration at burn up B as $X_{NdO1.5} \sim 0.0016$ B and the ideal behavior of $NdO_{1.5}$ in the fuel, the feasibility of the forward reaction can be evaluated as

$$\Delta G^0_{reaction} - RT\ln(X_{NdO1.5} \cdot a_{Mo} \cdot a_{BaZrO_3})^2 - 3\,\Delta G_{O2} < 0$$

$\Delta G^0_{reaction}$ obtained by using the reported stability data of the pyrochlore [92–94] and the thermodynamic data of Nd_2O_3 and the other constituents [41] is given by

$$-1808.2 + 0.556\,T\,(kJ\,mol^{-1})$$

$$\Delta G_{O2} = -559.4 + (0.117 + 0.0083\,\ln B)\,T\,(kJ\,mol^{-1}).$$

Taking for example, the values of a_{BaZrO3} and a_{Mo} as 0.47 and 0.3, respectively, the feasibility is written as,

$$T < 130/(0.3446 - 0.0415 \ln B)$$

It shows that even at burn up as high as 10 %, the pyrochlore and the molybdate formation is feasible below 522 K. Similar consideration of the reaction, 2 $[SrZrO_3]$ + 2 $[NdO_{1.5}]$ + 2 [Mo] + $3O_2$ (g) = $Nd_2Zr_2O_7$ (s) + 2 $SrMoO_4$ (s), that has $\Delta G^0_{reaction} = -1820.4 + 0.4802T$, will lead to the feasibility criterion as $T < 132.2/(0.2793 - 0.0415 \ln B)$ wherein a_{SrZrO3} and a_{Mo} values are substituted as 0.25 and 0.3, respectively. This criterion applied to the case of 10 atom% burn up indicates the feasibility of pyrochlore formation below 719 K. So for all practical purpose the pyrochlore formation is feasible toward the fuel periphery only through the conversion of $SrZrO_3$ component in the perovskite phase to $SrMoO_4$(s). Under the prevailing oxygen potential, it can be shown that the UO_2 oxidation, such as

$$2[UO_2] + O_2\,(g) + 2\,[SrO] = 2SrUO_4$$

leading to the formation of alkaline earth uranates is not feasible due to the low activities of the alkaline earth oxide as well as urania in the fuel.

The uranate formation via reduction of SrO component in the fuel phase as

$$[UO_2] + 2\,[SrO] = Sr\,(g) + SrUO_4\,(s),$$

depends on Sr atom transport from the reaction site. The reduction can also result in all gaseous products as,

$$[SrO] + [MO_2] = Sr\,(g) + MO_3\,(g) \quad (M = U, Mo)$$

The standard free energy changes of the reactions are, respectively

$$\Delta G^0_{reaction\,(M=U)} = 948.251 - 0.2973\,T\,kJ\,mol^{-1}$$

$$\text{and}\,\Delta G^0_{reaction\,(M=Mo)} = 1006.909 - 0.2956\,T\,kJ\,mol^{-1}$$

The direct interaction of SrO and MoO_2 components of $SrMoO_3$ lattice have kinetic advantage over randomly dissolved states of the binary components in a phase. Under the prevailing oxygen potential and Mo activity in the alloy phase, the MoO_3 pressure remains fixed as

$$RT \ln p_{MoO3} = -468.3 + 0.10\,T + 0.0125T \ln B \quad (kJ\,mol^{-1}),$$

and the Sr (g) pressure is obtainable from the stated redox equilibrium as

$$RT \ln (p_{Sr}) = -0948.251 + 0.2973\,T + RT \ln (a_{SrMoO3} / p_{MoO3}) \quad (kJ\,mol^{-1})$$

At 1500 K and 10 atom % burn up, the value of $p_{MoO3} = 2.6 \times 10^{-10}$ and the Sr pressure is, 1.2×10^{-8} bar for pure molybdite phase ($a_{SrMoO3} \sim 1$). The result indicates that there will be some volatilization of $SrMoO_3$ from the central part of the fuel and the transport of the vapor products will ultimately lead to the formation of $SrMoO_4$ (s) phase in the peripheral part of the fuel pin. This is in addition to the molybdate produced out of the molybdite disproportionation in the cooler region.

The Mo oxidation keeping pace with the net oxygen released out of the fission in thoria-based fuel, the oxygen rich phase formation will be limited to the molybdite, molybdate and uranate. If Mo oxidation was not fast enough, the oxygen potential would surpass the molybdenum/molybdite control and lead to the formation of a number of oxide phases, tellurites, tellurates, and molybdates of the fission products and clad materials. Summary of the available thermodynamic properties of the oxygen rich phases is included in Table14. In the conventional oxide fuels, these phases are relevant only for the analysis of failed pin. The fuel pin failure occurs when the thin clad wall ($\sim$0.5 mm thickness) develops defects providing permeation path of the coolant fluid, like water or liquid sodium to come in contact with hot fuel. The fluids with a higher oxygen potential due to thermal dissociation in case of water or less regulated oxygen in liquid sodium react with the fuel and fission products forming the oxygen rich phase.

The thoria-based fuels will develop internal stress from the formation of the bulky oxygen rich phase and from the large retention of the gaseous products that exist under highly supersaturated state as well as dispersed gas bubbles. As discussed before, the outer part of the fuel pin will be enriched with the molybdates precipitate as well as the gas bubbles migrated out from the center. The resultant stress will promote the development and propagation of peripheral cracks. Thus, as compared to the case reported for the conventional urania-based fuels, the extent of fuel swelling from the dissolved and undissolved components of the fission

Table 14 Thermodynamic data of oxygen rich compounds of the fission products

System	Method/thermodynamic property
$CdMoO_4$(s)	Solution calorimetry, Δ_fH^0 (298.15 K) = -1034.3 ± 5.7 (kJ mol^{-1}) [95]
	Knudsen effusion, Δ_fG^0 ($987 < T < 1033$) = $-1002 + 0.267$ T (kJ mol^{-1}) [96],
	Δ_fG^0 ($1044 < T < 1111$) = $-1101.9 + 0.363$ T (kJ mol^{-1}) [96]
$ThMo_2O_8$(s)	Solution calorimetry, Δ_fH^0 (298.15 K) = -2742.2 ± 4.5 (kJ mol^{-1}) [97]
	Transpiration, Δ_fG^0 ($1195 < T < 1292$) = $-2682.6 + 0.595$ T (kJ mol^{-1}) [98]
$UMoO_6$(s)	Transpiration, Δ_fG^0 ($1100 < T < 1250$) = $-1962.0 + 0.463$ T (kJ mol^{-1}) [99]
$La_2Te_3O_9$(s)	Solution calorimetry, Δ_fH^0 (298.15 K) = -2814.6 ± 12.9 (kJ mol^{-1}) [100]
$La_2Te_4O_{11}$(s)	Solution calorimetry, Δ_fH^0 (298.15 K) = -3116.5 ± 17.3 (kJ mol^{-1}) [100]
Ni_3TeO_6(s)	Transpiration, Δ_fG^0 ($1122 < T < 1202$) = $-1307 + 0.64$ T (kJ mol^{-1}) [101]
$UTeO_5$(s)	Solution calorimetry, Δ_fH^0 (298.15 K) = -1606.3 ± 3.5 (kJ mol^{-1}) [102]
	Transpiration, Δ_fG^0 ($1107 < T < 1217$)) = $-1614.2 + 0.45$ T, (kJ mol^{-1}) [103]
	Knudsen effusion, Δ_fG^0 ($1063 < T < 1155$) = $-1616 + 0.40$ T, (kJ mol^{-1}) [104]
UTe_3O_9(s)	Solution calorimetry, Δ_fH^0 (298.15 K) = -2287.2 ± 9.5(kJ mol^{-1}) [104]
	Transpiration, Δ_fG^0 (947–1011)K = $-2313.1 + 0.89$ T (kJ mol^{-1}) [103]
	Knudsen effusion, Δ_fG^0 (888–948 K) = $-2318 + 0.80$ T, (kJ mol^{-1}) [104]

products will be more in thoria. The swelling from the undissolved oxygen rich phases, A_2BO_4 (A = Cs/Rb, B = Mo, U), and A^1BO_4 (A^1 = Ba/Sr) out of the 2.8 atom percent of fission released oxygen can be worked out as given below.

4.6 Fuel Swelling by Fission Products Retention

The contribution to swelling by a solid fission product 'i' in a particular phase can be expressed considering the burn up (B in at %), molar value of the oxide fuel (V_{Th}) and fission product phase (V_i), and the yield of the fission product (Y_i). The overall swelling of the fuel with initial number of the atoms as N^0_{Th} is thus given as [105]

$$\left[\Omega_{Th}N^0_{Th}(1-B)+BN^0_{Th}\sum\Omega_i Y_i-\Omega_{Th}N^0_{Th}\right]/(\Omega_{Th}N^0_{Th}).$$
$$\equiv (\Delta\Omega/\Omega)_{solid\,FPS}=\left[\sum Y_i\,(\Omega_i/\Omega_{Th})-1\right]B$$

For zirconium the distribution into the fuel phase was worked out taking into account its total yield and its consumption in the perovskite phase formed out of the alkaline fission products. Yttrium and rare earths were considered to be dissolved in the fuel phase. The molybdenum content in the alloy phase was corrected for its partitioning into the perovskite phase. Considering the fission product yield and distribution in different phases, the swelling contribution are summarized in Table 15 at a burn up of 5 atom%. Overall swelling due to solid fission product phases works out to be 2.3 % at this burn up. One expects similar value of swelling due to solid fission products in urania fuel which is reported to have an overall swelling of (3–3.5 %) at 5 atom% burn up in PWR [1] The difference of (0.7–1.2) % is mainly the swelling contribution from gaseous fission products. The significance of surface energy in gas bubble formation is discussed explicitly in [1], where it is shown that for a given amount of gas dispersed in bubble form the extent of swelling, $(nRT/2\gamma)r$, depends on the temperature, surface energy (γ) and the radius (r) of the bubble. Once the bubble formation starts it occurs in cascade where the role of surface energy is quite insignificant. However, the role of surface

Table 15 Fractional contribution to fuel swelling by fission product phases

Fission products (i)	Phase incorporating the component (i)	Average molecular volume of the components $(Å)^3$	Swelling contribution $Y_i(\Omega_i/\Omega_{Th})$
Y, RE, Zr	Fluorite phase, $Th(U)O_2$	40.2	0.73
Ba, Sr	Perovskite phase (Ba, Sr)(Zr, Mo)O3	70.0	0.30
Mo and noble metals	Alloy phase	14.7	0.11
Cs, Rb	$(Cs, Rb)_2UO_4$	139.7	0.33

energy is significant in deciding the critical radius of nucleation. Since thoria has larger surface energy [106] as compared to urania [107], the gas bubbles will start nucleating with larger radii in case of thoria.

5 Conclusions

Thoria-based fuels though have the merit of less thermal dilation, excellent thermal transport properties, good mechanical strength, chemical inertness, and less thermal release of the gaseous and volatile fission products, there are certain demerits also. Thoria-based fuel is shown to have poor buffering ability and transport property for oxygen. This leads to subtle difference in the redistribution of the oxygen released during fission among the fission products and clad. Mo oxidation kinetics is shown to play decisive role on the potential buffering all through the burning process. Considering this, the present analysis suggests that the fuel will generally bear higher oxygen potential right from the early stage of burn up. In contrast, the oxygen potential is quite controlled in urania as fission generated oxygen can either oxidize the fuel or undergo faster transport to clad for its oxidation. Mo oxidation in urania occurs at a much later stage, when the fuel attains high O/M and is morphologically degenerated with dispersed gas bubbles that slow down the transport property. The higher oxygen potential in thoria fuel results in the formation of oxidized products like $SrMoO_3$. In the cooler part of the fuel pin, there is likely hood of the formation of more oxidized products like $SrMoO_4$ (M = Mo, U). The fuel containment problem from the clad corrosion by volatile fission products like iodine, tellurium, and cesium could be less in thoria-based fuels, but the matrix swelling will be more due to the larger retention of gaseous fission products and formation of oxygen rich phases.

References

1. Bailly H, Menessier D, Prunier C (1999) The nuclear fuel of pressurized water reactors and fast reactors-design and behavior. Intercept Ltd., Paris
2. Ali(Basu) M, Mishra M, Bharadwaj SR, Das D (2010) Thermodynamic and transport properties of thoria–urania fuel of advanced heavy water reactor. J Nucl Mater 403:204–215. doi:10.1016/j.jnucmat.2010.01.009
3. Matzke Hj (1992) Diffusion processes in nuclear fuels. In: Agarwala RP (ed) Diffusion processes in nuclear materials, Elsevier, North Holland
4. Brunn H, Hoppe R (1977) Über ordungsvarianten des NaCl-Typs. Neuekubische forman von $NaMO_2$ (M = Sc, Y, Dy, Tm, Yb, Lu), K_2MO_3, Rb_2MO_3 (M = Ce, Pr, Th) und Cs_2ThO_3. Z.Anorg. Allg Chem 430:144–154. doi:10.1002/zaac.19774300114
5. Pies W, Weiss A (1977): e256, XVI.4.1 Oxo-compounds of thorium (oxothorates), XVI.4.2 oxo-compounds of protactinium (oxoprotactinates). In: Hellwege K-H, Hellwege AM (eds). Springer materials—the 'Landolt-Börnstein database (http://www.springermaterials.com). doi:10.1007/10201569_16

6. Pies W, Weiss A (1977): f797, XIX.2.1 simple oxo-compounds of molybdenum (oxomolybdates). In: Hellwege, K.-H., Hellwege, A.M. (ed.). In: Hellwege K-H, Hellwege AM (eds). Springer materials—The Landolt-Börnstein database (http://www.springermaterials.com). doi:10.1007/10201577_18
7. Kaczorowski D (2004): Ternary actinide pnictides and chalcogenides containing s and p electron elements. In Wijn HPJ (ed) Springer materials—the Landolt-Börnstein database (http://www.springermaterials.com). DOI: 10.1007/10838073_10
8. López ML, Viega ML, Jerez A, Pico C (1992) Crystal and X-ray powder diffraction data for mixed oxides MTe_2O_6 (M = Ce, Th). Powder Diffr 7:32
9. Olander D (2009) Nuclear fuels: present and future. J Nucl Mater 389:1–22
10. Hofmann P, Spino J (1985) Conditions under which cd can cause scc failure of zircaloy tubing. J Nucl Mater 127:205–220
11. Schram RPC (2005) Analysis of the oxygen potential of $Th_{1-y}U_yO_{1+x}$. J Nucl Mater 344:223–229. doi:10.1016/j.jnucmat.2005.04.046
12. Cubicciotti D, Sanecki JE (1978) Characterization of deposits on inside surfaces of LWR cladding. J Nucl Mater 78:96–111
13. Ugajin M, Shiratori T, Shiba K (1979) Chemical form of the solid fission products in (Th, U)O_2 simulating high burnup. J Nucl Mater 84:26–38
14. Kinoshita H, Uno M, Yamanaka S (2004) Stability evaluation of fluorite structure phases in ZrO_2–MO_2 (M = Th, U, Pu, Ce) systems by thermodynamic modeling. J Alloy Comp 370:25–30
15. Holleck H, Wagner W (1968) Oxyde, Nitrures et Carbures Ternaires de U–Ce–Zr.Thermodyn Nucl Mater 1967(IAEA Vienna 1968):667–681
16. Ugajin M, Shiba K (1980) The chemical state of burnup-simulated (Th, U)O_2 at high oxygen potential. J Nucl Mater 91:227–230
17. Sibieude F, Foex M (1975) Phases et transitions de phases a haute temperature observees dan les systemes ThO_2–Ln_2O_3 (Ln = lanthanide et yttrium). J Nucl Mater 56:229–238
18. Kellar C, Berndt U, Engerer H, Leitner L (1972) Phasengleichgewichte in den systemen thoriumoxid-lanthanidenoxide. J Solid State Chem 4:453–465
19. Cordfunke EHP, Konings RJM (1990) Thermochemical data for reactor materials and fission products. Elsevier, Amsterdam
20. Mishra R, Ali (Basu) M, Bharadwaj SR, Kerkar AS, Das D, Dharwadkar SR (1999) Thermodynamic stability of barium thorate from a Knudsen effusion study. J Alloys Comp 290:97–102
21. Bharadwaj SR, Mishra R, Ali (Basu) M, Das D, Kerkar AS, Dharwadkar SR (1999) Gibbs energy of formation of barium thorate by reactive carrier gas technique. J Nucl Mater 275:201–205
22. Ali (Basu) M, Mishra R, Kerkar AS, Bharadwaj SR, Das D (2001) Gibbs energy of formation of $Ba(OH)_2$ vapor species using the transpiration technique. J Nucl Mater 289:243–246
23. Ali (Basu) M, Mishra R, Bharadwaj SR, Kerkar AS, Dharwadkar SR, Das D (2001) Thermodynamic stability of $SrThO_3$. J Nucl Mater 299:165–170
24. Prasad R, Dash S, Parida SC, Singh Z, Venugopal V (2003) Thermodynamic studies on $SrThO_3$(s). J Nucl Mater 312:1–9
25. Zharkova LA, Barancheeva NG (1964). In: Dash S, Singh Z, Prasad R, Sood DD (1993) (eds) The standard molar Gibbs free energy of formation of $BaMoO_3$(s). J Nucl Mater 207:350–352
26. Agarwal R, Singh Z, Venugopal V (1999) Calorimetric investigations of $SrMoO_3$ and $BaMoO_3$ compounds. J Alloys Compd 282:231–235
27. Shirsat AN, Kaimal KNG, Bharadwaj SR, Das D (2004) Thermodynamic stability of $SrCeO_3$. J Solid State Chem 177:2007–2013
28. Goudiakas J, Haire RG, Fuger J (1990) Thermodynamics of lanthanide and actinide perovskite type oxides IV. Molar enthalpies of formation of $MM'O_3$ (M = Ba or Sr, M' = Ce, Tb, Am). J Chem Thermodyn 22:577–587

29. Cordfunke EHP, Booij AS, Huntelaar ME (1998) Thermochemical properties of $BaCeO_3$(s) and $SrCeO_3$(s) from T = (5 to 1500 K). J Chem Thermodyn 30:437–447
30. Fava J, Flem GL, Devalette M, Rabardel L, Coutures JP, Foex M, Hagenmuller P (1971) Mise au Point D'un four de Haute Température –Application a L'étude des systems ThO_2–SrO et ThO_2–BaO. Rev Int Hautes Temper Et Refract 8:305–310
31. McIvar EJ (1966) Unit cell size of solid solutions of uranium dioxide and fission product oxides. Report AERE-M-1612
32. O'Hare PAG, Boerio J, Hoekstra HR (1976) Thermochemistry of uranium compounds VIII. Standard enthalpies of formation at 298.15 of the uranates of calcium ($CaUO_4$) and barium ($BaUO_4$). Thermodynamics of the behavior of barium in nuclear fuels. J Chem Thermodyn 8:845–855
33. Subsri R, Mallika C, Mathews T, Sastry VS, Sreedharan OM (2003) Solubility studies, thermodynamics and electrical conductivity in $Th_{1-x}Sr_xO_2$ system. J Nucl Mater 312:249–256
34. C. Keller (1976) Ternary and polynary oxides of thorium. Gmelin Handbuch der Anorganischen Chemie, Thorium, Teil C_2. Springer, Berlin
35. Kleykamp H (1985) The chemical state of the fission products in oxide fuels. J Nucl Mater 131:221–246
36. Desai KH, Grimes RW, Parfitt D, Wiss T, Uffelen PV (2009) Atomic-scale simulation of soluble fission products in UO_2. EUR Report ISSN 1018-5593, JRC EUR 23821 EN
37. Yokokawa H, Sakai N, Kawada T, Dokiya M (1991) Thermodynamic stability of perovskite and related compounds in some alkaline earth–transition metal-oxygen systems. J Solid State Chem 94:106–120
38. Ali (Basu) M, Mishra R, Kerkar AS, Bharadwaj SR, Das D, Dharwadkar SR (2000) Vaporization behavior and Gibbs energy of formation of Cs_2ThO_3. J Nucl Mater 282:261–263
39. Ali (Basu) M, Mishra R, Bharadwaj SR, Kerkar AS, Kaimal KNG, Das D (2001) Thermodynamic stability of Cs_2ZrO_3 by Knudsen effusion technique. J Alloys Comp 314:96–99
40. Ali (Basu) M, Shirsat AN, Kumar SC, Bharadwaj SR, Das D (2003) Vaporization behavior and Gibbs energy of formation of Rb_2ThO_3. J Nucl Mater 323:68–71
41. Barin I (1995) Thermochemical data of pure substance, 3rd edn. VCH, Weinheim
42. Berman R M (1976) Fission product distribution in oxide fuels.WAPD-TM-1236
43. Kleykamp H, Paschoal JOA, Pejsa R, Thümmler F (1985) Composition and structure of fission product precipitates in irradiated oxide fuels: Correlation with phase studies in the Mo–Ru–Rh–Pd and BaO–UO_2–ZrO_2–MoO_2 systems. J Nucl Mater 130:426–433
44. Cordfunke EHP, Booij AS, Smit-Groen V, van Vlaanderen P (1997) Structural and thermodynamic characterization of the perovskite—related $Ba_1 + yUO_3 + x$ and (Ba, Sr)1 + $yUO_3 + x$ phases. J Solid State Chem 131:341–349
45. Nakajima K, Arai Y, Suzuki Y, Yamawaki M (1997) Vaporization behavior of $SrPuO_3$. J Nucl Mater 248:233–237
46. Dash S, Singh Z, Parida SC, Venugopal V (2005) Thermodynamic studies on Rb_2ThO_3(s). J Alloys Comp 398:219–227
47. Cordfunke EHP, Ouweltjes W, Van Vlaanderen P (1987) The standard molar enthalpy of formation of Cs_2ZrO_3. J Chem Thermodyn 19:1117–1120
48. Lamoreaux RH, Hildenbrand DL (1984) High temperature vaporization behavior of oxides I. Alkali metal binary oxides. J Phys Chem Ref Data 13:151–173
49. Hoppe R, Seeger K (1970) Zur Kenntnis des Rb_2PbO_3 Typs: Über Rb_2ZrO_3, Rb_2SnO_3, Rb_2TbO_3. Z Anorg Allg Chem 375:264–269
50. Dash S, Sood DD, Prasad R (1996) Phase diagram and thermodynamic calculations of alkali and alkaline earth metal zirconates. J Nucl Mater 228:83–116
51. Villars P, Cenzual K; Gladyshevskii R, Shcherban O, Dubenskyy V, Kuprysyuk V, Savysyuk I (2010) Landolt-Börnstein—group III condensed matter, vol 43A8, p 519. Springer, Berlin. doi: 10.1007/978-3-540-70892-6_303, ISBN: 978-3-540-70891-9

52. Cordfunke EHP, Ouweltjes W, Prins G (1988) Standard enthalpies of formation of tellurium compounds III. Cs_2TeO_3, $Cs_2Te_2O_5$, $Cs_2Te_4O_9$, and Cs_2TeO_4. J Chem Thermodyn 20:569–573
53. Lindemer TB, Besmann TM, Johnson C E(1981) Thermodynamic review and calculations-alkali metal oxide systems with nuclear fuels, fission products and structural materials. J Nucl Mater 100:178–226
54. Shukla NK, Prasad R, Sood DD (1993) The standard molar enthalpies of formation at the temperature T = 298.15 K of barium molybdate $BaMoO_4$(cr) and strontium molybdate $SrMoO_4$(cr). J Chem Thermodyn 25:429–434
55. Feber RC(1977) Gas impurities in the primary coolant of high-temperature gas-cooled reactors. Los Alamos Sci Lab Report, LA-Nureg-6635
56. Tangri RP, Venugopal V, Bose DK, Sunderesan M (1989) Thermodynamics of vaporization of cesium molybdate. J Nucl Mater 167:127–130
57. Johnson I (1975) Mass spectrometric study of vaporization of cesium and sodium molybdates. J Phys Chem 79:722–726
58. Yamawaki M, Oka T, Yasumoto M, Sakurai H (1993) Thermodynamics of vaporization of cesium molybdate by mass spectrometry. J Nucl Mater 201:257–260
59. Semenov GA, Fokina LA, Mouldagalieva RA (1994) Mass spectrometric study of vaporization of cesium tellurate and tellurite. J Nucl Mater 210:167–171
60. Prussin SG, Olander DR, Lau WK, Hansson L (1988) Release of fission products (Xe, I, Te, Cs, Mo Tc) from polycrystalline UO_2. J Nucl Mater 154:25–37
61. Konashi K, Yamawaki M (1997) Evaluation of iodine pressure in oxide fuel pins under irradiation. J Nucl Sci Technol 29:1–10
62. Götzmann O (1983) A thermodynamic assessment of in-reactor iodine SCC of Zircaloy: Remarks concerning the article by D.R. Olander(1982) J Nucl Mater 110: 343–345, J Nucl Mater 118:349–351
63. Konashi K, Siokawa Y, Kayano H, Yamawaki M (1997) Radiation effect on the state of fission product iodine. J Nucl Mater 248:220–225
64. Chattopadhyay G, Juneja JM (1993) A thermodynamic database for tellurium-bearing systems relevant to nuclear technology. J Nucl Mater 202:10–28
65. Mishra R, Ali M, Bharadwaj SR, Das D (2003) Gibbs energy of formation of the Rh–Te intermetallic compounds Rh_3Te_2 and $RhTe_{0.9}$. J Nucl Mater 321:318–323. doi:10.1016/S0022-3115(03)00301-5
66. Ali (Basu) M, Shirsat AN, Mishra R, Kerkar AS, Kumar SC, Das D (2003) Thermodynamic stability of $RuTe_2$ solid by vapor pressure study. J Alloys Comp 352:140–142
67. Stolyarova TA, Osadchii EG (2011) Enthalpies of formation of tellurides of palladies from elements. Vestnik Otdelenia nauk o Zemle RAN 3: NZ6091, doi:10.2205/2011NZ000221
68. Johnson I, Johnson CE, Crouthamel CE, Seils CA (1973) Oxygen potential of irradiated urania–plutonia fuel pins. J Nucl Mater 48:21–34
69. Croff AG (1983) ORIGEN-2: Versatile computer code for calculating the nuclide composition and characteristics of nuclear materials. Nucl Technol 62:335–352
70. Bramman JI, Sharpe RM, Thom D, Yates G (1968) Metallic fission product inclusions in irradiated oxide fuels. J Nucl Mater 25:201–215
71. Kleykamp H (1979) The chemical state of LWR high-power rods under irradiation. J Nucl Mater 84:109–117
72. Kleykamp H (1989) Constitution and thermodynamics of Mo–Ru, Mo–Pd, Ru–Pd and Mo–Ru–Pd systems. J Nucl Mater 167:49–63
73. Octavio J, Paschoal A, Kleykamp H, Theummler F (1983) Phase equilibria in the quaternary Mo–Ru–Rh–Pd system. Z Metallk 74:652–664
74. Fukuzawa T, Tanaka M, Tanabe T, Imoto S (1984) Technology Report. Osaka University vol 34, p 219
75. Yamawaki M, Nagai Y, Kogai T, Kanno M (1980) Thermodyn Nucl Mater 1(IAEA, Vienna, 1980):249

76. Naito K, Tsuji T, Matsui T, Date A (1988) Chemical states, phases and vapor pressures of fission produced noble metals in the oxide fuels. J Nucl Mater 154:3–13
77. Matsui T, Naito K (1989) Vaporization studies on fission produced noble metal alloys by mass-spectrometric method. Thermochim Acta 139:299–312
78. Kaufman L, Bernstein H (1970) Computer calculation of phase diagrams. Academic Press, New York
79. Edwards HS, Rosenberg AF, Bittel JT (1963) Thorium oxide-diffusion of oxygen, compatibility with borides, and feasibility of coating borides by pyrohydrolysis of metal halides. ASD-TDR-63-635
80. Bayogln AS, Lorenzelli R (1984) Oxygen diffusion in FCC fluorite type nonstoichiometric nuclear oxides $MO_{2\pm x}$. Solid State Ionics 12:53–66
81. Matsui T, Naito K (1985) Chemical diffusion coefficient of oxygen in thoria-urania mixed oxide. J Nucl Mater 135:149–154
82. Mishra R, Shirsat AN, Das D(2006) Chemical diffusion of oxygen in thoria-urania solid solution. Proc Int Symp Mater Chem ISMC-06, 147–149
83. Sari C, Schumacher G (1976) Oxygen redistribution in fast reactor oxide fuel. J Nucl Mater 61:192–202
84. Carslaw HS, Jaeger JC (1947) Conduction of heat in solids. Clarendon, Oxford
85. Gulbransen EA, Andrew KF, Brassart FA (1963) Oxidation of molybdenum between 550° to 1700°C. J Electrochem Soc 110:952–959
86. Giacchetti G, Sari C (1976) Behavior of Mo in mixed oxide fuel. Nucl Technol 31:62–69
87. Barlett RW (1965) Molybdenum oxidation kinetics at high temperature. J Electrochem Soc 112:744–746
88. Wilson CN (1979) Oxygen transport limitation in using getters to control fuel/cladding chemical interaction. Thermodyn Nucl Mater IAEA Vienna 1980, 1:427–438
89. Goldberg I et al (1977) Fission gas release from ThO_2 and ThO_2–UO_2 fuels. Trans Am Nucl Soc 27:308–310
90. Paschoal JOA, Kleykamp H, Thümmler F (1987) Phase equilibria in pseudoquaternary BaO–UO_2–ZrO_2–MoO_2 system. J Nucl Mater 151:10–21
91. Ryu KH, Haile SM (1999) Chemical stability and proton conductivity of doped $BaCeO_3$-$BaZrO_3$ solid solutions. Solid State Ionics 125:355–367
92. Lutique S, Jaborsky P, Koenings RJM, Krupa JC, van Genderen ACG, van Miltenburg JC, Wastin M (2003) Low temperature heat capacity of $Nd_2Zr_2O_7$ pyrochlore. J Chem Thermodyn 35:955–965
93. Knacke O, Kubaschewski O, Hesselmann K (1991) Thermochemical properties of inorganic substances, 2nd edn. Springer, Berlin
94. Sedmidubsky D, Benes O, Konings RJM (2005) High temperature heat capacity of $Nd_2Zr_2O_7$ and $La_2Zr_2O_7$ pyrochlores. J Chem Thermodyn 37:1098–1103
95. Ali(Basu) M, Bharadwaj SR, Das D (2005) The standard molar enthalpy of formation of $CdMoO_4$. J Nucl Mater 336:110–112
96. Mishra R, Bharadwaj SR, Das D (2006) Determination of thermodynamic stability of $CdMoO_4$ by knudsen effusion vapor pressure measurement method. J Therm Anal Cal 86:547–552
97. Ali M, Bharadwaj SR, Mishra R, Kerkar AS, Das D (2000) The standard molar enthalpy of formation of $ThMo2O_8$. Thermochim Acta 346:29–32
98. Basu M, Mishra R, Bharadwaj SR, Kerkar AS, Dharawadkar SR (1998) Gibbs energy of formation of thorium molybdate ($ThMo2O_8$) by the transpiration technique. J Nucl Mater 257:185–188
99. Tripathi SN, Chattopadhyay G, Kerkar AS, Chandrasekharaiah MS (1985) Thermodynamic stability of $UMoO_6$ by the transpiration method. J Am Ceram Soc 68:232–235
100. Ali(Basu) M, Bharadwaj SR, Kumar SC, Das D (2005) Standard enthalpy of formation of $La_2Te_3O_9$ and $La_2Te_4O_{11}$. J Nucl Mater 347:69–72
101. Ali(Basu) M, Mishra R, Kerkar AS, Bharadwaj SR, Das D (2002) Gibbs energy of formation of solid Ni_3TeO_6 from transpiration studies. J Nucl Mater 301:183–186

102. Basu M, Mishra R, Bharadwaj SR, Namboodiri PN, Tripathi SN, Kerkar AS, Dharwadkar SR (1999) The standard molar enthalpies of formation of $UTeO_5$ and UTe_3O_9. J Chem Thermodyn 31:1259–1263
103. Mishra R, Namboodiri PN, Tripathi SN, Bharadwaj SR, Dharwadkar SR (1998) Vaporization behaviour and Gibbs' energy of formation of $UTeO_5$ and UTe_3O_9 by transpiration. J Nucl Mater 256:139–144
104. Krishnan A, Ramarao GA, Mudher KDS, Venugopal V (1998) Thermal stability and vapour pressure studies on UTe_3O_9(s) and $UTeO_5$(s). J Nucl Mater 254:49–54
105. Wirth BD, Olander DR (2005) Section 9: Fission product behaviour. Nuclear Engineering Department, University of California, Berkeley: NE120
106. Inoue T, Hj Matzke (1981) Temperature dependence of Hertzian indentation fracture surface energy of ThO_2. J Am Ceram Soc 64:355–360
107. Hall ROA, Mortimer NJ, Mortimer DA (1987) Surface energy measurements on UO_2—a critical review. J Nucl Mater 148:237–256

Transport Properties of Gaseous and Volatile Fission Products in Thoria-based Fuels

D. Das, M. Basu, S. Kolay and A. N. Shirsat

Abstract Transport properties of gaseous and volatiles that constitute a significant fraction of the total fission products (FPs) in high burnup matrix play important role in the performance of nuclear fuel. The transport decides release and retention behaviors of the FPs and in turn governs the matrix swelling and mechanical interaction with clad. A detailed understanding of the various factors involved in the overall transport process of the individual species inside a dense fuel matrix is necessary in predicting the performance of a fuel in high burnup. An introduction to the well-established factors governing the thermal and athermal transports of gas and volatile FPs in the conventionally used urania-based fuels in high burnup is made at the beginning of this chapter, before elaborating the case of thoria-based fuels. The experimentally measured diffusion properties of xenon, and corrosive volatile like iodine and tellurium inside thoria-based matrices, virgin as well as high burnup Simfuel, are included in the elaboration. The transport properties of the species are closely compared for the two fuels in order to understand their relative performances in high burnup situation. The knowledge of high burnup performance is particularly important for the thoria-based fuel that has limited recycling option due to the handling problems of high activity from hard gamma emitting daughters.

1 Introduction

For integrity analysis of the nuclear fuel pin, the knowledge of release characteristics of the gaseous and volatile fission products is important. Xenon and krypton gases and the volatiles such as iodine, tellurium, cesium, and rubidium generated from the fission process undergo significant redistributions inside fuel pellet by thermal and

D. Das (✉) · M. Basu · S. Kolay · A. N. Shirsat
Chemistry Division, Bhabha Atomic Research Centre, Trombay,
Mumbai 400085, India
e-mail: dasd@barc.gov.in

D. Das and S. R. Bharadwaj (eds.), *Thoria-based Nuclear Fuels*,
Green Energy and Technology, DOI: 10.1007/978-1-4471-5589-8_5,

athermal means. A PWR fuel with a power rating of $\sim$40 kW m^{-1} attains 50 GWd T^{-1} burnup in about 4 years and during this long period, there is progressive transport and redistribution of the fission products (FPs) in radial and axial directions and this results in concentration profile development of the FPs inside the fuel matrix. As a consequence of the transport process, a fraction of the fission-generated gases and volatiles get released out of the fuel pellets; volatiles mainly deposit on clad surface while the gases fill up the fuel–clad gap and plenum region. The retained and the released fractions depend on the transporting species as well as the matrix and also on a number of other parameters such as the average temperature profile, morphological history (grain size, porosity and microcracks) of the fuel, extent of burnup, and power (ramp) history. In consistence with higher melting point [1, 2] of thoria than that of urania and lower self-diffusivity of Th^{4+} than U^{4+} in the respective oxide lattices [3], the grain growth as well as impurity transport in thoria is comparatively slower. These properties lead to lesser release of the gases and volatiles in the thoria fuel. Further, in thoria and urania lattices, the differences in the energetics of vacancy migration (2.8 eV versus 2.6 eV, Table 1 of chapter "Introduction") and anion interstitial migration (3.27 eV versus 2.6 eV, Table 1 of chapter "Introduction") can result in comparatively more impediment of gaseous and volatile species diffusions in thoria. The strictly four valent Th cation will not act as favorably to reduce the barrier energy for interstitial occupancy of the electronegative elements like I and Br as in the case of uranium. In urania, the diffusions are influenced by O/M ratio; oxygen hyperstoichiometry is known to augment diffusion of gaseous species [4]. For the same power rating and thermal boundary condition, thoria with its higher thermal conductivity compared to urania will exhibit lower temperature profile and thus the release of gas/volatile is expected to be less in normal and in power ramp situations. The transport and release data are extensibly generated for urania which is widely used fuel in PWR, BWR, and PHWR. For thoria, similar data are less reported. Because of close similarity in the two actinide oxides, urania and thoria, the rich experiences in release behavior known for the former provide guideline for the behavior of the latter. In the subsequent sections, the observational data of urania fuel will be referred so as to serve as the nearest measure to understand the thoria fuel's behavior toward gas/volatile release.

2 A Survey on Release Behaviors of Gases and Volatile Fission Products in Oxide Fuels

Xenon and krypton accounting about 11–12 atom% of the FPs in PWR [5] (Table 1 of chapter "Introduction") have negligible solubilities in the oxide fuel and thus mostly remain dispersed as microbubbles within the grains and as growing hemispherical to lenticular bubbles on grain surface, and as interconnected bubbles along the grain boundaries of fuel matrix (Fig. 1). Nano-sized intergranular bubbles nucleating in fission spike events attain their steady number density through

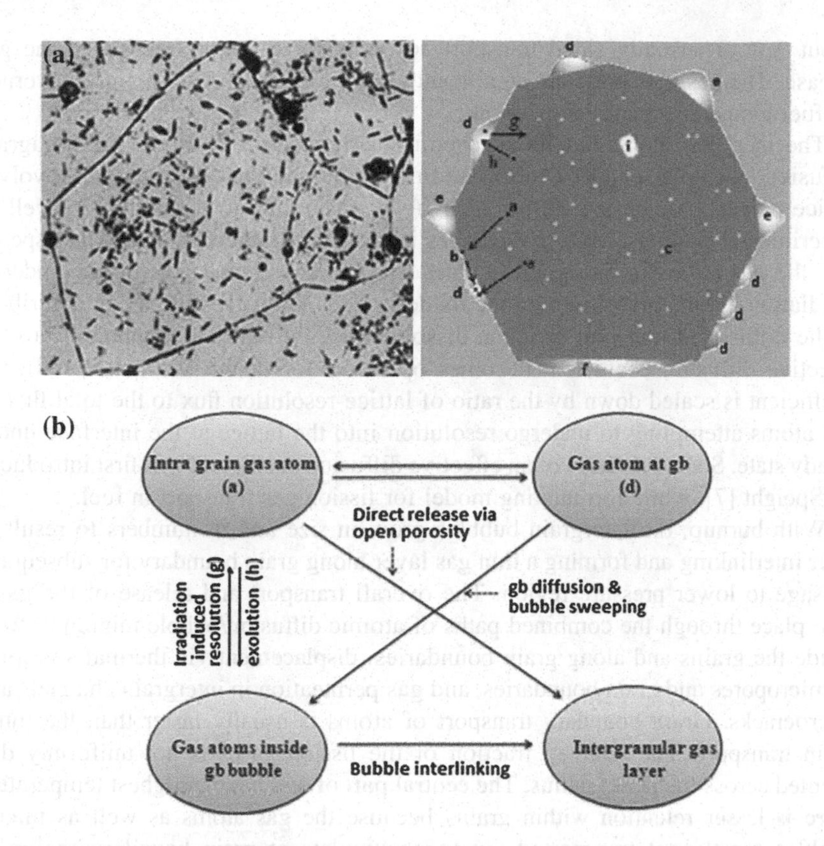

Fig. 1 (**a**) Intragrain gas atoms/microbubbles, and grain boundary (gb) bubbles in burnt fuel—a typical sectional micrograph of burnt fuel (*left*) and a modeled representation of a grain (*right*) [(At *right side*) Diffusing atom inside grain (*a*) and along gb (*b*); gas bubble inside grain (*c*) and at gb and grain edge (*d* and *e*); collapsing gas bubbles at gb (*f*); grain resolution of atoms (*g*) and exolution (*h*); and as fabricated pores (*i*)]. (**b**) Fission gas transport processes

continuous nucleation and resolution processes at a given power rating. The microdispersed bubbles then grow through addition of gas atoms from fission events and also from radioactive decay of parent nuclei originating from the FPs.

The gas atoms generated in the polycrystalline matrix undergo intragranular diffusion and in that process, a major fraction of them come in contact and dissolve into the pressurized microbubbles, while a small fraction always diffuses out and then undergoes grain boundary transport until they settle down as gas bubbles in voids and microcracks. In addition to the diffusional transport, irradiation-induced athermal processes contribute to the ejection of gas atoms out of the grain. The athermal transports, recoil, and knockout processes together account for less than 1 % of released fission gases [6] for fuel burnup up to 45 GWd T^{-1}. On further increase of the burnup to 60 GWd T^{-1}, the athermal release figure accelerates to about 3 %. Thermally aided transport accounts for a major fraction of the gas release. The gas transports are significant above 1300 K and are mainly governed by fuel temperature and its gradient.

The inert gas atoms having no chemical affinity to the matrix, the intragrain diffusion takes place under concentration gradient and the diffusion path involves lattice defects. Xenon gas diffusion involves, according to theoretical as well as experimental findings, MO_2 trivacancies in the oxide lattice [4]. Generally speaking, the gas atoms in intragrain bubbles may remain trapped therein or undergo irradiation-induced resolution in the oxide lattice. A resolved atom can contribute to the diffusional transport or it can dissolve back into the microbubble. Thus, an effective diffusion coefficient becomes operative for them. The lattice diffusion coefficient is scaled down by the ratio of lattice-resolution flux to the total flux of gas atoms attempting to undergo resolution into the lattice at the interface under steady state. Such definition of an effective diffusion coefficient was first introduced by Speight [7], while formulating model for fission gas transport in fuel.

With burnup, the intergrain bubbles grow in size and in numbers to result in their interlinking and forming a thin gas layer along grain boundary for subsequent passage to lower pressure region. The overall transport and release of the gases take place through the combined paths of atomic diffusion/bubble migration from inside the grains and along grain boundaries, displacement via thermal sweeping of micropores and grain boundaries, and gas permeation in intergrain channels and microcracks. Grain boundary transport of atoms is usually faster than the intragrain transport. The retained fraction of the fission gases is not uniformly distributed across the pellet radius. The central part of fuel having highest temperature there is lesser retention within grains because the gas atoms as well as microbubbles mostly get transported out to accumulate at grain boundary region as intergranular gas bubbles and as clusters in surface defects. Larger grain growth in the central region (Fig. 2) [8] of the fuel pellet sweeps away the intergranular gas leading to their release. High-resolution SEM shows the development of interconnected network of gas bubbles on grain surface and along grain boundaries (Fig. 3) [9]. Faster transports through thermal diffusion as well as grain sweeping from the hottest central part results in Xe depletion. The depletion process is facilitated in presence of power ramp. Post irradiation examinations (EPMA, XRF, SIMS) of moderately high burnup fuel (average burnup $\sim$40–50 GWd T^{-1}) show that the intragrain concentration of Xe in the central part is very low (Fig. 4) [10]. The extent of central dip in Xe profile is of course subject to the central temperature that is governed by the average value of linear power rating of the fuel pin. Both the Xe profiles and grain morphologies are seen to be distinctly different in fuel center and fuel periphery.

At temperature above 1900 K, the thermal gradient driven transport becomes prevalent. It is generally noted that the microdispersed bubbles that remain immobile under normal operational temperature of the oxide lattice, start mobilizing at very high temperatures (>2200 K) [11]. Therefore, under the power ramp when there is rapid rise in fuel enthalpy value, there is an abrupt increase in the overall transport and release of the fission gases and volatiles. Reported study with

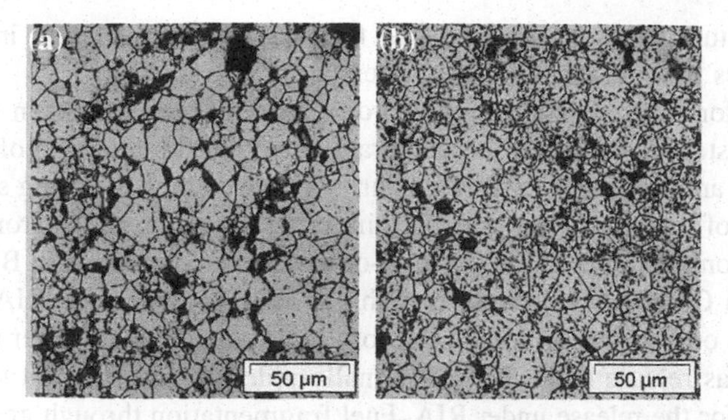

Fig. 2 Grain sizes at (**a**) central vis-a-vis (**b**) off-central zones of fuel pellet (65 GWd T^{-1}) ([8] copyright Elsevier)

Fig. 3 Grain morphology (at higher magnification) of the central zone (50 GWd T^{-1}) ([9] copyright Elsevier)

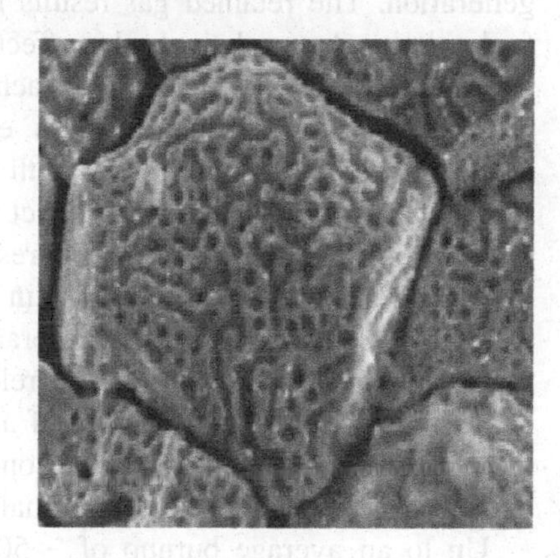

Fig. 4 Xe (EPMA) profile in PWR fuel (~50 GWd T^{-1}) ([10] copyright Elsevier)

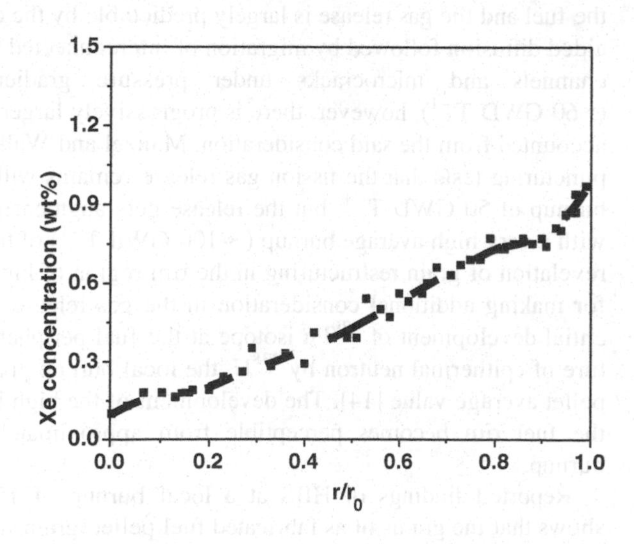

low temperature disk fuel showed that there is no augmentation in fission gas release unless the fuel pin experiences temperature transient.

Other factors like development and propagation of microcracks in fuel pellet in the event of steep rise and fall of temperatures add to the quantum of gas release. Microcracks are preferentially formed during power ramps or during start-up/shut-down times of the reactor. In reactivity-initiated accidents (RIA) from accidental ejection of control rod in PWR, or, drop-down of the control rod in BWR, or, loss of coolant in CANDU [12], there is high gas release invariably. RIA gas release also depends on previous irradiation history; a rod operated at higher power rating has higher gas release under RIA. Thermally aided transport is not the dominant mechanism for the release under RIA. Fuel fragmentation through grain boundary decohesion facilitates the prompt release.

The retention as well as release of the gases has deleterious effects to the power generation. The retained gas results in internal pressure development inside the fuel pellet and cumulatively this effect leads to swelling of fuel and its mechanical interaction with clad that can breach under the severity of stress corrosion by reactive volatiles (iodine, tellurium, etc.) and hydride precipitation. The released gases, on the other hand, mix up with the pressurized helium normally filled in the fuel–clad gap and plenum to conduct away the fission heat. The thermal transport property of the He filled layer progressively deteriorates with increased release of the fission gases. The resultant rise in temperature profile of the fuel pin augments the gas release and in turn deteriorates the gap conductivity further. The detrimental effects from retention and release of the fission gases pose limitation in achieving high discharge burnup of a fuel. In order to understand this limitation, the transport behavior of the fission gases at high burnup is important in the performance evaluation under normal and power ramp situations.

Up to an average burnup of $\sim$50 GWd T^{-1} the grain structure, barring the grain growth in the hottest central part, nearly represents the as fabricated state of the fuel and the gas release is largely predictable by the consideration of thermally aided diffusion followed by migration of interconnected bubbles in grain boundary channels and microcracks under pressure gradient. At higher burnups (>60 GWD T^{-1}), however, there is progressively larger gas release than could be accounted from the said consideration. Manzel and Walker reported [13] from rod puncturing tests that the fission gas release remains within 5–10 % at an average burnup of 50 GWD T^{-1}, but the release gets augmented to almost 25 % for fuel with a very high average burnup ($\sim$100 GWd T^{-1}) of the fuel pellet (Fig. 5). The revelation of grain restructuring in the rim region of high burnup pellet has called for making additional consideration in the gas release analysis. With the preferential development of ^{239}Pu isotope at the fuel periphery through resonance capture of epithermal neutron by ^{238}U, the local burnup grows to about 2.5 times the pellet average value [14]. The development of the high burnup structure (HBS) of the fuel rim becomes perceptible from approximately $\sim$150 GWd T^{-1} local burnup.

Reported findings of HBS at a local burnup of 150–300 GWd T^{-1} [8, 15] shows that the grains of as fabricated fuel pellet (grain size 5–10 μm) fragments to

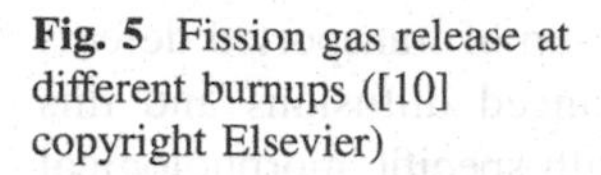

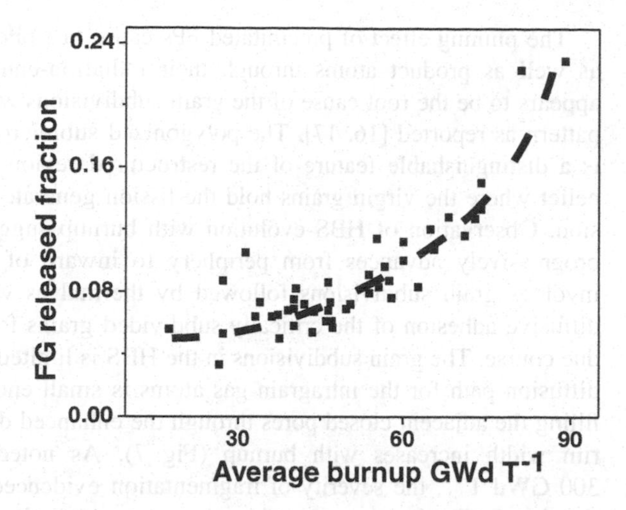

Fig. 5 Fission gas release at different burnups ([10] copyright Elsevier)

the submicron sizes (0.1–0.5 μm) as represented in Fig. 6. The cause of grain subdivisions in the high burnup rim of the fuel is not fully understood. The multiple subdivisions of the as fabricated grains are possibly triggered by localized stress from increased precipitations of FPs (gas/solid) in the lattice which is also accommodating high defects and dislocation network from radiation damage under the augmented fission events.

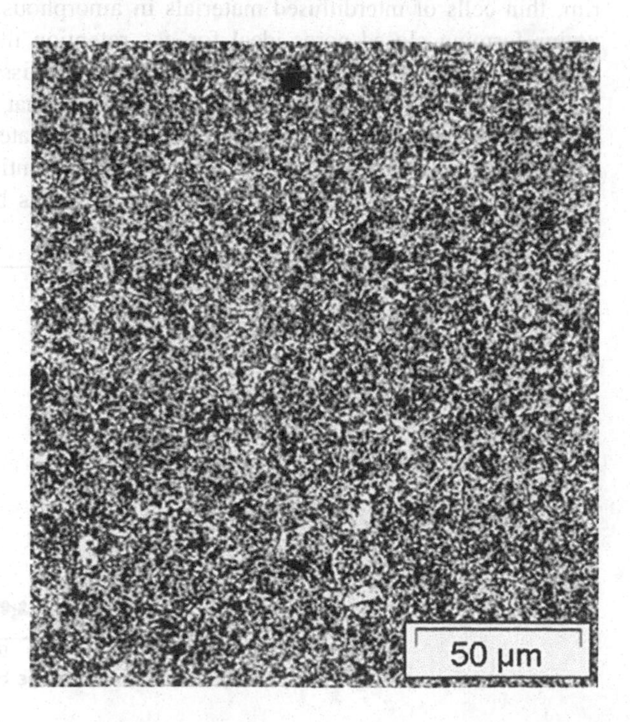

Fig. 6 Micrographic picture of restructured rim ([8] copyright Elsevier)

The pinning effect of precipitated FPs cause fast pile-up of transported defects as well as product atoms through their radiation-enhanced diffusions and this appears to be the root cause of the grain subdivisions with specific morphological pattern as reported [16, 17]. The polygonized submicron-sized grain morphology is a distinguishable feature of the restructured region as against the rest of the pellet where the virgin grains hold the fission generated gas without the subdivision. Observation of HBS evolution with burnup suggests that the restructuring progressively advances from periphery to inward of the rim and the process involves grain subdivisions followed by thermal as well as radiation-enhanced diffusive adhesion of the critically subdivided grains forming closed pores in the due course. The grain subdivisions in the HBS is limited to a critical size when the diffusion path for the intragrain gas atoms is small enough to be transported out filling the adjacent closed pores through the enhanced diffusions. The restructured rim width increases with burnup (Fig. 7). As noted at the local burnup of 300 GWd T^{-1}, the severity of fragmentation evidenced within ~150 μm of the rim gradually lessens inward and makes transition to the virgin structure approximately at a depth of 1000 μm [13, 18]. Study shows that restructuring requires lower temperature and the limit for the HBS formation is set below 1373 ± 100 K [19].

The Xe concentration in grains steeply falls once the burnup crosses the critical limit, which is noted to be within 60–75 GWd T^{-1}, the steep fall marking the onset of grain restructuring process. The indicated scatter in attaining the burnup limit (Fig. 8) may be connected with irradiation history. In the fully restructured rim, thin cells of interdiffused materials in amorphous state hold the subdivided grains forming closed pores ideal for gas retention in the HBS. The uniformly dispersed pores (2–5 μm diameters) are filled with fission gases at high pressure (>30 MPa at 300 K). SIMS analysis confirms that the highly porous HBS (~10–20 % porosity) holds most of the locally generated gases in the closed pores while its submicron-sized grains have poor gas retention (~0.2 wt%) [20]. The poor gas retentivity inside grains at steady state has been confirmed by EPMA

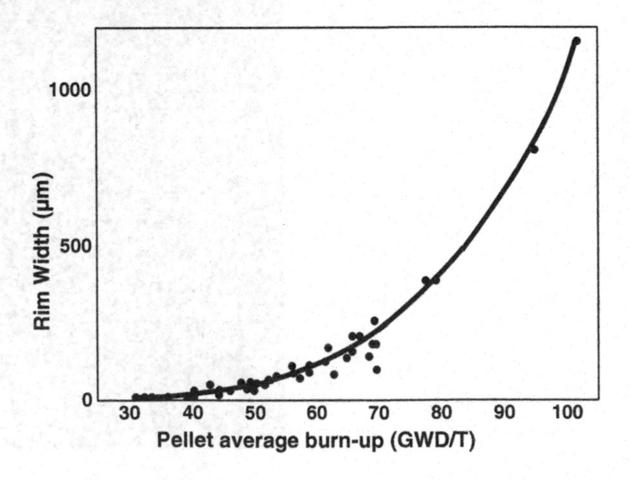

Fig. 7 Rim width increase with burnup ([13] copyright Elsevier)

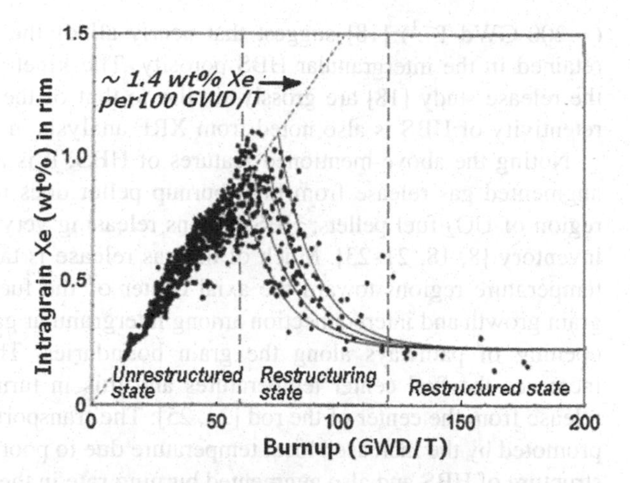

Fig. 8 Intragrain Xe concentration evolution at rim region of LWR pellet ([15] copyright Elsevier)

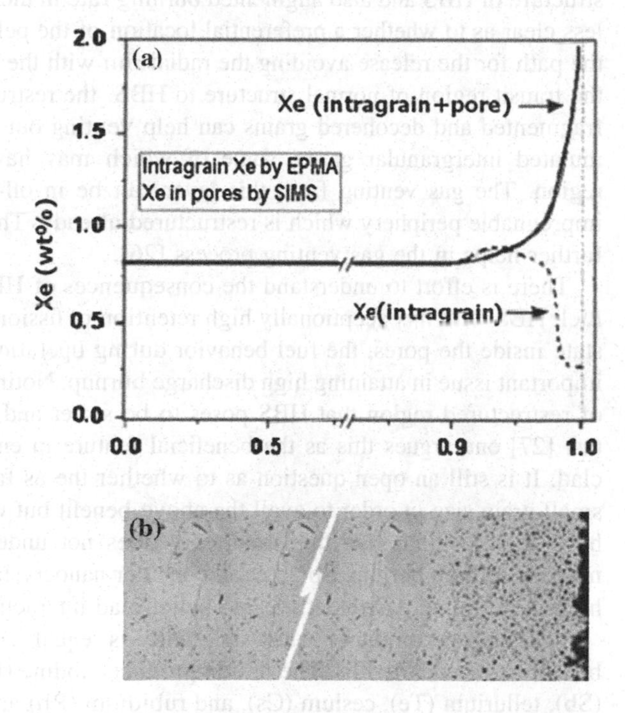

Fig. 9 Radial Xe-profiles of high burn-up fuel pin (a); micrograph sectional view (b)

analysis of high burnup fuels. The Xe profile across the restructured rim is typically shown in Fig. 9.

The undercritical gas retention (0.2 w/o) closely corroborates to dynamic solubility for the highly pressurized gas in HBS. In HBS, the confined gas pressure remains nearly flat except a little tapering in the outermost layer of the rim resulting from recoil and knockout release processes. The reported data on microstructural pores [20] and Xe/Kr thermal release of the rim-samples

($\sim$200 GWd T^{-1}) [18] suggest that nearly all of the locally produced gases is retained in the intergranular HBS porosity. The kinetic parameters derived from the release study [18] are grossly similar to that of the bulk phase. The high gas retentivity of HBS is also noted from XRF analysis of total Xe in HBS.

Noting the above-mentioned features of HBS, it is generally believed that the augmented gas release from high burnup pellet does not originate from the rim region of UO_2 fuel pellets; HBS regions releasing very little ($\sim$2 %) of the total inventory [8, 18, 21–23]. Much of the gas release is taken to be from the higher temperature regions toward the axial center of the fuel pellet, where significant grain growth and interconnection among intergranular gas bubbles can result in the opening of pathways along the grain boundaries. The high porosity in HBS increases the fuel center temperatures and this in turn enhances the fission gas release from the center of the rod [24, 25]. The transport kinetics are, respectively, promoted by the increased local temperature due to poor heat transfer in the porous structure of HBS and also augmented burning rate in the rim region. However, it is less clear as to whether a preferential location of the pellet's radial position clears the path for the release avoiding the radial rim with the impregnable HBS [25]. In the transit region of normal structure to HBS, the restructuring layer with freshly fragmented and decohered grains can help venting out much of the locally accumulated intergranular gases, some of which may have originated from hotter region. The gas venting from this layer can be in off-radial paths to evade the impregnable periphery which is restructured already. The presence of power ramp further helps in the gas venting process [26].

There is effort to understand the consequences of HBS in safe performance of fuel. HBS with its exceptionally high retention of fission gases in over pressurized state inside the pores, the fuel behavior during operational power transients is an important issue in attaining high discharge burnup. Noting, however, the properties of restructured region that HBS poses to be softer and tougher in microhardness test [27] one argues this as the beneficial feature in ensuring stress tolerance on clad. It is still an open question as to whether the as fabricated fuel should have small grain size in order to avail the above benefit but with a high gas release, or, have a large grain size that reportedly does not undergo HBS restructuring at medium to high burnup. Spino et al. consider nanocrystalline UO_2 as a potentially high performing, gas retentive, and pellet-clad-interaction resistant fuel [27].

For fuel performance analysis, there is equal concern about the release behaviors of volatile FPs. The fission products, iodine (I), bromine (Br), antimony (Sb), tellurium (Te), cesium (Cs), and rubidium (Rb) and also, cadmium (Cd), tin (Sn), and silver (Ag) have significant volatilities in free as well as chemically bonded states and are not chemically inert as the noble gases. They have nonzero affinity to the fuel matrix and to the other FPs. Barring Cs, the fission yields of the volatile products are low but they are of concern in fuel containment because of their significant mobilities to redistribute themselves inside the matrix reaching the fuel–clad interface. They interact with the clad and lead to stress corrosion cracking and fatigue failures particularly under pellet clad mechanical interaction. Though their transports take place through the combined paths as described for the gaseous FPs, the driving forces for their diffusions are influenced by chemical affinities with matrix and FPs. Most of the fission-generated halogens remain as alkali halides (MX, M = Cs/Rb, X = I/Br). The elemental iodine and iodides of the alkali metals, Cs and Rb, having very little solubilities in the oxide phase, remain dispersed in the lattice, grain boundaries, voids, and microcracks. Tellurium behaves in same way as iodine and chemically it remains partitioned in alkali tellurides and alloys. A significant fraction of Te and Sb remain dissolved in the alloy phases of the FPs Pd, Ru, Rh, Tc, and Mo. These FPs being incompatible with the oxide get concentrated at the grain boundaries as metal/alloy particles. Like the case of oxygen as described in the previous chapter (Sect. 3.3), the transport equation will involve the thermodynamic factor arising from chemical affinity of the species with the matrix and such equation can be set up by considering entropy increase in the transport under the concentration and temperature gradients. The volatile components generally remain distributed in the alloy and in vapor phases inside dispersed bubbles and accumulated gas layers at intergrain spaces. The transport and redistribution processes are predominant at the hottest central part of the fuel. Bubbles interlinking at intersection of grain boundaries eventually leads to venting of the volatile components to the fuel–clad gap. Like the case of fission gases, the thermal transport accounts for the major fraction of release of the volatiles. Athermal transports of the species will also be present and their contributions will be important only at the low temperature rim.

Similar to the fission gases, the volatiles will have burst release during reactivity-initiated high enthalpy deposition in fuel and augmented release in high burnup restructuring process at the rim.

3 Brief Descriptions of the Different Transport Processes

In the brief survey of different transport processes for the release of fission gases and volatiles from fuel pellet, it is seen that there is involvement of thermally assisted intragrain diffusions, grain boundary diffusions, radiation-enhanced diffusions, displacements through sweeping of grain boundaries and micropores, and migration of gases/vapors under pressure gradient along intergrain channels and propagating cracks. Additionally, there are the athermal routes of transports by fission recoil and knockout events. All these transport processes leading to the release of species take place in combination during irradiation. The fission recoil and knockout processes are primarily governed by the kinetic energies of the FPs and they result in stepwise displacements (recoil length <10 μm) of fission fragments. They play dominant role in the gas release at low temperatures from the peripheral region of fuel pin. Thermal diffusions, on the other hand, take place under steep temperature and concentration gradients, leading to sustained advancements of the species toward the boundary for their ultimate release from the matrix. The transports through sweeping of micropores and propagation of microcracks are promoted by thermal and thermomechanical paths, respectively.

Under the usual operating temperatures of fuel, the primary mechanism of transports and releases of fission gases and volatiles is the thermally assisted diffusion. The thermal and athermal transport mechanisms are briefly described for making better comparison of contributing kinetics in the overall release of gas/volatiles from urania and thoria matrices.

3.1 Thermally Assisted Transports

The diffusion of a species in lattice are thermally assisted under concentration/activity and temperature gradients. With the help of thermal energy, the species makes net displacement in a direction that results in overall increase of entropy. Noting that the chemical potential of the species (μ_i) corroborates to the partial molar entropy increase in isolated system as $(\partial S/\partial n_i)_{U,V} = -\mu_i/T$ the spontaneity in displacement in a predefined field of temperature and concentration can be expressed by the drift velocity as the product of mobility (D_i/RT) and the force derived out of the potential gradient $-T\bar{\nabla}(\mu_i/T) \equiv -[(\partial\mu_i/\partial c_i)\bar{\nabla}c_i + T\bar{\nabla}T\partial(\mu_i/T)/\partial T]$. Since μ_i is used in expressing net displacement of species, it is defined with respect to ideal solution state where the gradient, $[\partial({\mu^0}_i/T)/\partial T] \equiv -h_i^0/T^2$ (h_i^0 being molar enthalpy), is concentration independent and is inherently contributing solely to heat flux part of the spontaneity. The flux is rewritten as $\bar{J}_i = -c_i(D_i/RT)\,[RT\,F_{\text{thermod}}\,\bar{\nabla}\ln c_i - h_i^X\bar{\nabla}(\ln T)]$ [28] where c_i is the spatial concentration of the diffusing species i having diffusion coefficient D_i and relative (excess) molar heat content h_i^X in the solution phase with the matrix, and F_{thermod} is the thermodynamic factor, $F_{\text{thermod}} = \{1 + \partial(\ln\gamma_i)/\partial(\ln c_i)\}$. Fick's diffusional contribution and Soret's supplement in the transport as expressed by the respective terms constructively add for the exothermic solution energy $\left(h_i^X \equiv -q,\ q > 0\right)$ when both the gradients are in same direction. With ideal state of solution in general and Henrian ideality in infinitely diluted state in particular, the quantity F_{thermod} attains unity.

3.1.1 Atomic Diffusion

The macroscopic diffusion coefficient is connected with microscopic jump distance (λ_i) and overall jump frequency (Γ_i) by the Einstein formula, $D_i = \lambda_i^2\Gamma_i/6$. For cubic lattices, one connects the jump distance with the unit cell dimension (a) as $\lambda_i = a$A, A being the characteristic parameter of crystal structure and diffusion mechanism. Γ_i is proportional to the number of nearest neighbor sites (z), the probability (P_s) that a site is in proper state for the occupation of jumping species and the transmission rate (τ_i) of the species to such a site via the path of lowest barrier energy. P_s is evaluated by considering the free energy of formation of the proper site $\left(\Delta G_s^{\#}\right)$ as $P_s = \text{Exp}[-\Delta G_s^{\#}/RT]$ and therefore it depends on the diffusion mechanism as to whether the z-site atom migrates through vacancy, interstitial, or their combination. In $\Gamma_i = \tau_i z P_s$, the transmission rate τ_i is made of the two factors, the formation probability $\left(P_i^{\#}\right)$ of activated complex representing the barrier state and the transmission frequency ν_i. $\tau_i = \nu_i P_i^{\#} \equiv \nu_i \mathrm{Exp}[-\Delta G_i^{\#}/RT]$, $\Delta G_i^{\#}$ being the free energy of formation of the activated complex. Einstein relation thus takes the form $D_i = (\lambda_i^2 \nu_i z/6)\mathrm{Exp}[-(\Delta G_s^{\#} + \Delta G_i^{\#})/RT]$. Replacing the free energy terms in this relation by enthalpy and entropy for the respective cases, the resulting expression takes the Arrhenius form $D_i = D_{i0}\mathrm{Exp}[-Q_i/RT]$ where $Q_i \equiv \Delta H_s^{\#} + \Delta H_i^{\#}$, and $D_{i0} \equiv (\lambda_i^2 \nu_i z/6)\mathrm{Exp}[(\Delta S_s^{\#} + \Delta S_i^{\#})/R]$. The two isobaric enthalpies together account for the barrier energy, Q_i. The diffusion through the lattice can take place through vacancy/interstitial/combinatorial (interstitialcy) mechanisms, and/or through dislocations and voids. Dislocations are increased in fuel matrix under the irradiation and increased dislocations augment transport of the diffusing species. The mean diffusion coefficient for transports of species in the presence of dislocations is generally expressed in the linear form as $D_i = f D_{i,\mathrm{disloc}} + (1-f)D_{i,\mathrm{v}}$, where $D_{i,\mathrm{disloc}}$ and $D_{i,\mathrm{v}}$ are, respectively, the coefficients of transport through dislocations and point defects (vacancy/interstitial). The factor, f represents the dislocation fraction.

3.1.2 Radiation-Induced Diffusion Augmentation

The diffusion in lattice can get augmented by creation of defects through fast energy deposition during stoppage of energetic particles. The fission fragments after initial slowing down through electronic excitations of lattice atoms form spikes/tracks before coming to halt. Out of about 15,000 of Frenkel pairs instantaneously formed along several micrometers long spike, two-third of the pairs undergo fast annihilation and the energy dissipation results in the surge of local temperature along the spike. Across the most disturbed width of several nanometers over the spike there is utter disorder in the lattice: there is separation of vacancies from interstitials, mixing of atoms, and resolution of gas atoms from microbubbles. Due to the formation of a large number of irradiation-induced vacancies, there is an enhancement in diffusional transport besides the contribution from a pure athermal component. Total vacancy concentration (C_{v}^0) made of the two components, $C_{\mathrm{v}}^0 = C_{\mathrm{v}}^{\mathrm{radiation}} + C_{\mathrm{v}}^{\mathrm{Thermal}}$, the radiation-induced vacancy concentration has been shown to be proportional to the square root of displacement rate of lattice atoms and therefore to $\sqrt{\dot{f}_d}$, $\dot{f}_d$ being the fission rate density (fission m^{-3} s^{-1}) [29]. The thermal diffusion coefficient is thereby augmented to $D_i^0 = D_i^{\mathrm{Thermal}} + D_i^{\mathrm{Enhance}}$, $D_i^{\mathrm{Enhance}} = C_i\sqrt{\dot{f}_d}$, where $C_i(T)$ is in the form of Arrhenius function similar to D_i^{Thermal}; $C_i(T)$ depends on frequency of vacancy jump, atomic jump distance, and vacancy and interstitial parameters. The pure athermal part of radiation-induced transport is proportional to the fission density $D_i^{\text{athermal}} = C'_i \dot{f}_d$, C' being a constant. Wallin and Turnbull [30] from their study on gas release from single as well as polycrystalline UO_2 observed the athermal release of Xe at lower temperatures where the thermal release is negligible. Their study further noted that in the temperature range of 1273–1673 K, where thermally activated diffusion dominates, the diffusion parameters are indeed influenced by fission dose received by the oxide fuel. They found out the thermal and radiation-enhanced parts of the transport coefficient and these are, respectively, expressed as $D_{\text{Xe}}^{\text{Thermal}}(\text{m}^2\,\text{s}^{-1}) = 7.6 \times 10^{-10}\text{Exp}[-35000/T]$, $D_{\text{Xe}}^{\text{Enhance}}(\text{m}^2\,\text{s}^{-1}) = 5.78 \times 10^{-25} \sqrt{\dot{f}_d}\text{Exp}[-13800/T]$. The pure athermal part according to their analysis for UO_2 case is $D_i^{\text{athermal}}(\text{m}^2\,\text{s}^{-1}) = 2 \times 10^{-40}\dot{f}_d$ [31]. The athermal part of Xe transport contributes most effectively in the outer rim of the fuel pellet, where the temperature is lowest and Xe release due to this transport mechanism will compete with recoil and knockout removal of Xe from the rim.

3.1.3 Grain Boundary Diffusion

In polycrystalline materials, the diffusional transport in macroscopic scale occurs through combined paths involving diffusion inside grain and along grain boundary (gb). Because of more freedom on grain surface, the gb diffusion barrier $(Q_{i,\text{gb}})$ is lower than the intragrain diffusion barrier $(Q_{i,\text{gr}})$ and the atomic jump frequency on gb surface is orders of magnitude higher as compared to that in intragranular region. Thus, the gb diffusion in crystalline solid takes place much more rapidly than intragrain diffusion and it dominates at lower temperatures. The diffusion taking place mostly through vacancy mechanism, the jump frequency grows with temperature and the frequency becomes several orders of magnitude ($\sim 10^6$) higher than that in the intragrain case at 2/3rd melting temperature of the matrix resulting therein faster grain growth. The gb diffusion is believed to assist the transports of oxygen, and plutonium in the oxide fuels. There is, however, impeding factor to fast diffusion on grain surface. Uncompensated chemical activity and uneven electrostatic potential of ionic solid like urania/thoria make their grain surface vulnerable to hold gas bubbles and other phase precipitates, particularly at structural defects, dislocations, and pores. The surface diffusion slows down on encountering the heterogeneous phases. Chemical affinity to the precipitates results in further impediment in the transport process. For example, the metallic precipitate on gb can act on the tellurium diffusion, iodine diffusion is indirectly influenced by Cs potential over cesium uranate/molybdate phases, and diffusing Xe can get trapped in gas bubbles. Many models express the effective gas diffusion in the presence of gas capture and gas atom resolution kinetics taking place across interface of a trap. With the attainment of steady state in the rates of the two kinetic processes, the effective diffusion coefficient is expressed as $D_{\text{gas}}^{\text{eff}} = D_{\text{gas}}^{\text{singleatom}}[b/(b+g)]$, where b and g are resolution and capture rates, respectively [6, 7]. Trapped inert gas encounters an energy barrier of approximately 0.5–0.8 eV [32]. This is reflected in the gas diffusion behavior in the presence of trapping sites in fuel lattice; the activation energy of diffusion gets augmented by the indicated barrier value.

Olander and Uffelen evaluated the effectiveness of gb diffusion in the presence of surface traps [33]. The expected fastness in gb diffusion is noted in trace-irradiated fuel wherein fission gas bubbles/precipitates and fission dose-induced defects are nearly absent [34]. In high burnup fuel, the gb transport is less effective due to the trapping phenomenon and the gas transport is predominantly via the growth and interconnection of the gb surface bubbles and gas venting through intergrain channels. Considering the shrinkage and swelling kinetics of UO_2 fuel in in-pile experiments, White [9] has made comprehensive analysis of gb morphological relaxation and intergranular bubble growth and derived an estimate of the gb diffusion coefficient. Within observational uncertainties, the derived transport coefficients in the two kinetic phenomena follow similar temperature trend to be given by $\ln(D_{gb}\delta_{gb}/\mathrm{m^3\,s^{-1}}) = -(39196 \pm 2600)/T - (34.8 \pm 1.5)$. The indicated standard deviations of slope and intercept reflect the difference in temperature trends of the two kinetics. The derived diffusion enthalpy of 326 kJ mol^{-1} approximately corroborates to the Arrhenius slopes noted in sintering and creep behaviors of the oxide matrix to be represented, respectively, as $\ln(D_{gb}^{sinter}\delta_{gb}/\mathrm{m^3\,s^{-1}}) = -40600/T - 29.3$ and $\ln(D_{gb}^{creep}\delta_{gb}/\mathrm{m^3\,s^{-1}}) = -29960/T - 34.7$ [35]. Discrepancy is, however, evident in absolute value of the transport coefficient obtained by the different approaches.

3.1.4 Grain Boundary Sweeping

At high temperature, grains grow in size at the expense of adjointly placed smaller size grains. The process of swallowing smaller grains is because of their higher surface free energies compared to the growing grain. The surface energy driven material transport from smaller to larger grain is accomplished through grain boundary sweeping. The sweeping boundary can transport some fractions of dissolved solutes as well as microbubbles/aggregates, leaving out their major fractions. As the smaller grain starts disappearing, its nonsoluble contents like the gas bubbles/precipitates are mostly left out in the adjoining intergranular space, enabling these entities to be more mobile. As against this, the gases present inside the growing grain has to travel a larger distance to diffuse out to the intergrain space. Overall, the gb sweeping is thermally activated process that can facilitate the transport of gases and volatiles. This transport mechanism operates in equiaxed grain growth as observed in the hotter central part of the PWR fuel (1900–2100 K) and in the columnar grain growth noted at still higher temperatures of fast reactor fuels [36, 37]. The grain boundary sweeping are retarded by insoluble phases like bubbles and FP precipitates as these react while coming out from the shrinking grains.

Grain growth rate $\mathrm{d}R_{gr}/\mathrm{d}t$ is a measure of boundary sweeping. For the growth of an isolated grain, the growth velocity is the product of mobility m_{gb} and the inward pull by virtue of the surface force proportional to γ_{gb}/R_{gr}. The temperature dependence is contained in m_{gb}, which is expressible in terms of gb diffusion coefficient (D_{gb}) using Stokes-Einstein relation, $m_{gb} = D_{gb}/kT$. This consideration leads to the parabolic kinetics of the growth, that is, $dR_{gr}^2/dt = \text{constant}$. The growth rate for oxide fuel like urania deviates from the parabolic form because the grain size exponent differs from being 2. Observationally, it is $dR_{gr}^n/dt = \text{constant}, 2<n<3$ [38]. The parabolic form is arrived from oversimplistic assumptions that the grain is growing in isolation with a constant mobility, the surface force stems from a unique curvature (isotropic) and that there is no drag from bubbles/precipitates. In the presence of precipitated particle, the parabolic expression for the grain growth rate is generally modified with a constant drag (K') as $dR_g/dt = K/R_{gr} - K' \equiv K[1/R_{gr} - 1/R_{max}]$ [39]. The drag's presence has set an upper limit to the average radius of grain to $R_{max} \equiv K/K'$. R_{max} was implied to be infinity in the consideration of isolated grain. Hillert [40] has considered a more gradual growth and deduced the retarding growth as $dR_g/dt = (K''/R_{gr})\,[1 - R_{gr}/R_{max}]^2$ These formulations, however, do not reflect the basic feature of the grain growth that takes place at the expense of smaller grains. The force manifesting from curvature difference between the grain and its adjoining smaller one with radius below a critical value (R_{cr}) drives the growth rate. Thus, Hillert has expressed the growth rate and hence the sweeping speed of gb surface as $dR_{gr}/dt = M'(1/R_{cr} - 1/R_{gr})$. There, he considers that M' is made of mobility and surface energy of the grain boundary. The driving and retarding forces respectively defined by the relative curvature values with respect to the limiting radii R_{cr} and R_{max} are to be considered together in a formulation.

3.1.5 Bubbles Migration

Besides the transport of tiny bubbles and microdispersed precipitates via grain boundary sweeping, there can be direct migration of these entities in the presence of forces $(\bar{F})$ like high thermal gradient [41, 42]. The drift velocity $(\bar{v})$ can be represented with the help of mobility (m_{bub}) as $\bar{v} = m_{bub}\bar{F}$. As is the force expression used in atomic diffusion under temperature gradient $(\nabla T < 0)$, the magnitude of the thermal force experienced by a gas/vapor bubble will be proportional to $Q_{bub}|\nabla T/T|$ and also to the bubble size. Q_{bub} is the heat of transport in the drift rate controlling mechanism, which can be through volume diffusion, surface diffusion, or diffusion of gas/vapor concerned. The mobility, m_{bub}, according to Stoke-Einstein relation is D_{bub}/kT, where D_{bub} represents the transport coefficient of the bubble. D_{bub} can be correlated with the diffusion coefficient of the leading species involved in the drift mechanism. For the bubbles located at the central region of fuel pin, the drift is poor due to negligible temperature gradient. For those located near the steepest thermal gradient (approximately at 0.7 times pin radius), the drift is again negligible due to very small size of the bubbles in the low temperature region.

According to Matzke, the bubbles remain small in size due to fission-induced resolution, and they have negligible mobility at least up to 2073 K [11]. Further, in the fuel matrix, there are various defects in large concentrations and these defects are responsible for pinning of bubbles. The bubbles have negligible mobility until they attain a critical size where the temperature gradient is large enough to pull them free of the bubble-defect interaction. Olander and Nichols have estimated that for the cases of UO_2 and UC fuels, the critical bubble radii are within 100–1000 Å for dislocations and 5000–10000 Å for grain boundaries [41–43]. Bubble migration is nearly absent in LWR fuels. The migration arises at much higher temperatures (>2000 K) as in the case of fast breeder fuels, where the bubble migration contributes to columnar grain growth with central void formation and to fuel swelling via local accumulation of the bubbles.

3.1.6 Sublimation Paths Along Grain Boundaries, Channels, and Cracks

Transport and release of gas atoms always accompany with the transfer of sublimed layer of fuel to its nearest cooler surface in the presence of thermal gradient at high temperature region of the fuel. For urania, the sublimation followed by condensation of the oxide can start above 1900 K [44]. In fast reactor fuel, the formation of columnar grains is the result of material transfer radially from hotter to cooler side of the fuel pin and in the process there is migration of the freed pores full of fission gases and volatiles toward the central region and their ultimate release in the central void [45]. This way of gas transport in fast reactor fuel distinguishes itself from the diffusional transport discussed for the case of thermal reactor fuels. The former follows approximately linear kinetics in time, which contrasts with parabolic kinetics in the latter case.

3.1.7 Burst Release of Gases/Volatiles

In post-irradiation annealing study, an initial rapid release of gas/volatile species is invariably noted from fuel sample for a short period, before the characteristic diffusion controlled slower release could be discerned. The burst-like initial release has been addressed by several authors for understanding its occurrence, particularly because similar release occurrences have been noted in heating and cooling phases in some in-pile experiments. The initial reasoning that the burst release of fission gases is essentially from oxygen-enriched surface layer formed by impurity pick up during irradiation/heating of the sample, could not be substantiated by later findings [46]. Some authors ascribe the burst release to rapid escape of untrapped gas atoms present in the surface layers [11]. The presence of overpressurized gas bubbles at the grain boundaries could be deduced through XRF and EPMA examinations of irradiated fuel. The reported findings of higher irradiation dose effects [47], namely, critical temperature lowering in the onset of

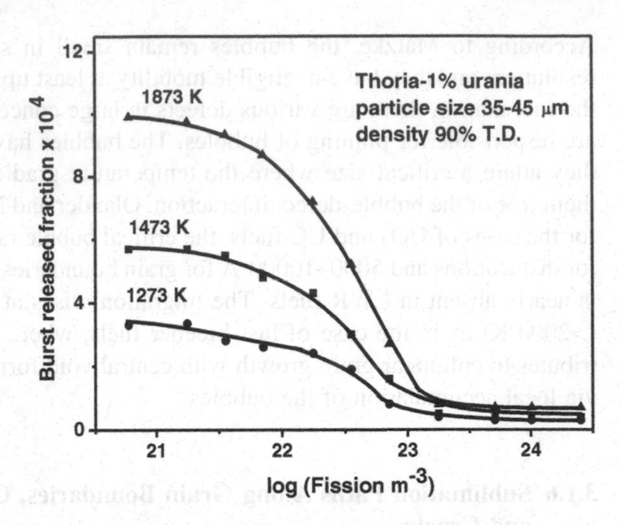

Fig. 10 Burst release trend with fission dose

burst release and augmentation in released fraction indeed corroborate to an increased gas inventory at grain boundaries of the surface layers. The state of this inventory has been variously attributed, such as to the formation of 'quasi crystallites,' overpressurized bubbles under hydrostatic pressure constraint [48], or to bubble phase as well as trapped form in surface defects [47]. Reported studies on burst release of Xe during post-irradiation annealing of thoria-1% urania at different temperatures showed that for a given fission dose in the samples (pellet/powder) the burst release is more, higher the set temperature, though by increasing the irradiation dose the isothermally released fraction progressively falls and finally converges to a constant value toward higher doses (Fig. 10) [49]. The falling pattern in the released Xe with the dose increase is connected with the loss of freedom of gas atoms due to their engagement in bubbles and surface defects.

The modeling of burst release has evolved through considerations of a number of features. Under thermal stress, bubble interconnection process followed by microcracking along grain boundaries is generally considered as the probable reason behind the observed burst release. The interconnection timing and the randomness involved in microcrack development in the brittle phase together decide the time constant in burst release. Burst release during power drop has been connected to gas venting out of mechanical relaxation of thermally stressed grains holding the accumulated gas in surface layers. In a recent attempt of the modeling, Kogai [50] has considered the bubble interconnection factor and the pellet cracking factor together by the use of product function of two sigmoidal evolutions respectively describing the involved kinetics. The areal fraction covered by critically loaded gas bubbles at grain boundary interlinks at a rate that is governed by grain boundary diffusion of surface-adsorbed gas atoms. In expressing the gas release kinetics from interlinked bubbles through emergent pores of the grain structure, the model refers to experimental observation that the kinetics is proportional to the square of bubble pressure. The cracking factor is defined proportional to the effective tensile stress of the grain structure. With the stated considerations, the modeled result corroborates to the observed burst period of gas release.

3.2 Athermal Components in Transport

3.2.1 Intergrain Gas Bubbles Interconnection and Release

Gas atoms transported out of lattice accumulate as bubbles on grain face and edge and the bubbles grow in numbers as well as in their sizes with burnup [51]. Beyond a surface coverage of critically sized bubbles, they undergo interconnection [29]. Because of lowered surface force of the resultant domain, the interconnection process extends to fill intergrain channel with thin gas layer pressurizing thereby the internal space of the polycrystalline matrix and resulting in gas venting through small grain edge tunnel and pore-openings. Both the athermal processes, namely, bubble interconnection and gas venting, are augmented in power surge by thermal as well as mechanical stress. Normally, it is envisaged that the bubble interconnection takes place as the gas atom concentration on the grain surface attains saturation, which can be evaluated from a simple consideration of mechanical equilibrium of gas-filled lenticular bubble under a given hydrostatic pressure (p_h) maintained inside the fuel [31]. The surface force of bubble maintains mechanical equilibrium with inside gas pressure (p_g) as $p_g + p_h = 2\gamma/R_{bub}$, where R_{bub} corroborates to radius of curvature of the bubble and γ the surface energy. Gas pressure increase results in bubble growth. Under ideal gas approximation, the number of gas atoms critically maintaining the pressure p_g inside a bubble is $N_{cr} = p_g V_{bub}/kT$. For lenticular bubble, the bubble volume is given by $V_{bub} = 4\pi R_{bub}^3 f(\theta)/3$, where $f(\theta)$ is the lenticular factor for correcting the spherical volume; 2θ is the dihedral angle between bubble surface and grain boundary. Thus, N_{cr} for ideal case is governed by the hydrostatic pressure, temperature, and the bubble characteristics including the surface energy. For the onset of bubble interconnection, if η fraction surface coverage is required then the bubble density (n_{bub} per unit area) and therefore, the critical number density (N_{grain}) of gas atoms per unit area on the grain surface are, respectively, $n_{bub} = \eta/\pi R_{bub}^2$ and $N_{grain} \equiv N_{cr} n_{bub} = 4R_{bub} f(\theta)\eta(p_g/3kT)$, $p_g = (2\gamma/R_{bub}) - p_h$. The use of ideal gas equation in the derivation is an oversimplification for the highly pressurized gas inside bubble. The nonideal behavior can be introduced in the expression for gas density inside bubble (N_{cr}/V_{bub}), which in that case will be modified from p_g/kT to $p_g/(kT + p_g \upsilon_{waal})$, where υ_{waal} is Van der Waal's constant.

The athermal gas transport via bubble interconnection is an inevitable process for ultimate release of the diffused out atoms from lattice wherein the gas atoms are generated by fission. This process has regular as well as sporadic components.

The regular seepage from pressurized intergrain channel and the intermittent burst release by the momentary relief of thermomechanical stress at grain edge, pores, and microcrack line, both have significant role in the overall gas release.

3.2.2 Recoil and Knockout

The fission energy mostly shared by the pair of fission fragments according to their masses is progressively lost over the traversed paths (typical range ~10 μm). In the initial part of their recoiled paths, the kinetic energy loss is predominantly through electronic interaction in the lattice; while in the last part, till they come to halt, the energy is abruptly used in creating atomic displacements. Toward the fuel's outer surface, the recoiling can result in release of the energetic FPs in the rod-free region. The athermal release governed by the fission rate (fission per unit time) f can be expressed as $fY_i r_i (S/V)_{\text{geom}}/4$, where Y_i is the fission yield of the fragment (i), r_i is the range, $(S/V)_{\text{geom}}$ is surface to volume ratio for the overall geometry of the fuel pellet [52]. Besides this recoil release, there can be release through knockout events of the matrix atoms. The knocked atoms, which can be several in numbers in such an event, while traversing with moderated kinetic energies can in turn transfer their momentum to other atoms in their paths resulting in knocking cascade. The knocked atoms can have sufficient kinetic energies to kick out FPs to the grain boundary. Generally, the knocked out atom having little energy cannot make re-entry in the adjacent grain and continues to stay on the surface until thermal means of its transport takes over. Such intergrain release of the FP atom will be governed by total surface to volume $(S/V)_{\text{total}}$ ratio of the polycrystalline fuel. The release to birth ratio of short-lived isotope with decay constant λ_i can be expressed as [52] $(R/B)_i = \alpha_U r_i' \dot{f}_d (S/V)_{\text{total}}/(4N_U \lambda_i)$, where α_U is the number of sputtered U atoms per recoiled fragment ($\alpha_U \sim 5$ in sintered UO_2), r_i' is the average range of the knocked isotope, $\dot{f}_d$ is fission density (fission rate in unit volume), and N_U is number density of U in the fuel matrix. Tiny gas bubbles knocked on by the fission fragments can result in its total resolution in the matrix. The knockout release is negligible compared to the recoiled release of short-lived isotopes. The athermal release due to recoil and knockout is limited to ~1 % at low to moderately high burnup. In high burnup situation (>60 GWd T^{-1}), this release goes up to ~3 % [53].

3.3 *Modeling of Polycrystalline Fuel Matrix for Release Studies*

As indicated in the previous section, the thermally assisted release of a fission gas/volatile atom from the polycrystalline fuel pellet takes place through a complex transport path traced inside grain and along grain boundaries. The statistical treatment of diffusional release kinetics of generated gas/vapor species from the polycrystalline matrix was made simpler by subtle assumption in Booth's model [54] described below.

3.3.1 Booth's Model

The Booth's model assumes that the overall release kinetics is mainly decided by the slowest step of intragrain diffusional release to the grain boundary, whereon faster steps are involved to result in the release. The model further assumes that the polycrystalline mass constituting irregular grain shape and size can be statistically taken as an assembly of unimodular distribution of spherical particles that reproduces the specific surface area (S) of the mass. Radius of the equivalent spheres is thereby written as $\mathbb{R} = 3/(\rho S)$, ρ being the grain density. In the hypothetical sphere, the specific surface to volume (S/V) signified by the ρS is conserved. The complex path of transport is simplified to the diffusional release from the Booth's spheres defined with the conserved surface property of the matrix. On the stated spherical geometry, transport analysis can be made for the cumulative release of a FP knowing, for example, its initial concentration distribution, generation/decay rates and the boundary conditions for the diffusion equation $\partial c/\partial t = (D/r^2)\partial(r^2\partial c/\partial r)/\partial r + \dot{S}$. The term $\dot{S}$ is constituted of generation and decay rates of the species.

If the intragrain diffusion solely controls the overall transport for the release process of a species, then the constant D in the Booth equation would be the single atom diffusion coefficient of the species. However, the intragranular as well as grain boundary transport being complex, the constant D is generally replaced by an effective diffusivity D_{eff} and the Booth radius is adjusted from $3(V/S)$ to somewhat higher value ($\mathbb{R}_{\text{eff}}$) in order to take care of the impeding factors in transport outside the grain. The diffusion equation is generally solved with assumed null concentration of the species on the surface of Booth's sphere. With this boundary condition as has been justified by the absence of chemical affinity of the species with matrix, the species can escape in free space once it diffuses out of the adjusted Booth sphere. Thus, for the simpler case of uniform initial concentration distribution of the species, $c(r, 0) = c_0$, and absence of its source/decay term ($\dot{S} = 0$), the fractional release under the boundary conditions, $c(\mathbb{R}, t) = 0$ and $c(0, t) \neq 0$ (finite), is expressible in a simple form in short time approximation as $f(t) \simeq 6(Dt/\pi\mathbb{R}^2)^{1/2} - 3(Dt/\mathbb{R}^2)$ [55]. $(D/\mathbb{R}^2)^{-1}$ is the characteristic release time from Booth's sphere. For shorter time approximation $Dt/\mathbb{R}^2 \ll 1$, the release is expected to be parabolic. Parabolic release kinetics of gas and volatile FPs are observed indeed in post-irradiation annealing study of trace-irradiated fuels.

For the case where the diffusing species i generated at the constant rate $y_i\dot{f}_d$ (y_i being the fission yield and $\dot{f}_d$ the fission rate in unit volume) has shorter average life as compared to the characteristics release time (i.e., $D_i/\mathbb{R}^2 \gg \lambda_i^{-1}$, λ_i being the

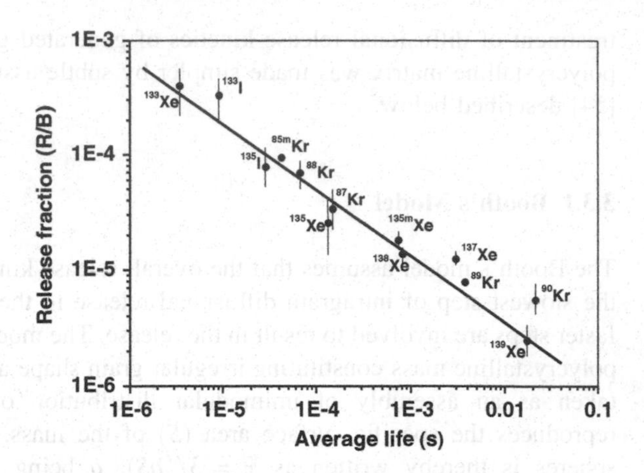

Fig. 11 Release trends of Xe, Kr, I isotopes of short average lives

decay constant), the diffusion equation with its last term $\dot{S}_i$ substituted with $y_i\dot{f}_d - \lambda_i c_i(r,t)$ can be solved for steady state release in the Booth's geometry (boundary conditions, $c(\mathbb{R},t) = 0$ and $c(0,t) \neq 0$ (finite)). The released fraction is expressible as [55] $R_i/B_i = 3\sqrt{D_i/(\lambda_i\mathbb{R}^2)} \equiv (S/V)\sqrt{D_i/\lambda_i}$. In the diffusional release, the release to birth ratio (R_i/B_i) has parabolic dependence on the average life (λ_i^{-1}) of the radioactive species. This is exemplified in figure below (Fig. 11). It may be recalled that in the case of knockout release, the R/B ratio is proportional to average life.

With the use of Booth model, one may work out the average release of a species from a fuel rod consisting of a large number of pellets (radius r_0) axially stacked inside close fitting cylindrical cavity of clad. Thermal hydraulic design caters to establishing regulated temperature profile in the rod at a given linear power rating (P). Quite uniformly generated fission heat of density, $P/(\pi r_0^2)$, is radially transported to coolant flowing outside the clad.

Booth spheres filling up the rod volume per unit length, πr_0^2, release out a cumulative fraction of the species during the time period, t, over which the rod is delivering the power, P. The value of diffusion constant radially decreasing down the temperature profile, the fractional release $f_i(r,t)$ of species i from a sphere located at larger radius of the fuel rod will be lesser. The released fraction is to be evaluated considering the constant rate of generation of the species. For species having very large decay time as compared to the time period of this evaluation, the $\dot{S}_i$ in the diffusion equation is $y_i\dot{f}_d$. Solution of the equation with the initial concentration, $c(r,0) = 0$ and the boundary conditions, $c(\mathbb{R},t) = 0$ and $c(0,t) \neq 0$ (finite), one has the approximate expression of fractional release as $f_i(r,t) = 4(D_i t/\pi\mathbb{R}^2)^{1/2}$ [56]. In the cumulative generation of a FP within the rod volume (πr_0^2), an annular element $2\pi r dr$ accounts for a certain fraction given by $2(r/r_0^2)\mathrm{d}r$ out of which the released fraction from this element is $f_{i,\mathrm{annular}}(t) = 2(r/r_0^2)$ $f_i(r,t)\mathrm{d}r$. Noting that in an infinitely long cylinder that is immerged in a regulated heat sink, the radial temperature gradient gets established in the presence of the uniform power density deposition $P/(\pi r_0^2)$. The gradient is given by $\mathrm{d}T/\mathrm{d}r = -(2r/r_0^2)[P/(4\pi\zeta)]$, ($\zeta(T)$ being the thermal conductivity of the fuel matrix), the released fraction can be rewritten in terms of temperature variable as $f_{i,\mathrm{annular}}(t) = -(4\pi\zeta/P)f_i(r,t)\mathrm{d}T$. Thus, the released fraction for the considered rod volume is the integral $f_{i,\mathrm{unitlength}}(t) = (4\pi/P)\int_{T_S}^{T_c}\zeta(T)f_i(r,t)\mathrm{d}T$. With the substitution of fractional release function $f_i(r,t) = 4(D_i t/\pi\mathbb{R}^2)^{1/2}$, and the Arrhenius dependence of the diffusion coefficient, $D_i(T) = D_{0,i}\mathrm{Exp}[-E_i/RT]$, one thus writes the released fraction as $f_{i,\mathrm{unitlength}}(t) = (16\sqrt{\pi}/P)(D_{0,i}t/\mathbb{R}^2)^{1/2}\int_{T_S}^{T_c}\zeta(T)\mathrm{Exp}[-E_i/2RT]\mathrm{d}T$. Using this result as is obtained over a particular length section of the fuel rod, the average fractional release from the entire rod can be expressed.

3.3.2 Limitations of Booth's Model

The predicted value of Xe release using Booth model can be compared with observed result of in-pile experiment. Usually, the predictivity is not good [57] (Fig. 12). Even with kinetic and geometric parameters adjusted in the transport equation, the Booth model works with severe limitations beyond trace irradiation level of the fuels. Though the intragrain transport could be fairly addressed by the use of D_{eff}, the complexity in grain boundary transport is hardly manageable with the use of the adjusted radius of Booth sphere. There is difference in analytical approaches of different authors in addressing the surface transport of the diffused out species from the Booth's sphere. The assumption of free escape of the species as it comes out of the grain is in gross contradiction with the reality arrived from gas release data from low to high burnup fuels. In low level irradiation, an incubation period is clearly noted before the release starts. There is occurrence of burst release in the heating and cooling cycle of the fuel. With burnup increase, the fuel becomes more susceptible to gas release as noted from the steep lowering of fuel enthalpy input, and therefore the central temperature (T_c) for attaining the onset of release occurence at different burnups (Bu). Within irregularity in release occurrence in some cases due to unpredictable microcracking of ceramic matrix in the thermal cycles and also burnup histories, the central temperature follows inverse proportionality with logarithmic burnup as $T_c = 9800/\ln(\mathrm{Bu}/0.005)$ [58]. The burnup dependence and the steep lowering of the central temperature at low burnup region is indicative of the rising gas confinement on and around grain surface of the dense polycrystalline assembly representing the fuel matrix.

The recent analyses with the Booth model [59, 60] take care of (a) coupling of the intragrain and grain boundary diffusions modes of the species for their localized accumulation in specific sites on/around the grain surface with burnup, (b) criticality of transition from surface arrested states to a common gas pool inside intergrain space posing a hydrostatic pressure against the polycrystalline boundary, and (c) criteria of regular and burst release of the species across the boundary and

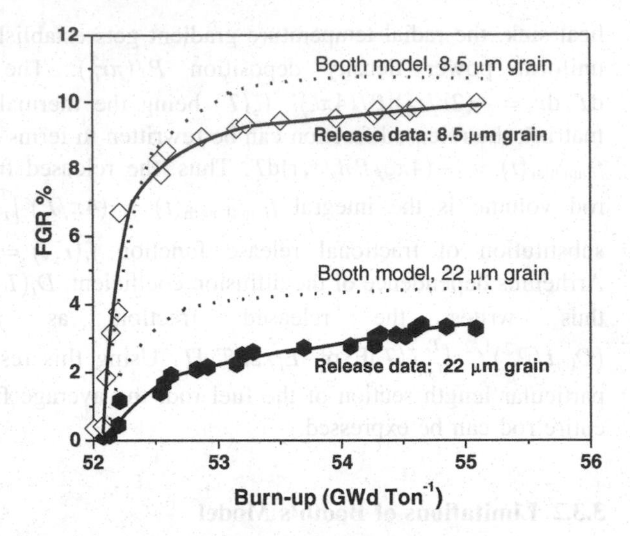

Fig. 12 Typical cases of in-pile release data vis-a-vis Booth modeled results for UO_2 fuels

reversibility in resumption of surface localized states for an incubation for the next release. The analysis of Kogai [61] considers combination of bulk diffusion of gas atoms in spherical grains and gb diffusion of surfaced out atoms toward neighboring gas bubbles that had nucleated in the past. In line with reported observations, he considers stationary number density of growing bubbles as they near the onset of interconnection on the grain surface. With a given input of surface dissolved atoms in the lenticular bubble, the gas pressure inside is re-established through its mechanical balance with curvature-dependent force under a hydrostatic pressure head, as discussed already. The growth rate of bubble radius is governed by gb mobility of uranium vacancies under the chemical potential gradient resulting from the pressure $\Delta p \equiv [p_g + p_h - (2\gamma/R_{bub})]$ and is expressed as [62] a $\partial R_{bub}/\partial t = (\delta_{gb} D_{gb,vac}/kT)\Delta p\, \upsilon_{vac} F(\eta)/(4R_{bub}^2 g(\theta))$, where δ_{gb} is gb width, $D_{gb,vac}$ is gb diffusion coefficient, υ_{vac} is volume of U vacancy, $g(\theta)$ is the volume ratio of lenticular to spherical bubbles of same radius of curvature, $g(\theta) \equiv 1 - 1.5\cos\theta + 0.5\cos^3\theta$, and $F(\eta) \equiv 8(1-\eta)/[(\eta-1)(3-\eta) - 2\ln\eta]$ is a function of areal fraction of grain covered by bubbles. As discussed already, the bubble interconnection sets in at a critical surface density of atoms. He treats the release from the interconnected network of bubbles as fluid flow under pressure gradient in porous medium following the well known Darcy's law [63]. For the conductivity of porous medium, Kogai has considered two kinetic factors, bubble interconnection and pellet cracking, with their sigmoidal evolutions as functions of the areal coverage of bubbles on grain surface and effective tensile stress on grain-interlinked boundary, respectively. The bubble interconnection kinetics is the rate controlling step. The observed facts of intermittency in gas release from fuel could be corroborated by the critical attainment periods of interconnection of surface arrested bubbles and renucleation of surface bubbles following gas venting out in the porous medium.

Certain drawbacks in the Kogai's approach in the coupling of diffused outflux of species from Booth's sphere to their gb kinetics has been addressed by others at a later stage [6]. The Kogai's approach did not cover trapping and resolution in intragrain transport, influence of finite concentration of FP species on the Booth's sphere, and contribution of gb sweeping in bubble growth. Further, the approach considered Smoluchowsky's 3D diffusion model to the 2D case of the precipitation kinetics of surface diffusing species into the bubble and by this it rendered sluggish precipitation than expected in 2D case. The intragrain flux directly intersecting the bubble sites on surface can straightforwardly enter into the gas assembly while the rest of the outflux establishes themselves in surface dissolved state and then make the 2D diffusion. Subsequent analysis [64] addressed to these shortcomings in the Kogai's approach.

For the restructured rim of fuel, the applicability of Booth model is questionable, though some authors [64] attempted to apply the model to explain the very low but constant gas retention (0.2 wt%) in the submicron grains in HBS. Restructured to restructuring regions within the thin rim ($\sim$1,000 μm) will have too sharp gradation in grain morphologies, and porosity size and density distributions for treating them with a Booth sphere. When most of the gas is located in pores outside the grains, the assumption of null surface concentration of the diffusing atoms is uncalled for.

4 Diffusion Studies of Fuels

The relevant properties and parameters involved in the release evaluation of gas and volatile products from fuel matrix are seen to be the lattice and grain boundary diffusion coefficients as functions of temperature and burnup. Theoretical evaluation of the single atom diffusion coefficient in lattice refers to data of formation energies and entropies of point defects and diffusion barrier complex involved in the matrix, and the lattice characteristics [65]. The controlling parameters involved in defining the release properties are grain size and distribution, desorptive surface area, open and closed porosities, pellet density, critical density of surface traps for bubble nucleations and growth, gas–solid interfacial property, bulk elasticity, grain growth behavior, and temperature profile of the fuel pellet. The sample under study needs to be characterized for these parameters. Experimentally, the diffusive kinetics of a species is studied isothermally in the samples either in release mode usually under null concentration at boundary for an initially defined inside concentration profile, or, in ingress mode with initially null concentration inside and a finite concentration at boundary surface. The time evolution in the transport is accordingly followed by the extent of release/retention, or, the extent of the ingress. The evolution can also be traced by concentration profiling for the diffusing species inside the fuel matrix at a regular interval. The FG release studies on a fuel pin are generally made by in-pile rod pressure and gamma ray spectrometry measurements [58], and also by post-irradiation examinations through ceramography and rod puncturing tests [13, 66]. For short-lived isotopes, with the knowledge of their decay properties, the monitoring of the released fractions help evaluating the respective transport coefficients. As noted earlier, the release to birth ratio for the short-lived isotopes is governed by $R_i/B_i = 3\sqrt{D_i/(\lambda_i\mathbb{R}^2)} \equiv (S/V)\sqrt{D_i/\lambda_i}$.

4.1 Release Kinetics from Post Irradiation Annealing Study

For evaluating the thermally assisted primary release process, namely the diffusion controlled outflux of gas/vapor atoms from the grain, one needs the lattice diffusion coefficients of the relevant species. Experimentally, it is most common to use post-irradiation annealing (PIA) technique to evaluate the transport characteristics of the gas and volatile products that are present in their complex states inside the grains. In PIA technique, the FPs are commonly generated in the sample by reactor irradiation. Ion implantation is used as an alternative [67], where the FPs could be introduced into lattice up to a limited depth decided by the range of ion energies used. In reactor, sample is normally irradiated around room temperature using uniform neutron flux so that initial concentration of the products can be fairly assumed to be uniform inside. The fuel matrix with well-defined physical characteristics such as density (ρ), specific surface area (S) is thus obtained with an initial concentration profile of the diffusing species. In many cases, particles from crushed powder are sieved out with selected size range close to the average grain size of the fuel. The surface to volume ratio of the granular specimen is evaluated as ρS. Generally, S is measured by BET adsorption technique using krypton/nitrogen gases and it includes the surfaces of open porosities, which are in fact additional let outs for the diffusing species. Working on the Booth's model, the fuel sample is thus characterized in terms of the diffusion geometries of equivalent spheres of radius $\mathbb{R} = 3/(\rho S)$. As will be elaborated in the next section, the isothermal runs in PIA technique for the reactor-irradiated samples conform to the boundary conditions $c_i(\mathbb{R}, t) = 0$ and $c_i(0, t) \neq 0$ (finite) with uniform initial concentration over the Booth's spheres. Thus, the intragrain diffusion coefficient of a species follows from the measured data of its cumulative released fraction as a function of the isothermal annealing time. As noted earlier that for short annealing period, the release is parabolic in time, $f_i(t) \simeq 6(D_i t/\pi\mathbb{R}^2)^{1/2}$, so that the $D_i/\mathbb{R}^2$ value follows from slope of linear trend in $f_i(t)$ versus $\sqrt{t}$ plot of the kinetic study. The data generated at different temperatures help establish the temperature trend of the lattice diffusion coefficient D_i, that is expected to be in Arrhenius form as $D_i = D_{0,i}\text{Exp}[-Q_i/RT]$. Thus in a linear trend in the plot of $\ln(D_i/\mathbb{R}^2)$ versus inverse temperature $(1/T)$, the slope gives the value of diffusion barrier energy Q_i of the species. The pre-exponential constant $(D_{0,i}/\mathbb{R}^2)$ obtained from intercept of the linearity essentially represents the frequency factor associated with the displacement over the characteristic dimension, $\mathbb{R}$. The transport coefficient arrived from trace-irradiated fuel corroborates to the value of single atom diffusion of species in virgin grain of the fuel. With burnup increase, the transport property deviates from this value. The increasing incorporation of FPs is an important factor for the deviation. The deviating behavior can be understood from PIA studies of specimens prepared with doped FPs simulating their concentrations for a burnup. Similarly, the influence of irradiation dose on the transport property can be followed. The general approach of these experimental techniques and the specific results reported for thoria-based fuel are given below.

4.1.1 Experimental Features in Post Irradiation Annealing Study

For the kinetic data acquisition, the virgin and simulated burnup fuels are made in the form of highly dense pellets with phase purity, compositional homogeneities (with respect to the ratios of U/Th, or, Pu/Th and O/M), and homogeneities with respect to the FP distributions in the case of SIMFUEL matrix. The bulk density is generally maintained at around 95 % of the theoretical value reproducing the reactor fuel density for better incorporation of the FPs with less swelling and without much sacrifice for the mechanical strength and thermal conductivity of the matrix. Since the pellet preparation involves long sintering cum densification step at high temperatures (>1873 K) in reducing environment with controlled oxygen potential, the FPs' addition in SIMFUEL preparation excludes highly volatile components like Rb, Cs, I, Te, etc. The added elements usually are: Nd as the representative of trivalent rare earth and Ce as tri/tetravalent rare earths, the transition metals (Y, Zr, Nb, Mo, Ru, Rh, Pd) and the alkaline earth (Sr and Ba) FPs. The densified pellet is examined for the chemical states as well as compositions of the added FPs [67]. Ideally, the chemical states should corroborate to those noted in reactor-irradiated pellets and the compositions should corroborate to a burnup value. The fuel pellets are grinded and sieved for collecting samples of narrow distribution of particle size, and the surface to volume ratio of the samples are evaluated by density and surface area measurements. The selected size range of the particles usually approach the average grain size observed in the polycrystalline pellet so that the overall release kinetics of gas/volatile atoms is representative of intragrain diffusion process. The powder samples with defined O/M ratio are irradiated at a given dose in nuclear reactor to generate the gaseous and volatile FPs inside. Uniformly irradiating flux results in evenly distributed products in the fuel particles when fission heat is properly dissipated out without significant temperature increase above ambience.

As mentioned earlier, the release kinetics of gas and volatile FPs from irradiated samples can be followed by measuring either the residual activities of decaying isotopes of the FPs after isothermal annealing over a known period, or, by measuring the activities in the released fraction. The former method needs annealing and quenching steps to repeat for generating the kinetic pattern of cumulative release over integrated annealing time. The latter one would involve

Fig. 13 Schematic of PIA setup for Xe release study

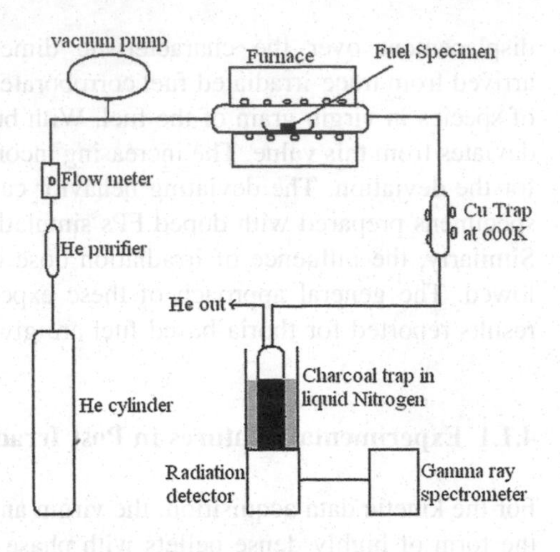

isothermal annealing run at a stretch during which the gases are continuously swept out by an inert carrier gas to the online activity monitoring system for a radioactive isotope of the released fission gas. The released volatiles because of their condensation in lower temperature region cannot be quantitatively carried by gas flow to the detector site. The released volatile FPs can, however, be chemically fixed or condensed on forward collecting targets in vacuum annealing run for over a known time period. The fission products of higher volatilities like iodine and tellurium can be chemically fixed on hot copper targets as Cu-iodide/telluride [68]. For the condensable vapors, the collection efficiency for the designed source to target geometry needs precalibration. To partially reduce handling problem and to avoid complexity in specific activity assay in freshly irradiated sample, the annealing experiments are generally carried out after an initial period of cooling the activity due to short-lived nuclei. Schematics of PIA setups, respectively, used for monitoring gas and volatile release in the isothermal annealing of sample are shown in Figs. 13 and 14. The released gas like Xe is carried away by flowing purified helium to get adsorbed in liquid nitrogen cooled activated carbon trap held on the top of a scintillation counter that makes online monitoring of ^{133}Xe for its 81 keV gamma ($t_{1/2} = 5.27$ days). For iodine and tellurium, the schematic shows vacuum annealing of the sample in an effusion cell and streaming of effusates through a hot duct of precalibrated streaming efficiencies for the respective volatiles for their forward collection on copper targets. For an isothermal run, a large number of copper targets are brought one after the other to the forward position of the effusing stream. The collected ^{131}I and ^{132}Te on the targets can be monitored offline for the respective gamma activities 364 keV ($t_{1/2} = 8.01$ days) and 228 keV ($t_{1/2} = 78$ h) by using HPGe–MCA monitoring system. For evaluating the fractional release over a time period, the total contents of the monitored isotopes in the irradiated sample are required and are generally obtained by the radiometric assays of a known amount of the samples used in the PIA studies. The

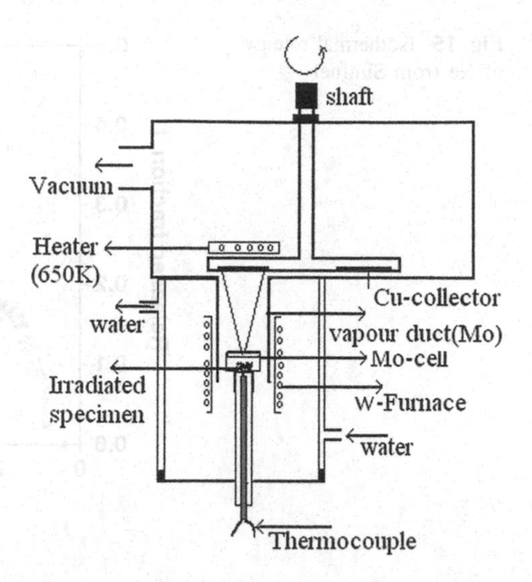

Fig. 14 Schematic of PIA setup for I and Te release study

released fraction evaluation takes note of decay property of the radiometrically monitored isotopes. The typical results of isothermal kinetic studies of xenon, iodine, and tellurium are shown in Figs. 15, 16, and 17. The sample characteristics are included in the Table 1 [68, 69]. The diffusion controlled parabolic growth of cumulative release are reflected in the linearly increasing part of $f_i(t)$ versus $\sqrt{t}$ plots for all the cases. The initial nonlinear growth in the plots generally reflects the burst release of the species, while the release attaining halt or toward halt as can be seen in the later part after the linear region is indicative of retention of species in trapped states. The nature of trapping and the saturation concentration in the trapped state is specific for the species. For species like iodine/bromine, the formation of electrovalent compound with cesium/rubidium in the trapped state poses additional impediment in their diffusions. Their diffusions in atomic forms are controlled by high chemical stability of alkali iodides and in ionic forms the electrostatic field with counterion that can have slower diffusion comes in the way. The diffusing species can be trapped in vacancy clusters like the trivacancies ($v_O v_M v_O$) in the MO_2 lattice, or in precipitated microbubbles, or in dislocations and voids as mentioned earlier. The competitive kinetics of trapping in diffusional release process is reflected either by a delay in release initiation or by its slowing down to a halt practically stopping the total release of species. The kinetic results as depicted for Simfuel at 1,773 K (Fig. 15) show that xenon release came to halt with its 60 % retention inside fuel, while iodine release showed sluggish start following the initial burst release and the sluggish growth continues until it attains 1.6 % whereafter there is steep linearity in the cumulative release [69]. On the contrary, the iodine release in the virgin fuel (at 1,803 K) has come to a halt after 2 % release. Resumption in iodine release in the virgin matrix [68] could be temporarily established by raising the temperature (Fig. 16). Tellurium mostly remaining in alloyed form, the halting effect in the release is less severe or absent in the virgin fuel and Simfuel cases, respectively.

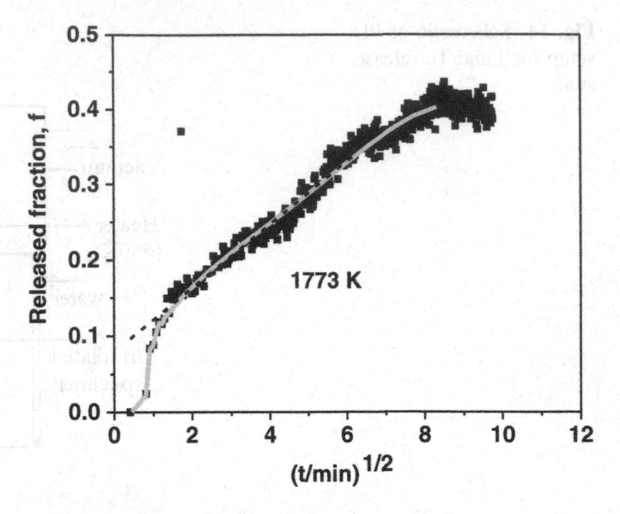

Fig. 15 Isothermal release of Xe from Simfuel

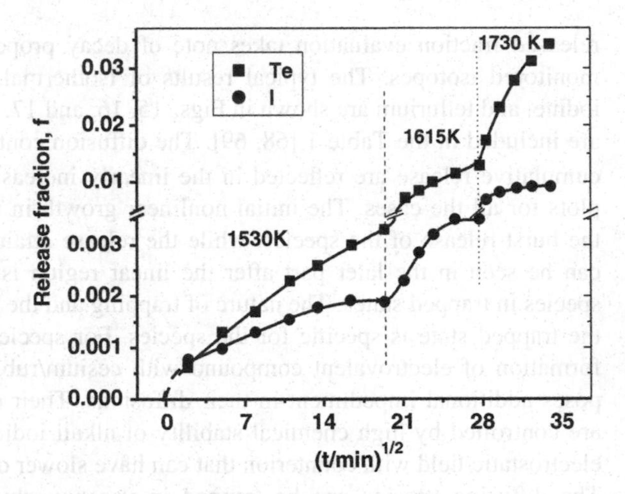

Fig. 16 Isothermal release characteristics of I and Te from virgin thoria fuel

4.2 Transport Properties of Fission Products in Thoria-Based Fuels

With the widespread use of urania-based fuels in nuclear reactors extensive studies have been reported on the transport behavior of gases and volatiles in urania and urania-plutonia fuels [4, 49, 70–74]. Studies on the thoria-based fuels are also reported [75–83]. The stated results of isothermal release kinetics of Xe, I, and Te from thoria-based virgin and Simfuels (Figs. 15, 16 and 17) are one among the scarcely reported data. As compared to the case of urania fuel, the fission yield of gases is 10 % higher in the thoria-based fuel, the excess gas is resulting from higher yield of Kr from the fission of ^{233}U daughter of ^{232}Th. However, as will be seen subsequently that results of fission gas and volatile transport studies carried

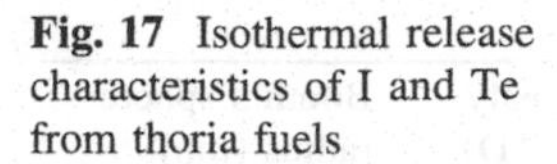

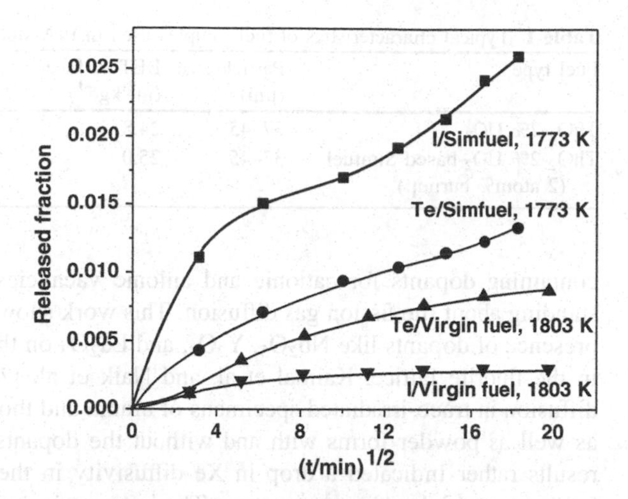

Fig. 17 Isothermal release characteristics of I and Te from thoria fuels

out in laboratory as well as reactor irradiated test fuel rods point to the fact that there is lesser release of the FPs from the thoria-based fuels.

Thorium unlike uranium remaining tetravalent in its compounds, a subtle difference in the transport properties is noted in thoria and urania matrices. In the latter matrix, the diffusion is significantly influenced by O/M ratio; the oxygen hyper-stoichiometry augments the diffusion, whereas this aspect is absent in thoria. In a diffusion study with air annealed thoria–urania samples, Olander has reported systematic augmentation in the diffusivities with increase in U/Th ratio [82]. The electronegative species such as Br and I are expected to show distinction in diffusion behavior in the two oxide lattices; the diffusion will be more impeded in thoria since the transition state complex of Br/I with the tetravalent Th cannot have favorable energetics as compared to that with the polyvalent U in urania. The reportedly higher value of anion interstitial migration energy in thoria (3.27 eV) than that in urania (2.6 eV) is indicative of this distinction. Experimental data for O diffusion show that the activation barrier for the diffusion is marginally higher in thoria (2.8 eV) than in urania (2.6 eV) [3]. Diffusion of the actinide atoms (Th/U) in the thoria lattice is again significantly lower than in urania; the activation barriers are reported to be 6.5 eV and 5.6 eV in the respective cases. The stated difference in cation diffusion property in the two fluorite lattices is generally taken to be the cause behind the lower transport property of FPs noted in thoria-based fuels [82, 83].

4.2.1 Fission Gas Transport Properties

Large atoms like Xe (atomic radius of 0.216 nm) cannot be accommodated in cationic site of U^{+4} or Th^{+4} that have smaller ionic radii (0.101 and 0.107 nm, respectively). Out of pile experiments of xenon transport in thoria- and urania-based samples suggest that Xe atom does not diffuse through cation or anion vacancies. The earliest work of Matzke [75] with Xe-implanted samples

Table 1 Typical characteristics of fuel samples used in PIA studies

Fuel type	Particle size (μm)	BET surface ($m^2\ kg^{-1}$)	Density (% TD)	Booth's sphere radius (μm)
ThO_2-2% UO_2	37–45	24.8	>95	12.5
ThO_2-2% UO_2-based Simfuel (2 atom% burnup)	37–45	25.0	97	12.1

containing dopants for cationic and anionic vacancies led to the above understanding about the fission gas diffusion. This work showed little or no effect of the presence of dopants like Nb_2O_5, Y_2O_3, and La_2O_3 on the diffusion property of Xe in the fluorite lattice. Kaimal et al. and Naik et al. [74, 76, 77, 79] studied Xe diffusion in trace-irradiated specimens of urania and thoria-1mol% urania in pellet as well as powder forms with and without the dopants, Nb_2O_5 and Y_2O_3. Their results rather indicated a drop in Xe diffusivity in the doped samples, the drop being more, higher the temperature. The increase in cation or anion vacancies not augmenting the transport, these authors made similar conclusion as that of Matzke et al. Experimental data supported by theoretical evaluation for the Xe transport in the fluorite lattice indicates that Xe atom diffuses through vacancy cluster such as the trivacancies of one metal atom and two O atoms in the MO_2 lattice (Schottky trios) [84]. The evaluation reveals that neutral trivacancy is the most favored site energetically for Xe atom. The theoretical revelation gets its support from result of channeling experiments with Rn in urania lattice. The calculated barrier energy in the trivacancy mechanism for Xe diffusion is 2.1 eV. There is, however, significant variation in the experimentally derived transport parameters for Xe in pure urania and thoria–urania fuel. Table 2 summarizes the results of annealing studies of ion-implanted [75] and trace-irradiated samples ($<10^{22}$ fission m^{-3}) [3, 49, 69–74, 77, 79, 80]. The results show that the activation energy for xenon diffusion derived from trace-irradiated thoria fuels varied from 2.48 to 4.96 eV, the reported lowest value being from Naik [49] while the highest one from Shiba [3]. For the sake of comparison, the reported results of Xe diffusion in urania are included in the same table.

Shiba [3] has studied the temperature dependence of Xe diffusivity in ThO_2-6.3 mol% UO_2. The diffusion barrier energy of 478 kJ mol^{-1} ($\sim$5 eV) noted in the trace-irradiated sample (0.9×10^{21} fission m^{-3}) has been attributed to Xe transport through tetra vacancy mechanism: a cationic vacancy joining with the normally envisaged neutral trivacancy out of a missing MO_2 molecule in the fluorite lattice [4, 84]. There, the cation vacancy formation and migration together control the movement of Xe-trivacancy complex and the energy involved is distinctly higher than that in the normal trivacancy mechanism. The barrier energy of 478 kJ mol^{-1} obtained by Shiba can be compared with those reported by different authors for Xe in urania: 335 kJ mol^{-1} by Matzke [4], 407 kJ mol^{-1} by Miekeley and Felix [71], 293 kJ mol^{-1} Turnbull et al. [72], and 644 kJ mol^{-1} by Prussin et al. [73] As compared to the results of Shiba [3], the lower barrier energy ($\sim$240 kJ mol^{-1}) reported by Naik [49] (Table 2) in trace-irradiated, thoria-based fuel could be

Table 2 Summary of transport properties of Xe in UO_2 and ThO_2-based fuel

Authors	System studied	Diffusivity parameters (D_0 and Q)	Authors	System studied	Diffusivity parameters (D_0 and Q)
Matzke [4]	Xe/UO_2, low dose implantation	4×10^{-7} m^2 s^{-1} 335 kJ mol^{-1} (3.47 eV) **1073 < T < 1873 K**	Naik [49]	Xe/ThO_2-1% UO_2 35–45 μm; 90 % TD; 24 m^2 kg^{-1}; 5.5×10^{21} fission m^{-3}	1.4×10^{-12} m^2 s^{-1} 239 kJ mol^{-1} (2.48 eV) **1273 < T < 1873 K**
Naik [49], Kaimal et al. [74]	Xe/UO_2 35–45 μm; 90 % TD; 24 m^2 kg^{-1}; ~10^{21} fission m^{-3}	4.9×10^{-11} m^2 s^{-1} 272 kJ mol^{-1} (2.82 eV) **1273 < T < 1773 K**	Naik [49, 79]	Xe/ThO_2-1% UO_2 35–45 μm; 90 % TD; 24 m^2 kg^{-1}; 1.4×10^{24} fission m^{-3}	4.5×10^{-15} m^2 s^{-1} 239 kJ mol^{-1}/2.48 eV **1273 < T < 1500 K**
Miekeley and Felix [71]	Xe/UO_2 (trace irradiated)	1.1×10^{-3} m^2 s^{-1} 407 kJ mol^{-1} **1200 < T < 1900 K**	Naik [49, 79]	Xe/ThO_2-1% UO_2 35–45 μm; 90 % TD; 24 m^2 kg^{-1}; 1.4×10^{24} fission m^{-3}	$D_0 = 5.3 \times 10^{-13}$ m^2 s^{-1} $Q = 289$ kJ mol^{-1} (3.00 eV) **1500 < T < 1873 K**
Turnbull et al. [72], Davies and Long [70]	Xe/UO_2 Urania fuel pellet; 10 μm av. grains: ~5.5 GWd T^{-1}	7.6×10^{-10} m^2 s^{-1} 293 kJ mol^{-1} **1273 < T < 1673 K**	Shiba [3]	Xe/ThO_2-6% UO_2 50 and 500 μm; ~18 and 2.4 m^2 kg^{-1}; ~99 % TD; 0.9×10^{21} fission m^{-3}	$D_0 = 2.6 \times 10^{-5}$ m^2 s^{-1} $Q = 478$ kJ mol^{-1} (4.96 eV) **1273 < T < 1573 K**
Prussin et al. [73]	Xe/UO_2 7–10 μm; >90 % TD; 10^{19}–10^{20} fission m^{-3}	$Q = 653$ kJ mol^{-1} (from apparent diffusivity) **1723 < T < 2023 K**	Shiba [3]	Xe/ThO_2-6% UO_2 50 and 500 μm; ~18 and 2.4 m^2 kg^{-1}; ~99 % TD; 1×10^{23} fission m^{-3}	$D_0 = 7.1 \times 10^{-11}$ m^2 s^{-1} $Q = 344$ kJ mol^{-1} (3.57 eV) **1373 < T < 1573 K**
–			Shirsat and Ali [69]	Xe/ThO_2-2% UO_2 Simfuel (20 GWd T^{-1}) 35–45 μm; 97 % TD; 25 m^2 kg^{-1}; 5×10^{20} fission m^{-3}	(15.9)10^{-11} m^2 s^{-1} 189 ± 11 kJ mol^{-1} **1700 < T < 1800 K**

arising from the usage of different grain densities (99 % TD and 90 % TD, respectively) by the two groups. The data variations for the barrier energy of Xe diffusion in thoria fuel are quite similar to those in the urania case. The Xe transport has been noted to be highly facilitated [69] when the sinteractive dopant like magnesia ($\sim$600 ppm) introduced in the thoria–urania lattice is finally removed by volatilization during high temperature sintering in reducing atmosphere. PIA study [69] of such highly sintered particulates (35–45 μm; 25 m^2 kg^{-1}; 97 % theoretical density) of thoria-2mol% urania-based Simfuel (20 GWd T^{-1}) showed high rate of fractional release of Xe (Fig. 15). The values of diffusion coefficient derived from the release kinetics at different temperature corroborate to low activation energy and high preexponential factor as compared to the reported results of Naik [49] and Shiba [3]. The results are intercompared in Table 2. The high transport property of Xe noted in the matrix has been interpreted to be due to the formation of excess trivacancies from magnesia removal in the sintering stage.

With the increase of irradiation dose, the Xe diffusivity at any temperature is found to drop progressively and beyond a dose of 10^{23} fission m^{-3} the diffusivity levelouts to a lower value [3, 4, 49, 75, 79]. The lowering of Xe diffusivity with irradiation dose has been noted in reactor-irradiated and ion-implanted samples of urania and thoria-based matrices. With powder samples (grain density >90 %, BET surface area 30 m^2 kg^{-1}), Naik showed that the measured Xe diffusivity in thoria-1mol% urania at 1,273 K dropped by more than two orders of magnitude before it levelouts [49]. A typical case is exemplified in Fig. 18. The Xe diffusion study of Shiba [3] reconfirmed the levelout trend of the diffusivity to a low value at high burnup.

The lowering has been interpreted to be due to the formation of radiation-induced defects (damages) where the inert gas atoms are 'trapped' [3, 4, 49]. The damage volume per fission event has also been reported by the authors from their studies on the changes of lattice parameter and electrical resistivity of the oxide matrix. Naik considered the observed decrease of Xe diffusivity with irradiation dose for their evaluation of the damage volume [49]. The defects essentially act as nucleation centers for the supersaturated gas atoms to accumulate and this ultimately leads to the formation of gas microbubbles and thereby to heterogeneous path for the impeded diffusion. At the dose of 1×10^{23} fission m^{-3}, the average

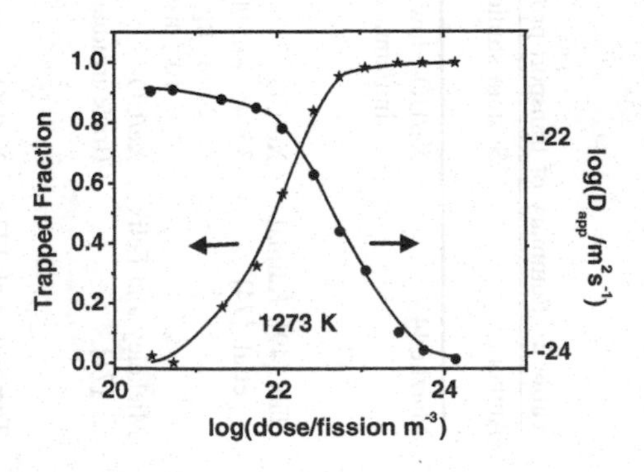

Fig. 18 Diffusivity and trapping fraction of Xe at different doses in ThO_2-1mol% UO_2

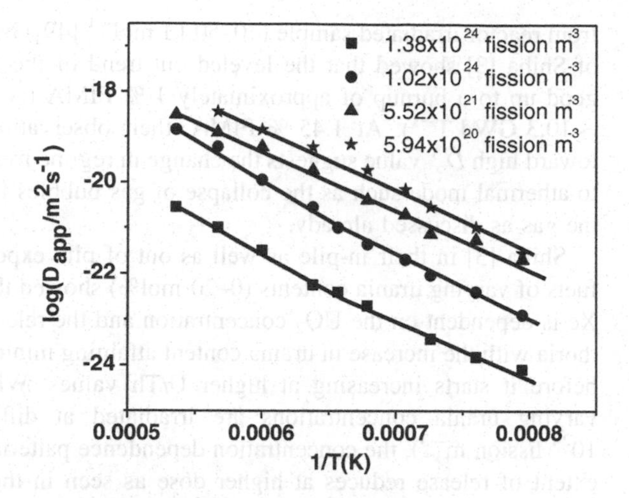

Fig. 19 Arrhenius plots of Xe diffusivity in ThO_2-1% UO_2 at different fission doses

Xe concentration of about 3×10^{-4} mol% is well above the Xe solubility in urania.

At the higher doses (>10^{23} fission m^{-3}), the levelout values of the diffusivity (D_{∞}) as a function of temperature show deviation from linearity in the Arrhenius plot, $\ln(D_{\infty})$ versus $1/T$. From the observed trend over the temperatures, 1273–1873 K, Naik et al. [49, 77, 79], could decipher two distinct regions of the Xe transport: below 1500 K the barrier height in the diffusion is 40–50 kJ mol^{-1} lower in comparison to that above 1500 K. This difference in the barrier heights is attributed to the thermal effect of the radiation defects at higher temperatures. The thermal effect refers to annealing of radiation damage centers, resolution of trapped atoms, and/or, mobilization of trapped bubbles. The typical temperature dependence of the apparent diffusivities (D_{app}) for Xe in thoria-1mol% urania sample at low and high doses as reported by Naik et al. [49, 77], are included in Fig. 19.

It may be noted that as compared to the single atom diffusion coefficient derived from trace-irradiated sample the leveled out value of the apparent diffusivity, D_{∞}, remains orders of magnitude lower. Naik et al. have quantified the trapped fraction (f) as $(1-f)^2 = D_{app}/D_{trace}$, where D_{trace} is the measured diffusion coefficient in trace-irradiated sample. In their sample, nearly 94 % Xe remains trapped in the high dose sample ($\sim 10^{24}$ fission m^{-1}) over high temperatures 1273–1673 K. In low dose sample ($\sim 5 \times 10^{21}$ fission m^{-3}), the trapped fraction remains low, varying from 16 down to 8 %. From a study of inert gas migration in high dose implanted metal and metal oxides, Matzke [4] has concluded that gas release takes place in various stages at different temperature intervals. In the case of Xe release from urania, he has attributed the noted 'stage IIA release' in the temperature interval of 1273–1573 K to true volume diffusion of gas atoms. The 'stage IIB release' noted in the higher temperature range has been attributed to the mobility of gas atoms from weakly trapped centers. The activation energy difference of 29–87 kJ mol^{-1} noted for the two stages is in close agreement with that derived from reactor irradiated sample (40–50 kJ mol^{-1} [49]) by Naik. The reported study of Shiba [3] showed that the leveled out trend of the Xe diffusivity (D_∞) holds good up to a burnup of approximately 1 % FIMA ($\sim 2.5 \times 10^{26}$ fission m^{-3} or ~ 10.3 GWd T^{-1}). At 1.45 % FIMA, their observation of significant deviation toward high D_∞ value suggests the change in regime from purely thermal diffusion to athermal mode such as the collapse of gas bubbles followed by venting out of the gas as discussed already.

Shiba [3] in their in-pile as well as out of pile experiments with thoria-based fuels of varying urania contents (0–20 mol%) showed that the released fraction of Xe is dependent on the UO_2 concentration and the release sharply falls from pure thoria with the increase in urania content attaining minimum value at U/Th ~ 0.05 before it starts increasing at higher U/Th values. When the samples with the varying urania concentrations are irradiated at different dose levels (10^{21}–10^{24} fission m^{-3}), the concentration dependence pattern remains the same and the extent of release reduces at higher dose as seen in the case of pure urania and levels out beyond the dose of 10^{23} fission m^{-3}. The transport coefficients measured at all the dose levels showed the UO_2 concentration dependency, though the attainment of minimum diffusivity at the stated concentration region of urania appears to be less apparent at around the dose of $\sim 1 \times 10^{22}$ fission m^{-3}. It is interesting to note that earlier reported study of capsule irradiation of $(Th,U)O_2$ pellet at high burnup [78] indicated no bearing of the U/Th ratio on the fission gas release.

4.2.2 Volatile Fission Products Transport Properties

For thoria-based fuel, the reported studies of transport behavior of the volatile FPs are a few as compared to those of the gaseous products. Early reports of out of pile and in-pile experimental results on iodine transport indicate that iodine diffuses at a comparable rate to xenon in the oxide fuel. The detailed studies of Davies and Long [70] and Turnbull et al. [72] for the release behaviors of Cs, Te, and I in low to high density urania fuels (60–96 % theoretical densities) over the temperature of 1200–2500 K showed that the volatiles have comparable release rates to Xe in low density sample. However, in high density samples, the same study showed that their release rates are higher than Xe. The transport and release behaviors of Rb, Cs, Te, Br, and I in urania, thoria, and their Simfuels were systematically studied by Matzke et al. and Verrall et al. by ion implantation followed by thermal annealing technique [67, 85–87]. Samples were ion implanted at different doses, 1×10^{15}–2×10^{20} ions m^{-2} to simulate the situations to low and high fission doses. Considering the range of the implanted species under the applied field energy (45–50 keV), the indicated high dose is representative of ~ 50 GWd T^{-1} burnup situation. At low doses ($<3 \times 10^{17}$ ions m^{-2}), the alkali metal FPs behaved in a similar way to Xe. Generally, the release from thoria is slow as compared that from urania. Nearly 80 % release is noted in the isochronal study up to a temperature of 1,800 K, and the release patterns from thoria and its SIMFUEL are nearly the same. The transports take place through volume diffusion of single atom. The early study [85] reports the transport parameters D_0 and Q as $10^{-4 \pm 1}$ m^2 s^{-1} and 418 ± 29 kJ mol^{-1}, respectively, for Cs in urania and thoria. When the ions are implanted at high dose (~ 1.5–2×10^{20} ions m^{-2}) the isochronal release study of Rb, Cs, Te, I, and Br in the fuel matrices revealed formation of segregates of the volatile species as microbubble/precipitate [88]. The studies covering the volatiles also showed that at the high dose, the transport behavior of the volatiles contrasted with that of Xe, though this was not so at very low implantation dose ($<1 \times 10^{15}$ ions m^{-2}). At the high implantation dose, the Cs/Rb release shows no retardation due to trapping phenomenon evidenced in the case of Xe. At the high dose, their release occurs at lower temperatures. The isochronal release study [67, 86, 87] of the implanted I showed similar result that the volatile comes out of the matrix at lower temperature when the doses of implantation were high.

The high release noted for the ion-implanted alkali metals sharply contrasts with the reported result of reactor irradiated fuels. The isochronal study with Cs in the thoria-6% urania by Akabori and Fukuda [89] showed that the alkali atom release is extremely low; a few atom percent is released in 400 h run at 1573 K. The transport parameter D_0 for Cs obtained using moderately irradiated samples (1×10^{24} fission m^{-3}) is about two orders of magnitude lower than the reported value from ion implantation experiments. The activation energy obtained in these two approaches is quite in agreement (see Table 3). At higher irradiation dose (3×10^{25} fission m^{-3}), the study of Akabori and Fukuda showed that the Cs release sharply falls and the derived activation energy rises by 200 kJ mol^{-1}. The authors invoked the trapping phenomenon at high dose in order to explain the lowering of Cs transport. The high yield of Cs can result its partitioning into the microbubbles/precipitates. The chemical states of Cs in the precipitates as well as within the fluorite lattice for the two cases, the reactor irradiated and the ion-implanted samples, can be quite different and this can make the difference of the release rates in the respective situations. Cs having a number of possible chemical states (gaseous atoms and oxides, ternary oxide phases, iodide and telluride), its resolution into the lattice and transport therein will be quite different than iodine for which the impediment in transport by trapping mechanism is not evidenced. Pechs et al. studied the redistribution behavior in LWR fuel that are operated at different power ratings. They showed that above the rating of 40 kW m^{-1} both Cs and I redistribute, which is influenced by power ramping [91].

Naik et al. [90] reported the iodine transport from reactor irradiated ThO_2-0.1% UO_2 pellet samples with bulk densities of 79–90 % theoretical value. Their result obtained in the temperature range of 723–1373 K indicated that I diffuses with activation energy of 236–248 kJ mol^{-1} (see Table 3). Considering that the activation energy is quite close to the energy involved in anion self diffusion in thoria (~ 270 kJ mol^{-1}) they concluded that I undergoes transport using the anionic sublattice. Their result further showed that by increasing the irradiation dose from 5.5×10^{20} to 1.1×10^{23} fission m^{-3}, there is two fold rise in the transport coefficient unlike the Xe case where the value of the coefficient dropped

Table 3 Summary of transport properties of I, Te, and Cs in UO_2- and ThO_2-based fuels

Authors	System studied	Diffusivity parameters (D_0 and Q)	Authors	System studied	Diffusivity parameters (D_0 and Q)
Matzke [85]	Cs/UO_2, low dose implantation	$10^{-4 \pm 1}$ m^2 s^{-1} 418 ± 29 kJ mol^{-1}	Naik [90]	I/ThO_2-0.1% UO_2 35–45 μm; ≤90 % TD; 24 m^2 kg^{-1}; 5 × 10^{20} fission m^{-3}	1.3 × 10^{-11} m^2 s^{-1} 236–248 kJ mol^{-1} **723–1373 K**
Prussin [73]	I/UO_2 7–10 μm; >90 % TD; 10^{19}–10^{20} fission m^{-3}	510 ± 80 kJ mol^{-1} (from apparent diffusivity) **1723–2023 K**	Kaimal [68]	I/ThO_2-2% UO_2 35–45 μm; >95 % TD; 24 m^2 kg^{-1}; 5 × 10^{20} fission m^{-3}	(10.2)10^{-12} m^2 s^{-1} 287 ± 19 kJ mol^{-1} **1290–1790 K**
Prussin [73]	Te/UO_2 7–10 μm; >90 % TD; 10^{19}–10^{20} fission m^{-3}	481 ± 29 kJ mol^{-1} (from apparent diffusivity data) **1723–2023 K**	Kaimal [68]	Te/ThO_2-2% UO_2 35–45 μm; >95 % TD; 24 m^2 kg^{-1}; 5 × 10^{20} fission m^{-3}	(2.2)10^{-4} m^2 s^{-1} 490 ± 23 kJ mol^{-1} **1290–1790 K**
Naik [90]	I/ThO_2 (urania doped); 35–45 μm; ≤90 % TD; 24 m^2 kg^{-1}; 5 × 10^{20} fission m^{-3}	1.1 × 10^{-10} m^2 s^{-1} 288 kJ mol^{-1} **1173–1373 K**	Shirsat and Ali [69]	I/ThO_2-2% UO_2 Simfuel (20 GWd T^{-1}) 35–45 μm; 97 % TD; 24 m^2 kg^{-1}; 5 × 10^{20} fission m^{-3}	(1.4)10^{-14} m^2 s^{-1} 162 ± 27 kJ mol^{-1} **1273–1773 K**
Akabori and Fukuda [89]	Cs/ThO_2-6% UO_2 11 and 52 μm; ~99.5 % TD; 1 × 10^{24} fission m^{-3}	1.16 × 10^{-6} m^2 s^{-1} 463 kJ mol^{-1} **1473–1773 K**	Shirsat and Ali [69]	Te/ThO_2-2% UO_2 Simfuel (20 GWd T^{-1}) 35–45 μm; 97 % TD; 25 m^2 kg^{-1}; 5 × 10^{20} fission m^{-3}	(3.9)10^{-12} m^2 s^{-1} 260 ± 25 kJ mol^{-1} **1273–1773 K**
Akabori and Fukuda [89]	Cs/ThO_2-6% UO_2 11 and 52 μm; 99.5 % TD; 1 × 10^{25} fission m^{-3}	4.46 × 10^{-1} m^2 s^{-1} 664 kJ mol^{-1} **1473–1773 K**			

significantly as discussed already. As noted already, the ion implantation experimental results also showed increased trend in I transport at higher dose. The contrasting behavior of increased transport of I at higher dose is expected from its nonzero affinity to the fuel matrix as compared to the inert gas atoms. The finite affinity for the anionic vacancy sites in the transport results in faster resolution from gas bubble/traps that are formed at significantly higher dose. Further, the radiation-induced defects/damages at high irradiation dose can aid the transport of the 'resolved' atom.

Prussin et al. [73] systematically studied the transport properties of Xe and the volatiles I, Te, Cs, etc. in irradiated high density urania by post-irradiation annealing technique where the kinetics of cumulative release of the species was measured by monitoring the residual activity in the sample during its isothermal annealing (1673–1923 K). His result showed (Table 3) very high values of activation energy in I and Te transports as compared to those reported by Naik et al. Moreover, the temperature trend of transport coefficient showed deviation from linearity in the Arrhenius plots. The wide difference in the activation barriers of I, Te, and Xe in urania/thoria derived from the PIA studies of the two groups, Naik et al. [90] and Prussin et al. [73], has been interpreted to be due to the difference in sintered densities of the fuels, namely 90 and 97 % for respective cases. The transport properties of I and Te was therefore reinvestigated in highly sintered Thoria-2mol% urania fuel (35–45 μm particles; >95 % TD; 24 $m^2\ kg^{-1}$; see Table 1) in a PIA setup designed for the release kinetics study of condensable vapors [68]. Using the experimental setup, the transport behaviors of the two volatile species were obtained in the trace-irradiated ($\sim 5 \times 10^{20}$ fission m^{-3}) fuel. The release characteristics in isothermal annealing at different temperatures (1290–1790 K) were carried out. Typical results of release kinetics are already shown in Fig. 16. The peculiarity of slowing down in release noted particularly in the case of I was interpreted to be due to the electrovalent bonding of iodine in CsI. The slowing down to a halt in I release and resumption of the release on stepping

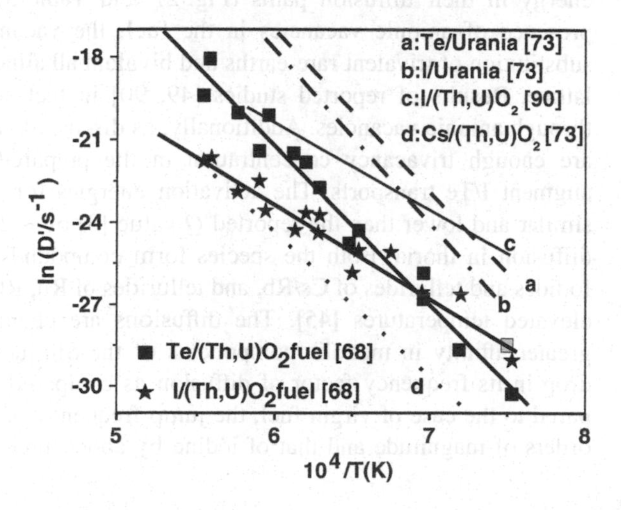

Fig. 20 Arrhenius plots of I and Te apparent diffusivity in thoria-2 mol% urania

up of the annealing temperature are indicative of ionic transport under the internal electric field. Unless the temperature is too high (Fig. 20), there is preferential transport of the electronegative element as compared to the cationic counterpart (Cs^{+}) that diffuses slower and result in drag by the electrostatic field. The temperature resetting process disturbs the internal electric field and thus the I release resumed for a while. In contrast with the release characteristic reported from ion-implanted samples [67, 85], the result of this PIA study shows insignificant release of I or Te, their overall release in the 4–5 h annealing period being less than 0.5 % at the highest temperature. The kinetic constant derived from the initial linear part of the plot of fractional release versus square root (annealing time) corroborates to the diffusional feature for the respective species (I and Te) in the thoria–urania-based virgin fuel. The activation energy and preexponential factor obtained from temperature trend of the kinetic constant are represented in the Arrhenius plot (Fig. 20) and also given in Table 3. The kinetic parameters of iodine approximately corroborate to the reported data of Naik et al.

Since the concentrations of I and Te in the trace-irradiated samples were very low, the obtained values essentially represent the single atom diffusion coefficients of the species at the different temperatures. For comparison sake, the reported behaviors of I and Te in pure urania fuel matrix [73, 90] are included in the table (Table 3) as well as in the figure (Fig. 20). The figure suggests that both the volatile species have slower transports in thoria than in urania. The frequency factors, D_0/a^2 for the respective species are seen to be significantly lower in thoria fuel. The lower frequency factor in thoria is indicative of poor affinities of the diffusing species to ThO_2 lattice. The low D_0 values are mainly responsible for the slower transports in thoria fuel. This is particularly true for iodine for which the diffusion barrier is lower in thoria.

There is little information on iodine and tellurium diffusion in burnup/simulated burnup fuel of thoria-based fuel. Compared to the virgin fuel case, it is seen that in simulated fuel lattice both the volatile species, I and Te, encounter less barrier of energy in their diffusion paths (Fig. 21 and Table 3). This can be due to the presence of anionic vacancies in the fuel, the vacancies being formed by the substitution of trivalent rare-earths and bivalent alkaline earths in the MO_2 fluorite lattice. Results of reported studies [49, 90] in fact suggest that iodine diffuses through anionic vacancies. Additionally, as discussed under the Xe diffusion, there are enough trivacancy concentration in the prepared Simfuel matrix that can augment I/Te transports. The activation energies for Te and I are, respectively, similar and lower than the reported Q value [3] of $\sim$270 kJ mol^{-1} for anion self diffusion in thoria. Both the species form compounds with many FPs elements. Iodides and tellurides of Cs/Rb, and tellurides of Ru, Rh, and Pd are quite stable at elevated temperatures [45]. The diffusions are chemically driven. Tellurium's greater affinity in metallic components of the Simfuel, is reflected in the larger drop in its frequency factor of diffusion as compared to that of iodine. As compared to the case of virgin fuel, the jump frequency of Te dropped by about eight orders of magnitude and that of iodine by about three orders only. The opposing

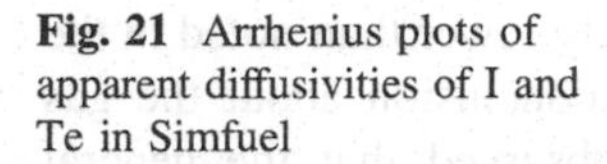
Fig. 21 Arrhenius plots of apparent diffusivities of I and Te in Simfuel

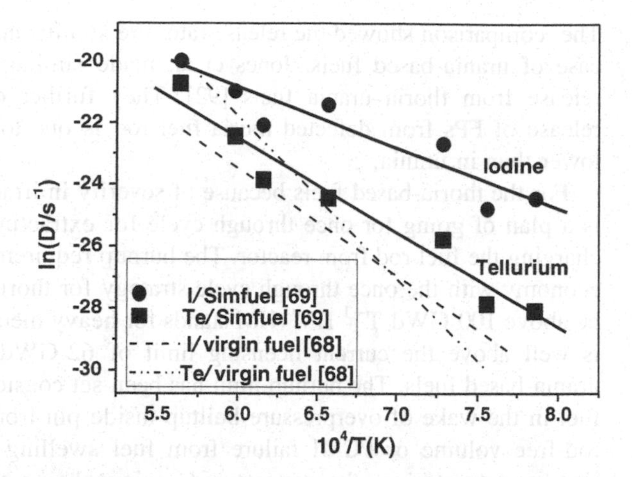

effects of jump frequency and activation barrier energy have been reflected in the kinetic coefficient. At higher temperature (>1600 K), the kinetic coefficient of tellurium has therefore fallen below that in virgin fuel (Fig. 21). As for iodine, this effect is not manifested within the experimental temperature range.

Knowledge of the transport properties of the metallic FPs such as Pd, Ag, Sn, Sb, and Cd that are quite volatile under the fuel temperatures is also important for understanding their segregation in cooler region of the pin. This is particularly important in fast reactor fuel where the temperatures are significantly high [45, 66]. For the case, it is known that Pd-based alloy concentrates in the fuel periphery and also in the central hole of the pin. In the open literature, there is however no data of transport properties of these metallic components in thoria-based fuels. The release behaviors of molecular species such as MoO_3, ReO_3, and Cs_2O are also important for understanding their redistribution particularly reaching in the fuel–clad gap. Generally, the molecular species are formed from the diffused out atomic/ionic species that are settled at grain boundary, microcracks, and micropore regions. The molecular species move mainly through grain boundary and crack, and their overall release kinetics is decided by the channels accessible for the chemical vapor transports.

4.2.3 Transport Properties in High Burnup Thoria Fuels

Consistent with the transport behaviors of FP species noted in the diffusion studies of trace-irradiated ThO_2 and UO_2 lattices, the reactor irradiated test pins of thoria-based fuels have shown lower fission gas release than urania under equivalent operating conditions [78, 92, 93]. The gas segregation in intergrain space and burst release under reactor irradiations are also known to be less in thoria as compared to those in urania. Goldberg have measured the in-pile release of fission gases on fuel rods with varying urania contents and compared the release for different power ratings and burnups with the established results of urania fuels [78, 93].

The comparison showed the release rates are significantly lower than noted in the case of urania-based fuels. Jones et al. made similar conclusion about the gas release from thoria–urania fuels [92]. They further observed that the general release of FPs from defected thoria fuel rod is one to two orders of magnitude lower than in urania.

For the thoria-based fuels because of severity in irradiated fuel handling, there is a plan of going for once through cycle for extracting more power before discharging the fuel rod from reactor. The burnup requirement in attaining an overall economy with the once through cycle strategy for thoria-based fuel is reported to be above 100 GWd T^{-1} HM (HM stands for heavy metals) [83]. This requirement is well above the current licensing limit of 62 GWd T^{-1} (in USA) for LWR urania-based fuels. The burnup limit has been set considering safe performance of fuel in the wake of overpressure builtup inside pin from excessive gas release in rod-free volume or PCMI failure from fuel swelling due to gas retention and matrix restructuring; the restructured regions being the hot central zone with excessive grain growth and the peripheral zone (fuel rim) with extensive subdivision of grains of HBS. As compared to urania-based fuel, the extent of gas release remaining low in thoria, the swelling from retained gases is the main issue in proving smooth performance up to the envisaged discharge burnup of 100 GWd T^{-1}. Models for evaluating the thoria-based fuels performance in the high burnup situation have been reported and their predicted results have been compared with the known results of test fuels' performances in LWRs. Such studies generally conclude that the thoria-based fuels show better performance than urania under normal as well as off-normal situation like the designed 'reactivity initiated accidents'.

Reported evaluation [83] also shows that the extent of HBS formation and gas retention in HBS are comparatively less in the case of thoria-based fuel. Considering the energy spectrum of epithermal neutrons and absorption cross sections of the neutrons for the two fertile nuclei, ^{238}U and ^{232}Th, it has been noted that the former nucleus has wider resonance range and stronger absorption as compared to the latter one. Moreover, ^{238}U has larger number of resonance absorption peaks in the most effective energy range of 1–100 eV neutrons. The less severity of rim effect in thoria-based fuel follows from the reported comparison of power profiles for the two types of fuels.

As for the gas/volatile release during off-normal situation of the nuclear reactor, the reported information of augmented transport and release [70, 71, 94] from urania due to increase in cationic vacancy concentration and augmentation of grain growth [95] would require reconsideration in the case of thoria-rich fuel. The oxygen incorporation in thoria-based fuel will be insignificant due its poor affinity for oxidation as well as oxygen transport. The gas/volatile release during off-normal situation of reactor operation will be mostly governed by the gas in plenum, where the accumulation takes place cumulatively over the irradiation period.

5 Summary

The diffusional transport properties of gas and volatile FPs in thoria matrix are quite different from those of the urania-based fuel matrix. The intragrain diffusion rates of species is slower in thoria as compared to those in urania. In the overall release process of species from fuel pellet, the participations of other thermally induced factors and contributions of athermal factors are approximately similar in the two oxide matrices. Rich experience available for thermal as well as athermal release factors in urania fuels set the guideline in the release.

Thoria-based fuel has less severe rim effect as compared to that in urania fuel. ^{238}U in small concentration in the ThO_2–UO_2 solid solution results in lower build up of ^{239}Pu. Further, there is less resonance capture in ^{232}Th that results in flatter distribution of the fissile daughter, ^{233}U, and flatter power profile. The reported enhancement in gas release from unrestructured region of urania fuel pellet during HBS development at rim is expected to be less in the case of thoria fuel.

References

1. Bakker K, Cordfunke EHP, Konings RJM, Schram RPC (1997) Critical evaluation of the thermal properties of ThO_2 and $Th_{1-y}U_yO_2$ and a survey of the literature data on $Th_{1-y}Pu_yO_2$. J Nucl Mater 250:1–12
2. Fink JK, Chasanov MG, Leibowitz L (1981) Thermophysical properties of uranium dioxide. J Nucl Mater 102:17–25
3. Shiba K (1992) Diffusion processes in thoria and thorium based oxides. In: Agarwala RP (ed) Diffusion processes in nuclear materials. Elsevier, North Holland
4. Matzke Hj (1992) Diffusion processes in nuclear fuels. In: Agarwala RP (ed) Diffusion processes in nuclear materials, Elsevier, North Holland
5. Basu (Ali) M, Mishra R, Bharadwaj SR, Das D (2010) Thermodynamic and transport properties of thoria–urania fuel of Advanced Heavy Water Reactor. J Nucl Mater 403:204–215
6. Uffelen PV (2002) Contribution to the modeling of fission gas release in light water reactor fuel. Dissertation, University of Liege
7. Speight MV (1969) A calculation on the migration of fission gas in material exhibiting precipitation and re-solution of gas atoms under irradiation. Nucl Sci Eng 37:180–185
8. Noirot J, Desgranges L, Lamontagne J (2008) Detailed characterisations of high burn-up structures in oxide fuels. J Nucl Mater 372:318–339
9. White RJ (2004) The development of grain-face porosity in irradiated oxide fuel. J Nucl Mater 325:61–77
10. Manzel R, Walker CT (2000) High burnup fuel microstructure and its effect on fuel rod performance. In: Proceedings of the international topical meeting LWR fuel performance, American Nuclear Society, LaGrange Park, UT, IL
11. Hj Matzke (1980) Gas release mechanisms in UO_2—a critical review. Rad Effects 53:219–242
12. Nuclear fuel behavior under reactivity initiated accident (RIA) condition, NEA/CSNI/R (2010)1, Nuclear Safety ISBN 978-92-64-99113-2
13. Manzel R, Walker CT (2002) EPMA and SEM of fuel samples from PWR rods with an average burn-up of around 100 MWd/kgHM. J Nucl Mater 301:170–182

14. Hj Matzke, Spino J (1997) Formation of the rim structure in high burnup fuel. J Nucl Mater 248:170–179
15. Lassmann K, Walker CT, van de Laar J, Lindstrom F (1995) Modelling the high burnup UO_2 structure in LWR fuel. J Nucl Mater 226:1–8
16. Hj Matzke, Kinoshita M (1997) Polygonization and high burnup structure in nuclear fuels. J Nucl Mater 247:108–115
17. Kinoshita M (1997) Towards the mathematical model of rim structure formation. J Nucl Mater 248:185–190
18. Hiernaut J-P, Wiss T, Colle J-Y, Thiele H, Walker CT, Goll W, Konings RJM (2008) Fission product release and microstructure changes during laboratory annealing of a very high burn-up fuel specimen. J Nucl Mater 377:313–324
19. Kinoshita M, Sonoda T, Kitajima S, Sasahara A, Kolstad E, Matzke Hj, Rondinella VV, Stalios AD, Walker C T, Ray ILF, Scheindlin M, Halton D, Ronchi C (2000) High burnup rim project (II): irradiation and examination to investigate rim-structured fuel. In: Proceedings of the ANS topical meeting on LWR fuel performance, LaGrange Park, UT
20. Spino J, Rest J, Goll W, Walker CT (2005) Matrix swelling rate and cavity volume balance of UO_2 fuels at high burn-up. J Nucl Mater 346:131–144
21. Spino J, Stalios AD, Santa Cruz H, Baron D (2006) Stereological evolution of the rim structure in PWR-fuels at prolonged irradiation: dependencies with burn-up and temperature. J Nucl Mater 354:66–84
22. Desgranges L, Valot C, Pasquet B, Lamontagne J, Blay T, Roure I (2008) A method for the quantification of total xenon concentration in irradiated nuclear fuel with SIMS and EPMA. Nucl Instrum Meth B 266:147–154
23. Walker CT, Bremier S, Portier S, Hasnaoui R, Goll W (2009) SIMS analysis of an UO_2 fuel irradiated at low temperature to 65 MWd/kgHM. J Nucl Mater 393:212–223
24. Walker CT, Brémier S, Pöml P, Papaioannou D, Bottomley PWD (2012) Microbeam analysis of irradiated nuclear fuel. Mater Sci Eng 32:1–15
25. Rondinella VV, Wiss T (2010) The high burn-up structure in nuclear fuel. Materials Today 13:24–32
26. Amaya M, Sugiyama T, Fuketa T (2004) Fission gas release in irradiated UO_2 fuel at burnup of 45 GWd T^{-1} during simulated reactivity initiated accident (RIA) condition. J Nucl Sci Technol 41:966–972
27. Spino J, Cobos-Sabate J, Rousseau F (2003) Room-temperature micro-indentation behavior of LWR-fuels, part 1: Fuel microhardness. J Nucl Mater 322:204–216
28. Shewmon P (1989) Diffusion in solids, 3rd edn. The Minerals, Metals & Materials Society, Warrendale, PA
29. White RJ, Tueker MO (1983) A new fission gas release model. J Nucl Mater 118:1–38
30. Wallin H, Turnbull JA (1994) IFA-563 isothermal wafer fuel test: irradiation history and fission gas release measurements up to 25 MWd/kg UO_2. Technical report HWR-411, OECD Halden Reactor Project
31. Blair P (2008) modelling of fission gas behaviour in high burnup nuclear fuel. Dissertation, École Polytechnique Fédérale de Lausanne
32. Hj Matzke (1986) Diffusion in ceramic oxide systems. Adv Ceram (Am Ceram Soc) 17:1–56
33. Olander DR, Uffelen PV (2001) On the role of grain boundary diffusion in fission gas release. J Nucl Mater 288:137–147
34. Turnbull A, Friskney CA (1975) The release of fission products from nuclear fuel during irradiation by both lattice and grain boundary diffusion. J Nucl Mater 58:31–38
35. Reynolds GL, Burton B (1979) Grain-boundary diffusion in uranium dioxide: the correlation between sintering and creep and a reinterpretation of creep mechanism. J Nucl Mater 82:22–25
36. Olander DR (1976) Fundamental aspects of nuclear reactor fuel elements, chapter 14. Technical Information Centre—Energy Research and Development Administration, University of California, Berkeley, TID-26711-P1 or ISBN 0-87079-031-5

37. White RJ (1994) Equi-axed and columnar grain growth in UO_2. In: Proceedings of the IAEA technical committee meeting on water reactor fuel element modelling at high burnup and its experimental support, Windermere, UK, pp 419–427
38. Atkinson HV (1988) Theories of normal grain growth in pure single phase systems. Acta Metall 36:469–491
39. Ainscough LB, Oldfield BW, Ware JO (1974) Isothermal grain growth kinetics in sintered UO_2 pellets. J Nucl Mater 49:117–128
40. Hillert M (1965) On the theory of normal and abnormal grain growth. Acta Metall 13:227–238
41. Nichols FA (1969) Kinetics of diffusional motion of pores in solids—a review. J Nucl Mater 30:143–165
42. Nichols FA, Ronchi C (1986) On the mobility of fission-gas bubbles. In: Hestings IJ (ed) Fission-product behavior in ceramic oxide fuel. Advances in ceramics, vol 17. The American Ceramic Society, pp 85–93
43. Olander DR (1976) Fundamental aspects of nuclear reactor fuel elements, chapter 13. In: Technical Information Centre—Energy Research and Development Administration, University of California, Berkeley, TID-26711-P1
44. Lawrence GT (1978) A review of the diffusion coefficient of fission-product rare gases in uranium dioxide. J Nucl Mater 71:195–218
45. Bailly H, Menessier D, Prunier C (1999) The nuclear fuel of pressurized water reactors and fast reactors-design and behavior. Intercept Ltd, Paris
46. Olander DR (1976) Fundamental aspects of nuclear reactor fuel elements. TID-26711-P1. ISBN 0-87079-031-5
47. Une K, Kashibe S (1990) Fission gas release during post irradiation annealing of BWR fuels. J Nucl Sci Technol 27:1002–1016
48. Walker CT, Knappik P, Mogensen M (1988) Concerning the development of grain face bubbles and fission gas release in UO_2 fuel. J Nucl Mater 160:10–23
49. Naik MC (1992) Diffusion controlled and burst release of gaseous and volatile fission products from UO_2 and ThO_2. In: Agarwala RP (ed) Diffusion processes in nuclear materials. Elsevier, North Holland
50. Kogai T(1992) A simple fission gas release/sweeling model. IAEA-TECDOC-697
51. Cornell RM (1971) Electron microscope examination of matrix fission-gas bubbles in irradiated uranium dioxide. J Nucl Mater 38:319–328
52. Lewis BJ (1987) Fission product release from nuclear fuel by recoil and knock-out. J Nucl Mater 148:28–42
53. Forsberg K, Lindström F, Massih AR (1994) Modeling of some high burnup phenomena in nuclear fuel. In: Proceedings of the IAEA committee meeting on water reactor fuel element modelling at high burnup and its experimental support, Windermere, pp 251–275
54. Booth AH, Rymer GT (1958) Determination of the diffusion constant of fission xenon in UO_2 crystals and sintered compacts. Report CRDC-720
55. Friskney CA, Speight MV (1976) A calculation on the in-pile diffusional release of fission products forming a general decay chain. J Nucl Mater 62:89–94
56. Chan P (2011) UNENE fuel engineering course, part 6: Fission product behaviour. http://www.unene.ca/ep704-fuel-2011/index.htm
57. Kolstad E (2007) On LWR fuel behaviour during normal operation. In: 7th international conference on WWER fuel performance, modelling and experimental support, Albena
58. Vitanza C, Graziani U, Fordest rommen NT, Vilpponen KO (1978) Fission gas release from in-pile measurements. OECD Technical report, HPR-221.10
59. Kidson GV (1980) A generalized analysis of the cumulative diffusional release of fission product gases from an "equivalent sphere" of UO_2. J Nucl Mater 88:299–308
60. Berna GA, Beyer CE, Davis KL, Lanning DD (1997) FRAPCON-3: a computer code for the calculation of the steady- state thermal-mechanical behavior of oxide fuel rods for high burnup. Technical report, NUREG/CR-6534

61. Kogai T (1997) Modelling of fission gas release and gaseous swelling of light water reactor fuels. J Nucl Mater 244:131–140
62. Matthews JR, Wood MH (1980) A simple operational gas release and sweeling model: II. Grain boundary gas. J Nucl Mater 91:241–256
63. Darcy H (1856) Wikipedia.org/wiki/Darcy's_law
64. Bremier S, Walker CT (2002) Radiation-enhanced diffusion and fission gas release from recrystallised grains in high burn-up UO_2 nuclear fuel. Rad Eff Defect 157:311–322
65. Nicoll S, Hj Matzke, Catlow CRA (1995) A computational study of the effect of Xe concentration on the behaviour of single Xe atoms in UO_2. J Nucl Mater 226:51–57
66. Kleykamp H (1985) The chemical state of the fission products in oxide fuels. J Nucl Mater 131:221–246
67. Hj Matzke, Verrall RA (1991) Release of volatile fission products from ThO_2 with a simulated burnup of 4 at%. J Nucl Mater 182:261–264
68. Kaimal KNG, Kerkar AS, Shirsat AN, Das D, Datta A, Nair AGC, Manohar SB (2003) Transport properties of iodine and tellurium in a thoria-2mol% urania matrix. J Nucl Mater 317:189–194
69. Shirsat AN, Ali (Basu) M, Kolay S, Datta A, Das D (2009) Transport properties of I, Te and Xe in thoria-urania SIMFUEL. J Nucl Mater 392:16–21
70. Davies D, Long G (1963) The emission of xenon 133 from lightly irradiated uranium dioxide spheres and powders. AERE report 4347, Atomic Energy Research Establishment
71. Miekeley W, Felix F (1972) Effect of stoichiometry on diffusion of xenon in UO_2. J Nucl Mater 42:297–306
72. Turnbull JA, Friskney CA, Findlay JR, Johnson FA, Walter AJ (1982) The diffusion coefficients of gaseous and volatile species during the irradiation of uranium dioxide. J Nucl Mater 107:168–184
73. Prussin SG, Olander DR, Lau WK, Hansson L (1988) Release of fission products (Xe, I, Te, Cs, Mo and Tc) from polycrystalline UO_2. J Nucl Mater 154:25–37
74. Kaimal KNG, Naik MC, Paul AR (1989) Temperature dependency of diffusivity of xenon in high dose irradiated UO_2. J Nucl Mater 68:188–190
75. Hj Matzke (1967) Xenon migration and trapping in doped ThO_2. J Nucl Mater 21:190–198
76. Naik MC, Paul AR, Kaimal KNG, Karkhanavala MD (1975) Effect of burnup on release of xenon from thoria. Rad Effects 25:73–77
77. Naik MC, Paul AR, Kaimal KNG, Karkhanavala MD (1977) Release of xenon from low-dose irradiated thoria pellets. J Nucl Mater 67:239–243
78. Goldberg I et al (1977) Fission gas release from ThO_2 and ThO_2-UO_2 fuels. Trans Am Nucl Soc 27:308–310
79. Naik MC, Paul AR, Kaimal KNG (1981) Migration and trapping of fission xenon in urania-doped thoria. J Nucl Mater 96:57–63
80. Shiba K, Itoh A, Ugajin M (1981) Fission xenon release from lightly irradiated (Th,U)O2 powders. J Nucl Mater 96:255–260
81. Shiba K, Itoh A, Akabori M (1984) The mechanisms of fission gas release from (Th,U)O2. J Nucl Mater 126:18–24
82. Olander DR (1987) Lattice and grain boundary diffusion in polycrystalline thoria-urania. J Nucl Mater 144:105–109
83. Long Y (2002) Modeling the performance of high burnup thoria and urania PWT fuel. Dissertation, MIT
84. Catlow CRA (1978) Fission gas diffusion in uranium dioxide. Proc R Soc Lond A Math Phys Sci 364:473–497
85. Hj Matzke (1967) The release of some non-gaseous fission products from CaF_2, UO_2 and ThO_2. J Nucl Mater 23:209–221
86. Verrall RA, Matzke Hj, Ogawa T, Palmer BJF (1986) Iodine release and bubble formation in oxide fuels. Canada report AECL-9475, p 558
87. Matzke Hj, Ray ILF, Verrall R (1987) Diffusion and behaviour of iodine and rubidium in UO_2 and ThO_2. IAEA report IWGFPT/27, p 183

88. Chkuaseli VF, Hj Matzke (1995) Volatile fission product bubble behaviour in uranium dioxide. J Nucl Mater 223:61–66
89. Akabori M, Fukuda K (1991) Release behavior of cesium in irradiated (Th,U)O_2. J Nucl Mater 186:47–53
90. Naik MC, Paul AR, Kaimal KNG, Karkhanavala MD (1977) Release of iodine from sintered thoria pellets at low temperatures. J Nucl Mater 71:105–109
91. Pechs M, Menzel R, Scheighofer W, Hass W, Hass E, Wurtz R (1981) On the behaviour of cesium and iodine in light water reactor fuel rods. J Nucl Mater 97:157–164
92. Jones RW, Lee HR, Hahn H, Walker JF, Celli A (1977) Thoria fuel technology for CANDU-PHW reactors. Trans Am Nucl Soc 27:303–304
93. Goldberg I, Waldman LA, Giovengo JF, Campbell WR (1979) Fission gas release and grain growth in ThO_2-UO_2 fuel irradiated at high temperature. WAPD-TM-1350 Addendum
94. Lewis BJ, Iglesias FC, Cox DS, Gheorghiu E (1990) A model for fission gas release and fuel oxidation behavior for defected UO_2 fuel elements. Nucl Technol 92:353
95. Mansouri MA, Olander DR (1998) Fission product release from trace irradiated UO_{2+x}. J Nucl Mater 254:22–33

Fabrication Technologies for ThO_2-based Fuel

S. K. Mukerjee, T. R. G. Kutty, N. Kumar, Rajesh V. Pai and Arun Kumar

Abstract Fuel fabrication technology, particularly once through utilization of uranium, has been mastered over the years. However, with the increasing demand of electricity at an affordable cost and depleting resources of uranium, introduction of thorium in the fuel cycle has become essential. The large-scale utilization of thorium requires adoption of closed fuel cycle scheme. Many of the fuel cycle technologies developed for uranium can be readily adopted for thorium, however, the man-rem problem associated with this fuel is a major concern. Therefore, fuel fabricators have, in recent past, initiated new R&D programs to solve this problem either through elimination of powder handling or making the unit operations of the production process amenable to remote handling and automation.

The status of development of these processes worldwide will be discussed in this chapter. Each process will be discussed in detail and its ability to achieve the desired objectives, particular related to reduction of man-rem exposure to the operator, will be critically reviewed.

1 Introduction

Fuel fabrication technology for industrial scale production of UO_2 has been mastered over the years. However, with the increasing demand for CO_2 free electricity at an affordable cost and due to rapid depletion uranium resources, introduction of thorium in the fuel cycle has become essential. The large-scale utilization of thorium requires adoption of closed fuel cycle scheme. Many of the fuel cycle technologies developed for uranium can be readily adopted for thorium, however, the man-rem problem associated with this fuel is a major concern. Since the natural thorium, ^{232}Th, is not a fissile nuclide, an external startup fissile

S. K. Mukerjee (✉) · T. R. G. Kutty · N. Kumar · R. V. Pai · A. Kumar
Bhabha Atomic Research Centre, Trombay, Mumbai 400085, India
e-mail: smukerji@barc.gov.in

D. Das and S. R. Bharadwaj (eds.), *Thoria-based Nuclear Fuels*,
Green Energy and Technology, DOI: 10.1007/978-1-4471-5589-8_6,

material has to be added to a thorium-based fuel. This can be natural uranium, enriched in ^{235}U, or plutonium from reprocessing of nuclear reactor fuel. During irradiation of thorium fuel, ^{233}U is formed as a result of neutron capture in ^{232}Th.

This ^{233}U is a highly fissile nuclide which significantly contributes to the power production, like ^{239}Pu in the uranium cycle. ^{233}U can be recovered through reprocessing of thorium fuel and can subsequently be used as a fuel. If there is a net production of ^{233}U at the end of the cycle, the amount of the external startup material to be added to the fresh fuel levels off to an equilibrium value after a number of cycles. In the ideal case, addition of external material is not necessary as the discharged fuel contains sufficient ^{233}U for a subsequent fuel loading, then one speaks of a self-sustaining equilibrium thorium cycle [1–4].

Th-U cycle offers several advantages over the U–Pu cycle. Some of them are: (a) high actinide burnup, (b) inherent proliferation resistance, (c) improved stability due to low fuel temperatures, (d) low volume of waste generation, (e) low fabrication cost, and (f) low fuel failure rate [1]. However, there are some problems associated with Th cycle, which are listed below [4–7]:

1. The reprocessed ^{233}U is always associated with ^{232}U, whose daughter products are hard gamma emitters and therefore require shielded operation. Also the alpha activity of ^{233}U is three orders of magnitude higher than that of HEU and just one order of magnitude less than that of weapons grade plutonium.
2. In reactor, ^{232}Th on neutron absorption produces ^{233}Th which decays first to ^{233}Pa before converting into ^{233}U as shown in Eq. (1).

$$^{232}\mathrm{Th}_{90} \longrightarrow (n,\gamma)^{233}\mathrm{Th}_{90} \longrightarrow \beta^{-1}\left(t_{\frac{1}{2}} = 22.3\ \mathrm{m}\right) {}^{233}\mathrm{Pa}_{91} \longrightarrow \beta^{-1}\left(t_{\frac{1}{2}} = 27\mathrm{d}\right) {}^{233}\mathrm{U}_{92} \quad (1)$$

^{233}Pa has a half-life of 27 days. This rather long half-life of ^{233}Pa results in a reactivity increase after reactor shuT.D.own due to ^{233}U production and this must be taken into account.

3. Thorium recovered from the irradiated spent fuel will contain ^{228}Th ($t_{1/2}$ = 1.91 years) isotope. Handling of recovered thorium will be difficult if processed after a short cooling period due to the presence of hard gamma emitters (^{212}Bi and ^{208}Tl) formed during ^{228}Th decay [8, 9]. Approximately, 20 years ($\sim$10 half-life of ^{228}Th) of cooling is required to bring down the gamma dose.

1.1 Problems in Fabrication of Fuels Containing $^{233}UO_2$

With the advent of the thorium fuel cycle wherein large quantities of ^{233}U will be generated and handled, methods for containment of this isotope become increasingly important. ^{233}U, which has a half-life of 1.62×10^5 years, becomes more

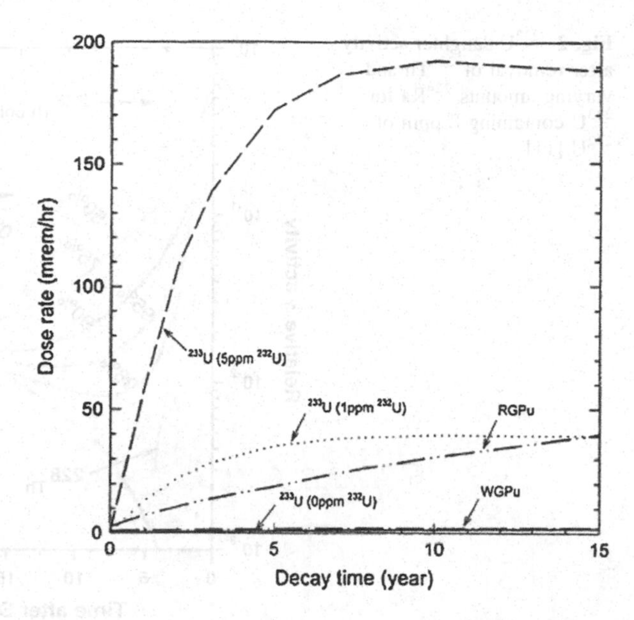

Fig. 1 Radiation dose rate buildup at 0.5 m from 5 kg spheres of freshly separated ^{233}U containing different amounts of ^{232}U and Pu of different grade ([9], Copyright Taylor and Francis)

hazardous because of its ^{232}U content, which varies from a few ppm to >500 ppm, depending upon the flux characteristics of the reactor in which Th was irradiated and the duration of the exposure [4]. ^{232}U has a half-life of around 68 years and first four members of its daughter chain are highly energetic α- emitters (refer Aqueous Reprocessing by THOREX Process, Fig. 3), with half-lives ranging from 0.15 s to 1.9 year. Therefore, it is obvious that ^{233}U containing even few ppm of ^{232}U, upon inhalation, would significantly increase the radiotoxicity [9].

The subsequent daughters of ^{232}U, namely, ^{212}Bi and ^{208}Tl, emit strong gamma radiations of 0.7–1.8 MeV and 2.6 MeV, respectively and makes heavy shielding mandatory for the protection of personnel handling aged ^{233}U. A general practice in handling ^{233}U containing ^{232}U in the range of about 10 ppm is to remove the ^{228}Th and ^{224}Ra, the longer lived daughters in the decay chain, as completely as possible and to allow the remaining short-lived daughters to decay before further processing [10–13]. Figure 1 shows the calculated buildup with time of the γ-dose rate at a distance of 0.5 m from 5 kg spheres of freshly separated ^{233}U containing 0, 1, and 5 ppm of ^{232}U. It may be noted that the dose rate is same for pure ^{233}U (with 0 ppm of ^{232}U) as that of weapon grade (WG) ^{239}Pu. For ^{233}U containing ^{232}U, the buildup with time indicates the ingrowth of ^{228}Th, which has a half-life of 1.9 years. It can also be seen from the figure that for ^{233}U containing 1 ppm of ^{232}U, the dose rate is nearly same as the reactor grade (RG) Pu [9].

Figure 2 shows the effect of removing all the ^{228}Th and various fractions of the ^{224}Ra from ^{233}U containing 8 ppm of ^{232}U. The safe radiation level that is tolerable in unshielded glove box operations is shown as the horizontal line. It may be seen that

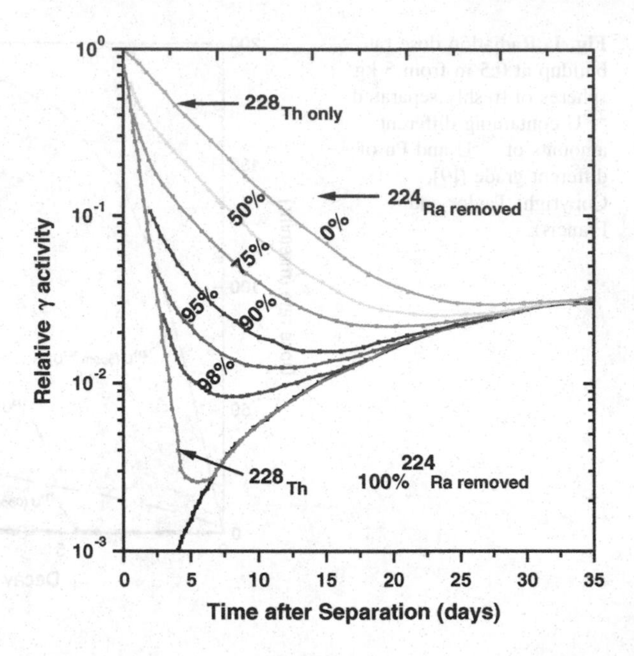

Fig. 2 ^{232}U daughter activity after removal of ^{228}Th and varying amounts ^{224}Ra for ^{233}U containing 8 ppm of ^{232}U [11]

removal of all ^{228}Th and >95 % of ^{224}Ra from the equilibrated stock (<10 ppm of ^{232}U) would result in U, which can be processed in unshielded glove boxes for a period of 1–3 weeks after interrupting the decay chain [11]. Thus, during the interval between about 4 and 25 days after separation, the various operations involved in converting the soluble ^{233}U compound into the desired fuel forms can be completed with a minimum of personnel exposure in lightly shielded facilities [12].

1.2 Historical Perspective

Thoria fabrication has been initiated in the US as early as in the 1960 s and 1970 s. Thoria fuel in Al-clad has been irradiated in Hanford and Savannah River reactors. The Hanford fabrication line was about 100 m^2 in size and fabrication was carried out in fume hood [4]. When fabrication of large quantities was involved, additional protection to workers like shielding with lead was employed.

1.2.1 Industrial Scale Production

There were two major campaigns in early 1960 s in US for the fabrication of ThO_2-UO_2 pellets:

1. Manufacturing fuel for the core of thermal reactor
2. Experimental Kilorod project.

Among the major fabrication lines, one should mention the semi-shielded Kilorod facility and remote thorium-based fabrication plant known as TURF of ORNL. The whole core demonstration of $(Th\text{-}U)O_2$ fuel happened in LWR in the 1960 s and 1970 s by two different approaches [14]:

1. Mixing of ThO_2 with highly enriched UO_2 in a uniform lattice at BORAX –IV, Indian Point I PWR and Elk River BWR power stations,
2. Using heterogeneous arrangement of seed and blanket regions, where blanket has less U in light water breeder reactor core (Shippingport reactor).

The methods of fabrication for the bulk thoria-based fuel were adapted from the large-scale manufacturing experience with UO_2 fuel pellets for LWRs.

Kilorod fabrication facility at Oak Ridge was operated during 1964 on a 10 kg/ day scale [4, 14]. A general practice employed in Kilorod project in handling ^{233}U containing ^{232}U in the range of about 10 ppm was to remove the ^{228}Th and ^{224}Ra, the longer lived daughters in the decay chain, as completely as possible and to allow the remaining short-lived daughters to decay before further processing [5]. Figure 7.4 in the next chapter can be referred for the ^{208}Tl activity development in ^{232}U decay for different post purification time in various purification processes. A "safe time window" of about one month is obtained for fabrication, when the various operations involved can be completed with minimum of personnel exposure in lightly shielded facilities. This time window depends upon the ^{232}U content and decreases with increase in ^{232}U content [4, 12].

In Kilorod process, ^{233}U solution ($\sim$5 days old) containing 38 ppm of ^{232}U was purified by solvent extraction and delivered to Kilorod facility at $\sim$40 days intervals. In each intervals, post purification time in sol–gel processing varied from about 5 to 25 days and in rod fabrication from about 9 to 28 days [10–12]. After each campaign about 7 days were required for post-campaign cleanout of equipment. The solvent extraction process removed the ^{228}Th, ^{224}Ra and ^{212}Pb daughters by factors of about 2,500, 5,000, and 100, respectively. Because of the low relative efficiency of removal of ^{212}Pb, the hard γ activity of the products decreased for about 2.5 days and then increased, approaching the activity of initial pure material after 5–10 days.

A totally remotized industrial scale fabrication plant was designed by ORNL engineers and built the TURF plant with remotely operated and manually maintained. This plant was designed to fabricate ThO_2–$^{233}UO_2$ fuel for water cooled reactors and carbide fuel for HTGRs. ^{232}U content in total U was expected around 500–800 ppm [4]. The shielding requirements of steel and concrete for such plants were based on the following assumptions [4]:

1. Time interval between solvent extraction and receipt at fabrication plant is $\sim$5 days.
2. After every 5 working days, a major cleanup of equipments and containers were undertaken.
3. No large quantities of materials were stored within 0.3 m of enclosure wall.
4. The amount of material retained in the equipment during processing was 3 kg.
5. The personnel exposure was limited to 40 mrem/week (0.4 mSv/week).

1.2.2 Remote Fabrication

Development of remote technology for the thorium fuel cycle is a major objective of the Thorium Utilization Program [11, 13]. The development of the ORNL sol–gel process for preparing ThO_2–UO_2 microspheres has established a firm starting point for this technology. The design, construction, and operation of the semi-remote Kilorod Facility for the fabrication of 1,100 rods of ThO_2-3 % $^{233}UO_2$ on a pilot plant scale was an important step toward this goal.

The shielding requirements for the fabrication of ^{233}U bearing fuels are shown in Fig. 3. It is observed that a thickness of 3.5 in. of steel is a practical limit for semi-remote fabrication because of the difficulty of working through a greater distance with gloved hands or tongs [4]. Also, radiation from sources requiring greater than 3.5 in. shielding greatly inhibit, contact maintenance of the equipment. In the case of equilibrium cycle HTGR fuel of burnup of approximately 100,000 MWD/t, one might expect from 40 to 70 ppm of ^{232}U in total heavy metal. Even at the lowest predicted ^{232}U level, it is observed from the figure that remote fabrication will be required at a plant capacity of any appreciable level [15, 16].

1.3 Fuel Fabrication Processes

There are many aspects which have to be addressed to when one selects a particular process flow sheet and the mode of fabrication to be employed for either first cycle or recycled fuel. A few important among them are radiotoxicity, quantity, and form of the fuel material which has to be fabricated. Fuels containing naturally occurring fissile ^{235}U in combination with fertile ^{238}U or ^{232}Th, emitting only alpha particles of relatively low-specific activity, can be manufactured by the so-called contact operations where the operator has direct contact with the fuel material. However,

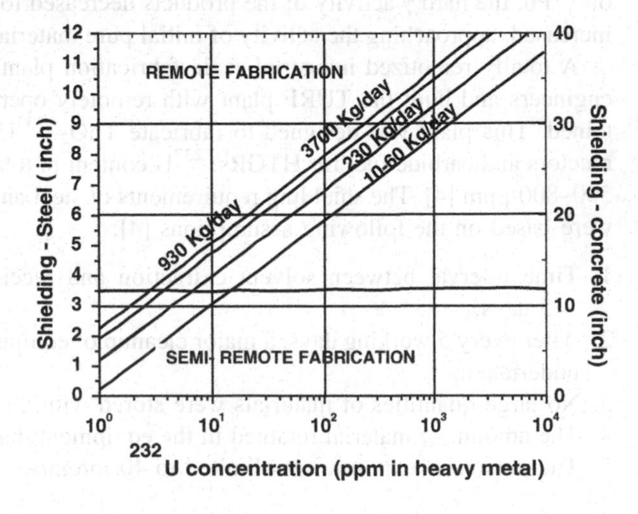

Fig. 3 Shielding requirements for plants fabricating ^{233}U fuel ([4], Courtesy: European Communities, 1997, Luxembourg)

process operations that involve generation and handling of fine powders of ^{235}U, ^{238}U, or ^{232}Th bearing fuels are carried out in ventilated enclosures for minimizing radioactive aerosol. The enclosures need not be hermetically sealed for handling ^{235}U, ^{238}U, or ^{232}Th bearing materials, if they are not pyrophoric. Glove box operations are those requiring hermetic sealing of equipment and are essential for handling highly radiotoxic plutonium and ^{233}U–bearing materials. During the past 4 decades, several countries have manufactured thorium-based oxide and non-oxide fuels in particulate (microspheres) and pellet forms by employing contact, hooded, glove box, semi-remote, and remote operations.

The following techniques have been developed so far for manufacturing ThO_2 and thoria-based mixed oxide fuels:

Powder-pellet route: for preparation of high density fuel pellets, using ThO_2, UO_2, and PuO_2 powders as starting materials; the fuel pellet stacks are encapsulated in cladding tubes.

Sol–gel microsphere pelletization: preparation of fuel microspheres using nitrate solutions of uranium, plutonium, and thorium as starting materials and adapting ammonia external gelation or ammonia internal gelation process for obtaining hydrated gel microspheres; these dust-free and free flowing sol–gel-derived oxide fuel microspheres are directly pelletized and sintered.

Coated agglomeration pelletization: ThO_2 mixed with a required amount of $^{233}UO_2$ powder is converted to free flowing agglomerate by powder extrusion route by mixing with an organic binder followed by extrusion through perforated rollers. These extrudes are shaped to spherical agglomerates of proper sizes in a spherodizer and then dried to remove the organic binder. These extrudes were coated with U_3O_8 powder, compacted and then sintered to obtain high density (Th, U)O_2 pellets.

Impregnation technique: where (a) partially sintered ThO_2 pellets of relatively low density ($\leq$75 % theoretical density) or (b) porous ThO_2 microspheres are vacuum impregnated in uranyl nitrate (U as ^{233}U) or Pu–nitrate solution followed by calcination & pelletization for microspheres followed by sintering to form high density ThO_2-based mixed oxide fuel pellets, which are encapsulated in cladding tubes.

2 Powder Preparation Methods

2.1 Preparation of Thoria Powder

Five different methods are reported for the production of thoria powders which are listed below [17, 18]:

1. *Thermal decomposition of oxalate precipitate*
2. *Direct denitration processes*

   (a) Thermal, (b) Hydrothermal, and (c) Microwave

3. *Spray calcine process*
4. *Combustion synthesis process*
5. *Freeze-dry process.*

Among the above, the oxalate precipitation was extensively used for the preparation of ThO_2 powder on commercial scale. The thermal decomposition of hydroxide and hydrothermal nitration were mainly used in the preparation of an oxide sol for the sol–gel process [17].

1. **Thermal decomposition of oxalate precipitate**

Thoria powder is prepared commercially by calcining the precipitate of hydrated thorium oxalate. The oxalate precipitation process consists of the following operations: (a) precipitation of thorium oxalate from nitrate solution by the addition of oxalic acid according to the equation,

$$Th(NO_3)_4 + 2\,H_2C_2O_4 + 6H_2O \rightarrow Th(C_2O_4)_2.6H_2O + 4HNO_3 \quad (2)$$

(b) digestion, (c) filtration, and (d) thermal decomposition of the oxalate to form the thorium oxide [14, 17].

The important variables which affect surface area, bulk density, particle size, and its distribution in the above precipitation reaction are:

(i) Precipitation temperature,
(ii) pH
(iii) Concentration of the reactants and their molar ratios
(iv) Sequence of combining reactants
(v) Temperature of calcination.

The properties of oxide powder have been found to differ depending on various parameters during precipitation and decomposition processes. Kantan et al. [19] showed that the precipitation and decomposition temperatures had a great influence on the final sintered density of oxalate-derived thoria powder. On varying precipitation temperatures from 1 to 60 °C and decomposition temperatures from 800 to 1,200 °C, the density of the sintered product varied from 86.7 to 94.7 % of T.D. The maximum density of 94.7 % T.D. was obtained using oxide powder precipitated at 20 °C and decomposed at 900 °C [17]. A lower precipitation temperature results in greater surface area, smaller particle size, and smaller bulk densities for the calcined ThO_2 powder [14]. White et al. [20] studied in detail the thorium oxalate precipitation variables such as temperature, agitation, and digestion time. They concluded that the precipitation temperature of 283 K with mechanical stirring and 15 min digestion time are the best combination that produces oxalate from which most sinterable ThO_2 powder can be derived. Lastly, the calcination temperature is the most important parameter determining the properties of both ThO_2 powder and the final product. Lower calcination temperature results in more active powder consisting of more porous particle with large surface area. If the calcination temperature is too low it leads to incomplete decomposition and entrapment of impurities such as carbon from the oxalate in thoria powder [14].

Balakrishna et al. [21] reported that addition of small quantities of Mg during oxalate precipitation improved the sinterability of thoria powder. They optimized the method of powder preparation for making high density pellets which consist of the following steps:

- Addition of $MgSO_4$ to the batch of oxalic acid (the residual Mg in powder ranged from 300 to 400 ppm).
- Continuous addition of pure thorium nitrate solution to this oxalic acid, to co-precipitate as thorium-magnesium oxalate.
- Calcination of oxalate at 900 °C in air to obtain sinterable thoria powder.

Although the calcination of thorium oxalate is usually carried out in air [22], the calcination in hydrogen atmosphere has also been reported [21, 22]. Balakrishna et al. [21] calcined the oxalate in air at 700, 750, 800, 900, and 1,000 °C in a tubular furnace. A few batches of the oxalate were separately calcined in hydrogen atmosphere at 750 °C. The samples after hydrogen calcination were reheated in air at 600 °C. BET specific surface area as a function of calcination temperature of thorium oxalate samples (calcined in air, each for 15 min) is shown in Fig. 4. The calcination of dried thorium oxalate in hydrogen at 750 °C resulted in black powder which had a carbon content of 0.36 wt%. On reheating the black powder at 600 °C in air restored its original white color of ThO_2. The carbon content decreased to 12 ppm after reoxidation. The carbon in the black powder is that left over from the starting oxalate. The calcination in H_2 prevents the oxidation of the left over carbon.

Thorium oxide powder obtained from calcination of thorium oxalate generally contained 0.05 wt% MgO as dopant. The calcined thoria powder, which varies in size from ~2 to 5 μm, has 'platelet' morphology as shown in Fig. 5. Compared to

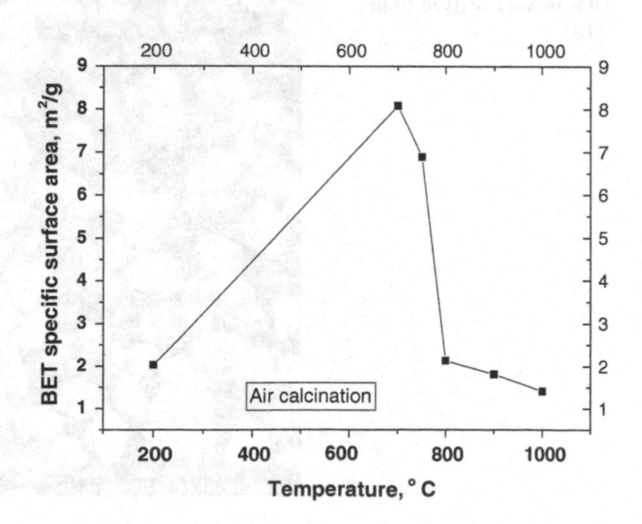

Fig. 4 Surface area as a function of calcination temperature of thorium oxalate samples in air ([21] Copyright Elsevier)

Fig. 5 SEM picture showing platelet morphology of thoria from thorium oxalate

this, the morphology of UO_2 powders, derived from ADU route is spherical in nature (Fig. 6). The premilling of the as-received thorium oxide for 8 h in a planetary ball mill is necessary to improve the surface area and sinterability of the powders. The above milling step breaks this platelet morphology of the powders by generating considerable amount of fines.

2. Direct denitration process

In this process, thorium nitrate is dissolved in deionized water at a 4 molar level and decomposed at 450 °C in quartz or alumina crucible. During this process, air

Fig. 6 The morphology of UO_2 powder derived from ADU

was bubbled through the solution which was stirred continuously through the drying operation [23–25]. The simplicity of this method has the advantage of using this method in a remote facility. However, the powder obtained by this process is coarse and pellets of density >90 % cannot be obtained using these powders [14].

3. **Spray calcine process**

The process consists of direct thermal conversion of a salt solution to an oxide powder by spraying the solution into a hot furnace with a collector at the bottom. This process is easy and requires rather simple equipment. Actinide oxide powders, both single and mixed, could be prepared in a similar manner using actinide nitrate as feed [17].

4. **Combustion synthesis process**

The combustion synthesis, also known as "self-propagating high temperature synthesis" is a novel technique used for the production of variety of materials. This process employs high exothermic heat of a chemical reaction between the fuel and oxidant to initiate the synthesis. The large exothermic reaction often results in the appearance of a flame whose temperature may exceed 1,000 K. All the volatile impurities are driven off at this high temperature resulting in purer products. The large volume of gases generated during the process, result in rapid cooling leading to nucleation without any substantial growth [26]. Also the gas evolution results in the disintegration of large particles or agglomerates resulting in very fine particles. The powders obtained by this route generally have high surface area and better sinterability. This process has earned significant interest due to its overall ease and less energy intensive. The combustion synthesis is advantageous over the solid state synthesis in terms of better compositional homogeneity and the purity of the final product.

Chandramouli et al. [27] prepared ceramic grade thoria powder by combustion synthesis using citric acid. The recipe for synthesis is to heat the aqueous solution of thorium nitrate and citric acid with citric acid/nitric acid ratio ≥ 1 preferably on a hotplate. Calcination of the product at 1,073 K gives a good quality thoria powder which can be compacted and sintered at 1,573 K to yield a density of 94 % T.D. Purohit et al. [28] made nanocrystalline thoria powder by combustion technique using glycine as a fuel and nitrate as an oxidizer. This technique involves the exothermic decomposition of viscous liquid prepared by thermal dehydration of aqueous solution containing thorium nitrate and glycine. For complete combustion of 1 mol of thorium nitrate, 2.2 mol of glycine are required as shown in the following stoichiometric redox reaction [28]:

$$Th(NO_3)_4 + 20/9(NH_2CH_2COOH) \rightarrow ThO_2 + 40/9CO_2 + 28/9N_2 + 50/9H_2O \tag{3}$$

Thoria powder of different crystallite size and surface area can be prepared by starting with different fuel to oxidant molar ratios. The exothermic decomposition of viscous liquid at about 200 °C containing thorium nitrate and glycine in the molar ratio 1:1.2 yields the well crystalline nano-sized thoria powder. Purohit et al. [28] reported that thoria powder prepared by this technique has high surface area (>50 m^2/g) which could be sintered to a density of >93 % T.D. at a relatively low temperature of 1,573 K for 3 h.

Ananthasivan et al. [29] used de-agglomeration of thorium oxalate for the synthesis of sinterable thoria powder. Thorium oxalate was obtained by slow addition of oxalic acid and filtered under suction. The precipitate is dispersed in water and in non-aqueous solvents and de-agglomerated by ultrasonication in both media. Thoria powder derived by this procedure could be sintered to a density of 9.7 g/cm^3 at a temperature as low as 1,673 K. The de-agglomeration technique is simple and does not pose the problems associated with self-sustained, uncontrolled burning encountered in the method based on combustion synthesis. Thoria powder obtained through non-aqueous precipitation using methanol, ethanol, propan-2-ol, and propan-2-one yielded nanocrystalline powders of 2–5 nm in size.

5. Freeze-dry process

The freeze-dry processing for powder production consists of:

- Mixing salt solution
- Flash freezing of the solution
- Sublimation of the ice phase
- Decomposition of anhydrous salt

The mixed nitrates of Th and U are ideally suited for producing powders of ThO_2-UO_2 crystallites by this process [14]. No material is added or lost in the process and therefore the composition of the final product would be that of the starting mixture. The powder produced by this method is granular and are highly sinterable and therefore amenable to simple fabrication process [30]. ThO_2, UO_2, and ThO_2-UO_2 powders have been fabricated in laboratory scale using this method. In this process, droplets of thorium uranyl nitrate solution are frozen by spraying it into a cooling bath. The frozen granules are vacuum dried, decomposed in Ar and calcined to produce thoria spheres having bulk density <1 g/cm^3 [14].

2.2 Preparation of ThO_2-UO_2 Powder

ThO_2 and UO_2 form a continuous series of solid solution. Hence the quick, economic and simplest method for making ThO_2-UO_2 pellets will be mixing and milling the components of the powder followed by compaction and sintering. ThO_2 is the only stable oxide of Th–O system in the condensed state and it has very little non-stoichiometry compared to UO_2 [31]. The crystal structure of ThO_2 is the fluorite type, isomorphous with UO_2, PuO_2, and CeO_2 and contains 4 Th atoms and 8 oxygen atoms per unit cell. Th^{4+} is the only valence state of thorium [14].

2.2.1 Preparation of ThO_2-UO_2 Powders by Mechanical Mixing

UO_2 powders are produced by any of the following methods:

1. Direct denitration of uranyl nitrate hexahydrate (UNH) to UO_3 followed by reduction to UO_2 by hydrogen.
2. Hydrolysis of UF_6 and precipitation of U as ammonium diuranate (ADU) or ammonium uranyl carbonate (AUC) which are then decomposed and reduced to UO_2.
3. Decomposition and reduction of UF_6 by superheated steam and hydrogen

An intimate mixing of ThO_2 and UO_2 powders followed by compaction will generally result in a homogeneous structure by diffusion in a reasonable time at temperatures greater than 0.5 T_m during sintering. The early process employed a double firing technique to ensure homogeneity between ThO_2 and UO_2. The ThO_2 and UO_2 powders were blended and milled in the required proportion and then compacted and then calcined at 1,973 K in H_2 to form a partial ThO_2-UO_2 solid solution [4, 14]. The resulting pellets are crushed, milled, again compacted, and then sintered for a second time in H_2 at 2,023 K to obtain pellets having complete solid solution formation. In the subsequent development, calcination step was eliminated and therefore, adopted a single firing process to initiate and complete the solid solution formation. The single step process is economical, time saving, and reduces personnel exposure. Two major techniques for the comminution and mixing of ThO_2-UO_2 powders were investigated which are listed below [14]:

1. For the process like micronizing, the energy for grinding was supplied by a fluid stream,
2. For the process like planetary ball milling, the energy was supplied by the motion of solid dense milling media.

Among the above, the micronization is the only method that constantly produced dense and homogeneous ThO_2-UO_2 pellets. In this process, the temperature is not raised appreciably during communition and no contamination pickup from the equipment has been noticed.

The above-mentioned single sinter method was used to manufacture ThO_2-UO_2 pellets (maximum UO_2 content <6 wt%) for LWBR core. The core contained about 4,29,000 kg of thorium and 500 kg of uranium in 1.6 million ThO_2-UO_2 pellets and 1.3 million ThO_2 pellets [17]. The key steps employed for the fabrication of high density ThO_2 and ThO_2-UO_2 fuel pellets are:

1. Blending
2. Micronizing
3. Secondary blending
4. Agglomeration
5. Compaction
6. Pretreatment and sintering

Here, the process consists of micronizing ThO_2-UO_2 blended mixtures twice followed by reblending. An organic binder was added to the powder mixture to maximize the green density. A CO_2 binder removal treatment at 1,198 K was incorporated which is followed by sintering at 2,023–2,063 K in H_2 atmosphere.

2.2.2 Preparation of ThO_2-UO_2 Powders by Co-Precipitation

Among the various techniques, the co-precipitation method gives an excellent route to make a very homogeneous mixture. The co-precipitation process had been given a low priority by fuel manufacturers since it involves handling of the liquid waste. However, if the co-precipitation process is incorporated in the reprocessing plant, then this method would become really advantageous [32–36]. The advantages of the wet process include [37]:

1. Very low generation of radiotoxic dust
2. Easy availability of cheap reactants
3. Reduction of the accessibility to pure plutonium or other fissile actinides, and reduction of risks of proliferation.
4. This method is easily amenable for glove box operation.

A number of reports are available on the production of uranium–plutonium mixed oxide by co-precipitation method but information on thorium–uranium mixed oxide using the above method is scanty [37]. In AECL, studies were made for the co-precipitation process using ammonia [1]. To the nitrate solution of U(VI) and Th(IV), ammonia solution was added to form ammonium diuranate and thorium hydroxide co-precipitate. The precipitate is calcined to form blended ThO_2 and UO_2 powder, which is subsequently pressed into fuel pellets. Radford et al. [37] prepared ThO_2-6 wt% UO_2 powders by the co-precipitation from mixed nitrate solution using NH_4OH. They changed the calcination temperature and studied the physical properties of the obtained pellets. Atlas et al. [38]. prepared ThO_2-20 wt% UO_2 pellets via co-precipitation of mixed oxalate from nitrate solutions by adding excess oxalic acid and studied various parameters which may affect the powder properties. White et al. [39] investigated precipitation temperature, agitation, and digestion time for the preparation of ThO_2-25 % UO_2. Argo [40] prepared ThO_2-20 % UO_2, ThO_2-35 % UO_2, and ThO_2-50 % UO_2 pellets (% in weight) by the co-precipitation method. Here, U(VI) nitrate solution was first reduced to U(IV) by adding a six-fold excess sodium formaldehyde sulfoxylate at room temperature. To form the co-precipitate, oxalic acid of six-fold excess of the stoichiometric amount, dissolved in distilled water was added to the mixed nitrate solution of Th(IV) and U(IV). The precipitate was separated by vacuum filtration and dried in air at room temperature.

The procedure for the fabrication of ThO_2-30 % UO_2 and ThO_2-50 % UO_2 powders by co-precipitation route followed in BARC consists of the following steps [36]:

(a) Preparation uranyl and thorium nitrate solution
(b) Reduction of U ions from (VI) to (IV) valency state
(c) Mixing of the solutions to the intended U to Th ratio
(d) Co-precipitation using oxalic acid
(e) Calcination

The starting solutions [36] are uranyl nitrate and thorium nitrate solutions. The concentration of uranyl nitrate solution used in the study conducted at BARC was 200 g/l (in 1 M HNO_3) while that of thorium nitrate was 300 g/l in water. The oxidation states of thorium and uranium were 4+ and 6+, respectively. Uranium was reduced to U(IV) in the initial stage to get a homogeneous product [41]. The reduction of uranyl nitrate solution was carried out by hydrazine with the help of platinum oxide as a catalyst. The product obtained after reduction contained more than 96 % U (IV) in ~0.1 M N_2H_4 and 0.4 M HNO_3. Then, U(IV) and Th(IV) nitrate solutions were mixed into the specified U:Th ratio.

The precipitation experiment was conducted with 10 % oxalic acid (0.79 M) solution. The oxalic acid of 0.1 M excess amount was used. The precipitation reaction can be represented by the following equation:

$$\begin{aligned}(1-y)\mathrm{Th(NO_3)_4} + y\mathrm{U(NO_3)_4} + 2\mathrm{H_2C_2O_4} \\ \rightarrow \left[(1-y)\mathrm{Th(C_2O_4)_2} + y\mathrm{U(C_2O_4)_2}\right] \cdot n\mathrm{H_2O} + 4\mathrm{HNO_3}\end{aligned} \quad (4)$$

where y is the mole fraction of uranium. The $(Th,U)(C_2O_4)_2 . nH_2O$ precipitate from the above equation was then allowed to settle, filtered, and washed with distilled water. It was then heated to 200 °C in 1 h and then held at that temperature for 1 h, which results in decomposition of the oxalates to carbonates by the following reactions [36],

$$\begin{aligned}\left[(1-y)\mathrm{Th(C_2O_4)_2} + y\mathrm{U(C_2O_4)_2}\right] \cdot n\mathrm{H_2O} \\ \rightarrow \left[(1-y)\mathrm{Th(C_2O_4)_2} + y\mathrm{U(C_2O_4)_2}\right] + n\mathrm{H_2O}\end{aligned} \quad (5)$$

$$\left[(1-y)\mathrm{Th(C_2O_4)_2} + y\mathrm{U(C_2O_4)_2}\right] \rightarrow \left[(1-y)\mathrm{Th(CO_3)_2} + y\mathrm{U(CO_3)_2}\right] + 2\mathrm{CO} \quad (6)$$

The resulted mixture was heated from 200 to 700 °C with a heating rate of 4 K/min and soaked at 700 °C for 3 h. This results in formation of mixed oxide phases by the following reaction.

$$\begin{aligned}\left[(1-y)\mathrm{Th(CO_3)_2} + y\mathrm{U(CO_3)_2}\right] \rightarrow \left[(1-y)\mathrm{ThO_2} + y\mathrm{UO_{2+x}}\right] + xy\mathrm{CO} \\ + (2-xy)\mathrm{CO_2}\end{aligned} \quad (7)$$

The major advantage of the above co-precipitation method is the decontamination from Fe which is normally encountered in the final product solution received from the reprocessing plant [41]. In the ADU route, a separate step was necessary to remove Fe by carbonate precipitation prior to the uranium precipitation.

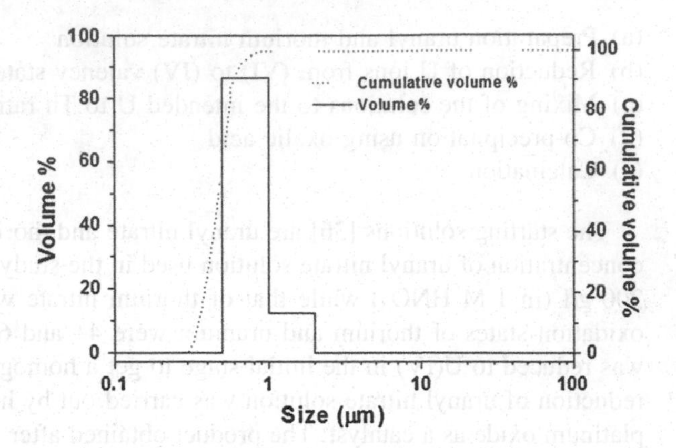

Fig. 7 Particle size distribution of ThO_2-50 % UO_2 powder from co-precipitation process ([36], Copyright Elsevier)

The average particle sizes of the ThO_2-30 % UO_2 and ThO_2-50 % UO_2 powders were found to be 0.63 ± 0.21 and 0.81 ± 0.17 μm, respectively. The particle distribution and the volume cumulative graphs for ThO_2-50 % UO_2 powders are shown in Fig. 7, which shows that about 90 % particles are below 1.0 μm. The surface area values for ThO_2-30 % UO_2 and ThO_2-50 % UO_2 powders were 12.10 and 7.16 m^2/g, respectively. A close examination on the shape of the above-mentioned powders showed that ThO_2-50 % UO_2 particles were more spherical, while the ThO_2-30 % UO_2 particles exhibited irregular surfaces with angular appearance.

2.2.3 Microwave Synthesis of $(Th,U)O_2$ Powder

It is reported that the microwave heating can lower the processing temperatures by several hundred degrees and shorten the processing time by several hours. Microwave heating not only increases the heating efficiency by concentrating the heating process within the material rather than in the furnace in which it is placed. Microwaves also overcome the problems associated with non-conduction of heat to areas containing unreacted/unaffected material by transferring energy homogeneously and efficiently on a molecular scale throughout the bulk. This ensures uniform heating throughout the body [42].

Homogeneous solid solutions containing uranium and thorium oxides with U/(U + Th) ratio of 0.15, 0.50, 0.65, and 0.80 have been successfully prepared by Chandramouli et al. [42] using the PVA-aided denitration method in a microwave oven. Excepting the case of U/(U + Th) ratio of 0.15, microwave calcining of the denitrated powders resulted in powders with lower surface area compared to the conventionally calcined powders. The density obtained with microwave calcined powders that were sintered using a conventional furnace was comparable with that obtained using conventional techniques. Microwave calcination resulted in relatively pure solid solutions perhaps due to shorter processing time namely 2 min as compared to 4 h at 1,073 K. As in the case of pure thoria prepared by PVA-aided denitration, the powder sizes were found to be in the nanocrystalline range agglomerated into larger particles. These larger agglomerates were of 0.5–1.0 mm size while the crystallites within were in the nanometer range.

2.2.4 Preparation of ThO_2-UO_2 Powders by Combustion Synthesis

The feasibility of combustion synthesis using citric acid as the fuel was studied by Anthonysamy et al. [43] in order to prepare a homogeneous mixture of ThO_2-UO_2 feed powder that can be compacted and sintered to high densities at a relatively lower temperature. A systematic study was carried out by them to examine the suitability of various combustion fuels such as urea, PVA, and citric acid for the preparation of thoria. They concluded that high density $U_yTh_{1-y}O_2$ solid solution with $y \leq 0.5$ can be prepared by combustion/denitration of a mixture of aqueous uranyl nitrate, thorium nitrate, citric acid with citric acid/nitrate ratio of 1.0 in a microwave oven followed by calcination of resultant powder in air at 973 K for 5 h. The microwave processed powders yield compacts that can be sintered to high densities (≥95 % T.D.). They also reported that citric acid is a better fuel than PVA for the preparation $U_yTh_{1-y}O_2$ solid solution through combustion route [44]. The powder obtained through this method was found to be nanocrystalline, highly free flowing, and porous in nature.

3 Powder Characterization

3.1 Density

Densities of powders are generally expressed as either bulk or tap densities. The term bulk density is applied to the weight of the remaining powder poured into measured volume after leveling and without applied compaction [14]. On the other hand, the tap density is applied to the same procedure but with some compaction by tapping the container. The measured bulk and tap densities of ThO_2 and ThO_2-UO_2 powders are low and in the range of 1–5.5 g/cc.

The influence of calcination temperature on bulk density has been studied by Clayton et al. [45]. For milled powder, bulk density increased from 1.97 to 2.64 g/cc which may be probably due to agglomeration of contacting particles during the heat treatment. Effect of calcination temperature on bulk and tap densities is shown in Fig. 8 which shows that both bulk and tap densities increase with increase in calcination temperature.

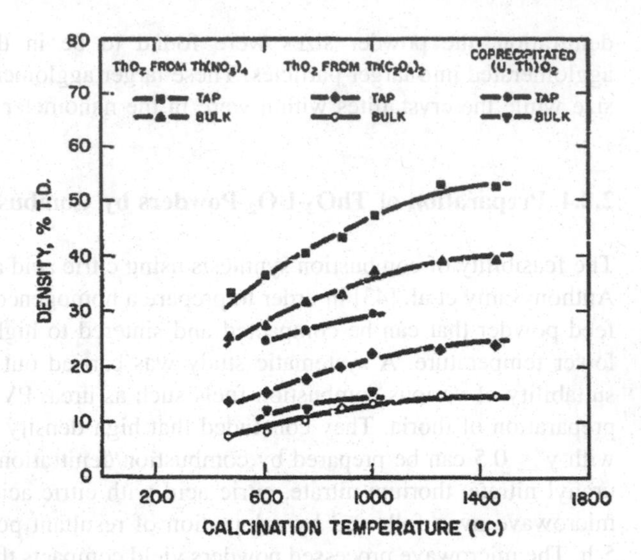

Fig. 8 Effect of calcination temperature on bulk and tap densities of Thoria-based powders ([23], Copyright Elsevier)

3.2 *Morphology*

An important factor that has significant influence on compaction and sintering is particle shape of ThO_2 and ThO_2-UO_2 powders. In fact particle shape affects the bulk density, particle size, surface area, etc. And particle contour determines the flow characteristics of the powder and the manner in which particle is packed affecting the green and sintered densities [14]. The exterior shape of the fine particle depends upon the method or condition of preparation. White et al. [20]. studied the shape of particles using SEM made by eight different oxalate precipitation conditions. He reported that the precipitation temperature has a strong effect on morphology.

Thoria particles precipitated at 343 K were square in shape and in the same case, particles digested for 6 h and ultrasonically agitated were round. The particles precipitated at 283 K and mechanically agitated were nearly cube in shape [14]. The morphology of thoria particles prepared by a continuous precipitation method at 283 K were agglomerates composed of intergrown platelets. The morphology of particles of ThO_2-UO_2 powders were reported to be identical and similar to that of ThO_2 particles.

3.2.1 Surface Area

The surface area, in general of ThO_2 and ThO_2-UO_2 depends on the surface areas of the material from which they are prepared, their thermal history and communition treatment if any [14, 23]. Surface area values ranging from 1 to 50 m^2/g are reported for various crystalline ThO_2 powder precipitations. Calcination temperature and

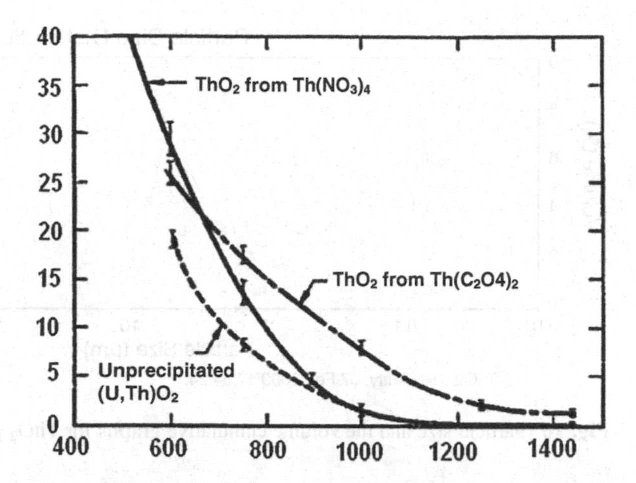

Fig. 9 Effect of calcination temperature on surface area for ThO_2 powder derived through different routes ([23], Copyright Elsevier)

time are important factors determining the surface areas of ThO_2 and ThO_2-UO_2 powders. Effect of calcination temperature on surface area for thoria-based powders is shown in Fig. 9. For oxalate-derived powders on increasing the calcination temperature from 873 to 1,723 K, the surface area decreased from 26.7 to 1.6 m^2/g.

Mathews et al. [46] studied the effect ball milling on particle size, shape, and surface area. It has been shown that the as-received ThO_2 powder could not be readily pressed without appropriate treatment. Flow ability and pressing problems were encountered during the fabrication and a large fraction of the green pellets had to be rejected due to cracking and breakage. After pre-compaction, granulation, and mixing with 0.2 % Sterotex, pellets could be pressed, and sintered densities as high as 90 % T.D. were achieved. The as-received powder particles were large with square platelet morphology and the resulting sintered pellet structure was coarse with random cracks and large pores. The milling process broke down these platelets into fine particles and resulted in improved pressing and sintering characteristics [46].

As mentioned earlier, the temperature of calcination is the most significant factor on surface area. Holding time has only a small effect. Very little decrease in surface area occurs after the first 10 h hold on calcination temperature. Clayton [45] has studied the effect of communition by micronizing. The results indicate that the largest increase in surface area occurs during the initial communition of powders calcined at lower temperatures [14]. It may be noted that powders that are calcined at higher temperatures makes them stronger and less friable and therefore difficult to break.

3.3 Particle Size

A narrow size particle distribution was observed for the oxalate-derived thoria powders. Very little difference in size distribution was seen on increasing the calcining temperature from 450 to 1,450 °C. About 99 % of the particles were

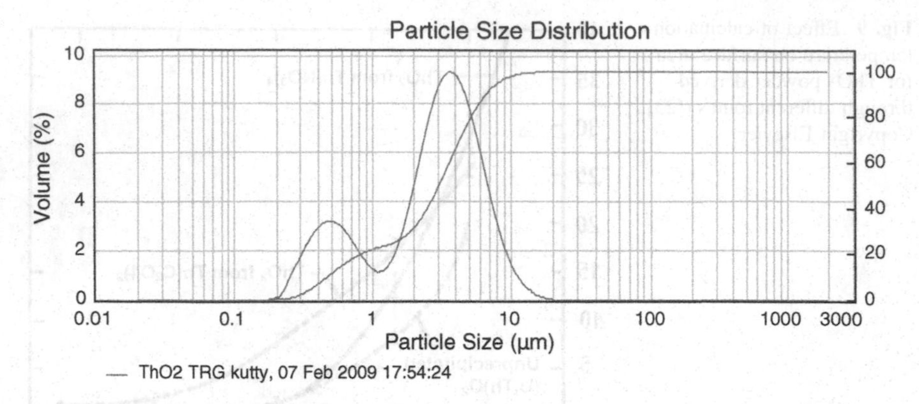

Fig. 10 Particle size and the volume cumulative graphs for ThO_2 powder made by oxalate route

<3 μm and no particles with size <0.2 μm were observed. A slight reduction in fines was observed and median particle size increased from 0.78 to 1.15 μm on increasing the calcination temperature from 450 to 1,450 °C. The particle derived from nitrate showed a far greater dependence on calcination temperature. On increasing the calcination temperature from 450 to 1,450 °C, the median particle size increased from 4 to 10.5 μm and at least 20 % particles were 10 μm or greater.

The particle distribution and the volume cumulative graphs for ThO_2 made by oxalate precipitation route are shown in Fig. 10. The average particle sizes of the ThO_2 powders were found to be 3.0 μm. From these figures, it may be noted that about 90 % particles for ThO_2 are below 8.0 μm.

3.4 Crystallite Size

Crystallite size, also referred as primary particle size, may be defined as a portion of solid which coherently scatters X-rays. The crystallite size is measured from X-ray peak broadening and sometimes using electron microscope. A diffuse X-ray pattern ray with broad peaks indicates low crystallinity whereas sharp diffraction pattern indicates high degree of crystallinity and crystallite growth [14]. This method is usually applicable to fine powders where the particle size is below 0.3 μm in diameter. Small crystallite size was observed for oxalate-derived ThO_2 powder. The crystallite size increases with increase in calcination temperature as shown in Fig. 11. It is important to understand the effect of crystallite size on sintering. According to Herring's scale laws, decreasing the particle size from micro- to nanoscale, has the potential to increase the sintering rate by about 12 orders of magnitude, depending on the sintering mechanism. Accordingly, nano-sized powders can be sintered at either lower temperature or the sintering time can be reduced for similar temperatures.

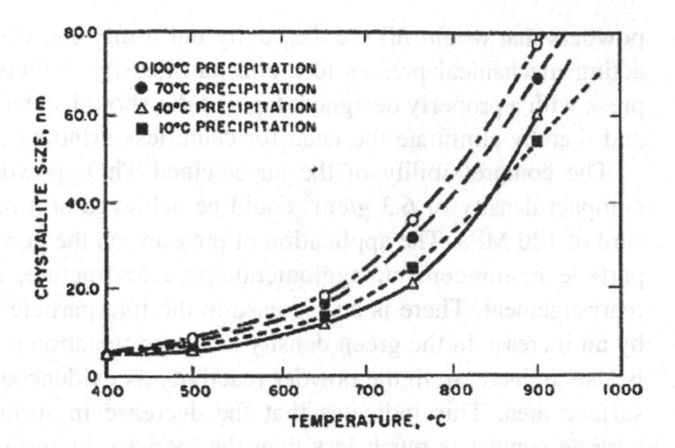

Fig. 11 Effect of calcination temperature on crystallite size for thoria powder ([23], Copyright Elsevier)

The powder obtained through oxalate route is reported to sinter to a much higher density than the powder-derived through the nitrate route. This is because the compact contains individual crystallites in direct contact with each other and individual porosity is readily eliminated during the sintering. On the other hand, the powder obtained through the nitrate route comprised of both large and small aggregates of tightly packed irregular crystallites that coarsens markedly with increase in calcination temperature [23].

4 Compaction of ThO_2

The fabrication of acceptable quality pellets from ThO_2 powders widely depend upon the powder properties. Some powders require binders, lubricants, and high compaction pressure to achieve the specified green density. Although ThO_2 may be pressed without binders, all the major fabrication campaigns reported in the literature used binders [23]. The most commonly used binder appears to be 1–4.5 wt% of Carbowax 8,000 [17]. Since the addition of binder affects porosity, density, and microstructure, the fabrication process that eliminates use of binder is desirable. The hard and abrasive nature of ThO_2 powder necessitates the use of die wall lubrication during pressing. Die wall lubrication is achieved by adding dry powder (0.2–0.3 wt%) such as stearates to the press feed. These lubricants are removed during the early stage of sintering and their effects on microstructure are minimal [21].

Compaction pressures as high as 965 MPa and as low as 172 MPa has been reported to press various types of thoria powders to obtain sintered densities greater than 90 % T.D. [17]. Generally, the pressing problems are caused by variations in powder particle size, bulk density, and flow characteristics. These variations can result in a non-uniform die fill which will result in variation of green and sintered densities. Hart et al. [17] reported that hydraulic presses can be used to compensate for variable die fill caused by poorly flowing powders. Free flowing powders that would fill the die cavity uniformly can be pressed in simple, single-acting mechanical presses to a constant density. Pellets pressed in a mechanical press with a properly designed, tapered die should sinter to right circular cylinders and thereby eliminate the need for centerless grinding [17].

The compressibility of the air-calcined ThO_2 powder was such that a green compact density of 6.3 g/cm^3 could be achieved at a moderate compaction pressure of 120 MPa. The application of pressure on the powder is believed to result in particle rearrangement, agglomerate/particle fracture, and once again fragment rearrangement. There is an increase in the interparticle contact area as evidenced by an increase in the green density as the compaction pressure is increased. There is also an increase in the powder reactivity as evidenced by an increase in specific surface area. This indicates that the decrease in surface area due to increased particle contact is much less than the increase in surface area due to fracture of agglomerates in the powder during compaction (see Table 1). Optimum compaction pressure is that pressure at which lateral and axial sintering shrinkages are equal and obey the relationship [21]

$$d/D = \left(\rho_g/\rho_s\right)^{0.33} \tag{8}$$

where, D = green diameter or length, d = sintered diameter or length, ρ_g = green density, ρ_s = sintered density.

As mentioned earlier, the powder produced by the common oxalate route has poor flowability and therefore difficult to get good compacts due to non-uniform die fill. Therefore, it is necessary to incorporate a pre-compaction granulation steps to achieve good quality green pellets. The pre-compaction pressure was found to have significant effects on the integrity of green compacts. The optimum pre-compaction pressure was found be in the range of 90–120 MPa which on granulation and further processing yielded good quality green pellets [21].

The final compacting pressure should be optimized with the following objectives in mind [23]:

- To have sufficient green strength for further handling
- To obtain high sintered density >94 % T.D.
- To minimize the variation in shrinkage after sintering which will eliminate or minimize the grinding to get the final dimension

Table 1 Surface area of ThO_2 green pellets as function of compaction pressure [21]

Compaction Pressure (MPa)	Surface area (m^2/g)
0	2.76
91	3.96
137	3.97
152	4.05
182	4.08
274	4.43
304	4.64

Balakrishna et al. [21] have taken powder from a single lot and compacted at seven different pressures and then sintered at 1,600 °C for 4 h in H_2. Their results suggest the followings:

- Green and sintered densities of thoria increase with increase in compaction pressure
- Variation in green densities decreased with increase in compaction pressure
- Variation in sintered densities was found to be comparatively high at low or high compaction pressure.

They concluded that pellets compacted in the pressure range of 90–120 MPa yielded most suitable product having green densities in the range of 6.2–6.3 g/cc. Comparatively large variations in both green and sintered densities at lower pressures and in sintered densities at higher pressures were evident [21].

Figure 12 shows the effect of calcination temperature on green density for various compaction pressures for the oxide derived through the oxalate route. At all temperatures, a linear relation between green density and compaction pressure in log scale exists. The higher calcination temperature needs greater compaction pressure for the same green density. The degree of compatibility was found to be almost same for the mixed oxide and thoria-derived from the oxalate route. Also the improvement in density with increasing pressure was found to be less for powders calcined at high temperatures. The green density of oxalate-derived thoria calcined at 600 °C showed ~17 % increase in green density when the compaction pressure was increased from 20 to 50 k psi (1 k psi = 6.89 MPa). The same material calcined at 1,450 °C showed only about ~6 % increase in green density for the same increase in compaction pressure [23].

Figure 13 shows the variation of green density as a function compaction pressure for ThO_2 powder. For comparison data of UO_2 are also shown in the same figure. It may be noted that at the same compaction pressure ThO_2 gives much higher green density. Figure 14 presents the axial and diametral shrinkages as a

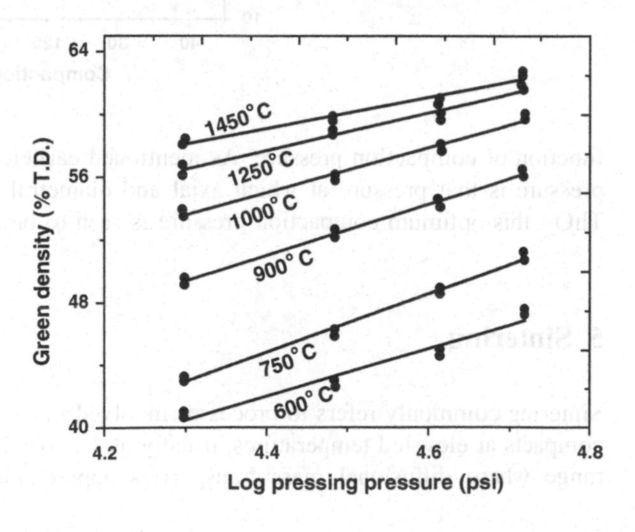

Fig. 12 Compactability of oxalate-derived thoria powder for different calcination temperatures ([23], Copyright Elsevier)

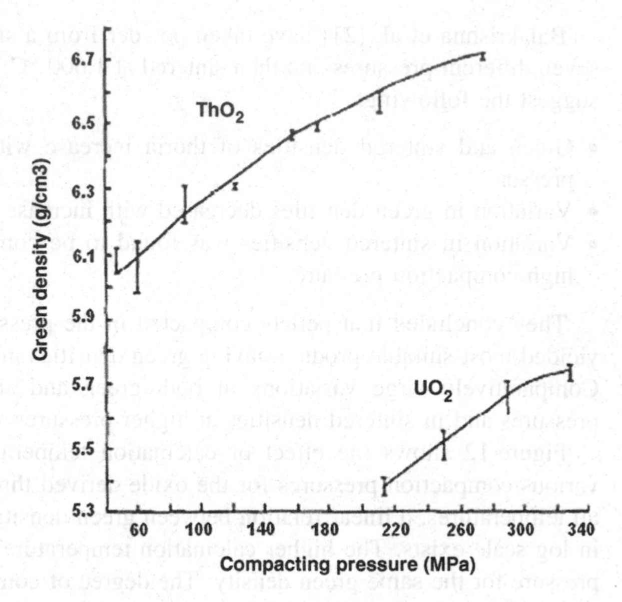

Fig. 13 Green density of ThO_2 at different compaction pressures. For comparison the values of UO_2 are also shown ([21], Copyright Elsevier)

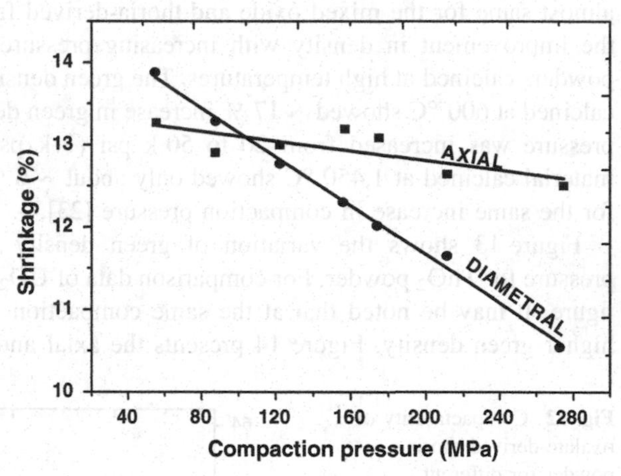

Fig. 14 Axial and diametral shrinkage of ThO_2 containing 0.5 mol% MgO versus compaction pressure when sintered in H_2 at 1,600 °C for 3 h ([21], Copyright Elsevier)

function of compaction pressure. As mentioned earlier, the optimum compaction pressure is that pressure at which axial and diametral shrinkages are same. For ThO_2, this optimum compaction pressure is seen to be about 120 MPa.

5 Sintering

Sintering commonly refers to processes involved in the heat treatment of powder compacts at elevated temperatures, usually at $T > 0.5\ T_m$ [K], in the temperature range where diffusional mass transport is appreciable. The driving force for sintering is the decrease in the surface free energy of powdered compacts obtained, by replacing solid–vapor interfaces with solid–solid interfaces. Thermodynamically, then, sintering is an irreversible process in which a free energy decrease is brought about by a decrease in surface area. For sintering to proceed, there must be local differences in chemical potential which is present due to differences in curvatures between the grain and the neck [31].

Curtis and Johnson [47] studied the densification behavior of thoria. They reported a density of 73 % for cold pressed thoria when sintering temperature was raised to 1,800 °C. By isostatic pressing they obtained a density of 86 % at the same temperature. But by small addition CaO or CaF_2, a density of 97 % is obtained at the same temperature. Clayton [48] studied about the chemical reactions that can occur during the various processing steps used in the fabrication of ThO_2 and $(Th,U)O_2$ pellets for the Shipping port light water breeder reactor. The oxalic acid used for the precipitation of thorium oxalate usually contains small quantities of residual sulfate (~500 ppm) in the form of alkali or alkaline earth metal sulfate. Some of this sulfate is precipitated along with oxalate. During calcining, the metal sulfates are converted to sulfides of Cr, Fe, and Ni. He also reported that during decomposition and reduction of ADU to UO_2, stable phosphate like $Ca_3(PO_4)_2$ is co-precipitated. The metal phosphide formation occurs during CO_2 pretreatment by the carbon reduction of phosphates. The efficiency of carbon removal from cold-compacted ThO_2 and $(Th,U)O_2$ pellets by CO_2 oxidation increases with temperature and at ~900 °C, almost complete oxidation of carbon to CO occurs. Carbon content is further reduced during the initial heating during sintering cycle in H_2 through the formation of methane. Clayton [48] also reported that addition of water vapor to H_2 sintering atmosphere also helps in removing carbon.

The processing methodology used by INEEL [7] for production of ThO_2 pellets consists of the following steps:

(a) Ball milling with WC balls of ~13 mm diameter and 0.5 wt% PEG-8000 for 24 h
(b) Pelletizing using special WC-lined die and hardened punches
(c) A binder burnout step at 500 °C for 1 h in air
(d) Sintering schedule consists of heating up to 1,750 °C using a heating rate of 5 °C/min, isothermal hold for 10 h and cooling to room temperature at the same rate.

Kang et al. [49] investigated the fabrication of ThO_2 and $(Th,U)O_2$ pellets. They milled ThO_2 and UO_2 powders by two different ways namely dry and wet milling. The wet milling was carried out using a ball mill for 24 h in a jar containing zirconia ball and alcohol and then dried in air at room temperature for >3 days. They observed that wet-milled powders had high density and uniform distribution of Th and U than dry milling process. The use of wet-milled powder and sintering at 1,700 °C yielded pellets of density ranging from 94 to 98 %.

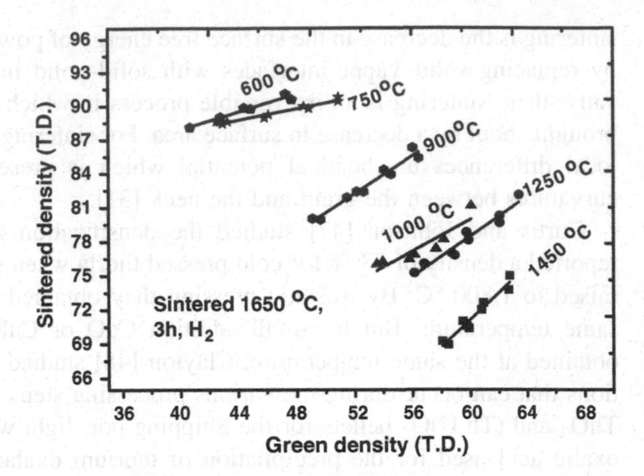

Fig. 15 Relation between green density, sintered density, and calcination temperature for oxalate-derived thoria powder [23]

Shiratori et al. [50] prepared high density thoria pellets without binders and lubricants from thoria powder derived from the oxalate process. They used ball milling at 120 rpm in dry air for 24 h. After ball milling, surface area of the powder increased by a factor of 4 and particle diameter decreased to 0.28 μm. Since the moisture in the powder causes cracking during sintering, the powder after milling was dried in open air at 200 °C for 8 h. The density of green pellets, which is directly related to surface area of the powder after milling, was a key factor to obtain the high density pellets. A density of maximum 98 % was obtained for ThO_2 at a low temperature of 1,550 °C in air.

Pope and Radford [23] have shown that for a particular calcination temperature, the sintered density was found to increase with increase in green density and the increase in calcination temperature resulted in a lower sintered density (Fig. 15). They reported the highest density obtained when surface area, green density, temperature were maximized.

5.1 Effect of Dopant

ThO_2 being a very stable oxide, can be sintered in atmosphere such as air, hydrogen, Ar, or vacuum. As previously stated, a sintering temperature as high as 1,800 °C or above is required to attain about 80 % T.D without any additive. But with the addition of suitable additives the sintering can be carried out at a temperature as low as 1,150 °C in air [21]. It is a known fact that the deviations from stoichiometry produce point defects. Similar effects can also be achieved by chemical doping by adding MgO/CaO to ThO_2. The effect of an additive depends mainly on its valency. When an additive, having a valency different from that of Th, is added to ThO_2 the following possibilities exist [31, 51–53]:

1. Oxygen interstitials/vacancies are created
2. Vacancies/interstitials in Th sites are created
3. Valency of some O ions are lowered/increased
4. Valency of some Th ions are increased/lowered.

Among the above possibilities, those conditions prevail for which energy requirement is minimum. The presence of an additive may assist sintering by one of the two possible mechanisms [21]. First, additive may create point defects in the ThO_2 lattice and thereby increases the diffusion of Th^{+4} ion by many orders of magnitude. A typical example for this is the case with Nb_2O_5 in ThO_2. Second, an additive may significantly retard grain growth so that pores are linked to the grain boundaries [36]. This is the case of CaO/MgO in ThO_2. The effect of higher valency additive is seen to be same as that of oxygen pressure in the atmosphere. If an oxidizing atmosphere is used along with a higher valency additive, the effect is added up [21]. If an oxidizing atmosphere is used with a lower valency additive, the effects nullify each other. When the range of stoichiometry is large as found in UO_2, advantage is taken by introducing oxygen pressure in the sintering atmosphere. When the range of stoichiometry is very small as found in ThO_2, one must take the advantage of introducing additives [51, 52]. While choosing an atmosphere or additive for achieving activated sintering, it should be remembered that effect of one must add to the other.

Ananthasivan et al. [44] have shown the effect of a higher valence additive, viz., V, Nb, and Ta on the densification of thoria. Nair et al. [53] reported that magnesia, calcia, and niobia bring about accelerated sintering in thoria. Balakrishna et al. [21] demonstrated that thoria could be sintered to densities greater than 9.76 g/cm^3 at 1,150 °C by doping with 0.25 mol% Nb_2O_5. Among the pentavalent dopants, a concentration of 0.5 mol% niobia is most effective in bringing about accelerated sintering in thoria. EPMA examination of pellets doped with niobia revealed that Nb is present in the different region of the pellets and amount is below the detection limit suggesting a homogeneous distribution of Nb in the sintered pellets. The microstructure of the pellet sintered at 1,350 °C reveals large grain growth suggesting the onset of melting. Ananthasivan et al. [44]. have reported the existence of a eutectic isotherm for Nb_2O_5-ThO_2 system at 1,333 °C for 27 mol% Nb_2O_5. Hence a temperature as high as 1,333 °C is necessary in order to bring about liqudation in thoria doped with niobia [51].

5.2 *Effect of Calcia*

In ThO_2, the diffusion of Th^{+4} ions is many times lower than that of O^{-2} ions. The addition of lower valency additive like CaO or MgO to ThO_2 is expected to create vacant oxygen sites in ThO_2 lattice. ThO_2 is always stoichiometric unless deviations

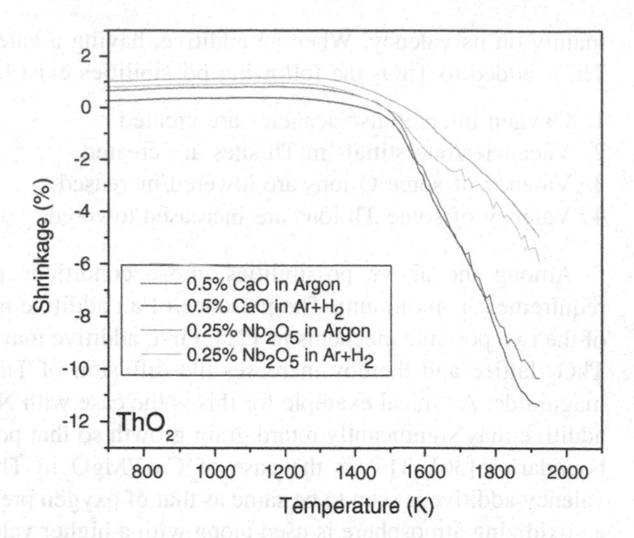

Fig. 16 Shrinkage curves for ThO_2 pellets in Ar-8 % H_2 and Ar atmospheres. The shrinkage curves for both the dopants are given in the figure

from stoichiometry are produced chemically, e.g., by adding CaO to produce oxygen deficient $(Th,Ca)O_{2-x}$ [54].

Since Ca has a lower valency than thorium, the addition of CaO creates oxygen vacancies or metal interstitials. The reducing atmosphere further reinforces the formation of metal interstitials or oxygen vacancies. Hence a lower valency additive and a reducing atmosphere will give a synergetic effect. From Fig. 16, it can be seen that the shrinkage is faster in CaO-doped ThO_2 in reducing atmosphere than the inert atmosphere. In PuO_2-ThO_2 system, both Th^{+4} and Pu^{+4} ions may be replaced by Ca^{+2} ions. However, in reducing atmosphere, Th^{+4} ions remain unaffected, but some of the Pu^{+4} ions are reduced to Pu^{+3} ions. The formation of Pu^{+3} ions results in generation of less number of interstitials. Hence the sintering is expected to slow down.

5.3 *Effect of Niobia*

When a Th^{+4} ion is substituted by Nb^{+5} ion in the ThO_2 lattice, an effective positive charge is imparted on the lattice. Hence the addition of Nb_2O_5 should decrease the concentration of anion vacancies, thereby increasing the concentration of cation vacancies through the Schottky equilibrium. An increase in Th lattice vacancy, increases its diffusion coefficient. If sintering is carried out in hydrogen, sintering-aid cation (Nb^{+5}) is expected to be reduced. If these reduced cations were to replace the Th^{+4} ions substitutionally, an increase in oxygen lattice vacancy should be expected, leading to a lower Th-ion Schottky vacancy equilibrium concentration, which will result in a decrease in Th diffusivity [21, 55].

Thus, Nb_2O_5 will not be much useful in reducing atmosphere which is evidenced in Fig. 16.

Two possibilities exist when Nb_2O_5 is dissolved substitutionally in ThO_2. First, one can have an oxygen interstitial model or thorium vacancy model [21, 55, 56]. For oxygen interstitial model, using the notation of Kroger and Vink, we have:

$$Nb_2O_5 \longrightarrow ThO_2 2Nb_{Th} + 4O_O + O_i \tag{9}$$

The unit cell corresponding to the model (9) is $Th_3NbO_{8.5}$. For thorium vacancy model, the following relation is applicable:

$$2Nb_2O_5 \rightarrow 4Nb_{Th} + 10O_O + V_{Th''''} \tag{10}$$

The unit cell corresponding to the model (10) is $Th_{2.75}NbO_8$. Here, $V_{Th''''}$ denotes Th vacancy having charge -4 with respect to the lattice. The same defects, namely O_i and $V_{Th''''}$ can also be created by an oxidative environment:

$$\tfrac{1}{2}\, O_2(g) \rightarrow O_i'' + 2h \tag{11}$$

$$O_2(g) \rightarrow V_{Th''''} + 4h + 2O_O \tag{12}$$

where h indicates the effective positive charge that is created in accommodating neutral atmospheric oxygen into ionic lattice. Since both Frenkel and Schottky defects are present simultaneously, the formation of oxygen interstitials should decrease the concentration of oxygen vacancies, thereby increasing the concentration of thorium vacancies through Schottky equilibrium [57]. The increase in the concentration of thorium vacancies leads to the increase in thorium diffusion coefficient and thus enhances sintering and grain growth [21].

In ThO_2-PuO_2 system, Nb^{+5} ions substitute for both Th^{+4} and Pu^{+4} ions thereby increasing the concentration of cation vacancies. If the sintering has been carried out in Ar and Ar-8 % H_2, Th^{+4} will not be affected since Th^{+4} being the only valence state of Th. But some of the Pu^{+4} ions will be reduced especially in Ar-8 %H_2 to Pu^{+3} ions at high temperatures. Hence the creation of more Pu^{+3} ions in the lattice should enhance the sintering since more cation vacancies are generated by the above process. However, in the reducing atmosphere like Ar-8 % H_2, Nb_2O_5 also gets reduced to lower oxide: $Nb_2O_5 \rightarrow NbO_2 \rightarrow Nb_2O_3 \rightarrow NbO$ [21, 56]. This means that Nb acts as a lower valency additive to ThO_2 in a reducing atmosphere. Its behavior will be then similar to CaO and hence sintering should be enhanced (see Fig. 16). Balakrisna et. al. [55] reported that high sintered density of 97 % can be achieved in $ThO_2 + Nb_2O_5$ pellets in reducing atmosphere at 1,700 °C.

The above findings suggest that niobia is a more effective dopant in oxidative atmosphere than in reducing atmosphere [36]. This can be further explained from the electrical conductivity measurements carried out on ThO_2. The variation of electrical conductivity (σ) with respect to absolute temperature (T) can be expressed by the following equation [58]:

$$\sigma T = A \exp(-E/kT) \tag{13}$$

where A is the pre-exponential factor and E is the activation energy. Bransky and Tallan [59] have measured the electrical conductivity of ThO_2 as a function of temperature at different oxygen pressures. They reported that the upper limit of oxygen pressure at which ThO_2, is stoichiometric is seen to be about 10^{-6} atm. If the oxygen pressure in the sintering furnace is greater than 10^{-6} atm, it may be considered as an oxidative atmosphere which would tend to generate oxygen interstitials in thorium oxide. In the oxidizing region, the increase in electrical conductivity occurred in both low and high temperatures. However, in the reducing atmosphere the increase in electrical conductivity was observed only at higher temperatures. Thus, at low temperatures the combination of higher valency additive and oxidizing atmosphere leads to give a higher defect concentration than does the combination of lower valency additive and reducing atmosphere [60–62]. The electrical conductivity of ThO_2 is decreased by doping with cations of higher valency like Nb^{5+}. The addition of Nb_2O_5 to ThO_2 causes to form significantly high concentrations of oxygen interstitial ions and an increase in thorium lattice vacancy, which results in increases in diffusion coefficient.

6 Techniques for Fabrication of $(Th,U)O_2$ Pellets

Studies on the use of thorium in existing reactors are of PHWR/PWR specific in most cases. The advanced heavy water reactor (AHWR) is being developed in India with the specific aim of utilizing thorium for power generation since India has vast reserves of thorium and its resource profile needs a closed cycle involving utilization of thorium [63]. AHWR is a vertical, pressure tube type, heavy water moderated and boiling light water cooled natural circulation reactor designed to produce 920 MW(th). The AHWR is fuelled with $(Th\text{-}^{233}U)O_2$ pins and $(Th\text{–}Pu)O_2$ pins. At equilibrium, the core of AHWR will consist of composite fuel assemblies each having 24 nos. of $(Th,^{239}Pu)$ MOX and 30 nos. of $(Th,^{233}U)$ MOX pins. The fuel is designed to maximize generation of energy from thorium and to maintain self-sufficiency in ^{233}U[64]. Since the ^{233}U required for the reactor is to be bred in situ, the initial core and annual reload for the initial few years will consist of $(Th\text{–}Pu)O_2$ clusters only. The reprocessed, ^{233}U is always associated with ^{232}U, whose daughter products are hard gamma emitters. The average concentration of ^{232}U is expected to exceed 1,000 ppm after a burnup of 24,000 MWD/t. The radioactivity of ^{232}U daughter products associated with ^{233}U starts increasing after separation. In view of this, a co-location of the fuel cycle facility, comprising reprocessing, waste management and fuel fabrication plant, with the AHWR is essential. The ^{233}U-based fuel needs to be fabricated in shielded facilities due to activity associated with ^{232}U daughters. This also requires considerable enhancement of automation and remotization technologies used in fuel fabrication. The spent fuel cluster, before reprocessing, would undergo disassembly for segregation of $(Th\text{-}Pu)O_2$ pins, $(Th\text{-}^{233}U)O_2$ pins, structural materials, and burnable absorbers. The $(Th\text{-}^{233}U)O_2$ pins will require a two stream reprocessing process, i.e., separation of thorium and uranium whereas the $(Th\text{-}Pu)O_2$ pins will require a three stream reprocessing process, i.e., separation of thorium, uranium, and plutonium [65, 66].

At present at BARC, R&D is being carried out on five major fabrication routes and its variants each having distinct advantages and disadvantages over the others [67, 68]. These are:

1. Powder route (mechanical mixing),
2. Co-precipitation technique
3. Sol–gel microsphere pelletization (SGMP)
   3.1. Gel impregnation
4. Coated agglomerate pelletization (CAP)
   4.1. Advanced CAP process
   4.2. Impregnated agglomeration process (IAP)
5. Pellet impregnation

We will briefly describe each of the above process and its merits and demerits.

6.1 Powder Route

The most common route for the fabrication of ThO_2-UO_2 pellets is by powder metallurgy technique. The key step in the production of the above mixed oxide fuels is the preparation of homogeneous oxide mixtures. The disadvantages of powder-pellet route are:

1. Handling of large amounts of fine powders (<1 μm) generating radiotoxic aerosols
2. Large number of fabrication steps in the flow sheet
3. Increase in personnel exposure due to buildup of fine powders on equipment surfaces

Therefore, alternative fabrication routes that are more amenable for remotization and automation procedures are being considered.

A study carried out at Bhabha Atomic Research Centre (BARC) show that high density ThO_2-UO_2 pellets could be fabricated from ThO_2-U_3O_8 green compacts without the addition of any dopants or sintering aids [69]. The U_3O_8 enhanced the sintering of ThO_2 compacts and the degree of enhancement depended upon the sintering atmosphere. Sintering of ThO_2-U_3O_8 green compacts under oxidizing condition yielded a density of 95 % T.D. at around 1,550 °C. The fabrication step is same, except, required amount of U_3O_8 is added instead of UO_2. Green density of the compacts was around 67 % of the theoretical density. To facilitate compaction and to impart handling strength to the green pellets, 1 wt% zinc behenate was added as lubricant/binder during the last 1 h of the mixing/milling procedure [69].

U_3O_8 is one of the most kinetically and thermodynamically stable forms of uranium. It is insoluble in water and it has a bulk density in the range of 1.5–4.0 g/cm^3 depending on the process used for production. Its particle density is 8.3 g/cm^3. U–O phase diagram gives the range of stability of U_3O_8 phase [6, 70, 71]. The crystal structure of U_3O_8 at room temperature is orthorhombic. Although U_3O_8 is a stable oxide of uranium at room temperature it dissociates into UO_2 above 1,100 °C when heated in air or inert gas like Ar. U_3O_8 can be completely reduced to UO_2 when heated in pure H_2 at about 700 °C [72]. Chervel et al. [73] have made pellets of UO_2 containing 15 and 25 % of U_3O_8 and annealed at 550 °C for 10 h. The XRD pattern of the above-mentioned annealed pellets did not reveal the presence of U_3O_8 phase.

Hund and Niessen [74] have studied the entire range of ThO_2-U_3O_8 system and reported the presence of U_3O_8 phase in addition to the solid solution for the higher U_3O_8 compositions. But for the composition containing lower U_3O_8 content, a single phase structure has been reported. The significance of U_3O_8 addition for enhancing sintering, especially in UO_2, has been discussed by many authors. Chevrel et al. [73] indicated that the composition of $UO_{2.025}$ appeared to be the most appropriate for the low temperature sintering which is obtained by the addition of U_3O_8 powder to UO_2. Harada [75] suggested a three stage sintering process for the fabrication of UO_2 pellets. His three stage process consists of sintering in reducing-oxidizing-reducing atmospheres at low temperatures between 1,200 and 1,500 °C. He suggested to adjust the oxygen partial pressure to the boundary between single phase UO_{2+x} and UO_{2+x}-U_3O_{8-z}. It is reported that in solid state the solubility of U_3O_8 in ThO_2 is negligible [67, 76].

From the study carried out in BARC, it was clear that high density ThO_2-UO_2 pellets could be fabricated from ThO_2-U_3O_8 green compacts without the addition of any dopant or sintering aids. Sintering under oxidizing condition yielded a density of 95 % T.D. at around 1,550 °C. But for the reducing and inert atmospheres even at 1,600 °C, the density was only around 87 % T.D. The onset of densification occurred above 1,150–1,200 °C for Ar and Ar-8 % H_2, while in oxidizing atmospheres it commenced about 100 °C lower than that in reducing and inert atmospheres. Figure 17 shows the shrinkage behaviors of ThO_2-2 % U_3O_8 pellets in Ar, Ar-8 % H_2, CO_2, and air. The dl/lo (lo: initial length) versus temperature curves can be classified into two distinct groups which show different behaviors. The curves for Ar (inert) and Ar-8 % H_2 (reducing) atmospheres almost coincide while those for oxidizing atmospheres like air and CO_2 also coincide but lie below the former at temperatures above 600 °C.

Figure 18 shows the shrinkage behaviors of pure ThO_2 and ThO_2-0.25 % Nb_2O_5 in air. For comparison, the shrinkage curve of ThO_2-2 % U_3O_8 in air is also shown in the figure. The effect of dopants on shrinkage is clear: the onset of sintering shifts

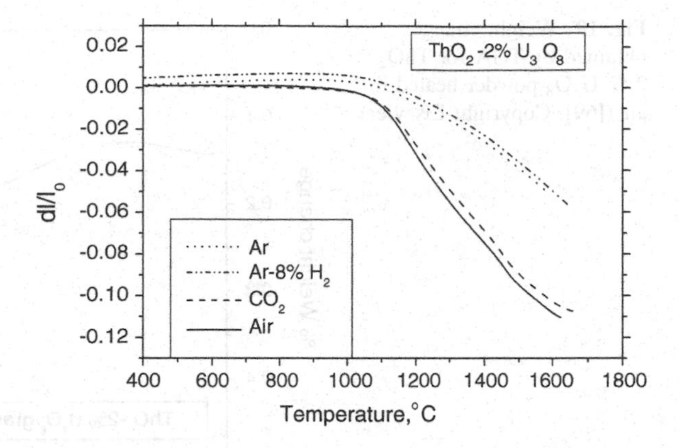

Fig. 17 Shrinkage behaviors of ThO_2-2 % U_3O_8 pellets in Ar, Ar-8 % H_2, CO_2 and air ([69], Copyright Elsevier)

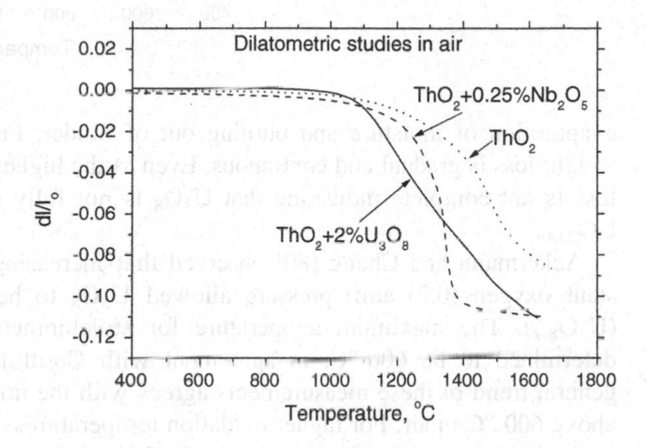

Fig. 18 Shrinkage curves for ThO_2-0.25 % Nb_2O_5 and ThO_2-2 % U_3O_8 obtained in air. The shrinkage curve for pure ThO_2 is also shown ([69], Copyright Elsevier)

toward to the lower temperature on the addition of a dopant. For pure ThO_2, the sintering commences only at temperatures above 1,250 °C, while it starts at about 1,000 °C for the ThO_2-2 % U_3O_8 pellet. The shrinkage occurs at a lower temperature by the addition of small quantities of additives like Nb_2O_5 or U_3O_8. At 1,400 °C, the shrinkage is 3 % for pure ThO_2 and 7.5 % for ThO_2-2 % U_3O_8 and 11 % for ThO_2-0.25 %Nb_2O_5. The effect of Nb_2O_5 is very significant especially in the temperature range of 1,300–1,400 °C, and above 1,400 °C, the rate of shrinkage decreases drastically with increase in temperature. On the other hand, the shrinkage rate is almost constant for ThO_2-2 % U_3O_8 from 1,100 to 1,600 °C.

Thus, the study carried out by Kutty et al. [69] confirms that air is the best atmosphere for sintering of ThO_2-U_3O_8 compacts. A maximum shrinkage of around 12 % is observed in air while it is only about 6 % for Ar and Ar-8 % H_2. This again shows that U_3O_8 has not been fully reduced to stoichiometric $UO_{2.00}$ on heating in air even at 1,600 °C [77–79]. To confirm this, thermogravimetry of ThO_2-U_3O_8 has been carried out in air up to 1,500 °C and the resultant thermogram is given in Fig. 19. The decrease in weight up to 400 °C is due to the

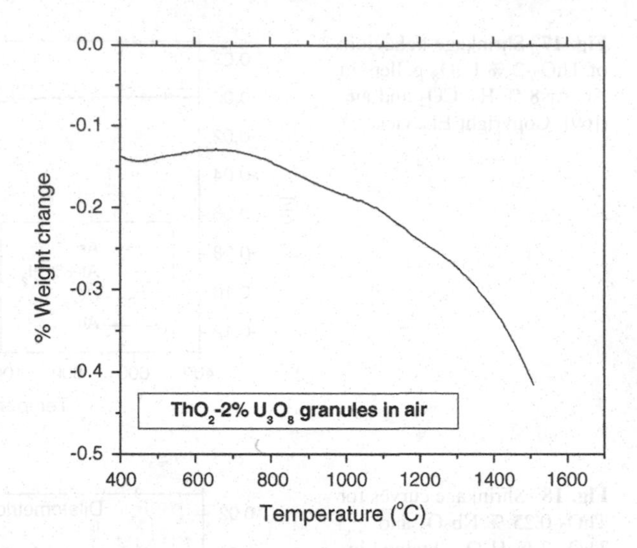

Fig. 19 Weight change obtained by TGA for ThO_2-2 % U_3O_8 powder heated in air ([69], Copyright Elsevier)

evaporation of moisture and burning out of binder. From 700 to 1,500 °C, the weight loss is gradual and continuous. Even at the highest temperature, the weight loss is not complete indicating that U_3O_8 is not fully reduced to stoichiometric $UO_{2.00}$.

Ackermann and Chang [80] observed that increasing the temperature at constant oxygen (0.21 atm) pressure allowed U_3O_8 to become non-stoichiometric (U_3O_{8-z}). The maximum temperature for stoichiometry at 0.21 atm of O_2 is determined to be 600 °C, in agreement with Cordfunke and Aling [81]. The general trend of these measurements agrees with the non-stoichiometric behavior above 600 °C in air. For higher oxidation temperatures, the true final composition of the U_3O_{8-z} oxide is determined by the following relation [72]:

$$\begin{aligned} \text{O/U} &= 1.3752 + 0.0046875T - 6.1855 \times 10^{-6}T^2 \\ &\quad + 3.5194 \times 10^{-9}T^3 - 7.3925 \times 10^{-13}T^4 \end{aligned} \tag{14}$$

where T is $600 < T < 1{,}223$ °C.

The O/U values of U_3O_{8-z} oxide at 1,000, 1,100, and 1,200 °C are calculated to be 2.637, 2.633, and 2.627, respectively. This means that on increasing the temperature, there is a considerable amount of deviation from stoichiometry generating more point defects. As the defect concentration increases, the driving force for sintering also increases. The defect structures of substoichiometric U_3O_8 at lower temperature and or high $p(O_2)$ appear to be randomly distributed doubly charged oxygen vacancies. However, at high temperature and or low $p(O_2)$ complex defect structures appear. Chervel et al. [73] have shown that the decomposition of U_3O_8 produces a network of fine pores enabling the samples richer in U_3O_8 to maintain high specific area at equivalent densification levels. The presence of high specific area for U_3O_8 containing samples constitutes a reserve of additional energy for sintering. Self-diffusion measurements of U in U_3O_8 have been carried out by Glasser Leme and Matzke [82] who reported activation energy for uranium migration of 2.4 eV by applying a thin layer of ^{233}U and observing the penetration profile following different annealing steps. This activation energy is much less than that for the self-diffusion of uranium in UO_2 (5.6 eV) [83]. Hence diffusion is expected to take place much faster in U_3O_8 phase.

6.2 Co-Precipitation Technique

The requirement for high homogeneity of the distribution of the actinides in the fuel is essential for most of new generation (Gen IV) reactors. The presence of fissile rich region causes a problem during irradiation and reprocessing, since these pellets will not dissolve completely in nitric acid without the addition of hydrofluoric acid. Hence, it is essential to avoid the formation of such fissile rich regions by adopting the proper manufacturing procedures. Among the various techniques, co-precipitation is an excellent route to obtain a very homogeneous mixture [51, 84]. The co-precipitation process is generally not preferred by fuel manufacturers since it involves handling of the enormous amount of liquid waste. If one can incorporate the co-precipitation steps in the reprocessing route, then this method will be really advantageous.

The procedure followed at BARC for the fabrication of ThO_2-30 % UO_2 and ThO_2-50 % UO_2 green pellets are same as that of powder route which consists of the following steps [36]:

(a) Milling of the co-precipitated ThO_2-UO_2 powder in a planetary ball mill for 4 h with tungsten carbide balls
(b) Double pre-compaction of the above prepared mixtures at 150 MPa
(c) Granulation of the pre-compacts
(d) Final cold compaction of the granulated powder at 300 MPa into green pellets

The green density of the compacts was around 62 % of the theoretical density. To facilitate compaction and to impart larger handling strength to the green pellets, 1 wt% zinc behenate was added as lubricant/binder at the last 1 h of the mixing/milling procedure. The resulting pellets were sintered at 1,400 °C in air for 6 h. The entire flow sheet of fabrication of ThO_2-UO_2 pellets by co-precipitation process is given in Fig. 20.

The O/M ratios of the calcined (in air 700 °C for 3 h) ThO_2-30 % UO_2 and ThO_2-50 % UO_2 powders produced by co-precipitation method were 2.218 and 2.301, respectively. The XRD powder patterns of these samples consist of two phase mixtures of CaF_2 type solid solution and U_3O_8. The lattice parameters calculated from the high angle scans ($2\theta = 100°$ to 145°) for ThO_2-30 % UO_2 and ThO_2-50 % UO_2 powders were found to be 0.55519 and 0.55205 nm, respectively.

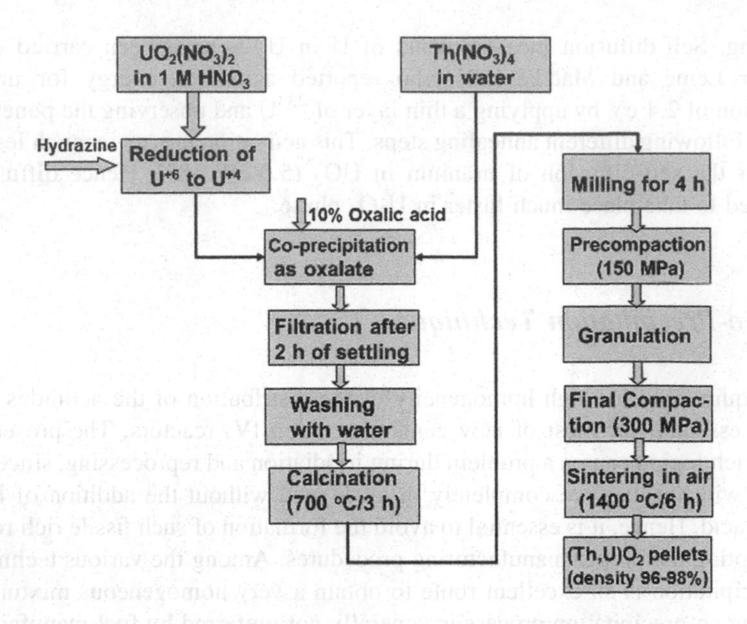

Fig. 20 Flow sheet of fabrication of ThO_2-UO_2 pellets by co-precipitation process

From the O/M ratios and the lattice parameter data, the amount of U_3O_8 present in the above powders was estimated. The estimation was made from the assumption that for high O/M values, ThO_2 forms a solid solution with $UO_{2.25}$. The remaining UO_2 is assumed to exist as U_3O_8. The amount of U_3O_8 present in ThO_2-30 % UO_2 and ThO_2-50 % UO_2 powders was calculated and found to be 9.88 and 14.19 mol%, respectively [36].

Figure 21 shows the XRD patterns of the ThO_2-30 % UO_2 and ThO_2-50 % UO_2 pellets sintered in air at 1,400 °C for 6 h after co-precipitation which showed that the U_3O_8 phase still exists together with the fluorite phase. The O/M ratios for the above composition were 2.170 and 2.230, respectively which are lower than those of the corresponding powders before sintering. But the amount of U_3O_8 present in both the pellets was less than that present in the corresponding powders. For ThO_2-30 % UO_2 composition, the amount of U_3O_8 present in the powder has come down on sintering in air at 1,400 °C for 6 h from 9.88 to 6.05 mol%. Similarly, for ThO_2-50 % UO_2 composition, it has come down from 14.19 to 7.50 %. These results clearly indicate that the U_3O_8 amount decreases with dissolution into UO_{2+x} during heating the sample forming solid solution with ThO_2. It clearly indicates that about 3–7 mol% U_3O_8 was decomposed and additionally dissolved in $(Th,U)O_2$ during the sintering. In the mixture of ThO_2-UO_{2+x} and U_3O_8, the densification at high temperatures probably proceeds as follows [21, 85, 86]:

1. Formation of defective structured U_3O_{8-z}
2. Dissolution of a part of U_3O_{8-z} to UO_{2+x}

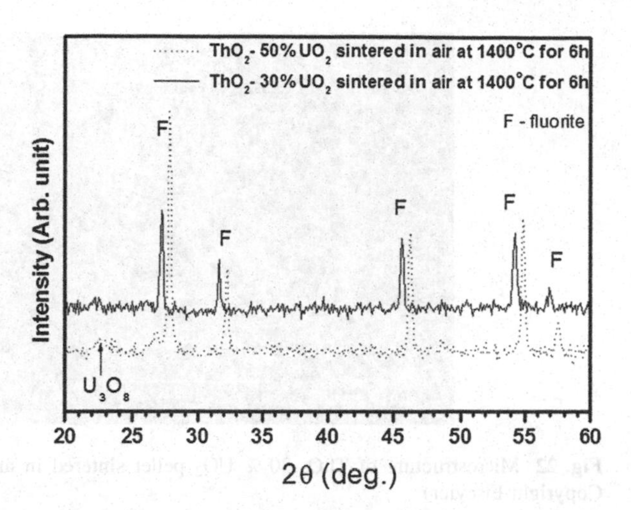

Fig. 21 XRD patterns of ThO_2-30 % UO_2 and ThO_2-50 % UO_2 pellets made by co-precipitation process and sintered in air ([36], Copyright Elsevier)

3. Solid solution formation between $(Th,U)O_2$ and UO_{2+x}
4. Simultaneous sintering of $(Th,U)O_2$.

From the above discussion, it is understood that the high densities obtained in this study are due to the following factors [84]:

1. High surface area of the starting powders
2. Higher O/M ratio of the powders
3. Presence of a small amount of U_3O_8 phase.

The larger surface area corresponds to higher surface energy. The driving force for sintering is the reduction in surface energy. The higher O/M ratio indicates the presence of higher concentration of oxygen interstitials. Lay and Carter [87] have shown that the self-diffusion coefficient of uranium in UO_{2+x} is proportional to x^2. This is predominantly due to the increased concentration of uranium vacancies in UO_{2+x}. It has been established that U_3O_8 acts as a sintering aid for ThO_2 [69]. As mentioned earlier, the O/U values of U_3O_{8-z} oxide in air at 1,000, 1,100 and 1,200 °C are calculated 2.637, 2.633, and 2.627, respectively [88]. This means that on increasing the temperature, there may be a considerable amount of deviation from stoichiometry generating more point defects. Therefore, the presence about 10–14 mol% U_3O_8 in the ThO_2-30 % UO_2 and ThO_2-50 % UO_2 powders helped in achieving higher density.

The average grain sizes of sintered ThO_2-30 % UO_2 (see Fig. 22) and ThO_2-50 % UO_2 pellets were 5.7 and 4.5 μm, respectively [36]. ThO_2-50 % UO_2 pellet showed one or two patches consisting of bigger grains of about 40 μm. In order to know the distribution of Th, U and O, electron beam scanning was performed by EPMA across the pellet from center to periphery [84]. A typical result of X-ray intensities of Th M_α, U M_α and O K_α is shown in Fig. 23. It shows the essentially uniform U distribution in the pellet.

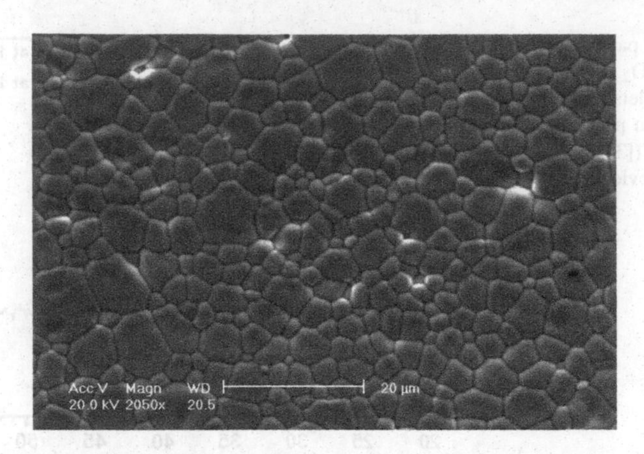

Fig. 22 Microstructure of ThO_2-30 % UO_2 pellet sintered in air and etched thermally ([36], Copyright Elsevier)

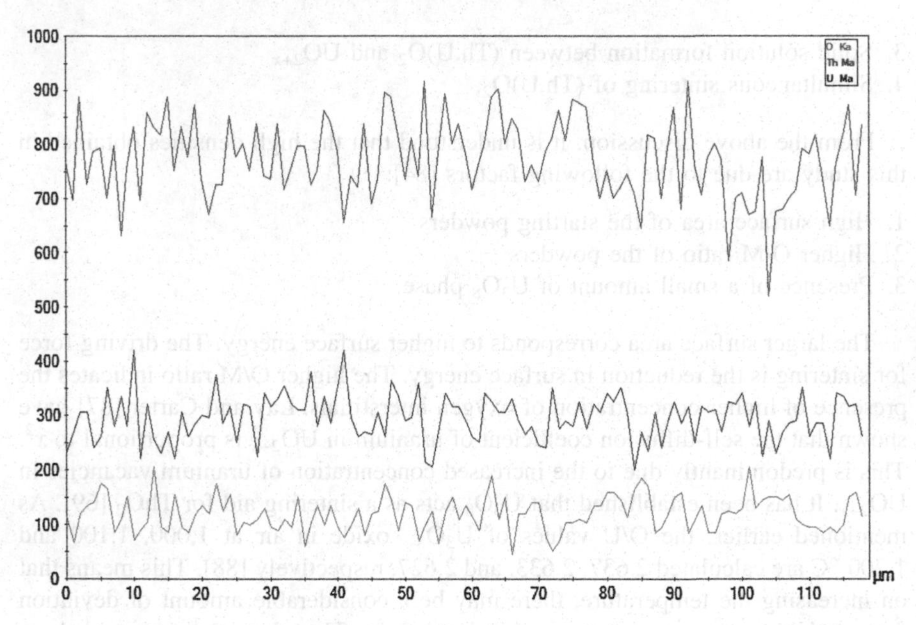

Fig. 23 X-ray line scan for Th M_α, U M_α and O K_α across the grain structure for ThO_2-30 % UO_2 ([36], Copyright Elsevier)

6.3 Sol–Gel Microsphere Pelletization Technique

The Sol–gel microsphere pelletization (SGMP) process not only avoids dust generation but also produces high density pellets of desired microstructure at compaction pressures and sintering temperatures, which may be even lower than for the conventional powder route. The other advantages of this technique include [89, 90]:

- Free flowing microspheres facilitate automation and remotization of the process.
- Process steps such as milling, pre-compaction, and granulation followed in conventional fabrication process are eliminated.
- Excellent micro homogeneity and desirable pore structure is achieved.
- Controlled density and tailored microstructure fuels can be fabricated.
- Suitable for manufacturing ^{233}U and Pu-based fuel fabrication.

SGMP technique employs free flowing dust-free sol–gel derived microspheres as starting material for the fabrication of pellets instead of powders. In order to obtain good quality pellets from microspheres, the microspheres should have [90]:

- Low crushing strength,
- Reasonably high tap density and,
- Higher surface area.

The microspheres should also have adequate tap density to get a reasonable compression ratio. Again the microsphere should be soft, but again too soft microspheres may result in the defect like lamination in the pellet.

6.3.1 EGT and EGU Route of Preparation of Sol–Gel Microspheres

External gelation of thorium (EGT) facility was used for preparation of gel microspheres of ThO_2-UO_3 [91]. The feed solutions were prepared by dissolving thorium and uranyl nitrate in the required molar ratio. The sols were then prepared by passing a controlled amount of NH_3 gas through a hollow, rotating dispenser shaft immersed in a jacketed and water heated vessel containing thorium uranyl nitrate solution between 80 and 100 °C. For the preparation of UO_3 gel by KFA External Gelation of Uranium (EGU) process, the feed solutions were prepared by adding urea and ammonium nitrate to uranyl nitrate solution. The optimum parameters for EGU process consist of a mixture of 1 mol/l of uranyl nitrate solution, 4 mol/l urea and 2.5 mol/l ammonium nitrate. The broth was then boiled for 10 min prior to gelation. Droplets of these sols were passed through an electromechanical vibrator with a horizontal nozzle into a gelation bath surrounded by ammonia gas. In the gelation bath the droplets gel settle at the bottom. The gel microspheres are then washed with ammonia solution at 60 °C and then dried at 200 °C.

The EGU or EGT process results in the microspheres which are very hard after drying and calcination. These particles are not suitable for direct pressing and sintering into good quality defect free pellets. To overcome these problems processes were modified by adding carbon black powder to the feed broth. The dispersed carbon was removed by controlled oxidation during heat treatment to obtain porous gel microspheres suitable for fabrication of high density pellets for direct pelletization and sintering [90].

6.3.2 IGP Route of Preparation of Sol Gel Microspheres

The internal gelation process (IGP) followed at B.A.R.C. uses nitrate solutions of uranium, plutonium or thorium, or its mixtures. The metal nitrate solution is mixed with urea and HMTA (Hexamethylenetetramine) solution, cooled to 0 °C. Droplets of this solution are contacted with hot silicone oil at 90 °C to get hydrated gel microspheres. These gel microspheres are then washed in CCl_4 and NH_4OH. Drying of the microspheres is carried out in air at 100 °C followed by calcination at 500 °C. The calcined microspheres are reduced in N_2–H_2 atmosphere at 600 °C [92].

Porous microspheres of ThO_2-2 % UO_2 containing ~0.4 % CaO as sintering aid prepared from 1.2 M heavy metal feed solution could be compacted at 350–560 MPa to green density in the range of 52–55 % T.D. [93]. The pellets were sintered at 1,400–1,500 °C in Ar-4 % H_2 atmosphere to achieve sintered density ≥95 % T.D.

6.3.3 Fabrication of (Th,U)O_2 Pellets from Gel Microspheres Prepared Without the Addition of Carbon as Pore Former

Although the use of carbon as the pore former gives very good pellets, the heat treatment procedure for obtaining the calcined and reduced microsphere requires a long time schedule and control of highly exothermic carbon oxidation step requires special attention. For fabrication of UO_2 pellets from gel microspheres prepared without the addition of carbon as pore former, the uranium concentration in the broth was kept higher (1.5 M) to get UO_3 crystallite size of more than 1,100 nm. The calcination and reduction of the microspheres were done at 800 °C and 600 °C in air and Ar-8 % H_2 atmospheres, respectively. The resultant microspheres could be compacted at 350 MPa and sintered to high density at 1,150 °C using low temperature oxidative sintering route [93]. Similarly, (U,Th)O_2 SGMP pellets were made using IGP process. (Th,U)O_2 microspheres containing 3 mol% uranium were prepared from feed composition of 1.25 M heavy metal feed solution, were calcined at 500 °C and compacted at 300 MPa and sintered to 97 %T.D. at 1,350 °C in air. The process flow sheet for obtaining dense (Th, ^{233}U) MOX pellets using SGMP technique has been shown in Fig. 24.

Figure 25 shows the variation of shrinkage with temperature of sintering for the ThO_2-3 % UO_2 composition in air and Ar-8 % H_2 atmospheres. The usefulness of air in sintering of thoria-based pellets are clearly demonstrated here. The maximum sintered density of 97 % of T.D. could be obtained in air at 1,350 °C. Figure 26 gives the intercomparison of the shrinkage behavior of ThO_2-3 % UO_2 composition made by powder and the sol–gel routes, respectively. It shows that sintering is comparatively faster for the pellets made by the sol–gel route [94–96]. The presence of large surface area of the sol–gel microspheres might have helped in enhancing the sintering for these pellets.

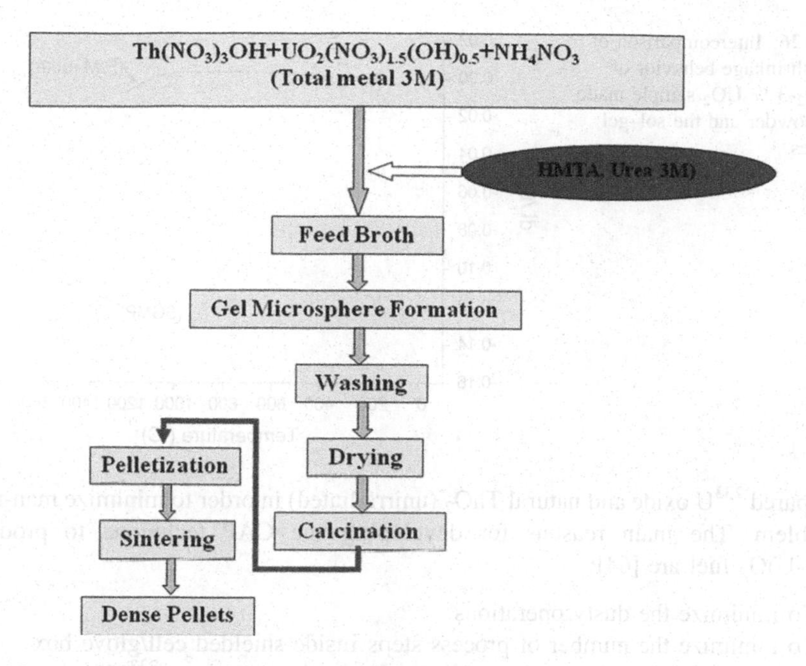

Fig. 24 Flow sheet for the fabrication of $(Th,U)O_2$ pellets through SGMP route

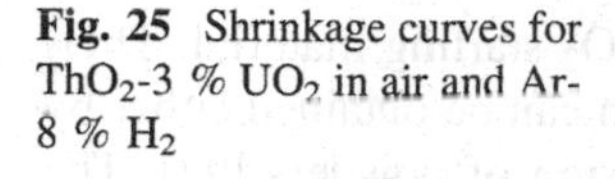

Fig. 25 Shrinkage curves for ThO_2-3 % UO_2 in air and Ar-8 % H_2

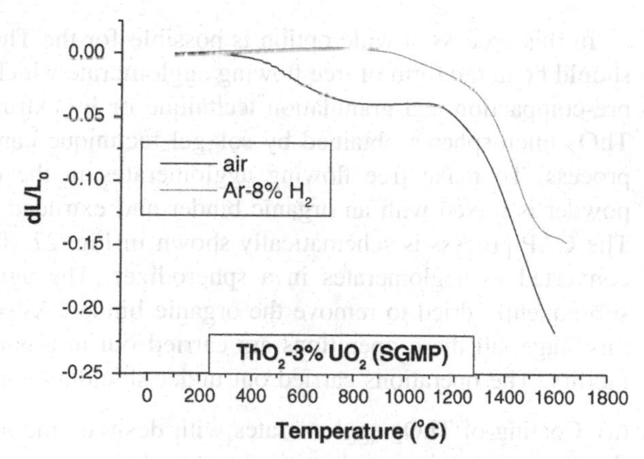

6.4 Coated Agglomerate Pelletization (CAP) Technique

The coated agglomerate pelletization (CAP) process was developed by Bhabha Atomic Research Centre (BARC) to replace the conventional powder metallurgy process that consists of pre-compaction and granulation [97]. The flow sheet of the CAP technique is made of the segmented flow sheet to be performed partly in the unshielded and partly in shielded facilities on the assumption to use freshly

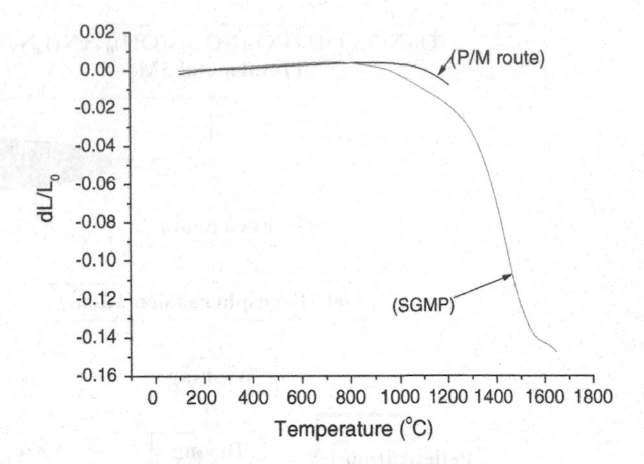

Fig. 26 Intercomparison of the shrinkage behavior of ThO_2-3 % UO_2 sample made by powder and the sol–gel routes

prepared ^{233}U oxide and natural ThO_2 (unirradiated) in order to minimize man-rem problem. The main reasons for developing the CAP technique to produce $(Th\text{-}U)O_2$ fuel are [64]:

- To minimize the dusty operations
- To minimize the number of process steps inside shielded cell/glove box
- To reduce the man-rem problems since the highly radioactive ^{233}U is confined to only certain steps in the fabrication route.

In this process, a wide option is possible for the ThO_2 starting material. ThO_2 should be in the form of free flowing agglomerate which can be obtained either by pre-compaction and granulation technique or by extrusion of powders [97]. The ThO_2 microspheres obtained by sol–gel technique can also be used in the CAP process. To make free flowing agglomerates in the extrusion route, the ThO_2 powder is mixed with an organic binder and extruded through perforated rollers. The CAP process is schematically shown in Fig. 27. The extruded ThO_2 paste is converted to agglomerates in a spherodizer. The agglomerates are sieved and subsequently dried to remove the organic binder. As only ThO_2 is handled up to this stage, all these operations are carried out in a normal alpha tight glove box facility. The operations carried out under shielding are [64]:

(a) Coating of ThO_2 agglomerates with desired amount of ^{233}U oxide
(b) Compaction in a multistation rotary press into green pellets
(c) Sintering in air
(d) Pellet loading and encapsulation into fuel rods.

Preliminary investigation was carried out in BARC to find out the optimum ThO_2 agglomerate size. For this purpose ThO_2 agglomerates were made by extrusion route. These agglomerates were segregated into various classes depending upon their size. This was done by sieving them through various mesh sieves, i.e., −20, −30, and −40 meshes. The agglomerates segregated were mixed with 4 % U_3O_8 powder in a planetary ball mill. The mixed powder was then

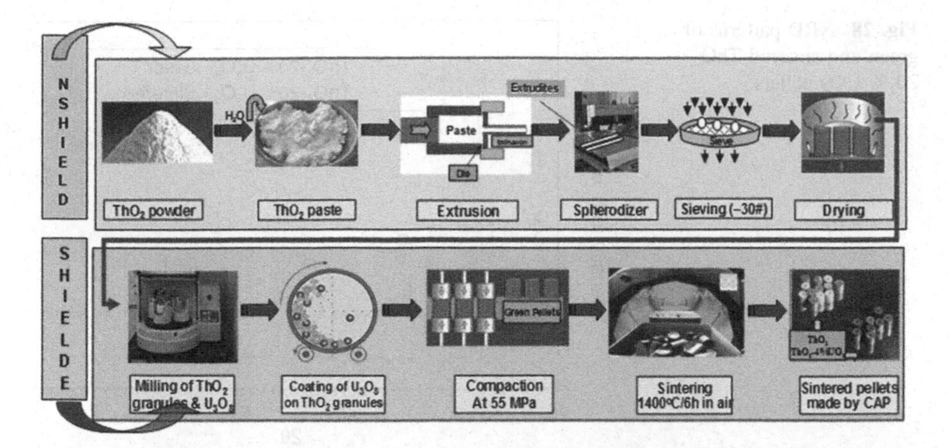

Fig. 27 Flow sheet for the fabrication of (Th, U)O_2 by CAP process ([109], Copyright, Elsevier)

compacted into pellets at 300 MPa pressure, which were then sintered in air at 1,450 °C for 8 h. The obtained density of the sintered pellet was found to be in the range 90–95 % T.D. The as-polished microstructure of the sintered pellet made by using −20 mesh granules was unacceptable since it showed delineation of particles at many places. The granule is seen as a separate entity with gaps all around which indicate that it has not diffused with other granules properly. The microstructure of −30 mesh sieved granules was better with occasional delineation at fewer places and pellet made by using −40 mesh granules shows a satisfactory microstructure with no sign of delineation of granules at any place in the pellet [98].

The green pellets of ThO_2-4 % UO_2 and ThO_2-20 % UO_2 pellets were prepared from ThO_2 agglomerates and U_3O_8 powder without the addition of any dopants or sintering aids. The green density of the compacts was in the range of 62–67 % of the theoretical density [97–100]. For ThO_2-4 % UO_2 pellets, sintering under oxidizing condition yields a density of around 95 % T.D. at around 1,550 °C. In the reducing atmosphere, however, a higher temperature of 1,600 °C was needed to have the same composition, and its density was only around 90 % T.D. The X-ray diffraction patterns for ThO_2 and ThO_2-4 % UO_2 confirm that the compounds are fcc single phased. ThO_2-20 % UO_2 should not be single phased, but U_3O_8 peaks cannot be detected in the diffraction pattern possibly because of a small amount (see Fig. 28).

ThO_2-20 % UO_2 pellet showed an inferior sintering behavior in reducing atmosphere in comparison with the other two compositions studied by Kutty et al. [97]. It has shown that the oxygen potential of $Th_{1-y}U_yO_{2+x}$ solid solution decreases with increase in uranium concentration (y) although it increases with the oxygen excess (x). For an O/M ratio of 2.005, Baker et al. [101] have shown that the oxygen potential of ThO_2-20 % UO_2 composition at 1,200 °C is about 40 % lower than that that of ThO_2-5 % UO_2. Since ThO_2-20 % UO_2 pellet has a lower oxygen potential than ThO_2 and ThO_2-4 % UO_2, it may be a factor for its lower

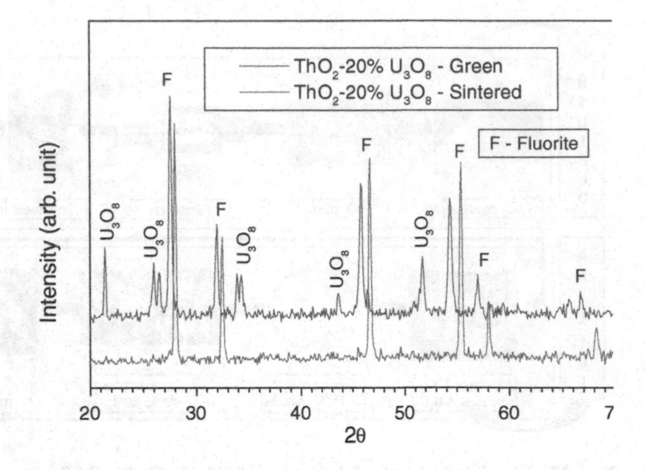

Fig. 28 XRD patterns of green and sintered ThO_2-20 % UO_2 pellets

shrinkage behaviour. Also, the microstructure of the above pellet showed the presence of inhomogeneous microstructure.

The microstructure of ThO_2-UO_2 pellet prepared by CAP process showed a unique "rock in sand" type structure [98]. There were colonies of fine grains which were surrounded by large-grained areas. In each of these colonies, grains were randomly distributed. The size of the each fine-grained colony is in range of 100–150 μm. These colonies were found to be extremely dense and some of them had no pores. These colonies represent initial ThO_2 granules, which were used for making green pellets. The identity of initial ThO_2 granules has disappeared since there was no boundary between the granules. All the granules were fused with each other (Fig. 29). No gap exists between the granules at any place in the pellet indicating that diffusion has occurred between them. It is not possible to form a solid solution between ThO_2 and U_3O_8 since U_3O_8 does not have any solubility in ThO_2 [102]. Therefore, the U_3O_8 coating on each granule is assumed to have been

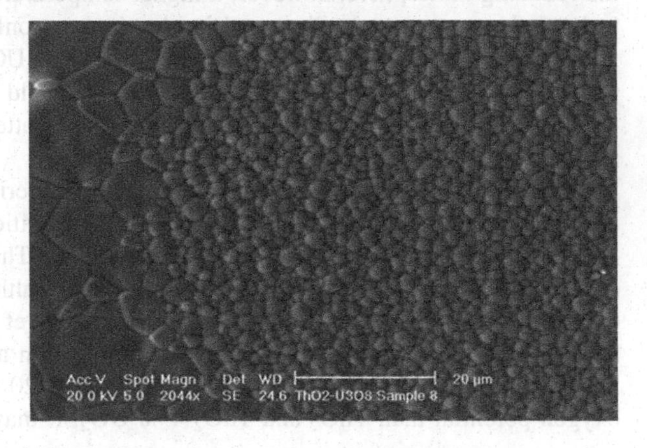

Fig. 29 The "rock in sand" type microstructure of ThO_2-UO_2 pellets ([98], Copyright Elsevier)

decomposed to UO_{2+x} during sintering and diffused to ThO_2 matrix forming solid solution [97]. This was confirmed by EPMA.

As mentioned earlier, the grain size in the colonies was of very fine (1.7 μm). In the initial U_3O_8 used for coating, grains were substantially bigger. This can be explained by considering the defect structure. The defects in pure ThO_2 are comparatively less. Since the grain growth is a diffusion-related phenomenon, it depends upon the defect concentration like O_2 interstitials or metal vacancies. Therefore, the grain growth is not enhanced inside the colonies. Also the temperature of the sintering ($\sim 0.45T_{m,}$ where T_m is the melting point in K) is comparatively less. Since the temperature of the sintering used by Kutty et al. [98] was $<0.5Tm$, this may be a factor responsible for the lower grain sizes. These two factors resulted in developing small grains in the initial ThO_2 colonies. The significance of U_3O_8 addition for enhancing sintering in ThO_2, has been discussed by Kutty et al. [69]. Thermogravimetric studies carried out in air on ThO_2-2 % U_3O_8 granules indicate that U_3O_8 has not been reduced to stoichiometric UO_2 even at 1,500 °C. This suggests that U_3O_8 present in the green compacts will exist as UO_{2+x} at sintering temperature. Since the diffusion coefficient of U, D^U, is proportional the square of the oxygen excess (x^2) in the lattice, sintering as well as grain growth is enhanced thus resulting in bigger grains.

The basic requirements for the high performance of the fuel are [98]:

(a) "Soft pellets"—To reduce pellet clad mechanical interaction (PCMI)
(b) Large grain size—To reduce fission gas release (FGR)

The strength of the pellet at room temperature is related to grain size by the Hall–Petch relation. Accordingly, the smaller grain-sized pellets will have higher strength. But at high temperature (above equicohesive temperature) the grain boundaries become weaker than grain matrix. Since the pellets of smaller grain size have wider grain boundary areas, these pellets become softer than pellets with larger grain size. Also as the grain size decreases, the creep rate of the fuel increases. Therefore, pellets with smaller grain size have higher creep rate and better plasticity at high temperatures. These pellets will reduce the PCMI. On the other hand, the pellets with larger grain size are beneficial to reduce the fission gas release. In developing thermal reactor fuels for high burnup, this factor should be taken into account [98].

The EPMA data confirm that uranium concentration was slightly higher in large-grained areas as shown in Fig. 30a, b. In summary, ThO_2-UO_2 fuels made by CAP process have the benefits of both small and large grains. The microstructure of comparatively fertile rich (Th) region has small grain size as indicated. Hence this region will show relatively high plasticity at high temperatures. On the other hand, the fissile rich region has larger grain size which will be of help in reducing the fission gas release. Since more fission is likely to occur in this region, more fission gas is expected to generate here. Therefore, the large grains in this region can be used as storage for the fission gases. Hence we can conclude that the fuel

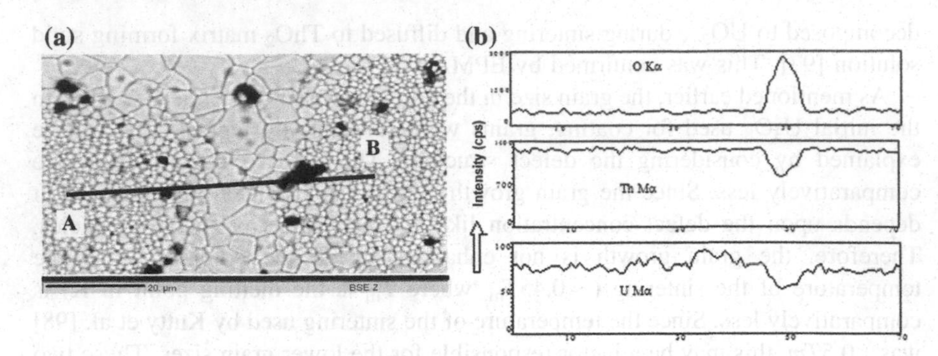

Fig. 30 a EPMA scan on ThO_2-UO_2 pellet section ([98], Copyright Elsevier), b U concentration variation along the scan ([98], Copyright Elsevier)

made by CAP process has properties to reduce the fission gas release and also as well to reduce PCMI. Hence this unique microstructure will be useful in attaining high burnup in thermal reactors.

6.4.1 Advanced CAP Process

In a variation of this technique, the advanced CAP (A-CAP), uses (Th-U)O_2 powders instead of U_3O_8 powder for coating. (Th-U)O_2 powders for the above process were made by co-precipitation technique [103]. The use of U_3O_8 in the CAP process naturally leads to a consequence that the uranium concentration in the solid solution is higher in the larger grained areas. It is often argued that this concentration inhomogeneity is desirable for fuel because a larger amount of heat is generated at the periphery of the particles. If the designer prefers a process that can deliver pellets having improved microstructures with better microhomogeneity, it is worth to modify the CAP process. A-CAP process can deliver pellets having microstructure which are either similar to CAP process or with the improved ones by selecting the appropriate sintering atmospheres. ThO_2-30 % UO_2 and ThO_2-50 % UO_2 powders are prepared by co-precipitation route. The major steps involved in the fabrication of ThO_2-30 % UO_2 and ThO_2-50 % UO_2 powders have already been described. The procedure for the fabrication of ThO_2-4 % UO_2 green pellets consists of the following steps:

(a) Mixing the ThO_2 granules with the required quantity of ThO_2-30 % UO_2 or ThO_2-50 % UO_2 powders in a planetary ball for 4 h with tungsten carbide balls,
(b) Cold compaction of the mixed powder at 300 MPa into green pellets,
(c) Sintering in reducing or oxidizing atmosphere.

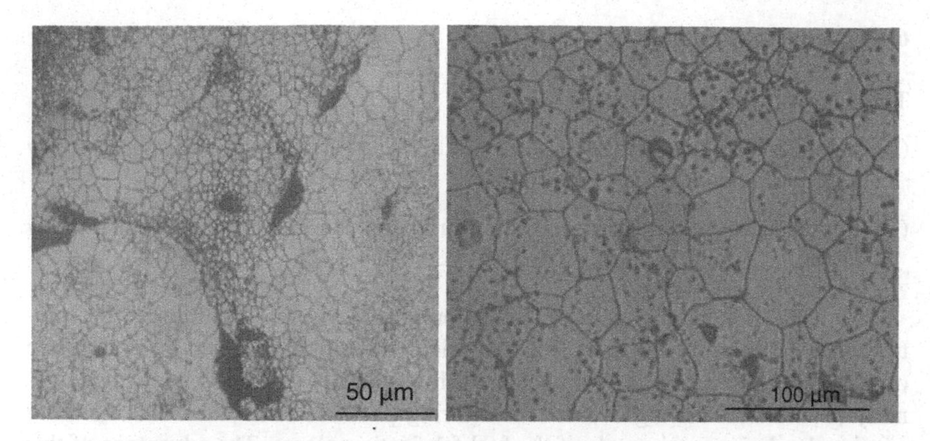

Fig. 31 Microstructure of ThO_2-4 % UO_2 pellet sintered in air (*left*) and Ar-H_2 (*right*) atmospheres ([103], Copyright Elsevier)

The sintering of the ThO_2-4 % UO_2 pellets made by A-CAP process was carried out in air at 1,400 °C for 6 h. A few green pellets were sintered separately in Ar-8 % H_2 at 1,650 °C for 6 h.

It was found that pellet sintered in air led to the formation of duplex grain structure and those sintered in Ar-8 % H_2 resulted in very uniform grain structure with excellent homogeneity (Fig. 31). The microstructure of the ThO_2-4 % UO_2 pellet sintered in air showed a "rock in sand" structure with small grains in the center of granules and large grains along the periphery. However, no delineation of granules could be found. The microstructure of the above pellet sintered in Ar-8 %H_2 atmosphere showed uniform grain size with large grains measuring 45 µm. The EPMA data on the sample sintered in air showed that the uranium concentration was slightly higher in the large-grained areas. But for the same composition sample sintered in Ar-8 % H_2 atmosphere, the uranium concentration was found to be uniform. The difference between the CAP and A-CAP processes is the amount of fine-grained areas which is considerably smaller for the pellets processed by the A-CAP route. The average size of the grains in the colonies for the pellets made by CAP and A-CAP were 2 and 4 µm, respectively [103].

6.5 Impregnation Technique

Impregnation technique is an attractive alternative for manufacturing highly radiotoxic ^{233}U bearing thoria-based mixed oxide fuel, remotely in a hot cell or shielded glove box facility. In this process, fresh ThO_2 green matrix is impregnated with ^{233}U containing nitrate solution to the required enrichment.

6.5.1 Pellet Impregnation

This involves mainly two steps;

(i) Preparation of "low density" (< 75 % T.D.) ThO_2 pellets having open porosity in an unshielded facility and

(ii) Impregnation and further processing in a shielded glove box/hot cell facility [104].

The ThO_2 pellets thus prepared are impregnated in uranyl nitrate (^{233}U) solution of molarity in the range of 1–3, in a shielded facility, followed by sintering to obtain ThO_2-based mixed oxide pellets of high density and good micro homogeneity [1]. Thus, handling of fine ^{233}U bearing powders is avoided and these are restricted only in certain parts of the fuel fabrication plants. The advantage of this process is that it can be so coupled with the reprocessing plant that the purified uranyl nitrate from the plant may be straightway used as the infiltrant. The impregnation process can eliminate conversion step and eliminate several expensive stages from operations. For example, the process steps like precipitation of ammonium diuranate, filteration, calcination, mixing, grinding, granulation, etc., which are associated with 'radiotoxic dust hazard', are eliminated. As no precipitation or washing steps are required within the shielded area, the radioactive wastes produced in the process are negligible [105–108].

The impregnation technique of Bhabha Atomic Research Centre (BARC) is very similar to Infiltration of Radioactive Materials (INRAM) [109–113] which has recently been applied for Am target fabrication. The process relies on the action of capillary forces to draw the solution into the pores of the host material. The amount of the second material introduced into the pellet can be controlled by adjusting the concentration of the infiltrant solution. The only requirement for the application of this process to the fabrication of fuel pellet is that the pellet should be insoluble in the solution containing the infiltrant and that the infiltrant can be easily convertible into the desired chemical form [114]. It is important that many of the porosities are interconnected and distributed uniformly across the pellet otherwise the impregnation will not be effective. Use of microwave during impregnation for local heating of the partially sintered low density pellets has been tested for expulsion of entrapped gas to accelerate impregnation. For uniform distribution of actinide in the sintered pellets, annular pellets are more suitable than the conventional ones. The concentration of the added material can be increased by multiple impregnations or by the use of the solution containing higher concentration of the material.

The flow sheet of fabrication of ThO_2-4 % UO_2 pellets by impregnation process is given in Fig. 32. The facility set-up for impregnation along with the pellets made by impregnation process are shown in Fig. 33. The procedures for the fabrication of ThO_2-UO_2 pellets followed by Bhabha Atomic Research Centre (BARC) involve the following steps:

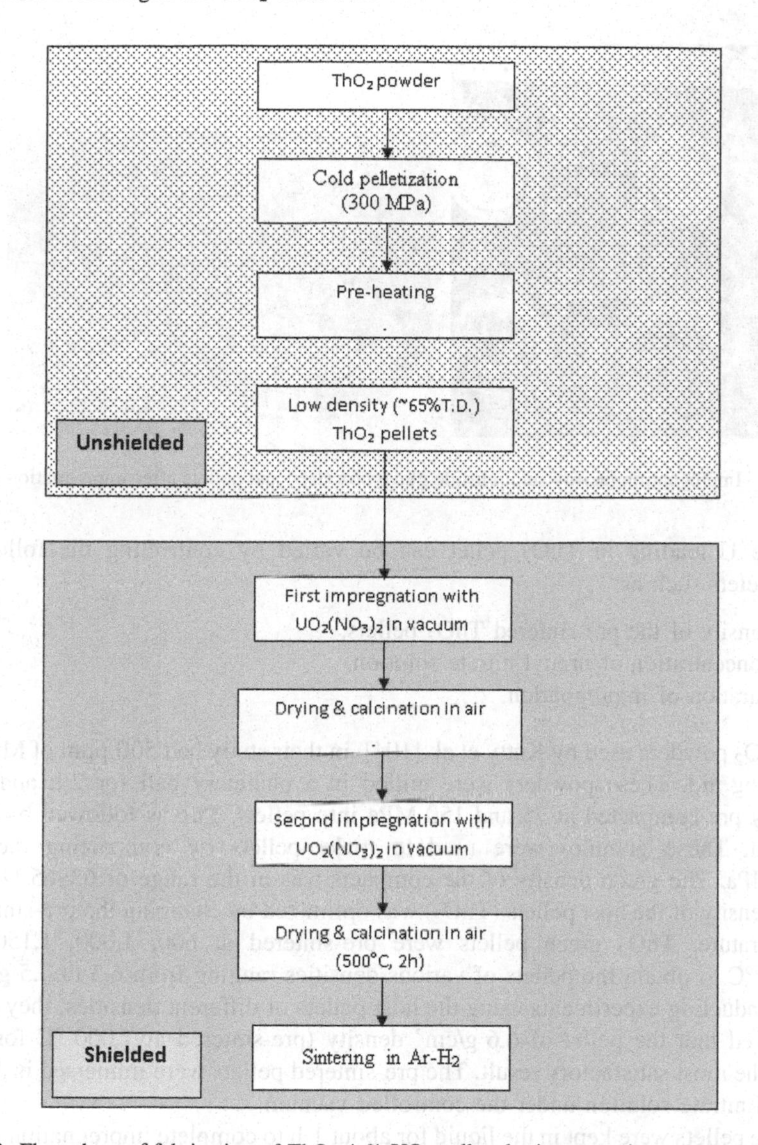

Fig. 32 Flow sheet of fabrication of ThO_2-4 % UO_2 pellets by impregnation process

1. Fabrication of low density (~66 % T.D.) ThO_2 pellets by powder route in an unshielded facility.
2. Impregnation of the above pellets by uranyl nitrate solution under vacuum in a shielded facility.
3. Drying and finally sintering at 1,700 °C in reducing atmosphere.

Fig. 33 Impregnation facility set up inside glove box (*left*) and pellets after impregnation (*right*)

The U loading in ThO_2 pellet can be varied by controlling the following parameters such as:

(a) Density of the pre-sintered ThO_2 pellets,
(b) Concentration of uranyl nitrate solution,
(c) Duration of impregnation.

ThO_2 powders used by Kutty et al. [104]. in their study had 500 ppm of MgO as sintering aid. These powders were milled in a planetary ball for 2 h and then doubly pre-compacted at 75 and 150 MPa into pellets. This is followed by granulation. These granules were used to make pellets by compacting them at 300 MPa. The green density of the compacts was in the range of 63–65 % T.D. The density of the host pellets (ThO_2) was optimized by changing the pre-sintering temperature. ThO_2 green pellets were pre-sintered at 600, 1,000, 1,150 and 1,200 °C to obtain the pellets of various densities ranging from 6.3 to 7.5 g/cm^3. By conducting experiments using the host pellets of different densities, they [104] observed that the pellet of 6.6 g/cm^3 density (pre-sintered at 1,000 °C for 2 h) gave the most satisfactory result. The pre-sintered pellets were immersed in 1.5 M uranyl nitrate solution under the controlled vacuum.

The pellets were kept in the liquid for about 1 h to complete impregnation. After impregnation, the pellet was dried and then calcined at 500 °C for 2 h. The amount of uranium impregnated in ThO_2 was found to be around 2 % from the weight gain of the pellet after the calcination. Attempts to increase the uranium amount by increasing the porosity of ThO_2 pellet using pore formers (methyl cellulose) did not give good results because such pellets were insufficient in strength to withstand the liquid impregnation. A series of experiments were conducted with batch sizes ranging from 250 g to 5 kg with 1.5 M uranyl nitrate solution. In all the above experiments, the uranium concentration in ThO_2 was around 2 %. For having ThO_2-4 % UO_2 pellet, then, the above pellet was again impregnated with 1.5 M uranyl nitrate solution and calcined at 500 °C for 2 h [104].

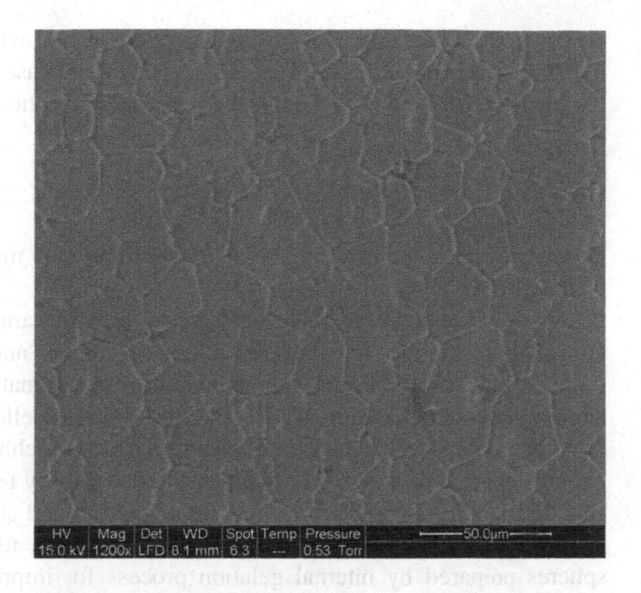

Fig. 34 Microstructure of ThO_2-4 % UO_2 made by impregnation process

The average grain sizes of ThO_2 and ThO_2-UO_2 pellets were found to be in the range 10–12 μm as shown in Fig. 34. The XRD data of the above pellets show they are of single phased. The XRD peaks are found to be sharp indicating that the material is crystalline. In order to know the distribution of Th, U and O, electron beam scanning was performed by EPMA across the pellet from center to periphery which shows the essentially uniform U distribution in the pellet. However, X-ray intensities measured by fixed time counting at various locations show that the U concentration is slightly higher at the periphery than that at the center for both ThO_2-2 % UO_2 and ThO_2-4 % UO_2 pellets. This is one of the drawbacks of the impregnation process. A better homogeneity can be attained by using annular pellet. The reason of non-uniform distribution and its effects on the fuel performance can be considered as follows. The impregnation process uses capillary force to draw up the solution into the pores of the host material [107–109]. It may be noted that the maximum concentration of the infiltrate exists at the exterior surface of the substrate. The concentration continuously diminishes from the periphery to the center of the pellet. Therefore, a continuous concentration gradient of fissile material in the pores of the composite continues to exist after the liquid is removed from the pellet. This concentration gradient continues to exist even after the heat treatment.

The concentration gradient thus obtained during the impregnation process may be beneficial to the nuclear industry by the following reasons [109, 115]. The greater concentration of fissile material at the periphery of the pellet will give greater neutron efficiency during operation of a reactor. The neutrons will be able to escape from the pellet more readily as the self-shielding effect is significantly reduced. Also the heat generation will be greater at the pellet periphery and decreases toward the center of the pellet [104]. For nuclear ceramics having low thermal conductivity values, this phenomenon will be beneficial in the case of an operating incident. The stored thermal energy in the impregnation fuel is significantly lower than that in the conventional fuel where the fissile material is uniformly distributed [116]. This is because the increase in the fuel temperature at the center of the pellet is significantly depressed in the present fuel.

6.5.2 Gel Impregnation

Dense $(Th,U)O_2$ pellets can be fabricated through microspheres impregnation technique.

Several investigations have been made with the aim to develop the gel pelletization technique for the preparation of ThO_2 and $(Th,U)O_2$ pellets. For reducing the time of fuel fabrication, a few alternatives such as the coated agglomerate pelletization (CAP) and the particle/pellet impregnation methods have been considered. In the pellet impregnation technique, all the pellet fabrication steps can be carried out in an unshielded facility for natural uranium and the shielded facility is needed for ^{233}U impregnation and sintering. The microsphere impregnation technique being developed in BARC uses porous thoria microspheres prepared by internal gelation process for impregnation of uranium. Pai et al. have fabricated dense $(Th,U)O_2$ pellets using microsphere impregnation technique [114]. In this process they have impregnated U into calcined ThO_2 microspheres. The loading of uranium in the ThO_2 microspheres depends on the morphology of the microspheres, i.e., pore size, shape, volume, and their inter connectivity. For impregnation studies of uranyl nitrate solution in calcined ThO_2 microspheres, the following parameters are need to be examined:

(i) Extent of evacuation and degassing achieved before impregnation.
(ii) Time of impregnation.
(iii) Concentration of uranyl nitrate solution used.
(iv) Effect of morphology of the ThO_2 microspheres.

Using the standardized process flowsheet for the preparation of ThO_2 microspheres and impregnation of uranyl nitrate solution, homogeneous loading of uranium within 3 % of the desired value could be obtained.

6.5.3 Impregnated Agglomeration Process (IAP))

The IAP route for the fabrication of $(Th,U)O_2$ pellet was developed for the following purpose which are given below:

- To further reduce powder handling and therefore to reduce man-rem
- To overcome disadvantages of CAP route (sticking of $^{233}UO_2$ powder to equipment and glove box).
- To improve the homogeneity and microstructure of fuel pellets

In CAP process, there is a possibility of powder specially $^{233}UO_2$ sticking to equipment and glove boxes. An attempt has been made recently to develop a new technique called Impregnated Agglomeration Process (IAP) at AFFF, Tarapur [117]. In this process, uranium oxide is dissolved in concentrated HNO_3 and diluted to 0.25 M. ThO_2 spheroids were coated with uranyl nitrate solution with the help of universal mixer. The dried spheroids are compacted and further processed as in CAP process. In IAP, powder handling is minimized and it is advantageous when $^{233}UO_2$ is handled. The flow sheet for the fabrication of $(Th,U)O_2$ pellet by IAP process is given in Fig. 35.

In IAP route, spheroids obtained through the extrusion route were spary coated with uranyl nitrate solution. The use of uranyl nitrate solution for coating in IAP route showed improvement in microstructure and homogeneity of fuel pellets in contrast to CAP route where coating was carried out using U_3O_8 powder. It was observed that by decreasing the molarity of uranyl nitrate solution the homogeneity can be improved. Pellets prepared by using 3 and 5 M nitric acid showed lower density with low yield due to the presence of cracks in the pellet. Higher molarity also caused corrosion of die tools used for pressing. Hence attempts were made to

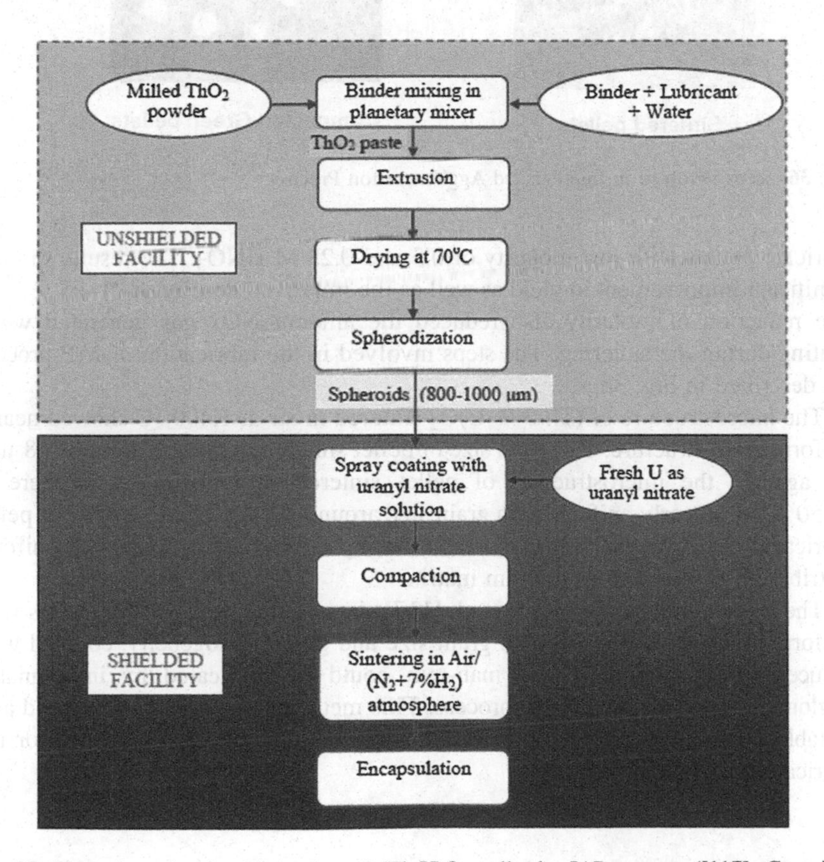

Fig. 35 Flow sheet for the fabrication of $(Th,U)O_2$ pellet by IAP process ([117], Copyright Elsevier)

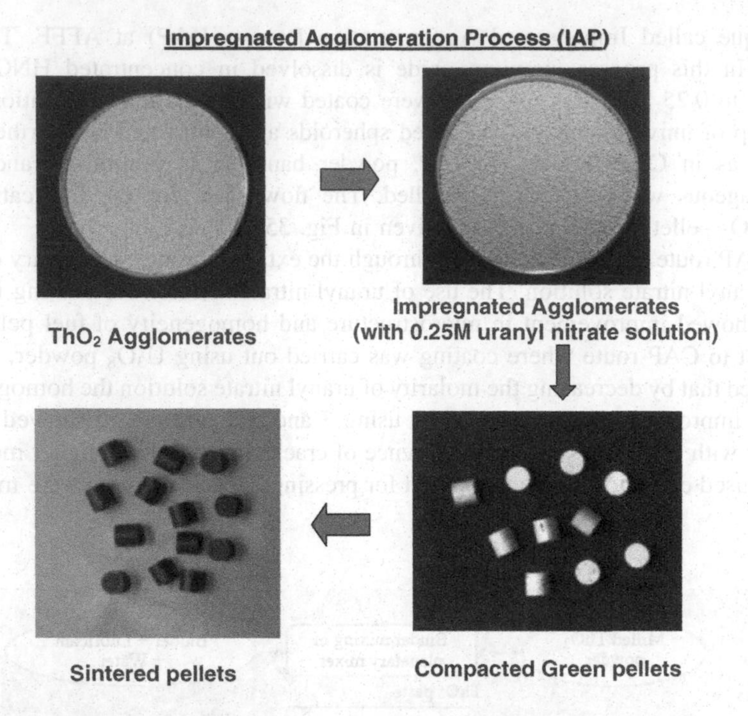

Fig. 36 Steps involved in Impregnated Agglomeration Process

fabricate pellets with low molarity of 0.5 and 0.25 M HNO_3. The results showed significant improvement in yield as well as the improved density of 91–95 % T.D. The reduction of molarity also reduced the amount NOx gas generated while heating during the sintering. The steps involved in the fabrication of IAP process are described in Fig. 36.

The microstructure of (Th-3 % U)O_2 sintered in air at 1,400 °C showed nearly uniform grain structure. The grain size of pellet sintered in air was around 7–8 μm. As against, the microstructure of pellet sintered in reducing atmosphere at 1,650 °C was nearly uniform with grain size around 50 μm. Homogeneity of pellet fabricated by IAP route was better than that by CAP route as it showed uniform distribution of uranium in thorium matrix.

The study conducted by Khot et al. [117] showed that high density pellets with uniform microstructure, optimum grain size and good homogeneity, coupled with reduced powder handling and man-rem could be fabricated by impregnated agglomerate pelletization (IAP) process. This method, therefore, has emerged as a suitable alternative to the conventional powder oxide pellet (POP) route for the fabrication of MOX fuel.

7 Intercomparison

The large-scale utilization of thorium requires the adoption of closed cycle and many of the fuel cycle technologies of uranium can be adopted for thorium. There are, however, a few major challenges in refabrication and reprocessing posed by the stable nature of thoria matrix and radiological issues associated with thorium fuel cycle. Thorium utilization calls for addressing to a lot of technological challenges. Several technologies are under development for the refabrication of $(Th\text{-}^{233}U)O_2$ fuel to reduce man-rem problems associated with ^{232}U daughter nuclides. A comparison of five different routes has been made in terms of radiotoxicity, microstructure, and waste generation and has been presented in Table 2. A judicious choice of the process for fabrication of pellets has to be made by taking into consideration of the above factors.

8 Other Methods

Other techniques like hot pressing, isostatic pressing, and extrusion had been tried for the limited amounts of ThO_2 and ThO_2-UO_2 pellets [14]. Kaufman [118] has studied the hot pressing behavior of low-sintered ThO_2 and ThO_2-UO_2 pellets over a temperature range of 1,873–2,423 K. The tests were carried out in a graphite punch and die assembly under a reduced pressure of CO. The end products were hypostoichiometric oxide and hence resulting the formation of metallic U and Th inclusions. Gardiner [119] could fabricate low density (~84 %) ThO_2 pellets by hot pressing with a load ranging from 0 to 20.7 MPa and temperature range 1,273–1,873 K.

Table 2 Comparison of five different routes for the fabrication of $(Th\text{-}^{233}U)$ MOX

	Powder Route	SGMP	Impregnation	CAP Process	Co-ppt process
Radiation dose	◆◆◆◆◆ (highest)	◆◆◆	◆◆ (lowest)	◆◆◆	◆◆◆
Uranium Distribution	Uniform	Uniform	For pellet impregnation: Higher U concentration at periphery For gel impregnation: Uniform	Lower U concentration in ThO_2 granules In A-CAP, better distribution of U	Extremely uniform
Grain size	Uniform	Uniform	Uniform	Duplex	Uniform
Waste	Solid waste	Liquid waste, organic waste	Less liquid waste	Less waste	Liquid waste

The successful extrusion of high density (94 %) ThO_2 and ThO_2-UO_2 rods have been reported by many investigators. Fitts et al. [120] reported high strength, 99 % density ThO_2 rods based on sol–gel process by extrusion. In general, extrusion of ThO_2 and ThO_2-UO_2 consists of preparation of a source powder incorporating a plasticizer such as methyl cellulose, to form a mass capable of being extruded. They are then subjected to extrusion, controlled drying, and high temperature sintering in air or inert/reducing atmospheres. The factors that affect the extrusion are [14]

- Binders used
- Powder particle size and distribution
- Wetness of the mass to be extruded
- Extrusion force.

Another non-conventional method is slip casting process which consists of making a stable water suspension of insoluble particles. The suspension is then poured into a plaster mold which gets solidified by means of a liquid extraction mechanism [14]. After the setting time, the solid has sufficient handling strength for further operation. Thin-walled ThO_2 has been made by slip casting procedure and then sintered at 2,373 K to get almost 100 % dense with a wall thickness of 0.6 mm.

9 Techniques for the Fabrication of (Th, Pu)O_2 Pellets

Very few reports are available in the literature on ThO_2-PuO_2 system. Today MOX (U, Pu) fuels are used in some conventional reactors, with ^{239}Pu providing the main fissile ingredient. An alternative is to use Th/Pu fuels with Pu being consumed and fissile ^{233}U being bred [121–123]. In the Energy Amplifiers, Th–Pu mixtures [124] are being considered as the candidate fuel which is more effective in eliminating Pu at acceptable concentrations than the conventional mixture of U and Pu. The device operates as an effective Pu to ^{233}U converter. The latter can be subsequently mixed with ordinary or depleted U and it constitutes an excellent fuel for PWRs.

In BARC, ThO_2-PuO_2 pellets of controlled density and microstructure have been manufactured by powder route using ex-oxalate ThO_2 and PuO_2 powders as starting materials. Two numbers of zircaloy clad 6 pin clusters of ThO_2-PuO_2 containing 4 % PuO_2 and 6.75 % PuO_2 have been manufactured and successfully irradiated in CIRUS reactor simulating the operating conditions of PHWR. Also flow sheet have been developed for manufacturing ThO_2-30 % PuO_2, ThO_2-50 % PuO_2, and ThO_2-75 % PuO_2 pellets using small addition of CaO or Nb_2O_5 as admixed dopants.

9.1 Th–Pu–O System

Freshly and Mattys [125, 126] have shown that ThO_2 and PuO_2 forms an ideal solid solution in the whole composition range like ThO_2 and UO_2. The lattice parameter of fluorite type cubic phase was found to decrease regularly from 0.5601 nm for ThO_2 to 0.5396 nm for pure PuO_2. A continuous series of solid solution has been reported by Mulford and Ellinger [127].

As mentioned earlier, ThO_2 is the only stable oxide of Th–O system in the condensed state and it has very little non-stoichiometry compared to UO_2. On the other hand, the phase diagram of plutonium–oxygen system shows the presence of four compounds namely Pu_2O_3, $PuO_{1.52}$, $PuO_{1.61}$, and PuO_2. PuO_2 is the stable oxide of Pu [128]. The lattice dimension and density of PuO_2 are 0.5396 nm and 11.46 g/cm^3, respectively. PuO_2 loses oxygen readily at elevated temperatures in either vacuum, reducing, or inert atmospheres. The deviation from stoichiometry is accompanied by the formation of Frenkel defects on the oxygen ion sublattice of the crystal [129]. This oxygen deficiency results in the formation of larger Pu^{+3} ions, which causes the unit cell to expand [130]. Unlike UO_2, PuO_2 does not readily incorporate excess oxygen in interstitials and therefore PuO_{2+x} formation has not been reported.

The fabrication processes for ThO_2-PuO_2 pellets are similar to those of ThO_2-UO_2. The green ThO_2-PuO_2 pellets for this study were prepared by the conventional powder metallurgy technique which consists of the following steps [121]:

- Milling of the as-received thorium oxide powder in a planetary ball for 8 h using tungsten carbide balls to break its platelet morphology
- Mixing/milling of the above-milled ThO_2 powder with the required quantity of PuO_2 powder and additive (0.5 wt% CaO or 0.25 wt% Nb_2O_5) for 4 h in a planetary ball mill with tungsten carbide balls
- Double pre-compaction at 150 MPa
- Granulation of the pre-compacts
- Final cold compaction of the granulated powder at 300 MPa into green pellets

Depending upon the composition, the density of the green compacts was found varying between 55 and 67 % of T.D. Kutty et al. [121, 122] studied the shrinkage behaviour of ThO_2, ThO_2-30 % PuO_2, ThO_2-50 % PuO_2, and ThO_2-75 % PuO_2 compacts in the various atmospheres in BARC using a push rod type dilatometer.

Figure 37 shows the intercomparison of the shrinkage curves for CaO-doped pellets in Ar-8 % H_2 for ThO_2, ThO_2-30 % PuO_2, ThO_2-50 % PuO_2, and ThO_2-75 %PuO_2. For ThO_2-30 % PuO_2 pellet, the shrinkage is faster in Ar-8 % H_2. The onset of shrinkage occurs at 1100 °C for CaO-doped pellet, which is about 100 °C lower than that for the Nb_2O_5-doped pellet. In the case of ThO_2-50 % PuO_2 pellet, the onset of shrinkage occurs earlier for Nb_2O_5-doped pellets than that for the CaO-doped pellets. It is also noted that for ThO_2-50 % PuO_2 and ThO_2-75 % PuO_2, the sintering rate is retarded in Ar-8 % H_2 at around 1,000–1,200 °C.

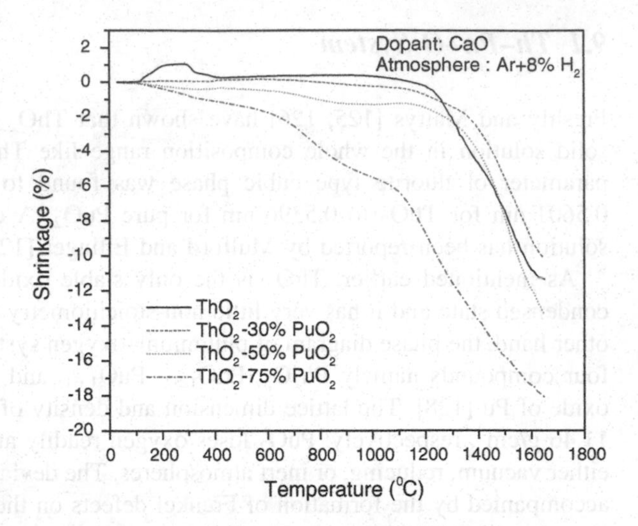

Fig. 37 Intercomparison of the shrinkage curves for CaO-doped pellets in Ar-8 % H_2 for ThO_2, ThO_2-30 % PuO_2, ThO_2-50 % PuO_2 and ThO_2-75 % PuO_2 ([121], Copyright Elsevier)

This effect is more prominent for ThO_2 -75 % PuO_2 pellet when a change in the slope in the shrinkage curve is noted at around 1,000–1,100 °C in Ar and Ar-8 % H_2 atmospheres for both CaO and the Nb_2O_5-doped pellets. Figure 38 illustrates effects of atmosphere on the shrinkage ThO_2-75 % PuO_2 pellets.

The significant observations of their study [121] are summarized below:

1. CaO was found to be a better dopant than Nb_2O_5 for Ar and Ar-8 % H_2 atmospheres.
2. Shrinkage was found to be marginally superior in Ar-8 % H_2 atmosphere than in Ar.
3. An expansion was observed for ThO_2-50 % PuO_2 and ThO_2-75 % PuO_2 especially in Ar-8 % H_2 atmosphere at around 1,100 °C.

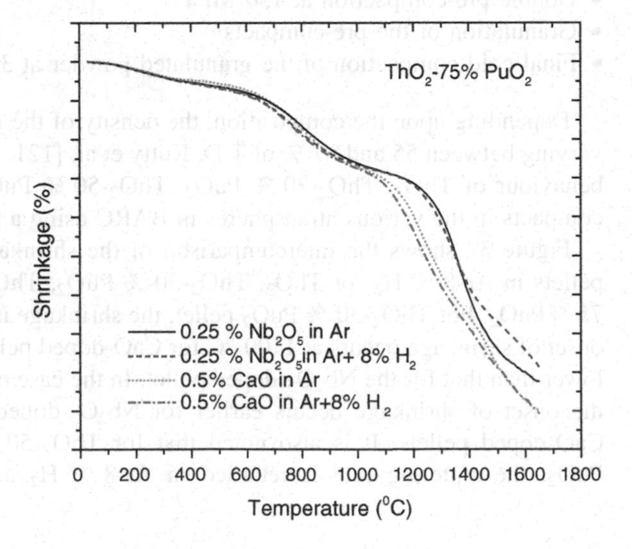

Fig. 38 Effects of atmosphere on the shrinkage ThO_2-75 % PuO_2 pellets ([121], Copyright Elsevier)

X-ray diffraction (XRD) pattern of ThO_2-50 % PuO_2 and ThO_2-75 % PuO_2 pellets sintered in Ar and Ar-8 % H_2 showed the presence of two phases—one is isostructural with PuO_2 (fluorite) and other is isostructural with bcc α-Pu_2O_3. Some portion of PuO_2 has combined with ThO_2 to form a fluorite type solid solution and the rest of the PuO_2 might have got reduced to bcc α-Pu_2O_3 and combined with ThO_2 to form bcc α-Pu_2O_3 type solid solution. Table 3 gives the typical values of O/M and their density. The microstructure of ThO_2-50 % PuO_2 and ThO_2-75 % PuO_2 pellets showed two phased structure which is in full agreement with XRD data (see Fig. 39). The amount of second phase was found to be higher in ThO_2-75 % PuO_2 sample. The average grain size of the major phase (fluorite) was estimated to be about 3.4 μm and that of the second phase region was found to be much smaller (<1 μm).

It is evident from the study of Kutty et al. [122] that the shrinkage occurs more rapidly for CaO-doped pellets in both the atmospheres. And also for the same dopant, shrinkage increases with the increase in PuO_2 content. At 1,300 °C, the shrinkage was negligible (<1 %) for pure ThO_2 in Ar-8 % H_2, but it was more than 11 % for ThO_2-75 % PuO_2 in the same atmosphere. It was also noticed for the same dopant (CaO), Ar-8 % H_2 is a marginally better medium for sintering than Ar for the all the compositions covered in their study.

Interdiffusion coefficient, D in ThO_2-PuO_2 system is expected to strongly depend on the oxygen potential, Pu content, and temperature as D in UO_2-PuO_2 system [129]. It is well known that the tracer diffusion coefficient of U, D^U and of Pu, D^{Pu} in UO_{2+x}, and MO_{2+x} vary by 4–5 orders of magnitude at constant temperature if the oxygen potential is varied [31, 129]. Since ThO_2 does not exist as ThO_{2+x}, the only possible deviation from stoichiometry for ThO_2-PuO_2 system is

Table 3 Typical values of green and sintered density, (O/M) ratios of ThO_2-PuO_2 pellets [121]

Pellet composition	Dopant	Sintering atmosphere	Green density (% T.D.)	Sintered density (% T.D.)	Oxygen to metal ratio (O/M)
ThO_2	CaO	Ar	67	93	2.000
		Ar-8 % H_2	67	95	2.000
	Nb_2O_5	Ar	67	96	2.000
		Ar-8 % H_2	67	96	2.000
ThO_2–30 % PuO_2	CaO	Ar	61	88	1.965
		Ar-8 % H_2	61	91	1.947
	Nb_2O_5	Ar	61.5		
		Ar-8 % H_2	61.5	–	–
ThO_2–50 % PuO_2	CaO	Ar	58	89	1.934
		Ar-8 % H_2	58	90	1.921
	Nb_2O_5	Ar	58	87	
		Ar-8 % H_2	58	88	–
ThO_2–75 % PuO_2	CaO	Ar	55	86	1.926
		Ar-8 % H_2	55	85	1.887
	Nb_2O_5	Ar	56	85	
		Ar-8 % H_2	56	87	–

Fig. 39 Microstructure of ThO_2-50 % PuO_2 pellet showing two phased structure

to exist as MO_{2-x}. The expected strong dependence of D^M on x in MO_{2-x} was confirmed by experiments on Pu in MO_{2-x}. Three possible mechanisms are suggested for MO_{2-x} which are listed below [129, 130]:

(a) A vacancy mechanism when O/M > 1.98,
(b) An interstitial mechanism for a lower O/M values (1.95 < O/M < 1.98),
(c) A cluster mechanism for O/M < 1.95.

The PuO_2 powder used in these experiments has a higher BET surface area than ThO_2. Hence, PuO_2 is expected to sinter more readily than ThO_2. Pritchard and Nance [131] have studied the sintering behavior of PuO_2 in different atmospheres. PuO_2 has been found to sinter better in H_2 than in Ar. Chikalla [132] has reported that PuO_2 can be reduced to ~50 % α-Pu_2O_3 on heating in H_2 at 1,650 °C or to 25 % Pu_2O_3 on just heating at 1,450 °C. PuO_2 loses little amount O_2 on heating up to 1,100–1,200 °C in inert or reducing atmosphere but loses O_2 readily at high temperature [133, 134]. The oxygen deficiency results in the formation of Pu^{+3} ions which help in enhancing sintering. Pu^{+3} ions were shown to diffuse faster in ThO_2 than Pu^{+4} ions. Therefore formation of Pu_2O_3 in the sample helps in achieving a faster shrinkage rate [134]. The superior sintering behavior of ThO_2-PuO_2 in Ar-8 % H_2 observed in this study may be associated with the presence of defect structure as some of the PuO_2 gets reduced to Pu_2O_3.

Sintering was found to be slightly superior for ThO_2-PuO_2 compositions in Ar-8 % H_2 than that in Ar. The similar observation has also been reported for UO_2-PuO_2 system [133]. Since O/M of the pellets sintered in Ar-8 % H_2 was found to be less than 2.00 especially for ThO_2-50 % PuO_2 and ThO_2-75 % PuO_2 (see Table 3), deviation of stoichiometry has occurred considerably in these pellets. Another important point noticed during the densification is the retardation of shrinkage at around 1,000 °C in Ar and Ar-8 % H_2 for ThO_2-50 % PuO_2 and ThO_2-75 % PuO_2. The retardation of shrinkage correlates with the onset of the solid solution formation. The solid solution is formed by the interdiffusion of Pu^{+4}

ion into ThO_2 lattice and Th^{+4} ion into PuO_2 lattice. These interdiffusion processes decrease the sintering rate and shift shrinkage to a higher temperature. The formation of Pu_2O_3 during the sintering results in the expansion of the unit cell. The ionic radii of Pu^{+3} and Pu^{+4} ions are 0.108 and 0.093 nm, respectively. Thus, expansion in the shrinkage curves at around 900–1,100 °C may be attributed to the formation of large Pu^{+3} ion [129].

10 Development of Master Sintering Curve for Thoria-Based Fuels

An important processing goal for the nuclear ceramics is to obtain a uniform microstructure with the desired grain size. It is also important to attain the desired dimension after sintering since the tolerance on the pellet diameter is small. The oversized pellets are to be ground to the desired size and the undersized pellets are to be reprocessed. It is therefore desirable to predict the final size and density of the pellets [135–137]. In the conventional sintering procedure, the parameters such as time and temperature of the sintering are arbitrarily decided on the 'trial and error' basis. It could be beneficial for the nuclear industry to predict the densification behavior from the sintering data that are readily available. If the prediction is made, the sintering strategy could be established on that basis [138], which leads to produce the pellets of good quality with less number of rejects. Therefore, there is a need for better understanding of whole sintering procedure. The theory of master sintering curve (MSC) provides a new insight into the understanding of sintering [139]. The master sintering curve (MSC) enables to predict the densification behavior under arbitrary time–temperature excursions with the help of a minimum of preliminary experiments. This curve is sensitive to such factors as starting morphology of the powder, fabrication route, dominant diffusion mechanism and heating condition used for sintering [140].

10.1 Theory

The master sintering curve can be derived from the densification rate equation of the combined-stage sintering model [141, 142]. The sintering of a powder compact is traditionally divided into three stages in which different mechanisms may be operative [143–145]. Most of the sintering models focus on a specific idealized geometry that is represented by only one of the three stages in the above sintering process.

To overcome the above-mentioned drawback, it is desirable to have a model that describes the entire sintering process. The combined-stage sintering model, proposed by Hansen et al. [141], describes the densification through the entire stages of sintering. By observing the similarities in the three stages of sintering, a single equation was derived which describes the densification through all stages of sintering. Johnson and Su [138] showed that the observed sintering rates in the combined-stage sintering model were within the range expected for the diffusion controlled sintering at the heating rates used. In this model the microstructure is characterized by two separate parameters representing geometry and the average grain size.

For the development of master sintering curves, the parameters in the sintering rate equations are separated into (a) those related to the microstructure and (b) those related to time and temperature terms, on the opposite sides of the equation [138]. These two sides are then related to each other experimentally. The combined-stage sintering model relates the linear shrinkage rate of a compact at any given instant to the grain boundary and volume diffusion coefficients, the surface tension, and certain aspects of the instantaneous microstructure of the compact. In this model, the instantaneous linear shrinkage rate is given as [142]:

$$-\mathrm{d}L/L\mathrm{d}t = (\gamma\Omega/kT)\,[(\Gamma_v D_v/G^3) + (\Gamma_b \delta x D_b/G^4)] \tag{15}$$

where dL/Ldt is the normalized instantaneous linear shrinkage rate, γ is the surface energy, Ω the atomic volume, k the Boltzmann constant, T the absolute temperature, G the mean grain diameter, D_v and D_b the coefficients of volume and grain boundary diffusion, respectively, δ the width of the grain boundary, Γ_v and Γ_b are the collections of microstructure scaling parameters for volume and grain boundary diffusion, respectively.

For isotropic shrinkage, the linear shrinkage rate can be converted into the densification rate by

$$-\mathrm{d}L/L\mathrm{d}t = \mathrm{d}\rho/3\rho\mathrm{d}t \tag{16}$$

where, ρ is the bulk density.

Substituting equation (16) in (15) and assuming that only one of the diffusion mechanisms (either volume or grain boundary diffusion) dominates the sintering process, one can rewrite equation (15) as [142]:

$$\mathrm{d}\rho/3\rho\mathrm{d}t = (\gamma\Omega/\mathrm{kT})\Gamma_{(\rho)}\mathrm{D}_0/(\mathrm{G}_{(\rho)})^{\mathrm{n}}\exp(-\mathrm{Q}/\mathrm{RT}) \tag{17}$$

where, Q is the activation energy, D_0 is the pre-exponential factor and R is the gas constant. The above equation can be rearranged and integrated as follows:

$$\int(\mathrm{G}_{(\rho)})^{\mathrm{n}}/(3\rho\mathrm{G}_{(\rho)})\,\mathrm{d}\rho = \int(\gamma\Omega\mathrm{D}_0)/(\mathrm{kT})\exp(-\mathrm{Q}/\mathrm{RT})\,\mathrm{d}t \tag{18}$$

In the above equation, the atomic diffusion process and the microstructural evolution terms are separated [140]. All the terms on the right hand side (RHS) of equation (18) are related to atomic diffusion process and are independent of characteristics of powder compacts. The terms on the left-hand side (LHS) are the quantities that define the microstructural evolution and are independent on the thermal history of the powder compacts [138].

The time and temperature dependent side (RHS) of the equation can be represented as theta parameter, Θ, as follows [140]:

$$\Theta = \int 1/T \exp(-Q/RT)dt \tag{19}$$

where, *t* is the instantaneous time, which is usually a function of temperature. Similarly, one could integrate the LHS of equation (18) if the evolution of the microstructure is known in detail.

Equation (19) can be simplified for isothermal portion of the sintering runs to [138]:

$$\Theta = t_i/T \exp(-Q/RT) \tag{20}$$

where, t_i is the duration of the isothermal portion of the run. For constant heating rate, Eq. (19) can be written as:

$$\Theta = (1/c) \int_{T_0}^{T_1} 1/T\exp(-Q/RT)dT \tag{21}$$

where, c is the heating rate used and T_0 is the temperature below which no sintering takes place.

The relationship between the density (ρ) and Θ is defined as the master sintering curve. For the construction of MSC, a series of runs at different temperatures (isothermal) or constant heating rates over a range of heating rates is needed. If the activation of energy of sintering is unknown, it has to be estimated in order to obtain the master sintering curve.

10.2 Construction of MSC

For the construction of MSC, the integral of Eq. (19) and the experimental density should be known. The dilatometry can be conveniently used to determine the density since the instantaneous density at all times can be obtained from the dilatometric data. Figure 40 shows the dL/L_0 versus temperature plot of ThO_2 under different heating rates in Ar-8 % H_2 atmosphere. It can be seen from the figure that the onset of shrinkage is shifted to higher temperatures on increasing the heating rates. The dilatometric curves of Fig. 40 were replotted as percentage of theoretical density (% T.D.) versus temperature. The dL/L_0 values were converted into % T.D. using the following relation:

$$\rho = [1/(1 - dL/L_0)]^3 \rho_0 \tag{22}$$

where, ρ and ρ_0 are the densities of the sintered and green pellets, respectively. The curves have the familiar sigmoidal shape and generally shifted to higher temperatures with increasing heating rate. It can be noted that the sintered densities obtained at any temperature showed a modest but a systematic dependence on heating rate.

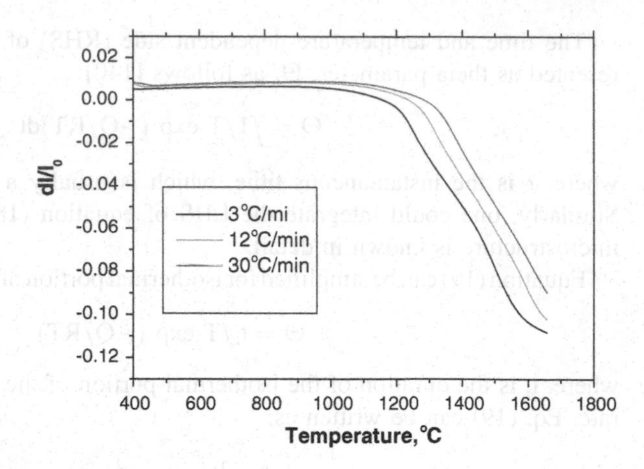

Fig. 40 Shrinkage curves for ThO_2 pellets containing 0.5 wt% CaO as dopant in Ar-8 % H_2 atmosphere for the different heating rates ([135], Copyright Elsevier)

For the calculation of Θ, the activation energy for the sintering process must be known. If the activation energy is unknown, it can be estimated with good precision from Θ versus density (ρ) data [138]. For this purpose, a particular value of activation energy is chosen and ρ-Θ curves are constructed for each heating rate. If the curves fail to converge, a new value of activation energy is chosen and the calculations are repeated. This procedure should be continued until all the curves are converged showing that the activation energy is the acceptable one for sintering. A curve can be then fitted through all the data points, and then convergence of data to the fitted line can be quantified through the sum of residual squares of the points with respect to the fitted line. The best estimate of Q will be the value of the minimum in the plot of activation energy versus mean residual squares [138].

Several assumptions are made in developing the above MSC which are listed below [138]:

1. The microstructural evolution is dependent only on density for any given powder and the fabrication procedure.
2. One of the diffusion mechanisms either volume or grain boundary diffusion dominate the sintering process.
3. Contributions from surface diffusion and vapor transport are negligible in the sintering process.
4. This procedure should be applied only to the powder compacts made from the same source powder and the same fabrication procedure.

As mentioned earlier, one of the essential data for obtaining the master sintering curve is the activation energy. For this, the density data for ThO_2 obtained from the dilatometric measurements, and Θ values obtained from Eq. (19) are employed. A ρ-Θ curve is then constructed for all the heating profiles for a chosen value of activation energy. The best convergence occurs at around 550 kJ/mol. Figures 41 gives the mean residual squares for the various values of activation energy and the minimum has been found to be for 540 kJ/mol. The activation

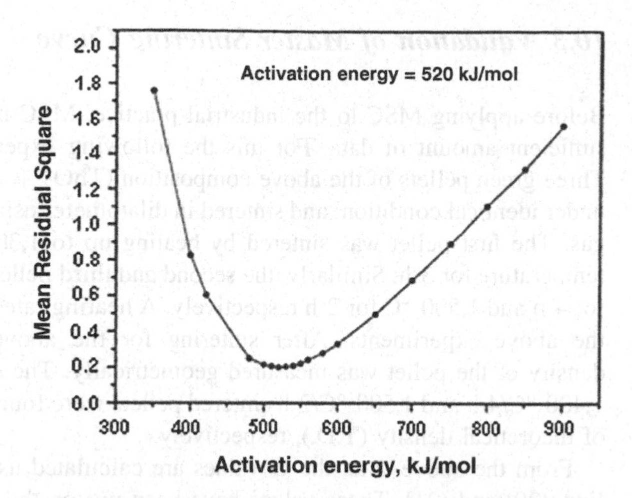

Fig. 41 Mean residual squares for the various values of the activation energy ([135], Copyright Elsevier)

energy thus obtained was found to be in reasonable agreement with the values reported in the literature [129, 130, 145]. Matzke [129] reported 650 kJ/mol for pure ThO_2. According to Shiba [146], it is 625 kJ/mol. The activation energy obtained in this study is about 20 % less than the value obtained by Matzke. The difference may be because the sample used in this study contains about 0.5 wt% CaO as sintering aid.

From the knowledge of the activation energy of sintering, MSC for ThO_2 has been constructed and is shown in Fig. 42. It can be seen that the value of Θ changed dramatically from 10^{-50} at the beginning to 10^{-35} at the higher density end. Despite a ten folds rise in heating rate, the individual sintering curves have merged reasonably close to a single curve. This result suggests that there must be a general curve, regardless of sintering path, which is what had been defined as the MSC [135].

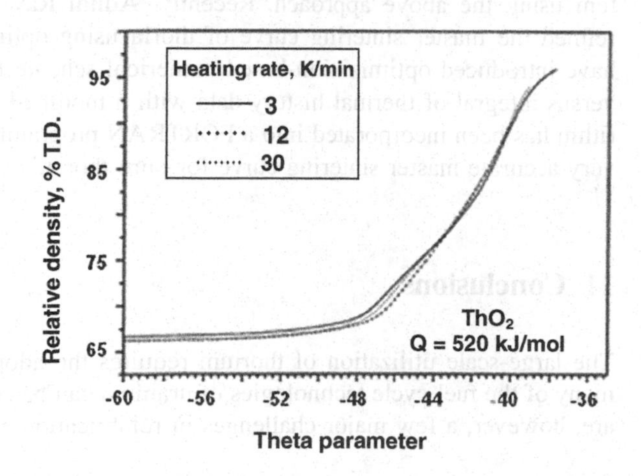

Fig. 42 Master sintering curve for 0.5 % CaO-doped ThO_2. The curve is constructed using an activation energy of 520 kJ/mol ([135], Copyright Elsevier)

10.3 Validation of Master Sintering Curve

Before applying MSC to the industrial practice, MSC has to be validated with a sufficient amount of data. For this the following experiments were carried out. Three green pellets of the above composition (ThO_2 + 0.5 wt% CaO) were made under identical condition, and sintered in dilatometer using Ar-8 % H_2 as the cover gas. The first pellet was sintered by heating up to 1,300 °C and holding at that temperature for 3 h. Similarly, the second and third pellet was sintered at 1,400 °C for 4 h and 1,500 °C for 2 h respectively. A heating rate of 20 K/min was used for the above experiments. After sintering for the above-mentioned periods, the density of the pellet was measured geometrically. The densities of 1,300 °C/3 h, 1,400 °C/4 h and 1,500 °C/2 h sintered pellets were found to be 78, 83.5 and 95 % of theoretical density (T.D.), respectively.

From the above data, the Θ values are calculated using the relations given in Eqs. (20) and (21). These values have been put on the master sintering curve. It can be seen that the values for all the three temperatures with different periods of time are lying on the MSC, validating the concept of MSC. Thus, it can be regarded that integration of a proposed sintering time–temperature profile yields a point on the MSC curve. The expected density can be obtained by finding the ordinate value at that point. On the other hand, if the final desired density is known, it is possible to find out the corresponding Θ value from the abscissa of the master sintering curve and thereafter to plan the sintering schedule [135, 136]. In the second set of experiment, the effect of dopant on MSC has been evaluated. To study this, ThO_2 pellets were fabricated under the identical condition with Nb_2O_5 as dopant instead of CaO. This curve does not coincide with that of ThO_2 doped with CaO. This result clearly shows that MSC is dependent on dopant. In other words, the sinterability of powder compacts made from different powders, dopants and fabrication procedures could be characterized with the help of the master sintering curve.

Kutty et al. [136] constructed master sintering curve for ThO_2-2 % U_3O_8 system using the above approach. Recently, Adithi Ray et al. [147] have further refined the master sintering curve of thoria using optimization technique. They have introduced optimization-based numerical scheme for fitting relative density versus integral of thermal history data with a modified sigmoid curve. The algorithm has been incorporated into a FORTRAN programme for the construction of very accurate master sintering curve for pure thoria.

11 Conclusions

The large-scale utilization of thorium requires the adoption of closed cycle and many of the fuel cycle technologies of uranium can be adopted for thorium. There are, however, a few major challenges in refabrication and reprocessing posed by the stable nature of thoria matrix and radiological issues associated with thorium fuel cycle. Thorium utilization calls for addressing to a lot of technological challenges. Several technologies are under development for the refabrication of $(Th\text{-}^{233}U)O_2$ fuel to reduce man-rem problems associated with ^{232}U daughter nuclides.

A comparison of five major different routes, namely powder-pellet, co-precipitation, sol–gel microsphere pelletization (SGMP), coated agglomerate pelletization (CAP), and impregnation has been made in terms of radiotoxicity, microstructure, and waste generation. The microstructure of the pellet made by CAP process showed a duplex grain structure while the other methods yielded a uniform grain size. The microhomogeneity was found to be slightly inferior for the pellets made by pellet- impregnation and CAP processes. A judicious choice of the process for fabrication of pellets has to be made by taking into consideration of the above factors.

References

1. IAEA-TECDOC-1450 (2005) Thorium fuel cycle-potential benefits and challenges. International Atomic Energy Agency (IAEA), Vienna
2. Lung M, Gremn O (1998) Perspectives of the thorium fuel cycle. Nucl Eng Des 180:133–146
3. Nuclear Energy Series No NF-T-2.4 (2012) Role of thorium to supplement fuel cycles of future nuclear energy systems. IAEA, Vienna
4. Lung M (1997). A present review of the thorium nuclear fuel cycles. European Commission Report EUR-17771, Luxembourg
5. Kazimi MS, Driscoll MJ, Ballinger RG, Clarno KT, Czerwinski KR, Hejzlar P, LaFond PJ, Long Y, Meyer JE, Reynard MP, Schultz SP, Zhao X (1999) Proliferation resistant, low cost, Thoria-Urania fuel for light water reactors. Annual Report, Nuclear Engineering, MIT, Cambridge
6. Kazimi MS, Pilat EE, Driscoll MJ, Xu Z, Wang D, Zhao X (2001) Enhancing proliferation resistance in advanced light water reactor fuel cycles. In: International conference: back-end of the fuel cycle: from research to solutions, Global 2001, Paris
7. MacDonald PE (1999) advanced proliferation resistant, lower cost, uranium-thorium dioxide fuels for light water reactors. NERI 99-0153
8. Kazimi MS, Czerwinski KR, Driscoll MJ, Hejzlar P, Meyer JE (1999) On the use of thorium in light water reactors. MIT-NFC-0016
9. Kang J, Von Hippel FN (2001) U-232 and the proliferation-resistance of U-233 in spent fuel. Sci Global Secur 9:1–32
10. Brooksbank RE, Parrott JR, Youngblood EL, McDuffee WT (1974) The containment of ^{233}U in the processing facilities of the ORNL pilot plant. In: Conf-740523-1, VII International Congres of Societe Francaise de Radioprotection, Versailles, France
11. Parrott Sr JH, McDuffee WT, Nicol RG, Whitson WR, Krichnisky AM (1979) The preparation of kilogram quantities of $^{233}UO_2$ for LWBR demonstration programme. ORNL/CF-79/279
12. Brooksbank RE et al (1966) The impact of kilorod facility operational experience on the design of fabrication plants for $^{233}U/Th$ fuels. In: 2nd international conference of thorium fuel cycle, Gatlinburg

13. Haws CC, Matheme JL, Miles FW, Van Cleve JE (1965) Summary of the Kilorod project. ORNL-3681
14. Belle J, Berman RM (1984) Thorium dioxide: properties and nuclear applications. Naval Reactors Office, DOE. Government Printing Office, Washington, DC
15. Sease JD, Pratt RB, Lotts AL (1966) Remote fabrication of thorium fuels. ORNL-TM-1501
16. Brooksbank RE, McDuffee WT, Rainey RH (1978) A review of thorium fuel reprocessing experience, ORNL. In: Conf. 780223-3
17. Hart PE et al (1979) ThO_2-based pellet fuels, their properties, methods of fabrication, and irradiation performance: a critical assessment of the state of the technology and recommendations for further work. PNL-3064
18. Hart PE (1979) Thoria development activities. PNL-2973
19. Kantan SK, Raghavan RV, Tendolkar GS (1958) Sintering of thorium and thoria. In: Proceedings of 2nd UN international conference on peaceful use of atom, Geneva, vol 6, pp 132–138
20. White GD, Bray LA, Hart PE (1980) Optimization of thorium oxalate precipitation conditions relative to thorium oxide sinterability. PNL-3263
21. Balakrishna P, Varma BP, Krishnan TS, Mohan TRR, Ramakrishnan P (1988) Thorium oxide: calcination, compaction and sintering. J Nucl Mater 160:88–94
22. Mohan A, Moorthy VK (1971) Studies on sintering of nuclear fuel materials -sintering behavior of Urani–Thoria mixtures. BARC-568, Bombay, India
23. Pope JM, Radford KC (1974) Physical properties of some thoria powders and their influence on sinterability. J Nucl Mater 52:241–254
24. Johnson RGR (1966) Fabrication of fuel pellets from pot process denitrated ThO_2 powder. WAPD-TM-577
25. Harada Y, Baskin Y, Handwerk JH (1962) Calcination and sintering study of thoria. J Am Ceram Soc 45:253–257
26. Chick LA, Pederson LR, Maupin GD, Bates JL, Thomas IE, Exarhos GJ (1990) Glycine-Nitrate combustion synthesis of oxide ceramic powders. Mater Lett 10:6–12
27. Chandramouli V, Anthonysamy S, Vasudeva Rao PR (1999) Combustion synthesis of thoria, a feasibility study. J Nucl Mater 265:255–261
28. Purohit RD, Saha S, Tyagi AK (2001) Nanocrystalline thoria powders via Glycine–Nitrate combustion. J Nucl Mater 288:7–10
29. Ananthasivan K, Anthonysamy S, Singh A, Vasudeva Rao PR (2002) De-agglomeration of thorium oxalate—a method for the synthesis of sinter active thoria. J Nucl Mater 306:1–9
30. Burke TJ (1982) Fabrication of high-density ThO_2 fuel pellets from freeze-dried granular feed. WAPD-TM-1524
31. Olander DR (1976) Fundamental aspects of nuclear reactor fuel elements. TID-26711-P1: p 145. DOE
32. Bairiot H (1999) MOX fuel cycle technologies for medium and long term deployment. In: IAEA proceedings of symposium on C&S papers, 2000, Vienna, vol 3, pp 81–101
33. Hugelmann D, Greneche D (1999) MOX fuel cycle technologies for medium and long term deployment. In: IAEA proceedings of symposium on C&S papers, 2000, Vienna, vol 3, pp 102–108
34. Wilson PD (1996) The nuclear fuel cycle: From ore to waste. Oxford University Press, Oxford
35. Zakharkin BS (1999) MOX fuel cycle technologies for medium and long term deployment. In: IAEA proceedings of symposium on C&S papers, 2000, Vienna, vol 3, pp 146–149
36. Kutty TRG, Khan KB, Achuthan PV, Dhami PS, Dakshinamoorthy A, Somayajulu PS, Panakkal JP, Kumar A, Kamath HS (2009) Characterization of ThO_2–UO_2 pellets made by co-precipitation process. J Nucl Mater 389:358
37. Radford KC, Bratton RJ (1975) Properties, blending and homogenization of (U, Th)O_2–UO_2 powder. J Nucl Mater 57:287–302
38. Atlas Y, Eral M, Tel H (1997) Preparation of homogeneous $(Th_{0.8}U_{0.2})O_2$ pellets via coprecipitation of (Th, U)$(C_2O_4)_2 \cdot nH_2O$ powders. J Nucl Mater 249:46–51

39. White GD, Bray LA, Hart PE (1981) Optimization of thorium oxalate precipitation conditions relative to derived oxide sinterability. J Nucl Mater 96:305–313
40. Argo L (2003) Experimental determination of the dry oxidation behavior of a compositional range of Uranium-Thorium mixed-oxide pellet fragments. Dissertation, University of Florida
41. Dhami PS, Kannan R, Rao KS, Shyam Lal R, Kumarguru K, Ajithlal RT, Sinalkar N, Dakshinamoorthy A, Jambunathan U, Dey PK (2005) Preparation of Thorium–Uranium mixed oxide fuel, NUCAR 2005, Amritsar, p 215
42. Chandramouli V, Anthonysamy S, Vasudeva Rao PR, Divakar R, Sundararaman D (1998) Microwave synthesis of solid solutions of urania and thoria—a comparative study. J Nucl Mater 254:55–64
43. Anthonysamy S, Ananthasivan K, Chandramouli V, Kaliappan I, Vasudeva Rao PR (2000) Combustion synthesis of urania–thoria solid solutions. J Nucl Mater 278:346–357
44. Ananthasivan K, Anthonysamy S, Sudha C, Terrance ALE, Vasudeva Rao PR (2002) Thoria doped with cations of group VB–synthesis and sintering. J Nucl Mater 300:217–229
45. Clayton JC (1976) Thorium oxide powder properties which are important to ThO_2 and ThO_2-UO_2 fuel pellet fabrication. WAPD-TM-1230
46. Mathews RB, Davis NC (1979) Fabrication of ThO_2 and ThO_2-UO_2 pellets for proliferation resistant fuels. Pacific Northwest Laboratory PNL-3210
47. Curtis CE, Johnson JR (1954) Interim report properties of thorium oxide Ceramics. ORNL-1809
48. Clayton JC (1994) Chemical reactions during ThO_2 and ThO_2-UO_2 fuel fabrication. WAPD-TM-3020
49. Kang KW, Yang JH, Kim KS, Song KW, Lee CH, Jung YH (2003) (Th, U)O_2 pellets: fabrication and thermal properties. J Kor Nucl Soc 35:299–308
50. Shiratori T, Fukuda K (1993) Fabrication of very high density fuel pellets of thorium dioxide. J Nucl Mater 202:98–103
51. Kutty TRG, Hegde PV, Banerjee J, Khan KB, Jain GC, Sengupta AK, Majumdar S, Kamath HS (2003) Densification behaviour of ThO_2–PuO_2 pellets with varying PuO_2 content using dilatometry. J Nucl Mater 312:224–235
52. Balakrishna P (1987) In: Sintering of uranium dioxide: an introduction. Nuclear Fuels Complex, Hyderabad, p 28
53. Nair MR, Basak U, Ramachandran R, Majumdar S (1999) Sintering of ThO_2, ThO_2-UO_2, ThO_2-PuO_2 fuel pellets with additives. Trans Powder Metall Assoc India 26:53
54. Hj Matzke (1981) p 156. In: Sorensen T (ed) Non-stoichiometric oxides. Academic Press, New York
55. Balakrishna P, Somauajulu GVSRK, Krishnan TS, Mohan TRR, Ramakrishnan P (1991) p 2995. In: Vincenzini P (ed) Ceramics today–tomorrow's ceramics. Elsevier, Netherlands
56. Kutty TRG, Hegde PV, Keswani R, Khan KB, Majumdar S, Purushotham DSC (1999) Densification behaviour of UO_2–50%PuO_2 pellets by dilatometry. J Nucl Mater 264:10–19
57. Matzke HJ (1966) On the effect of TiO_2 additions on defect structure, sintering and gas release of UO_2. AECL-2585
58. Sbbharao EC, Sutter PH, Hrizo J (1965) Defect structure and electrical conductivity of ThO_2-Y_2O_3 solid solutions. J Am Ceram Soc 48:443–446
59. Bransky I, Tallan NM (1970) Electrical properties and defect structure of ThO_2. J Am Ceram Soc 53:625–629
60. Upadhyaya DD, Sunta CM (1985) Defect interactions and TL behaviour in thoria. J Nucl Mater 127:137–140
61. Balakrishna P, Ananthapadmanabhan PV, Ramakrishnan P (1994) Electrical conductivity of sintered niobia-doped and magnesia-doped thoria. J Mater Sci Lett 13:86–88
62. Kutty TRG, Khan KB, Kumar A, Kamath HS (2009) Densification strain rate in sintering of ThO_2 and ThO_2-0.25%Nb_2O_5 pellets. Sci Sinter 41:103–115
63. Kakodkar A (2002) The twin challenges of abundant nuclear energy supply and proliferation risk reduction—a view. In: 46th general conference on IAEA, Vienna

64. Kamath HS (2003) Development and microstructural characterization of ThO_2-UO_2 Fuels. In: 14th annual conference on Indian nuclear society, IGCAR, Kalpakkam
65. Sinha RK, Kakodkar A (2006) Design and development of the AHWR-the Indian thorium fuelled innovative nuclear reactor. Nucl Eng Des 236:683–700
66. Anantharaman K, Shivakumar V, Sinha RK (2002) Design and fabrication of AHWR fuels. In: International conference on characterization and quality control of nuclear fuels (CQCNF2002), Hyderabad, India
67. Kutty TRG, Somayajulu PS, Mukherjee SK, Panakkal JP, Vaidya VN, Majumdar S, Kamath HS (2002) Densification behaviour ThO_2-UO_2 pellets fabricated by four different routes. In: 68th annual technical meeting of Indian ceramic society ICCP-04 2004, Mumbai
68. Kutty TRG, Kumar A, Panakkal JP, Kamath HS (2010) Development and microstructural characterization of ThO_2-UO_2 fuels. BARC Newsl 314:28
69. Kutty TRG, Hegde PV, Khan KB, Jarvis T, Sengupta AK, Majumdar S, Kamath HS (2004) Characterization and densification studies on ThO_2-UO_2 pellets derived from ThO_2 and U_3O_8 powders. J Nucl Mater 335:462–470
70. Hoekstra HR, Siegel S, Fuchs LH, Katz JJ (1955) The uranium–oxygen system: $UO_{2.5}$ to U_3O_8. J Phys Chem 59:136–138
71. Karkhanavala MD, George AM (1966) δ-U_3O_8, A high temperature modification: Part I: preparation and characterization. J Nucl Mater 19:267–273
72. Labroche D, Dugne O, Chatillon C (2003) Thermodynamic properties of the O-U system. Part II: critical assessment of the stability and composition range of the oxides UO_{2+x}, U_4O_{9-y} and U_3O_{8-z}. J Nucl Mater 312:50–66
73. Chevrel H, Dehaudt P, Francois B, Baumard JF (1992) J Nucl Mater 189:175–182
74. Hund F, Niessen G (1952) Anomalous solid solution in the system. Thorium oxide-uranium oxide. Z Elecrochem 56:972–979
75. Harada Y (1997) UO_2 sintering in controlled oxygen atmospheres of three-stage process. J Nucl Mater 245:217–223
76. Paul R, Keller C (1971) Phasengleichgewichte in den systemen UO_2-$UO_{2,67}$-ThO_2 und UO_{2+x}-NpO_2. J Nucl Mater 41:133–142
77. Rand MH (1975) Thermochemical properties. In: Kubaschewski O (ed) Thorium: physico-chemical properties of its compounds and alloys. Atomic energy review, vol 5. IAEA, Vienna, p 7
78. Mathews JR (1987) The technological problems and the future of research on the basic properties of actinide oxides. J Chem Soc, Faraday Trans 83(2):1273–1285
79. Belle J, Lustman B (1958) In: Properties of UO_2, Fuel elements conference, Paris, TID-7546, p 442
80. Ackermann RJ, Chang AT (1973) Thermodynamic characterization of U_3O_{8-Z} phase. J Chem Thermodyn 5:873–890
81. Cordfunke EHP, Aling P (1965) System $UO_3 + U_3O_8$: dissociation pressure of γ-UO_3. Trans Faraday Soc 61:50–53
82. Glasser Leme D, Hj Matzke (1983) The diffusion of uranium in U_3O_8. J Nucl Mater 115:350–353
83. Hj Matzke (1990) Atomic mechanisms of mass transport in ceramic nuclear fuel materials. J Chem Soc, Faraday Trans 86:1243–1256
84. Khan KB, Kutty TRG, Somayajulu PS, Sengupta AK, Panakkal JP, Majumdar S, Kamath HS (2005) Fabrication and characterization of ThO_2-UO_2 pellets made by co-precipitation process. In: characterization and quality control of nuclear fuels (CQCNF-2005), Hyderabad
85. Assmann H, Doerr W, Peehs M (1986) Control of UO_2 microstructure by oxidative sintering. J Nucl Mater 140:1–6
86. Song KW, Kim KS, Kang KW, Jung YH (2003) Grain size control of UO_2 pellets by adding heat-treated U_3O_8 particles to UO_2 powder. J Nucl Mater 317:204–211
87. Lay KW, Carter RE (1969) Role of the O/U ratio on the sintering of UO_2. J Nucl Mater 30:74–87

88. Kutty TRG, Khan KB, Hegde PV, Sengupta AK, Majumdar S, Kamath HS (2003) Determination of activation energy of sintering of ThO_2-U_3O_8 pellets using the master sintering curve approach. Sci Sinter 35:125–132
89. Ganguly C (1993) Sol-gel microsphere pelletization—a powder-free advanced process for fabrication of ceramic nuclear fuel pellets. Bull Mater Sci 16:509–522
90. Ganguly C, Basak U, Vaidya VN, Sood DD, Balaramamoorthy K (1992) SGMP-LTS process for fabrication of high density UO_2 and $(U,Pu)O_2$ fuel pellets. In: Proceedings of 3rd internatioanl conference on CANDU fuel, Chalk River
91. Ganguly C, Langen H, Zimmer E, Merz E (1986) Sol-gel microsphere pelletisation process for fabrication of high density ThO_2-2%UO_2 for advanced pressurized heavy water reactors. Nucl Technol 73:84–95
92. Vaidya VN, Mukerjee SK, Joshi JK, Kamat RV, Sood DD (1987) A study of chemical parameters of the internal gelation based sol-gel process for uranium dioxide. J Nucl Mater 148:324–331
93. Suryanarayana S, Kumar N, Bamankar YR, Vaidya VN, Sood DD (1996) Fabrication of UO_2 pellets by gel pelletization technique without addition of carbon as pore former. J Nucl Mater 230:140–147
94. Pai RV, Mukerjee SK, Vaidya VN (2004) Fabrication of $(Th, U)O_2$ pellets containing 3 mol% of uranium by gel pelletisation technique. J Nucl Mater 325:159–168
95. Kumar N, Pai RV, Joshi JK, Mukerjee SK, vaidya VN, Venugopal V (2006) Preparation of $(U, Pu)O_2$ pellets through sol–gel microsphere pelletization technique. J Nucl Mater 359:69–79
96. Basak U, Nair MR, Ramchandran R, Majumdar S (2000). Fabrication of high density ThO_2, ThO_2-UO_2 and ThO_2-PuO_2 fuel pellets for heavy water reactor. In: Annual conference of Indian nuclear society (INSAC-2000), p 170
97. Kutty TRG, Khan KB, Somayajulu PS, Sengupta AK, Panakkal JP, Kumar A, Kamath HS (2008) Development of CAP process for fabrication of ThO_2-UO_2 fuels. Part I: fabrication and densification behaviour. J Nucl Mater 373:299–308
98. Kutty TRG, Kulkarni RV, Sengupta P, Khan KB, Bhanumurthy K, Sengupta AK, Panakkal JP, Kumar A, Kamath HS (2008) Development of CAP process for fabrication of ThO_2-UO_2 fuels. Part II: characterization and property evaluation. J Nucl Mater 373:309–318
99. Kutty TRG, Somayajulu PS, Bhanumurthy K, Panakkal JP, Kumar A, Kamath HS (2009) Characterization of ThO_2-UO_2 fuels made by CAP process. In: International conference on the peaceful uses of atomic energy, New Delhi
100. Kutty TRG, Khan KB, Somayajulu PS, Sengupta AK, Sah DN, Panakkal JP, Majumdar S, Kamath HS (2005) Characterization of ThO_2-UO_2 fuels made by CAP Process. In: Characterization and quality control of nuclear fuels (CQCNF-2005), Hyderabad
101. Bakker K, Cordfunke EHP, Konings RJM, Schram RPC (1997) Critical evaluation of the thermal properties of ThO2 and Th1 − yUyO2 and a survey of the literature data on Th1 − yPuy02 J. Nucl Mater 250:1–12
102. Catlow CRA, Lidiard AB (1974) In: Proceedings of symposium on themodynamics of reactor materials, vol II. IAEA, Vienna, p 27
103. Kutty TRG, Somayajulu PS, Khan KB, Kumar A, Kamath HS (2009) Characterization of $(Th, U)O_2$ pellets made by advanced CAP process. J Nucl Mater 384:303–310
104. Kutty TRG, Nair MR, Sengupta P, Basak U, Kumar A, Kamath HS (2008) Characterization of $(Th-U)O_2$ fuel pellets made by impregnation. J Nucl Mater 374:9–19
105. Feraday MA, Cotnam KD, Preto F (1979) Alternative method of making recycle fuel: impregnation of low density pellets. Am Ceram Soc Bull 58:12
106. Croixmarie Y, Abonneau E, Fernandez A, Konings RJM, Desmouliere F, Donnet L (2003) Fabrication of transmutation fuels and targets: the ECRIX and CAMIX-COCHIX experience. J Nucl Mater 320:11–17
107. Haas D, Somers J, Renard A, Fuente AL (1998) Actinide and fission product partitioning and transmutation. In: Proceedings of 5th OECD/NEA information exchange meeting, Belgium

108. Fernández A, Richter K, Closset JC, Fourcaudot S, Fuchs C, Babelot JF, Voet R, Somers J (1999) Innovative materials in advanced energy technologies, Part C. In: Proceedings of 9th Cimtec-World forum on new materials, symposium VII, p 539
109. Fernández A, Richter K, Somers J (1998) Fabrication of transmutation and incineration targets by infiltration of porous pellets by radioactive solutions. J Alloy Compd 271:616–619
110. Richter K, Fernández A, Somers J (1997) Infiltration of highly radioactive materials: a novel approach to the fabrication of targets for the transmutation and incineration of actinides. J Nucl Mater 249:121–127
111. Boucharat N, Fernández A, Somers J, Konings RJM, Haas D (2000) Fabrication of zirconia based targets for transmutation. In: 6th IMF workshop, Strassbourg. (Prog Nucl Energy(2001) 38:291–294)
112. Richter K, Fernandez A, Somers J (1997) Infiltration of highly radioactive materials: a novel approach to the fabrication of transmutation and incineration targets. J Nucl Mater 249:121–127
113. Fernandez A, Haas D, Konings RJM, Somers J (2002) Transmutation of actinides: qualification of an advanced fabrication process based on the infiltration of actinide solutions. J Am Ceram Soc 85:694–696
114. Pai RV, Dehadraya JV, Bhattacharya S, Gupta SK, Mukerjee SK (2008) Fabrication of dense (Th, U)O_2 pellets through microspheres impregnation technique. J Nucl Mater 381:249–258
115. Melville AF (1975) Preparation of mixed oxide nuclear fuel. US Pat no. 40201311975
116. Harvey RL (1978) Method of fabricating nuclear fuel. US Pat no. 4110159
117. Khot PM, Nehete YG, Fulzele AK, Baghra C, Mishra AK, Afzal M, Panakkal JP, Kamath HS (2012) Development of impregnated agglomerate pelletization (IAP) process for fabrication of (Th, U)O_2 mixed oxide pellets. J Nucl Mater 420:1–8
118. Kaufman SF (1969) The hot-pressing behavior of sintered low-density pellets of UO_2, ZrO_2-UO_2, ThO_2, ThO_2-UO_2. WAPD-TM-751
119. Gardiner DA (1964) A study of the response contours of the hot-pressed thorium oxide pellets. ORNL-3608
120. Fitts RB, Moore HG, Olsen AR, Sease JD (1968) Sol-gel thoria extrusion. ORNL-4311
121. Kutty TRG, Hegde PV, Banerjee J, Khan KB, Jain GC, Sengupta AK, Majumdar S, Kamath HS (2003) Densification behaviour of ThO_2-PuO_2 pellets with varying PuO_2 content using dilatometry. J Nucl Mater 312:224–235
122. Kutty TRG, Khan KB, Hegde PV, Pandey VD, Sengupta AK, Majumdar S, Kamath HS (2002) Microstructure of ThO_2-PuO_2 pellets with varying PuO_2 content p504. In: Ganguly C, Jayaraj RN (eds) Characterization and quality control of nuclear fuels. CQCNF-2002, Hyderabad
123. Rodriguez P, Sundaram CV (1981) Nuclear and materials aspects of the thorium fuel cycle. J Nucl Mater 100:227–249
124. Rubbia C, Buono S, Gonzalez E, Kadi Y, Rubio JA (1995) A realistic plutonium elimination scheme with fast energy amplifiers and thorium-plutonium fuel. CERN/AT/95-53(ET)
125. Freshley MD, Mattys HM (1962) Properties of sintered ThO_2-PuO_2, Hanford Power Products Division, HW-76300
126. Freshley MD, Mattys HM (1963) Fast fuel development (classified), Hanford Power Products Division, HW-76302
127. Mulford RN, Ellinger FH (1958) ThO_2–PuO_2 and CeO_2–PuO_2 solid solutions. J Phys Chem 62:1466–1467
128. Wriedt HA (1990) The O-Pu (oxygen-plutonium) system. Bull Alloy Phase Diagram 11:184–202
129. Hj Matzke (1981) In: Sorensen T (ed) Non-stoichiometric oxides. Academic Press, New York, p 156
130. Catlow CRA (1987) Recent problems and progress in the study of UO_2 and mixed UO_2–PuO_2. J Chem Soc, Faraday Trans 83:1065–1072

131. Pritchard WC, Nance RL (1965) Studies on the formation of Pu_2O_3 in the sintering of PuO_2. Los Alamos report LA-3493
132. Chikalla TD, McNeilly CE, Skavdahl RE (1964) The plutonium-oxygen system. J Nucl Mater 12:131–141
133. Kutty TRG, Hegde PV, Khan KB, Majumdar S, Purushotham DSC (2000) Sintering studies on UO_2–PuO_2 pellets with varying PuO_2 content using dilatometry. J Nucl Mater 282:54–65
134. Kutty TRG, Khan KB, Hegde PV, Sengupta AK, Majumdar S, Purushotham DSC (2001) Densification behaviour and sintering kinetics of PuO_2 pellets. J Nucl Mater 297:120–128
135. Kutty TRG, Khan KB, Hegde PV, Banerjee J, Sengupta AK, Majumdar S, Kamath HS (2004) Development of a master sintering curve for ThO_2. J Nucl Mater 327:211–219
136. Kutty TRG, Khan KB, Hegde PV, Sengupta AK, Majumdar S, Kamath HS (2003) Determination of activation energy of sintering of ThO_2-2%U_3O_8 pellet using master sintering curve approach. Sci Sinter 35:125–132
137. Khan KB, Kutty TRG, Jarvis T, Hegde PV, Sengupta AK, Majumdar S, Kamath HS (2003) Master sintering curve for AHWR Fuel. In: 14th Annual conference, INS, Chennai, India
138. Johnson DL, Su H (1997) A practical approach to sintering. Am Ceram Soc Bul 76:72–76
139. Henrichsen M, Hwang J-H, Dravid VP, Johnson DL (2000) Ultra rapid phase conversion in beta-alumina tubes. J Am Ceram Soc 83:2861–2862
140. Su H, Johnson DL (1996) Master sintering curve: a practical approach to sintering. J Am Ceram Soc 79:3211–3217
141. Hansen JD, Rusin RP, Teng M, Johnson DL (1992) Combined-stage sintering model. J Am Ceram Soc 75:1129–1135
142. Su H, Johnson DL (1996) Sintering of alumina in microwave- induced oxygen plasma. J Am Ceram Soc 79:3199–3201
143. Thummler F, Thomma W (1967) The sintering process. Metall Rev 115:69–108
144. Coble RL, Burke JE (1963) Sintering in ceramics. In: Burke JE (ed) Progress in ceramic science, vol 3. Pergamon, Oxford, pp 197–251
145. Kuczynski GC (1949) Self-diffusion in sintering of metallic particles. Trans Am Inst Min Met Eng 185:169–178
146. Shiba K (1992) Diffusion processes in thoria and thorium based oxides with emphasis on fission fragments, irradiation effects. In: Ararwala RP (ed) Diffusion processes in nuclear materials. North Holland, Amsterdam
147. Ray A, Banerjee J, Kutty TRG, Kumar A, Banerjee S (2012) Construction of master sintering curve of ThO_2 pellets using optimization technique. Sci Sinter 44:147–160

Aqueous Reprocessing by THOREX Process

P. V. Achuthan and A. Ramanujam

Abstract In nature, thorium exists mainly in a monoisotopic form with no fissile isotope for its use as fuel in nuclear reactors. Though a fissile isotope of uranium, ^{233}U is formed by neutron irradiation of fertile thorium in reactors; the subsequent reprocessing of irradiated thorium for recovery, purification, and further handling of ^{233}U product (accompanied by ^{232}U) has remained a challenge because of the complex radiological problems associated with irradiated thorium. During irradiation in the reactor, ^{233}Pa with a 27 day half-life is formed by (n, γ) reaction of ^{232}Th. Its complete decay to ^{233}U is to be ensured prior to reprocessing for maximum recovery. The $(n, 2n)$ reactions encountered during the irradiation of Th give rise to long-lived ^{231}Pa and rather short-lived ^{232}U ($\sim$70 years). The ^{232}U and its hard beta gamma emitting short-lived daughters in the separated ^{233}U and the ^{229}Th and ^{228}Th in thorium contribute to the radiation dose of these products. TBP has been the most widely used extractant in the nuclear industry for hydro metallurgical and reprocessing applications and hence, TBP-based THOREX process, in its various forms has become the natural choice to treat irradiated thorium in plants meant for PUREX process, on a campaign basis. Different flow sheets have been used and fine-tuned to meet the specific processing requirements of irradiated Th from different reactor systems based on the type of thoria target/fuel and the cladding under treatment, their irradiation, and cooling history and the end objectives, namely, separation and purification of Th/^{233}U/^{233}Pa in short cooled fuels, or Th and ^{233}U in long cooled fuels/targets or ^{233}U alone from the fuel target matrix and the final product decontamination factor aimed at. Thus the process has evolved over the years from the experience of individual plants that had conducted

P. V. Achuthan
Nuclear Recycle Board, Bhabha Atomic Research Centre, Trombay,
Mumbai 400085, India
e-mail: apv@barc.gov.in

A. Ramanujam (✉)
No. 30, Basant Gardens, Sion-Trombay Road, Chembur,
Mumbai 400071, India
e-mail: aiyaswamy.ramanujam@gmail.com

D. Das and S. R. Bharadwaj (eds.), *Thoria-based Nuclear Fuels*,
Green Energy and Technology, DOI: 10.1007/978-1-4471-5589-8_7,

these pioneering earlier campaigns. This chapter details the major steps involved in this generic process and the several possible variations in the process flow sheet that can be utilized to meet the different end objectives. The different head end treatment options, fluoride ion catalyzed nitric acid dissolution of the fuel, the different TBP-based cycles for the extraction, individual separation and purification of Th/^{233}U, extraction behavior of Th/^{233}U/^{233}Pa and fission products during this step, the solvent management in the cycles, third phase formation tendencies of Th with TBP, the solvent degradation and its consequences in the process are some of the topics covered. The global status on thorium utilization along with country-wise practices and experiences on thorium reprocessing, and recent developments in this domain have been briefly reviewed. Just prior to ^{233}U product reconversion and fuel fabrication, as of now, a chemical separation of the longer lived daughters of ^{232}U (like ^{228}Th and ^{224}Ra) is required to control the personnel exposure during these steps. The activities of short-lived, but strong gamma emitters in the chain, ^{208}Tl and ^{212}Bi, depend on ^{228}Th, the nuclide with longest half-life in the chain. Techniques based on ion exchange, solvent extraction, or precipitation routes are followed for purification of ^{233}U product from ^{228}Th and for removing the residual thorium accompanying product. These are summarized along with the practices followed at different facilities to treat bulk quantities of aged ^{233}U and their recycle during fuel fabrication. Third-phase formation encountered in TBP extraction of Th and Pu has been discussed in detail as it has serious repercussions on process performance and safety. In this context, many homolog neutral organophosphorus esters in the TBP family and alkyl amides have been tested as substitutes for TBP and some of them are reported to have better performance in terms of Th/U separation factors and higher thorium loading. Their status is presented. Major areas that need further developmental efforts for the process to succeed on industrial scale have been identified.

1 Introduction

At high burnup levels ThO_2 fuel contains significant quantities of ^{233}Pa and its fissile daughter ^{233}U along with accumulated fission products resulting from the fissile components present in the fuel. As with ^{239}Pu in the conventionally used uranium-based fuel, ^{233}U is partly burnt inside reactors and the residual Th/^{233}U/^{233}Pa contained in the discharged fuel need to be recovered for reuse. The process of separating ^{232}Th and ^{233}U/^{233}Pa from fission products and other impurities present in the spent fuel or irradiated thoria target after allowing for decay of short-lived fission products is termed as fuel reprocessing. Historically, the thorium-based fuel reprocessing technology emerged from the serious consideration in the 'Atoms for Peace Programme' supplementing limited uranium reserves for developing the nuclear industry [1]. In the mid 1950s, several projects dealing with demonstration reactors and study of the thorium fuel cycle were initiated, first in USA, and then

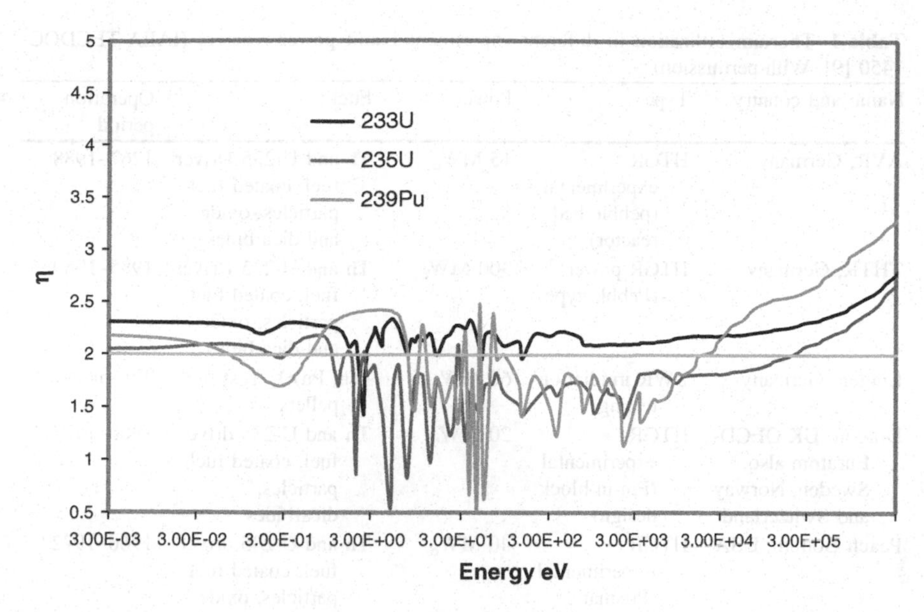

Fig. 1 Eta versus neutron energy plot for fissile nuclides

in Europe. The effort showed that thorium can provide an alternative to uranium fuel cycle if enough fissile isotopes are available from other sources to initiate the "thorium fuel cycle." One of the principal advantages noted in thorium cycle being that the generated fissile isotope, ^{233}U, is the best fissile isotope of all existing fissile isotopes for thermal neutrons (Fig. 1) with an average eta value of (number of neutrons produced for every neutron absorbed) 2.27 for ^{233}U in a standard PWR compared to 2.06 for ^{235}U and 1.84 for ^{239}Pu [2]. As early as 1958, about 55 kg of ^{233}U was available with the USA [1]. Over the years about 1,500 kg of ^{233}U was separated in the USA from 900 tons of thorium [3]. There have also been sizable activities in Germany, UK, India, and France [4–12]. Many reactor prototypes were built, operated, and thorium extraction plants set up. Table 1 [9] shows the major reactor systems used to experiment with thorium/^{233}U fuels.

Although socio-political factors have brought the thorium projects practically to a halt in many countries, India with its increasing energy demand against the limited reserves of uranium and other conventional fuels has seriously considered exploiting the potential of large thorium deposits (519,000 tons) [9]. Following a three-stage nuclear energy program India has been developing the thorium fuel cycle for future energy independence [13].

In the fuel reprocessing technology several specific features of the thorium-based spent fuels need considerations. Unlike the formation of Pu in uranium fuel, ^{233}Pa formed by (n, γ) reactions of ^{232}Th (Fig. 2) decays with a half-life of 27 days to fissile ^{233}U, necessitating a longer cooling period for maximum recovery of ^{233}U in one step. Further, the $(n, 2n)$ reactions encountered during the irradiation of

Table 1 Thorium utilization in different experimental and power reactors (IAEA-TECDOC-1450 [9]. With permission)

Name and country	Type	Power	Fuel	Operation period
AVR, Germany	HTGR experimental (pebble bed reactor)	15 MW_e	Th and U-235 Driver fuel, coated fuel particles, oxide and dicarbides	1967–1988
THTR, Germany	HTGR power (Pebble type)	300 MW_e	Th and U-235 Driver fuel, coated fuel particles, oxide and dicarbides	1985–1989
Lingen, Germany	BWR irradiation-testing	60 MW_e	$(Th, Pu)O_2$ Test fuel, pellets	Terminated in 1973
Dragon, UK OECD-Euratom also Sweden, Norway and Switzerland	HTGR experimental (Pin-in-block design)	20 MW_t	Th and U-235 driver fuel, coated fuel particles, dicarbides	1966–1973
Peach Bottom, USA	HTGR Experimental (Prismatic Block)	40 MW_e	Th and U-235 driver fuel, coated fuel particles, oxide and dicarbides	1966–1972
Fort St Vrain, USA	HTGR power (prismatic block)	330 MW_e	Th and U-235 driver fuel, coated fuel particles, dicarbides	1976–1989
MSRE ORNL, USA	MSBR	7.5 MW_t	U-233 Molten fluorides	1964–1969
Borax IV and Elk River Reactors, USA	BWRs (pin assemblies)	2.4 MW_e 24 MW_e	Th and U-235 driver fuel, oxide pellets	1963–1968
Shippingport and Indian Point, USA	LWBR PWR (pin assemblies)	100 MW_e 285 MW_e	Th and U-233 driver fuel, oxide pellets	1977–1982 1962–1980
SUSPOP/KSTR KEMA, Netherlands	Aqueous homogeneous suspension (pin assemblies)	1 MW_t	Th and HEU oxide pellets	1974–1977
NRU and NRX, Canada	MTR (pin assemblies)		Th and U-235 Test fuel	Irradiation-testing of few fuel elements
KAMINI, CIRUS[a] (Shut down in 2010) and DHRUVA, India	MTR thermal	30 kW_t 40 MW_t 100 MW_t	Al and U-233 Driver fuel, 'J' rod of Th and ThO2 'J' rod of ThO_2	All three research reactors in operation

(continued)

Table 1 (continued)

Name and country	Type	Power	Fuel	Operation period
PURNIMA –II, India[a]	MTR thermal	100 mW$_t$ (nominal)	$^{233}UO_2(NO_3)_2$ Solution	1984–1986
KAPS 1 and 2, KGS 1 and 2, RAPS 2, 3 and 4, India	PHWR (pin assemblies)	220 MW$_e$	ThO_2 Pellets for neutron flux flattening of initial core after start-up	Continuing in all new PHWRs
FBTR, India	LMFBR (pin assemblies)	40 MW$_t$	ThO_2 blanket	In operation

[a] Modifications in the table

thorium in this fuel cycle lead to the formation of long-lived ^{231}Pa and relatively short-lived (68.9 years) ^{232}U with its hard beta gamma emitting daughter products. Thus the ^{233}U produced in the reactor is always contaminated with ^{232}U and the level of contamination depends on the isotopic composition of initial thorium fuel, the burnup, and the neutron spectrum encountered in the reactor. The radiation dose contribution from ^{232}U contamination in the separated ^{233}U product and the contamination from ^{229}Th and ^{228}Th in the separated thorium product will have to be taken into consideration while handling these products. Because of the presence of ^{232}U and its highly radioactive daughter products, a fully integrated and remotely controlled fuel reprocessing and fuel re-fabrication facility is required for ^{233}U product handling. Figure 2 shows the various nuclides formed during thorium irradiation.

2 Radiological Features Associated with Reprocessing of Irradiated Thorium/Thoria Fuels and Targets for ^{233}U/^{233}Pa Recovery

During reprocessing, ^{231}Pa is the main long-lived actinide that needs to be assessed for its long-term environmental impact on high level liquid waste (HLLW) treatment and disposal. The radiological hazard in handling the reprocessed ^{233}U product arises mainly from the alpha and gamma activities associated with the accompanying ^{232}U isotope and its short-lived decay chain products as shown in Fig. 3. The best way to tackle this problem is to isotopically purify the ^{233}U. The conventional centrifuge and gas diffusion techniques are not suitable because of the unit mass difference and the rather low yield of ^{232}U. This can be best achieved by use of laser methods, which require sophisticated equipment yet to be proven on a large scale. Under these circumstances, the only near-term options left are to control the personnel exposure during ^{233}U reconversion and fuel fabrication by resorting to semi-automated handling techniques and by chemical separation of the gamma emitting daughters of ^{232}U (like ^{228}Th and ^{224}Ra) just prior to fuel fabrication.

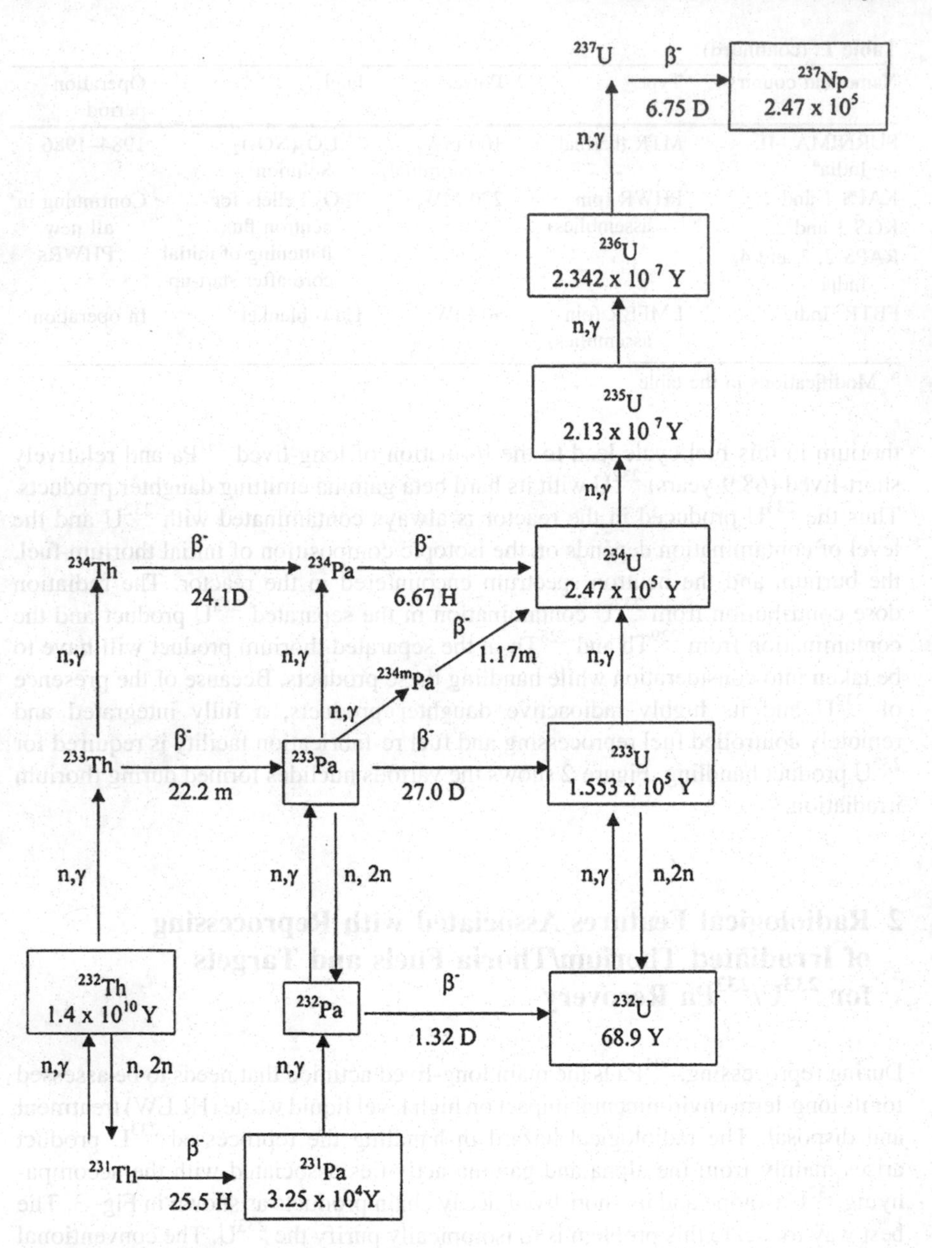

Fig. 2 Nuclear reactions of importance in thorium fuel cycle

(Table 2 lists the radiations associated with the daughters of ^{232}U). It can be seen from the table that the strong gamma emitters in the chain are ^{208}Tl and ^{212}Bi. These nuclides are very short-lived and their activities will depend on the parents ^{212}Pb, ^{224}Ra, and in turn on ^{228}Th, the nuclide with longest half-life in the chain. Efficient removal of these nuclides will govern the growth rate of radiation dose of the product

Fig. 3 ^{232}U decay chain

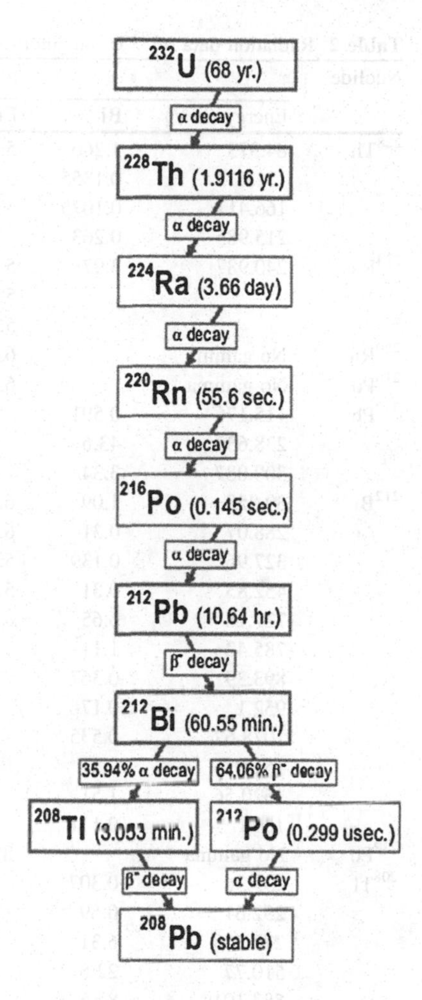

after purification. The growth of ^{208}Tl in the product as a function of post purification time by different purification techniques and varying aging time of product is given in Fig. 4 [10]. The growth rate will reach saturation on ^{228}Th growth reaching a maximum at about 10 years.

3 Reprocessing of Irradiated Thorium/Thoria Fuels and Targets: THOREX Process

Globally, vast experience is available on reprocessing and recovery of plutonium and uranium from spent uranium fuel by well-known solvent extraction-based Plutonium Uranium Extraction (PUREX) process. In this process, the recovery and purification of U and Pu from U/Pu spent fuels is done using aqueous route, by the

Table 2 Radiation data of ^{232}U daughters

Nuclide	γ		α		β	
	Energy (keV)	BI %	Energy (MeV)	BI %	Energy (MeV)	BI %
^{228}Th	84.373	1.266	5.42	72		
	131.613	0.1355	5.338	28		
	166.411	0.1075				
	215.985	0.263				
^{224}Ra	240.987	3.97	5.681	95		
			5.448	4.6		
			5.194	0.4		
^{220}Rn	No gamma		6.282	100		
^{216}Po	No gamma		6.774	100		
^{212}Pb	115.176	0.591			0.36	80
	238.633	43.6			0.57	20
	300.087	3.34				
^{212}Bi	39.858	1.09	6.084	27.2	2.25	66.3
	288.07	0.31	6.047	69.9		
	327.96	0.139	5.765	1.7		
	452.83	0.31	5.5	1.2		
	727.18	6.65				
	785.42	1.11				
	893.39	0.367				
	952.1	0.176				
	1078.62	0.535				
	1512.75	0.31				
	1620.56	1.51				
	1806	0.111				
^{212}Po	No gamma		8.785	100		
^{208}Tl	233.36	0.307				
	252.61	0.69				
	277.358	6.31				
	510.77	22.6				
	583.191	84.5				
	763.13	1.81				
	860.564	12.42				
	982.7	0.203				
	1093.9	0.4				
	2614.533	99.16				

dissolution of the fuel in nitric acid, followed by several cycles of solvent extraction using tributylphosphate (TBP) diluted in n-dodecane as solvent to extract and separate U and Pu. Fuel reprocessing for the recovery of fertile and fissile materials differs from conventional chemical processes due to radioactive nature of the materials processed. The equipment mostly made of corrosion-resistant SS304L is installed and operated remotely with minimum maintenance behind massive concrete shielding with stringent air ventilation and exhaust

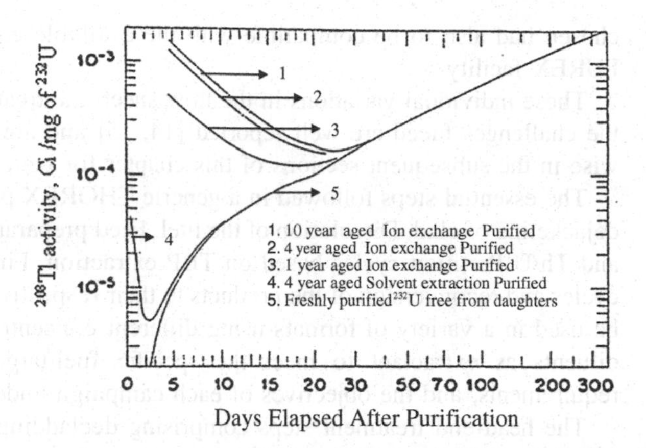

Fig. 4 ^{208}Tl activity from ^{232}U decay as a function of elapsed time after purification, purification method, and aging prior to purification

requirements to protect operating personnel and environment from radiation and contamination hazards. In case of equipment failure, the design should provide for duplicate equipment and remote or direct maintenance. Further, the reprocessing plant design and chemical flow sheet should preclude the possibility of accidental criticality, a self-sustained nuclear fission chain reaction. The prevention techniques are based on the escape of fission neutrons by controlling mass, volume, and concentration of fissile materials or geometry control of the equipment used. Addition of soluble neutron poisons to the solution and incorporation of neutron absorbers in structural materials are also adopted.

Compared to the PUREX Process for processing irradiated uranium fuels, the Thorium Uranium Extraction (THOREX) process for the recovery and recycle of ^{233}U and thorium from irradiated thorium fuels and targets using TBP as extractant has several special features and requirements that need attention. However, the major steps generally followed while processing thoria by THOREX process, are similar to PUREX process and are described here in detail.

THOREX process was developed and operated mainly by the USA in sufficiently large scale in its regular PUREX facilities to treat irradiated thoria targets or spent thoria-based fuels [14, 15]. Over the early years, the process in its various formats, was mostly evolved and standardized by the USA and UK who have had the maximum exposure in processing irradiated thoria-based fuels and targets in terms of tonnage. The process underwent modifications based on their operational experience and the challenges faced by them during those pioneering campaigns conducted in already existing PUREX facilities with minimum changes and alterations in available equipment.

The THOREX flow sheets followed by individual facilities were modified and fine-tuned to meet the specific requirements of the type of thoria target/fuel and the cladding under treatment, their irradiation and cooling history, and the end objectives, viz., separation and purification of Th/^{233}U/^{233}Pa in short cooled fuels, or Th and ^{233}U in long cooled fuels/targets or ^{233}U alone from the fuel target matrix and the final product decontamination factor (DF) aimed at. The flow sheet chosen had also to be compatible with the available equipment in the operating PUREX facility.

These individual variations in the flow sheet, the treatment steps followed, and the challenges faced are well reported [14, 15] and are briefly described nation-wise in the subsequent sections of this chapter for better continuity.

The essential steps followed in a generic THOREX process are: Decladding or dejacketing of clad, Dissolution of the fuel, Feed preparation, Co-decontamination, and Th/U Partition cycles based on TBP extraction, Final U and Th purification cycles and reconversion of the products to their respective oxides. The process can be used in a variety of formats using different concentrations of TBP in suitable diluents as extractant to meet the specific fuel/target composition, end use requirements, and the objectives of each campaign under execution.

The head-end treatment steps comprising decladding/dejacketing followed by dissolution of the fuel and feed preparation are meant to bring the fuel into aqueous solution suitable for THOREX process. The aim of the co-decontamination and partitioning step is to extract Th and ^{233}U together from the aqueous phase leaving the fission product impurities behind, followed by their individual separation. With appropriate concentration of TBP in a suitable diluent as extractant and by manipulating the solvent extraction conditions like aqueous phase acidity, salt strength, etc., good recoveries and separation factors are achieved. In the subsequent steps, Th and U are further purified individually. The concentrated fission product waste solution is stored as HLLW.

The major variations in the generic process include, (a) selective separation of ^{233}U alone leaving behind bulk of thorium and fission products in the aqueous raffinate waste, followed by delayed processing of thorium for its recovery from long cooled fission product waste, (b) separation of ^{233}Pa alone from short cooled fuels and targets for its delayed decay to ^{233}U free from ^{232}U contamination along with simultaneous recovery of Th and ^{233}U, (c) separation of Th/U/Pu from three component thoria fuel, (d) selective extraction of ^{233}U from its alloying components, etc. Thus, in the case of long cooled fuels, depending on whether only ^{233}U or both thorium and ^{233}U (or Th, U and Pu) are to be recovered, the TBP content in the diluent (usually Shell Sol-T, n-dodecane or n- paraffin) varies. The general THOREX flow schemes used for different purposes are given in Fig. 5a, b.

In the early days, the use of Al as clad for thoria fuel and powder was also a factor that influenced the aqueous dissolved feed composition (For example, simultaneous nitric acid dissolution of Al along with Th using mercury as catalyst). Aqueous feeds also used Al as salting out agent for the extraction thorium and uranium in TBP. With the subsequent prevalence of zircaloy and stainless steel as clad materials for fuels exposed to higher burnups and the need for a waste with least salt content to achieve high concentration factors for wastes before vitrification, the utilization of aluminum as salting agent is no longer in vogue. Whenever Al is used as clad for special low irradiation targets, it is first removed by chemical decladding and sent as separate intermediate level waste.

The extent of purification of Th or U achieved in the processing is expressed as DF which is the ratio of beta or gamma activity of fission products associated with

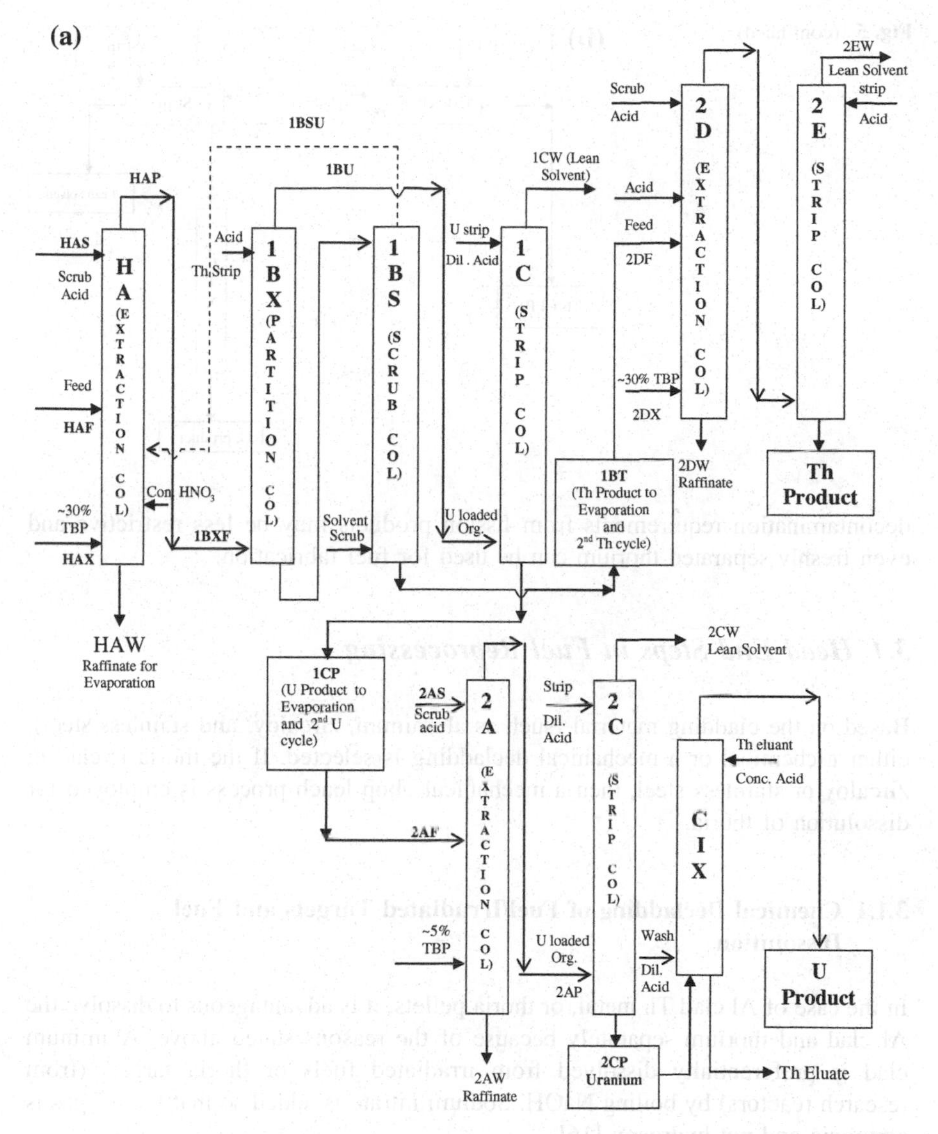

Fig. 5 **a** Schematic of a typical THOREX flow sheet, **b** Schematic of a flow sheet for only ^{233}U separation

unit mass of Th or U in the feed to that in the product. The larger the DF, the greater the purification efficiency of the process. In general, a DF of 10^6–10^7 for both Th and U with respect to fission product activities is aimed in the process.

In the case of final U and Th products, their respective accompanying ^{232}U and ^{228}Th isotopes and their daughter products contribute to the overall radiological dose. If remote handling of the product is warranted because of this factor, then

Fig. 5 (continued)

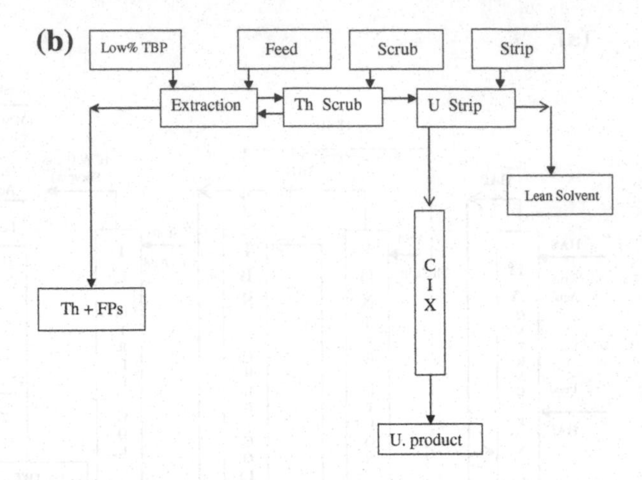

decontamination requirements from fission products may be less restrictive and even freshly separated thorium can be used for fuel fabrication.

3.1 Head-End Steps in Fuel Reprocessing

Based on the cladding material (such as aluminum, zircaloy, and stainless steel), either a chemical or a mechanical decladding is selected. If the thoria is clad in zircaloy or stainless steel, then a mechanical chop-leach process is employed for dissolution of thoria.

3.1.1 Chemical Decladding of Fuel/Irradiated Targets and Fuel Dissolution

In the case of Al clad Th metal, or thoria pellets, it is advantageous to dissolve the Al clad and thorium separately because of the reasons stated above. Aluminum clad is preferentially dissolved from irradiated fuels or thoria targets (from research reactors) by boiling NaOH. Sodium nitrate is added so that the off-gas is ammonia and not hydrogen [16].

$$20\,Al + 17\,NaOH + 21\,NaNO_3 \rightarrow 20\,NaAlO_2 + 18\,NaNO_2 + 4\,H_2O + 3NH_3 \quad (1)$$

Unlike the dissolution of uranium in PUREX process, the dissolution of the irradiated thoria in THOREX process is a more difficult proposition. To overcome the poor dissolution of thoria in HNO_3, a small amount of HF is added to enhance the rate of dissolution [17]. The presence of fluoride ion leads to the corrosion of dissolver made of stainless steel. This problem is controlled by careful addition of aluminum nitrate to complex and control the free fluoride ion concentration in the dissolver.

After alkali dissolution and removal of Al, the fuel thorium is dissolved using a mixture of 13 M HNO_3 with 0.03–0.05 M HF and 0.1 M $Al(NO_3)_3$. This is the "THOREX dissolvent" recommended for ThO_2 dissolution [18]. Boiling temperature is employed to enhance dissolution but as the dissolution progresses, the dissolution rate decreases slowly. The thorium metal is easier to dissolve than thoria. High density sintered thoria pellets are most difficult to dissolve. Dissolution rates of ThO_2 are dependent on pellet density, extent of irradiation, particle size and dissolvent solution composition, and temperature employed. Along with optimum dissolution conditions, the method of preparation of ThO_2 (like addition of MgO) has also a bearing on dissolution pattern [15]. In place of HF, its fluoride salts NaF, KF have been used. Some studies indicate that the use of fluoride salts contribute to slightly increased ThO_2 dissolution rate at least in the early stages of the dissolution although there is no difference in the overall dissolution rate and time [19, 20].

The rate of thoria dissolution also depends on efficient recirculation in the dissolver as thoria in powder/pellet/wafer forms have a tendency to accumulate and settle at the bottom of the dissolver as sludge. As thoria dissolution is very slow and time-consuming, after a pre-selected time duration of dissolution at boiling temperature (about 10–18 h), arbitrarily based on processing experience, bulk of the dissolved product is removed. The undissolved heel is left in the dissolver for continued dissolution with the next batch. As prolonged heating for several hours at boiling temperature is required for a satisfactory rate of dissolution, some of the volatile fission products and Ru in its oxidized RuO_4, ruthenate form, are carried by the off gases and need to be efficiently scrubbed out before release to atmosphere through stacks.

The THOREX solvent containing fluoride is corrosive despite the excess (2–4 times) Al to fluoride mole ratio being maintained, especially, if the solutions are held at high temperature for prolonged periods of time. The air sparging and abrasive nature of ThO_2 particles can also erode the passive metal. Plugging of jets are also reported [14]. If short cooled fuels with significant amount of ^{233}Pa are to be processed, then Pa removal is required to be carried out. For this, the early classical THOREX processes employed in the USA had made use of acid-deficient feeds during extraction to achieve better DF and solid sorbents for Pa removal [15]. For preparing acid-deficient thorium feed, the dissolved Th solution is concentrated and denitrated until it becomes acid deficient. The product solution of 0.5–0.7 M Th from the dissolver is concentrated to a final value of 3 M Th and after that, water addition is done at boiling condition to remove the excess nitric acid, till the required acid deficiency (−0.2 M) is achieved. This concentrated acid-deficient feed is then transferred at 75 °C to a receiver tank with water jet as solidification problems may be encountered at lower temperatures.

For Pa removal by sorption, the solution is contacted with solid sorbents like MnO_2 or vicor glass. Since the current practice is to treat long cooled fuels, these steps may not be warranted for normal operation.

3.1.2 Chop-Leach Process for Decladding and Fuel Dissolution

If the thoria fuel is clad in zircaloy or stainless steel (as in the case of fuels from power reactors), chop-leach method is used in the decladding cum dissolution operation to bring the ThO_2 into solution. In this step, the fuel assemblies are mechanically chopped and the fuel core from the chopped fuel is leached with standard "THOREX Reagent," nitric acid—F–Al mixture, leaving behind the undissolved clad materials as solid waste (hull). As mentioned earlier, the "THOREX Reagent" employed for ThO_2 dissolution shows acceptable dissolution rates at boiling conditions, producing 0.8–1 M solutions of $Th(NO_3)_4$, with a final acidity of about 8 M HNO_3. ^{233}U dissolves much easily along with thoria under the dissolution conditions employed. Since thorium dissolves in HNO_3 by a simple reaction, the reaction produces much lower amounts of NO_2

$$ThO_2 + 4\ HNO_3 \rightarrow Th(NO_3)_4 + 2\ H_2O \quad (2)$$

However, oxides of nitrogen liberated during the dissolution are reoxidized and returned to the dissolver using down draft condensers. The off-gases are treated to remove the traces of nitric acid and volatile fission products, cooled and filtered before being exhausted through tall stacks. During dissolution at boiling temperature, the chopped cladding materials, zircaloy and stainless steel, also show enhanced solubility because of the presence of fluoride ion. After chop-leach dissolution, it is essential to monitor the rinsed Hull wastes containing the chopped and well-leached zircaloy or SS clads for the presence undissolved Thoria heel prior to their final disposal.

Feed clarification to remove the zircaloy fines, undissolved thoria heels, and other particulate impurities are needed prior to solvent extraction to avoid interfacial accumulation of these particles along with colloidal fission products, silica, and degraded organic products of the extractant. Feed conditioning to adjust the final acidity and thorium concentration is required before the feed is subjected to TBP extraction to optimize the decontamination from fission products in the extracted thorium and uranium.

3.2 Solvent Extraction with TBP

3.2.1 Extraction Properties of TBP and Its Utilization in THOREX Process

The versatile extractant TBP in hydrocarbon diluent has been the most widely used solvent for the extraction of both ^{233}U and thorium or for the selective extraction of ^{233}U alone. Depending on the requirement whether both thorium and ^{233}U are to be recovered or only ^{233}U is to be recovered, the TBP content in the diluent (usually Shell Sol-T, n-dodecane, or n-paraffin) to be used as extractant varies.

As an extractant, TBP meets most of the irradiated uranium and thorium reprocessing requirements. It is highly selective for U/Pu and to a lesser extent for Th and has very low extractability for fission products. Many of its chemical and physical properties recommend itself for its intended role as extractant for Th, U, and Pu. It is commercially available and can be easily purified, has high boiling point (284 °C), and low mutual solubility in aqueous phase. It has been used for the past several decades as the extractant for hydrometallurgical separation and purification of natural U and Th to obtain nuclear grade materials and in the industrial scale reprocessing of uranium-based fuels by PUREX process. The PUREX process experience indicates a satisfactory chemical, thermal, and radiation stability for TBP under the process conditions employed.

The drawbacks are its high density (0.973 g/mL) and viscosity which are compensated by dilution with inert diluents like n-dodecane.

Unlike the extraction of U in PUREX process, the main drawback for its use in THOREX process is the limited solubility of its complexes with Th in the diluents used, which leads to third-phase problems during extraction. This puts a limit on the amount of thorium that can safely be extracted under normal operating conditions. Such a situation exists for Pu as well. This can be mitigated by using appropriate diluents, like aromatic or branched hydrocarbons. This tendency by TBP to form third phase with Th (and Pu) at higher loading levels is discussed in the subsequent Sect. 3.5. Many higher and branched homologs of neutral organophosphorus esters have been tried as substitutes for TBP and some of them have shown better performance in terms of Th/U separation factors and higher thorium loading of organic phase without third-phase formation. Yet another reservation in the use of TBP and its other phosphorous-based higher homologs is the difficulties encountered in the ultimate disposal of these organophosphorus reagents in an environmentally safe manner because of their residual phosphorus components.

To avoid the above stated constraints, specifically designed organonitrogen-based (easily destroyed "CHON" type) compounds like amides with better separation, extraction, and loading properties are being proposed and evaluated. Unlike TBP, these reagents still need further plant scale exposure for better evaluation. Hence, as of now, TBP still remains the best choice for this task, at least in the immediate future. However, these alternate extractants as substitutes for TBP are discussed briefly in Sect. 3.6.

The diluent used with TBP should be unreactive toward HNO_3 and HNO_2 and stable to radiation and should be compatible with the solubility requirements of Th complexes with TBP. In addition, it should have a high flash point and low vapor pressure. The nonpolar straight chain aliphatic hydrocarbons containing 12–14 carbon atoms or n-paraffin mixtures (boiling point in the range:180–210 °C) are usually preferred. Shell Sol-T, Sol trol, etc., have also been widely used, especially in PUREX process.

Generally, 30–45 % (v/v) TBP in n-dodecane or n-paraffin diluent is used for initial extraction of both Th and ^{233}U and for final purification of Th. In the THOREX process variant where ^{233}U alone (or $^{233}U/Pu$) is to be extracted preferentially from THOREX feeds and during the final ^{233}U purification cycles, often a lower percentage of TBP (usually 3–8 %) is used.

3.2.2 Extraction of Acid and Metal Ions by TBP in THOREX Process

The good extractability of nitrate complexes of U and to a slightly lesser extent those of Th^{4+}, coupled with the poor extractability of fission products by TBP forms the basis for the separation of these components in the THOREX process. As Th is weakly extracted by TBP as a function of the aqueous phase acid, nitrate or metal (salt) nitrate concentration, its further separation from U is achieved by first extracting both Th and ^{233}U, near quantitatively from aqueous phase having high acid/metal nitrate content and then preferentially back extracting or stripping the extracted Th by a fresh aqueous stream having lower nitrate strength which is still high enough to retain the U in TBP phase.

Nitric acid is extracted as TBP.HNO_3 and is salted back to the aqueous phase during the loading of TBP with metal nitrates. Among metal ions, the main species extracted are $MO_2(NO_3)_2$.2TBP for M(VI) ions, $M(NO_3)_4$.2TBP for M(IV), and $M(NO_3)_3$.3 TBP for M(III) where M stands for the metal with extractability following the same order. The important extraction equilibria between U(VI), Pu(IV), Th(IV), etc., and TBP are given below.

$$UO_2^{2+} + 2\,NO_3^- + 2\,TBP \rightleftharpoons UO_2(NO_3)_2{\cdot}2\,TBP, \quad (3)$$

$$Pu^{4+} + 4\,NO_3^- + 2\,TBP \rightleftharpoons Pu(NO_3)_4{\cdot}2TBP, \quad (4)$$

$$Th^{4+} + 4\,NO_3^- + x\,TBP \rightleftharpoons Th(NO_3)_4{\cdot}xTBP, \quad (5)$$

$$M^{3+} + 3NO_3^- + 3TBP \rightleftharpoons M(NO_3)_3{\cdot}3TBP. \quad (6)$$

M^{3+} is a trivalent fission product or trivalent actinides such as Ce, Rare Earths, Am, etc., are poorly extracted by TBP. MO_2^+, pentavalent actinides are also poorly extracted in TBP.

When nitrates of U, Pu(IV) and Th are extracted by TBP, the D values increase with aqueous nitric acid, pass through a maximum, and decrease a little thereafter. The initial rise is due to increase in the $[NO_3^-]$ which acts as a salting agent. But at higher acidities, the HNO_3 itself competes with metal ions resulting in a fall in D values. In the presence of large amounts of extracted species, the available free TBP decreases retarding further extraction. At low concentration levels, the distribution ratio (D) for U and Pu(IV) is directly proportional to the square of free TBP concentration in the organic phase and to the square and the fourth power of nitrate ion concentration, respectively, in the aqueous phase.

However, in the case of Th(IV), variations in TBP dependence are reported [21] and a third power dependence used in modeling extraction systems was found to yield better fits with solvent extraction plots [22]. Additional complex species with 2–4 mol of TBP per Th mole are reported whose formation depends on the concentrations of Th, TBP, total nitrate, and nitric acid present in the system. At high thorium nitrate concentrations in the organic phase, the TBP/Th ratio becomes less than four. Under these conditions, if straight chain aliphatic hydrocarbons are used as diluent for TBP in the absence of a phase modifier, often the solubility limit of the solvated complex in the diluent is breached. This leads to a split in the organic phase, one containing mainly the diluent with very low TBP content and another heavier fraction containing most of the TBP saturated with Th. This phenomenon is described in detail in Sect. 3.5 dealing with third phase formation during thorium extraction by TBP.

This phenomenon of third phase formation complicates the extraction column operations in the process and should be avoided under all circumstances by carefully controlling the solute concentration, acidity, temperature, or by modifying the solvent composition. The temperature dependence of the extraction behavior of U, Pu Th, Pa, and fission products in TBP is complex but it is useful in some column operations of the process. Usually the high temperature operation is beneficial for scrubbing out fission products and for stripping U from TBP.

While processing long cooled fuels, ^{90}Sr-^{90}Y, ^{95}Zr-^{95}Nb, ^{106}Ru-^{106} Rh, ^{137}Cs, ^{144}Ce-^{144}Pr, are major fission products and ^{231}Pa and ^{233}Pa (in short cooled fuels) are the other major actinides that affect the product purity. Among the fission products, the behavior of ^{106}Ru-^{106}Rh and ^{95}Zr-^{95}Nb often determines the overall process performance. Most of the fission product extraction behaviors reported here have been observed under PUREX process conditions, but they are still valid for THOREX process conditions and can be used as guidelines for dealing with them.

The complicated extraction behavior of Ru in the process is due to the presence of several nitrato-nitrosyl ruthenium complex species and the slow rate of interconversion between the multiple species. Fast extraction at lower acidities, followed by prolonged scrubbing at higher acidity and temperature are recommended for better decontamination from Ru. As mentioned earlier, during the dissolution with prolonged heating for several hours at boiling temperature, some of the Ru in its oxidized RuO_4 form, has a tendency to escape with the off gases. This needs to be efficiently scrubbed out as it can settle in the off gas ducts and can escape through stack releases to atmosphere. Degraded process solvents show higher levels of Ru contamination even after solvent wash/cleanup.

Fission product Zr has a tendency to hydrolyze and polymerize at lower acidity and high temperature. Like Th(IV), it has a strong affinity for the TBP degradation products (di- and mono butyl phosphates, DBP, and MBP) that lead to a reduction in the decontamination and to formation of precipitates. Long-cooled fuels (>3 years) pose less problems as ^{95}Zr has a short half-life (65 days). Some inactive Zr is also present in feed streams, due to the dissolution of small amounts of zircaloy clad. Because of this presence, due to isotope dilution effect, the extraction of radioactive ^{95}Zr would get reduced. DF for Nb can be improved by efficient filtration and by passing the relevant aqueous streams through silica gel for Nb sorption. In French PUREX plants, F^- ion in combination with Al had been used to complex Zr and to prevent its extraction [23]. A similar effect can be expected with THOREX feeds as they too have these additives. Fission product technetium (^{99}Tc) present as TcO_4^- facilitates its own extraction and that of Zr by forming extractable complexes.

Among the long-lived actinides, ^{237}Np exists mainly as Np(V) which extracts poorly in TBP and follows the raffinate. In comparison, Np(IV) and Np(VI) follow U streams. The concentration of nitrous and nitric acids, U, and temperature influence the conversion of Np from one state to another. Similarly, if Pu(IV) is present in the system, its removal from the products is achieved by its reduction to Pu(III) with a suitable reductant, in which form it is least extracted by TBP.

^{231}Pa is a long-lived irradiation product formed during Th irradiation and it can be isolated in the head end or tracked to the HLW in THOREX process, to control its long-term hazard potential. ^{231}Pa and ^{233}Pa (short-lived) can be removed from the THOREX feeds by carrier precipitation or by other suitable techniques as a pre-treatment step. Protactinium exists in two valence states; (IV) and (V). As Pa(V) is the most stable state, it is expected to be in this state in process solutions. It is a homolog of niobium and tantalum, resembling many of their chemical properties like tendency to hydrolyze in neutral and slightly acid solutions, to precipitate as hydrous oxide at high concentrations, and as colloids at low concentrations. Extraction of Pa by TBP has been found to be significant and increases with both acid and TBP concentrations. Hence, to avoid the co-extraction and spread of Pa in various streams, the recovery of protactinium from thorium nitrate solutions prior to solvent extraction by adsorption on hydrous manganese dioxide has been reported. Separation diminishes at high concentration of nitric acid and thorium nitrate. Unfired Vycor glass and silica gel are also reported to be good candidates for removal of ^{233}Pa from acidic THOREX feed solutions. Investigation of the solvent extraction of Pa by various solvents showed that di-isopropyl corbinol gives the best distribution ratio for the extraction from nitric acid and the highest ratio occurs around 3 M HNO_3. Fluoride ion concentration in the presence of thorium does not change the distribution ratio significantly, but high temperatures decrease the distribution ratio. This solvent was used in one of the early THOREX variants for the selective removal of Pa prior to extraction of Th and U by TBP and discarded subsequently for reasons explained in a subsequent section of this chapter. During extraction, reduction in acid concentration and addition of phosphoric acid in the feed or scrub improves decontamination from Pa(V). However, these steps add to clogging problems in the extraction columns, and are best avoided.

3.2.3 Co-Decontamination and Partitioning Cycles for Thorium and Uranium-233 Recovery and Their Individual Purification in THOREX Process

The first co-extraction cycle is operated with acid-deficient feed or with an acidic feed (depending on the decontamination required from Pa) but under acidic scrub conditions. The aqueous feed is introduced near or slightly above midpoint of the compound extraction—scrub column contactor (HA). The aqueous feed (HAF) from the middle and the scrub (HAS) from the top of the HA column being heavier flow down the column. The extractant 30 % TBP in n-paraffin (HAX) is introduced from the bottom of the column and being lighter flows upward. First, thorium and uranium are co-extracted in the lower extraction section of the compound column (HA) by 30 % TBP in n- paraffin to get primary separation from fission products and other radioactive and corrosion products. Uranium has a strong affinity for TBP and can be extracted easily under the acid and salting conditions chosen. Thorium, though having lesser affinity, is also co-extracted under these conditions.

The flow rates of feed HAF, scrub HAS, and organic extractant HAX are adjusted in such a way to get a maximum throughput with good product recovery and decontamination from fission products and other radioactive products. The losses to the waste raffinate are kept to a minimum. To keep the thorium losses low (especially in the case of acid-deficient feeds), conc. acid introduction near the bottom of the column can be done, if required, which enhances the thorium extraction. The Th/U losses to the aqueous waste raffinate (HAW) are usually low (about 0.1 % of feed).

The loaded organic is given a suitable acid scrub (HAS) in the upper part of the HA to further purify the products before they leave the column from the top as loaded organic stream. In some cases, the scrub HAS also contains other suitable chemical additives as necessary to remove the accompanying residual fission products and Pa contamination. The additives include, low concentrations of phosphoric acid to complex Pa, ferrous sulfamate, or other reductants to reduce and scrub out Pu and to reduce chromate as chromium, etc. These chemical additives are plant selective and are used only when they are absolutely essential to get the required DF and are introduced at appropriate points in the extraction column. Their uncontrolled use can cause column disturbances in the extraction and other columns due mainly to insoluble precipitate or gel/crud formation. Phosphates and DBP form precipitates with Pu(III), Pu(IV), Th(IV), Fe(III), Zr(IV), etc. Sulfamate ends up as sulfate in the final waste.

An additional scrub column (HS) is often provided to improve the decontamination from fission products. During operation, variations in acid profile are possible for better decontamination from fission products Zr and Ru and to keep the Th/U losses low. A high loading of thorium (but well below third phase formation levels) in solvent phase reduces the extraction of fission products.

The Th/U bearing loaded organic (HAP) from extraction column is used as partition column feed (IBXF). This is introduced at the bottom of the partition column (IBX) where Th and U are partitioned from each other. The partitioning is done by exploiting the relatively lower extractability of Th in TBP at lower acidities and salting strength of the aqueous partitioning stream. The partition column process performance depends on strict control of conditions like acidity and salt strength such that the conditions favor the preferential stripping or back extraction of Th as against uranium which still follows the organic stream. This is usually done using 0.2 M HNO_3 as partitioning agent (IBX Strip) for back extracting Th from organic phase. The stripped out thorium in the aqueous phase also acts as salting out agent to keep most of the uranium in the organic phase.

The uranium bearing organic with residual contamination of still accompanying thorium exits the top of the partition (IBX) column as IBU and is stripped or back extracted with very dilute (0.01 M) nitric acid in the strip column (IC). The stripped U product is concentrated (after diluent wash and steam stripping to remove entrained and soluble TBP and its degradation products) to meet the feed requirements of second uranium purification cycle.

The thorium bearing aqueous partitioned product stream with accompanying residual uranium exits the partition column and is subjected to an organic TBP scrub in a separate scrub column (IBS) to re-extract the residual uranium. This U bearing organic scrub IBSU is recycled to HA co-extraction column described above along with fresh organic as extractant (HAX). The thorium bearing aqueous product stream (IBT) exiting the scrub column IBS is then concentrated (after diluent wash and steam stripping to remove entrained and soluble TBP and its degradation products) to meet the feed requirements of second thorium/final purification cycle (2D).

The HAW leaving the (HA) extraction column is washed free of entrained/dissolved TBP by diluent wash and steam stripped and concentrated by evaporation to form the HLLW. If required, extra aluminum nitrate is added to complex the free fluoride ions present (due to the absence bulk metal ions) to minimize corrosion especially due to the presence of HF in vapor phase. The enhancement in acidity during evaporation is reduced by destroying or 'killing' the acid using non-salt forming chemicals like formaldehyde which is added slowly under carefully controlled reaction conditions of acid, temperature, etc. Uncontrolled addition of formaldehyde can lead to violent reactions as this reagent reacts with nitric acid [24] with a time lag/incubation period.

$$4\,HNO_3 + H_2CO \rightarrow 4\,NO_2 + CO_2 + 3\,H_2O\ (8\text{–}6\ M\ HNO_3), \tag{7}$$

$$4\,HNO_3 + 3\,H_2CO \rightarrow 4\,NO + 3\,CO_2 + 5\,H_2O\ (2\text{–}8\ M\,HNO_3), \tag{8}$$

$$2\,HNO_3 + H_2CO \rightarrow 2\,NO_2 + HCOOH + H_2O\ (< 2\,M\,HNO_3). \tag{9}$$

The concentrated thorium feed (2DF) entering the second thorium purification cycle extraction column (2D) is treated with ferrous sulfamate or any other suitable reducing agent to reduce Pu and the chromate ions, if present. Sometimes, H_3PO_4 is added to high acid scrub entering the 2D extraction column above feed point in the scrub section to improve Pa DF. As mentioned earlier, this step leads to gel type precipitate formation of Th and is best avoided, especially for long cooled fuels having negligible ^{233}Pa. The high acid scrub is to maintain the required salting strength to favor the extraction of Th and a low acid scrub from the top of the scrub section ensures the reduction of acid content in the exiting thorium bearing organic product (2DP). This product is stripped in the strip column 2E with dilute nitric acid (0.01 M). The Th lean organic stream (2EW) leaving the column is subjected to alkaline solvent wash treatment before recycle. The required DF is mostly achieved by control of A/O ratio and aqueous feed and scrub stream acidities. The stripped Th product is concentrated after diluent wash and steam stripping to remove entrained and soluble TBP and its degradation products.

The stripped and concentrated aqueous U product (1CP) from partition cycle strip column (1C) is introduced as feed 2AF in 2A column above the midpoint. At the lower end of the column concentrated nitric acid is added, if required, to prevent excessive loss of U to waste raffinate (2AW). The low acid scrub introduced from top of the column contains reducing agent, to improve decontamination from Pu, if required. In the strip (2C) column, the loaded organic (2AP) is stripped off U with dilute nitric acid (0.01 M). The acidity of the incoming organic feed contributes to the acidity of the aqueous phase. The lean solvent (2CW) is sent for recycling.

Depending on the plant requirements and the concentration levels of ^{233}U to be handled as feeds, the TBP percentage in the solvent used for this purification cycle can be either 30 % or lower (about 5–8 %) with appropriate adjustments in the aqueous to organic flow ratios of the incoming streams to the columns.

3.2.4 Extraction and Separation of ^{233}U Alone with Dilute TBP from THOREX Feeds

Generally, 3–7.5 % TBP is used when ^{233}U alone is to be recovered from high acid THOREX feeds. During extraction in HA column, about 8 g/l Th is also co-extracted in this step when 5 % TBP is used. The Th uptake reduces the extraction of fission products significantly. The co-extracted Th is scrubbed out with 1–2 M HNO_3 and the final Th contamination depends on the number and the efficiency of scrub stages provided. Finally, ^{233}U is stripped from the organic phase with very dilute acid.

The Th remaining in the raffinate can be recovered by a higher percentage TBP flow sheet, either immediately or after allowing for the further decay of contaminants like ^{228}Th. To guard against the third phase formation possibility during the extraction, the Th saturation of TBP is kept much below the solubility limits. In general, 30–42.5 % TBP in paraffin diluents has been used for extraction of Th. For Th feeds in 1-3 M HNO_3, the organic Th loading is restricted to about 28–35 g/l in the case of 30 %TBP and to about 55 g/l in the case of 42.5 % TBP.

3.2.5 Solvent Degradation and Its Mitigation in THOREX Process

TBP is prone to hydrolysis giving rise to DBP, MBP, and butyl alcohol. Nitric acid also reacts to give DBP and butyl nitrate with complete dealkylation leading to orthophosphoric acid. The degradation can take place in both the phases and is promoted by higher temperatures. To start with, initially, the presence of DBP in TBP increases the extraction of Th, U, and Zr synergistically as DBP-TBP adducts and their stripping becomes difficult, thus affecting the separation and decontamination factors. MBP is formed to a lesser extent than DBP and is water soluble.

Further, radiolytic and chemical (acid) attack of TBP leads initially to the formation of DBP and in more acute exposure to MBP and even phosphoric acid. These products interact with Th, Pu, Zr, and Fe to form precipitates, especially at lower acidities as prevalent in strip columns. Being insoluble in both phases, they tend to disperse through the column and also collect at the inter phase. These precipitates also absorb radioactive fission products, silica, and fines of dissolver origin and form cruds/sludges which act as radiation sources, subjecting the organic phase to much more radiation than what would have normally been encountered. On prolonged accumulation, they cause clogging and choking of the cartridges, ultimately leading to column flooding and escape of precipitate bearing mixed phases through the exit streams.

Hence, plant experience dictates as reported at Hanford Plant [14], that irrespective of the cycle, all the columns need periodic flushing to remove Th-DBP solids. This is done by soaking the column internals with hot concentrated nitric acid followed by TBP pulsing for dissolving/dislodging the precipitates and for flushing them out. The need and periodicity varies from cycle to cycle and from column to column based on the fluctuations in the individual column performance. Flushing the equipment with sodium carbonate solutions has also been used as a cleanup procedure.

To mitigate these solid formation problems, a decrease in the organic hold-up to reduce the residence time between columns and a reduction in the organic exposure time to acid and radiolytic attack arising from the extracted acid and fission products and ^{233}Pa are recommended based on Plant experience. Further, as mentioned earlier, precipitate forming additives in scrub and other streams are best avoided. Equipment wise, use of short residence contactors like centrifugal contactors along with reduced inter column organic hold-up would be of help.

Another area where there is a possibility for TBP to degrade and interfere with the process is in the inter cycle product recycle evaporators and high level waste evaporators. To prevent the escape of entrained organics from the pulse columns, sufficient volume and residence time is given for phase settling at the disengagement section. In spite of this, some organic escapes with aqueous stream due to entrainment and inherent but limited solubility of the TBP in the aqueous phase. Hence the feeds to all the evaporators are scrubbed with organic diluents and steam stripped to remove these entrained and dissolved organic matters which otherwise will decompose in evaporators leaving solid residues and heavy organics. These residues have a tendency to absorb and retain uranium/plutonium/thorium. They enter the extraction column along with concentrated feeds from evaporators and accumulate as cruds at the interface affecting the column performance as described before.

At times, due to maloperation of the columns, some inadvertent escape of organic extractant in bulk can take place via aqueous stream as mixed phase. If uncontrolled bulk amount of TBP/diluent enters the evaporator in this manner, on prolonged evaporation, the TBP content of the organic phase increases due to the escape of diluents and it gets fully loaded with the metal ions present (U, Th, Pu, etc.). The density of the organic phase becomes more than that of the aqueous phase and the organic sticks to vessel surfaces and the bulk settles down as heavy organic at the bottom, disturbing the heat transfer and dissipation processes in the ongoing evaporation step. At boiling temperature, simultaneously, the presence of nitric acid also causes severe nitration of the organic phase leading to reddish brown organic oily layer commonly termed as red oil. If the evaporation continues at temperatures above 130 °C and in presence of metal nitrates and nitric acid (>8 M), the TBP may decompose with explosive violence. If the metal ion in the heavy organic phase is a fissile element then the criticality hazard potentials should be evaluated for prevention measures.

Hence, a strict control is maintained on the quality of the aqueous stream entering the evaporator and the samples are checked for double layers and any deviation in material balance of the metal ion content in the aqueous phase is analyzed and accounted for. If any organic inflow is suspected then the evaporation is stopped and both the layers are flushed out and after separation, the organic layer is subjected to recovery of the metal ions and washed and disposed or recycled depending on its quality. To avoid explosion hazards, interlocks are provided on low pressure steam to control the evaporator temperature below 130 °C and the acidity is also controlled at lower levels. Sufficient free board volume and negative pressure is maintained in the evaporators to accommodate any surges in the release of gaseous products due to this.

Generally, the volume reduction factor (VRF) is limited to about 10 for HLW-evaporators. When highly evaporated products are to be used as aqueous feeds to extraction column, a feed clarification step would help to keep the column free of cruds during operation

3.2.6 Solvent Treatment and Recycle in THOREX Process

The major degradation product DBP, being acidic, can be washed off from TBP with Na_2CO_3 and NaOH solutions, (also with $KMnO_4$ as additive for improved cleanup from radioactive contaminants). Usually this is done on a continuous basis as part of the each extraction cycle before recycling the organic after each extraction and strip cycle. As these washing steps contribute to salt and organic bearing wastes, the operation needs lot of care and control. Various sorbants and vacuum distillation techniques developed for TBP purification as a part of PUREX process can be adapted for this process as well [25–27]. In the plant practice, the solvents used in the first cycle and for U purification are treated together as one bulk, whereas the solvent used for Th purification is treated separately.

In the wash column, the organic is first treated with dilute nitric acid. The subsequent alkaline wash contains $KMnO_4$ as additive, if required, for better cleanup performance. But any escape of this reagent along with organic phase due to improper settling may lead to precipitation of MnO_2 and formation of Ce(IV), Pu(VI) and Cr(VI). The solvent treatment step can be a major source of solvent loss if organic escapes with aqueous waste stream as entrainment due to poor settling and disengagement of phases. The solvent quality after solvent wash is tested by P or Z test (Pu or Zr retention test) and compared with fresh solvent [11].

In the case of diluents, chemical attack by nitric and nitrous acids combines with radiolytic attack to produce a spectrum of nitrogen compounds, ketones, esters, and unsaturates resulting in increased viscosity and density. They also interact with TBP and its degradation products to produce long chain extractants, more potent than TBP, with great affinity for Zr, Pu, and Ru and also ^{129}I, if present. These are difficult to be washed off from TBP and are to be taken into account while judging the quality of solvent for reuse.

3.3 Purification and Conversion of ^{233}U

Several techniques have been in use for purification of ^{233}U product from Th and other impurities based on the facility end use requirements, equipment availability, and contaminants present. They range from solvent extraction (mainly TBP as reported above), ion-exchange to precipitation processes. Anion exchange processes had been used in some facilities as uranium forms strong anionic chloro and acetato complexes in the respective acid media which are sorbed preferentially (over Th) on the anion exchangers. These early approaches had several disadvantages like the need for hydroxide precipitation of U/Th for their conversion to chloride or acetate media, careful control of the acid strength in the media used, corrosion of equipment (Chloride) and iron contamination (due to sorption of $FeCl_4^-$) of the product, gasification in the exchanger column, etc.

Alternately, a cation exchange procedure in nitric acid for preferential sorption and separation of tetravalent Th from weakly absorbed uranyl ion has several advantages over the earlier methods and is amenable for continuous plant operation as demonstrated in later dedicated THOREX facilities. A precipitation procedure for the separation of Th from ^{233}U as its oxalate and the recovery of U from the oxalate supernatant as ADU is also attractive mainly as a batch technique useful in laboratory campaigns, due to its similarity to Pu product reconversion operations.

Conventional routes are followed for the conversion of ^{233}U and Th products into their respective oxides. As ^{233}U builds up dose due to ^{232}U, the final end product fabrication should be done forthwith. Otherwise, ^{233}U would require purification to reduce the dose. This is done by removing ^{228}Th, the longest lived daughter of ^{232}U and thus breaking the decay chain (Usually by cation exchange technique). The ^{229}Th formed from ^{233}U decay also follows ^{228}Th during this step. Thorium nitrate product is converted into oxide following established routes.

3.4 Solid, Gaseous, and Liquid Waste Management

During reprocessing, wastes are generated in solid, liquid, and gaseous forms. The clad material, at the end of the dissolution (in mechanical chop/leach process) is monitored for completion of dissolution before disposal. The HLLW carries bulk of the fission product activity in highly concentrated form. It is temporarily stored in stainless steel underground storage tanks with provision for cooling to remove fission products decay heat and is finally vitrified for long-term storage. The medium or intermediate level waste (ILW) is stored in underground storage tanks for subsequent waste management. The low level waste (LLW) effluents from the plant are treated and released to the sea at levels much below the maximum permissible limits. The medium and low level solid wastes are segregated as combustible, noncombustible, α-bearing, non-α-bearing, etc., and treated independently to reduce the volume for final disposal.

The spent TBP-diluent waste is decomposed by alkaline hydrolysis to separate the mixture into hydrocarbon and active DBP/MBP acid fractions for final disposal [28].

Long-lived fission gases and volatile products like T_2, THO, ^{85}Kr, Xe and $^{129}I_2$, $^{14}CO_2$ and $^{106}RuO_4$ are liberated during dissolution and other operations. The process equipment like dissolver, evaporators, and other vessels are, therefore, kept at a few centimeters of water vacuum to prevent the escape of activity, and the vessel off-gases are subjected to separate treatment and filtration to remove the fission products and condensable vapors. Finally, the gases along with area ven tilation air are released through a tall stack after filtration through high efficiency particulate activity (HEPA) filters. The environmental releases from reprocessing have decreased considerably over the years due to national and international regulatory controls.

3.5 Extraction of Thorium by TBP and Third Phase Formation Phenomenon

Pure TBP containing thorium nitrate has a density nearly the same as that of aqueous solutions; therefore the separation of phases is slow when highly concentrated solution is used. Different organic diluents have been used with TBP to increase the density difference between the phases to eliminate this phase separation problem. Carbon tetrachloride has been used to make the organic phase denser than the aqueous phase; and hydrocarbon diluents, as well as butyl ether, have been used to make the organic phase less dense.

The general reactions governing the extraction of both nitric acid and thorium by dilute TBP have been studied. For Th, the mechanism of extraction has been reported [21] using 7.5 % by volume TBP solutions in hexane. The overall extraction reaction is given below:

$$Th^{4+}_{aq} + 4\,NO^{-}_{3\ aq} + 4TBP_{org} \rightleftarrows Th(NO_3)_4{\cdot}4TBP_{org} \qquad (10))$$

The complex $Th(NO_3)_4{\cdot}4TBP$ is considered as the primary complex which is extracted in dilute TBP and is assumed to have a coordination number of 8 with a cubic configuration with TBP and the nitrate occupying alternate corners of the cube. With all the coordination positions of thorium ion occupied by either nitrate ions or TBP dipoles, complex is assumed to be stabilized. The TBP in the complex permits its solubility in an organic solvent. At higher thorium or TBP concentrations other reactions are reported to take place. The extraction of nitric acid follows the sequence of reactions [21]:

$$H^{+} + NO^{-}_{3} \rightleftarrows HNO_3, \qquad (11)$$

$$\text{and} \quad TBP + HNO_3 \rightleftarrows TBP.HNO_3, \qquad (12)$$

giving an overall reaction,

$$TBP + H^{+} + NO^{-}_{3} \rightleftarrows TBP \cdot HNO_3. \qquad (13)$$

Studies by Gresky et al. [21] indicate that the composite equation holds in the region of dilute TBP, e.g., 7.52 % by volume in hexane, and at moderate nitric acid concentrations in the aqueous phase, e.g., 1–2.5 M. At higher equilibrium acid concentrations, e.g., >8.0 M, apparently uncomplexed nitric acid exists in the organic phase. At higher TBP concentrations, the extraction reaction becomes more complicated. Thus the thorium distribution coefficient as a function of nitric acid concentration is not a smooth curve but shows definite inflections. This points to the fact that in the Th-TBP system at equilibrium with aqueous nitrate concentrations above 2.5 M NO^{-}_{3}, there is a change in the mechanism of extraction [21].

At high thorium nitrate concentrations in the organic phase, at which the TBP/Th ratio becomes less than 4, and when a normal hydrocarbon diluent is used, two organic phases may appear. The dense organic (third) phase contains practically all the TBP and the thorium nitrate, while the light organic phase contain only small amount of TBP and thorium nitrate. To account for a TBP/Th ratio of less than 4, that of the primary complex, a polymerization reaction is assumed to occur. The co-ordination number of 8 of thorium ion is expected to remain constant in this reaction. Two of the primary complexes, having a cubic configuration, are apparently coupled together by nitrate ions at two adjacent corners of each cube, liberating two TBP molecules, as follows [21]:

$$2\left[Th(NO_3)_4.4TBP\right] \rightleftarrows 2\left[Th(NO_3)_4{\cdot}3TBP\right] + 2\,TBP \qquad (14)$$

The polymerization reaction may continue, forming polymers of a higher order. As the polymerization number increases, the TBP/Th ratio approaches 2.0 as limit. The tetramer, $4Th(NO_3)_4{\cdot}10TBP$, is reported. Polymer chains which are close to each other in a normal hydrocarbon solvent may exert attractive forces analogous to van der Waals's forces. These forces may overcome the solvation by the normal hydrocarbon diluents and bring about the formation of dense phase by a process analogous to the condensation of a vapor.

If an aromatic solvent such as benzene (phase modifier) is present in the diluents, an induced dipole can be formed in the ring structure of the aromatic molecule [21]. This aromatic molecule with its induced dipole can then attract the thorium nitrate-TBP polymer chain and solubilize the polymer chains, preventing the formation of a dense phase. Thus in solvent extraction processes, the high loading of the solvent with metal salts or acids can sometimes cause a third phase to form. The organic phase splits into two phases of different compositions and densities. This phenomenon is a major drawback in terms of industrial process implementation.

Third phase formation problems are a major concern in fast reactor fuel reprocessing by PUREX process and thorium spent fuel reprocessing by THOREX process as both Pu(IV) and Th(IV) are prone to third phase formation at higher loading in the TBP phase. This has resulted in the use of a narrow range of operating conditions affecting both safety and throughput considerations. These processes are more prone to third phase formation problems because of the use of n-alkanes as preferred diluents over others due to their otherwise excellent properties. But as the extractants and the extracted complexes are polar, the phase compatibility problems are often encountered in these diluents under certain operating conditions. Thus, THOREX campaigns using high percentage TBP (30–45 %) use narrow operating regions to obviate the problems of third phase formation.

The aggregation of molecules was treated as polymerization in the past based on co-ordination requirements. Considerable work has since been reported on third phase formation in liquid–liquid extraction systems. Numerous qualitative reports are available on the experimental [29–34] parameters leading to third phase formation and their empirical relations. Mitigation of the third phase formation by fine-tuning the lipophilicity of the extractant molecule or modifying the diluent composition has been reported. But there is lack of a predictive model for third phase formation. Considering the organic phase as a complex fluid containing micro emulsions or reverse micelles as explained by Osseo-Asare [35] has provided a better understanding of third phase formation. Till recently, the third phase was principally investigated from co-ordination chemistry viewpoint without considering any supramolecular organization of the species. Only extensive organic phase aggregation of the acid or metal salts was an observation during third phase formation with TBP and other organophosphorous extractants without any quantitative explanation. Since 1998, extensive information on molecular aggregation by direct measurements is reported with evidence for short-range attractive interactions between reverse miscelles responsible for third phase formation.

Chiarizia et al. [36–38] based on the small-angle neutron scattering data examined the third phase formation in TBP-alkane system loaded with HNO_3-U(VI), HNO_3-Th(IV), HNO_3-Zr(IV), HNO_3-Pu(IV), and different inorganic acids. Nave et al. [39] also studied the TBP-DD system by varying the nitric acid concentration in the aqueous phase. The mechanism of third phase formation in the

TBP-alkane solutions is driven by attractive interactions between reverse miscelles and attempts have been made to describe the phenomenon by models. A detailed review on third phase formation in liquid–liquid extraction is given in a recent literature [40].

3.6 Alternate Solvents as Substitutes for TBP in Thorium Reprocessing

As has been repeatedly emphasized, tributyl phosphate has been the universal choice as extractant for spent fuel reprocessing due to its various desirable properties. But, despite its success, it also has certain drawbacks as described in the earlier sections. Thus time and again, studies are reported on the search for better extractants as alternate to TBP without sacrificing its advantages. The first choice is compounds of the same class as TBP but with longer or branched alkyl groups. The higher molecular weight homologs have lower aqueous phase solubility and show increased organic phase solubility of their metal complexes [41]. Tri (n-hexyl) phosphate, tri (2-ethylhexyl) phosphate, triamyl phosphate, triisoamyl phosphate etc., are reported [41, 42] to have low aqueous solubility and no third phase formation problem with plutonium. Physicochemical properties such as density, viscosity, phase-disengagement time (PDT), and hydrolytic and radiolytic degradation of TAP in normal paraffinic hydrocarbon in the presence of nitric acid have been investigated [43]. The variations in these parameters are not much different from those obtained with degraded TBP.

Siddall, Mason, and Griffin have observed that, in the absence of steric effects in organophosphorus ligands, the extraction of both U(VI) and Th(IV) increases as the basicity of the coordinating $P = O$ of the neutral extractant increases [44, 45]. However, among the homologous series of neutral phosphate, phosphonate, phosphinate, and phosphine oxide, phosphate has the least basic phosphoryl oxygen of the series and gives the largest separation factor for U/Th separation. Several trialkylphosphates have been developed and tested for the extraction of U(VI) and Th(IV) ions [42, 45]. It is observed that the distribution ratio for the extraction of Th(IV) is drastically suppressed by the introduction of branching at the first carbon atom of the alkyl group. Suresh et al. [46, 47] investigated the extraction of uranium and thorium by TsBP and TiBP (isomers of TBP with branched carbon chain) as alternative choices for TBP. The esters with bulkier substitutes in place of the butyl group have been proposed for process applications in uranium and thorium separation [47]. The limiting organic concentration (LOC) of thorium in equilibrium with aqueous nitric acid-thorium nitrate was reported to decrease in the order THP > TAP > TBP. Pathak et al. [48] have shown that TEHP can be a better choice for U/Th separation compared to TBP and TsBP. Brahmmananda Rao et al. [49] have synthesized TcyHP, having three closed bulky aliphatic rings and compared the extraction of U(VI) and Th(IV) with those of

TBP, TsBP, and THP. The distribution ratios for extraction of U(VI) and Th(IV) by TcyHP at all acidities are almost double the value for THP. One of the major drawbacks of the extraction of U(VI) and Th(IV) by TcyHP is the formation of a third phase during extraction at much lower metal concentrations.

The behavior of degradation products from these extractants and their role in the process, their solubility (and of their alkaline salts during solvent washes) in both phases and the ease of their removal would influence the final choice.

Since the early work of Siddall, *N,N*-dialkyl amides are extensively evaluated as alternate extractants to TBP [50, 51] in PUREX process. The amide extractants are favored due to the following advantages. (i) Low volume of secondary waste generated as they are completely incinerable, (ii) chemical and radiolytic degradation products are of innocuous nature, hence give better decontamination from fission products with easier regeneration/cleanup, (iii) Solvents have low aqueous phase solubility, (iv) final U and Pu products streams are free of P contamination, and (v) ease of synthesis. Major drawbacks are third phase formation even with U and unfavorable viscosity compared to TBP.

Musikas et al. [52–54] have reported extensively on the extraction behavior of inorganic acids and U/Pu extraction chemistry with *N,N*-dialkyl amides. Certain dialkyl amides have been identified for the reprocessing of irradiated nuclear fuels in nitric acid media. Most of the earlier amides used either aromatic or substituted aliphatic hydrocarbons as diluents. However, these diluents are not suitable for commercial-scale reprocessing due to their poor radiation and chemical stability in the presence of nitric acid. Most of the solvent physicochemical properties can be tuned by a judicious choice of substituent alkyl groups. Hence, this group of extractants has received attention recently as alternative to TBP. An increase in the chain length, particularly adjacent to the carbonyl group, improves the extraction ability and increases the level of metal ion loading without third phase formation. However, it adversely influences the phase disengagement time, hydraulic behavior, and the aqueous solubility of degradation products.

Recent systematic studies have suggested the use of linear dialkyl amides as an alternative to TBP in PUREX process and branched dialkyl amides as alternatives to TBP in THOREX process [55–59]. Among these, *N,N*-dihexyl derivatives of hexanamide, octanamide, and decanamide are found to be promising alternatives to TBP in PUREX process. These ligands readily dissolve in *n*-dodecane and do not form third phases with nitric acid (up to 7 M). *N,N*-dihexyl octanamide (DHOA) has been extensively studied as a promising extractant for PUREX and compared with TBP.

Branching at the carbon atom adjacent to the carbonyl group greatly suppresses the extraction of quadrivalent actinides and fission products as compared to the hexavalent metal ions. This could be due to steric hindrance experienced by the amide molecules within the co-ordination sphere of Th(IV) ion already surrounded by four nitrate ions. Hence, branched amides are potential candidates for the separation of U from irradiated thorium. Extensive distribution and counter current extraction studies with branched-chain dialkyl amides have been carried out which indicate that the branched-chain *N,N*-di(2ethylhexyl) isobutyramide (D2EHIBA) could be a promising replacement of TBP for the separation of U alone from Th. However, D2EHIBA has limitations when used for the recovery of U and Pu during the reprocessing of spent fuel having thorium, uranium, and plutonium due to the poor extractability of Pu. In this context, DHOA is proposed for the selective extraction of U and Pu leaving thorium in the raffinate [60].

4 The Status and Developments in Thoria/Thorium Reprocessing

In the initial days of THOREX process development, thoria/thorium slugs or powder canned in Al were used for the irradiation in reactors. Hence, two decladding options were available: selective removal of Al by alkali dissolution without any significant effect on thorium, and co-dissolution of the clad along with thorium. In both the options, core dissolution could not be carried out without the addition of fluoride, hence Al as a complexant was used to keep the dissolver corrosion under control. Moreover, better decontamination from fission products and Pa was obtained when acid-deficient feed was used for extraction with aluminum nitrate as a salting agent in both feed and scrub. Thus, the difficulties in removing the residual alkali after alkaline decladding when heavy particles of thoria powder settled as slurry in the dissolver and the advantages of co-dissolution of Al and thorium prompted the adoption of this route for use in the early pilot and plant scale operations. However, nonvolatile clad metal nitrate in the concentrated aqueous waste solution resulted in large waste volume unlike the nitric acid-salted PUREX process.

Early attempts to develop a flow sheet having both aluminum and acid in the feed and acid alone in the scrub resulted in a second organic phase and also poor decontamination factors. Efforts for the development of an acid THOREX process were continued as the use of nitric acid as a "salting" agent in the extraction step would result in an aqueous waste volume about 1/10 that of the aluminum-salted flow sheet [61]. This resulted in the acceptance of the standard process of alkaline decladding as followed for Al-clad uranium fuel. The problem of second organic phase formation in the presence of nitric acid when Amsco was used as diluent was solved by the use of Decalin as diluent for tributyl phosphate which increased the solubility of the thorium-TBP complex. However, the decontamination of the products from fission products and Pa continued to be poor in acid medium. This was solved by a feed conditioning step in which the feed was made acid deficient by boiling, making the Ru and Pa in-extractable in TBP. Low acid scrub was used to further decontaminate the loaded organic.

The replacement of zircaloy and SS as clad in place of Al in power reactors gave an impetus for the development of alternate decladding technologies like Darex, Sulfex, and Zirflex. These methods did not find much favor with the operators and as of today, the chop-leach method employed in PUREX process for power reactor fuels is well accepted.

Dissolution of irradiated Al clad thoria has been carried out on an industrial scale using standard THOREX reagent at both Hanford and Savannah River sites in USA [14, 15]. However, the sequence of addition and the leftover heel from previous dissolution batch had significant effect on dissolution. These studies also revealed the need to study the dissolution behavior of thoria prepared by a specific route. Addition of MgO in the pellet was reported to have a significant effect on the rate of dissolution. Similarly, the chop leach process of SS clad thoria irradiated at Consolidated Edison Reactor at Indian Point was demonstrated in the commercial reprocessing NFS facility, USA.

Studies on the reprocessing of fuel particles from high temperature reactors were also initiated in USA, Germany, UK, and others to reprocess the fuel from high temperature gas cooled reactors employing thoria-urania particles. Grind-burn-leach and Grind leach methods were developed in the early days of thorium utilization. Equipment and the processes were also developed and tested in hot cell facilities.

In the recent past, not much work has been reported on thorium fuel cycle except from India. India has continued with its efforts to develop a viable thorium fuel cycle at a steady pace as this cycle constitutes the major part of the Indian nuclear energy program in its third and final stage.

4.1 THOREX Process Experience in USA

Laboratory development work was started in the middle of the last century at ORNL, USA to establish chemical conditions required for plant scale processing of irradiated thoria, which was designated as "THOREX Process". The goal was to establish the conditions needed to separate and decontaminate ^{232}Th, ^{233}U and ^{233}Pa from short cooled irradiated thoria and thorium targets [21]. The recovered ^{233}U was useful for the power generation, ^{233}Pa on decay after separation led to isotopically pure ^{233}U and the ^{232}Th was useful to breed further ^{233}U. Solvent extraction process based on TBP diluted in inert diluent was the natural choice due to the success of TBP-based PUREX process developed earlier for the separation of U and Pu. The knowledge base accumulated could be easily integrated into the new system. However, there were a few additional points, like recovery/removal of high active ^{233}Pa, which necessitated the use of other solvents and sorbants to resolve the problem. These studies resulted in three different flow sheets, viz., Interim 23, THOREX-1 and THOREX-2 Processes.

4.1.1 Interim-23 Process

This process was developed to recover and decontaminate ^{233}U based on 1.5 % TBP extraction of ^{233}U from a feed solution prepared by dissolving both Al and thorium slugs in 13 M HNO_3 containing fluoride and mercury catalyst. Silica gel was used for the final Pa removal and Dowex 50 cation exchanger resin was used for product concentration. An overall loss of 0.5 % was estimated with thorium separation factor $>10^5$. Solvent extraction provided a beta separation factor $>10^5$ and $>10^7$ after silica gel and resin concentration step.

4.1.2 THOREX-1 Process

This process to separate and decontaminate ^{233}Pa-^{233}U-^{232}Th had the following steps: (1) ^{233}Pa extraction in diisobutyl carbinol, (2) ^{233}U extraction in 5 % TBP, and (3) ^{232}Th extraction in a 45 % TBP-15 % benzene-40 % Amsco solvent. Although the product had adequate recovery and decontamination, it had engineering problems like management of three different solvents, their chemical treatment for cleanup, and six first-cycle columns out of which at least four were hot. The chemical problems were incomplete stripping of Pa from carbinol, need of benzene to avoid third phase formation, and incomplete protactinium decontamination from thorium owing to its extractability in TBP from acidic system.

Pa accounted for more than 90 % of the gross beta activity in THOREX feeds and was well extracted in the acidic TBP system. A method for converting Pa into inextractable species was thus required to permit the use of a single solvent and to reduce the number of columns for separation. Laboratory studies on Pa behavior indicated an acid-dependent ionic-polymeric equilibrium shifting toward the polymer species in an acid-deficient system that decreased its extraction.

4.1.3 THOREX-2 Process

This alternate THOREX-2 process which found favor on the basis of its engineering feasibility, utilized a single solvent (TBP) and had the following major steps: (1) thorium and ^{233}U extraction and separation from protactinium and fission products by use of concentrated TBP (41–55 %) followed by Pa adsorption from HAW on a silica gel column, (2) thorium partition from ^{233}U by preferential stripping with dilute nitric acid of suitable concentration, and (3) ^{233}U stripping with very dilute nitric acid.

Satisfactory extraction conditions for the first step were thus established to maintain a slightly acid-deficient extraction section so as to limit the protactinium and fission product extractability that did not lead to any thorium hydroxide precipitation or losses, followed by highly acid-deficient aluminum nitrate scrub section to hold protactinium extraction factors less than one. Sufficient aluminum nitrate salt strength was maintained throughout the system for satisfactory thorium extractability. Appropriate flow ratios and free solvent capacity ensured the prevention of Th bearing third phase formation without benzene addition. The silica gel bed sorption of Pa did not require any further feed adjustment and its capacity was governed by the flow rates. The Th/U partitioning step was controlled by proper adjustment of flow ratios and strip solution acidity to remove all thorium and yet maintain ^{233}U in the organic phase. The final step of ^{233}U stripping with a very dilute nitric acid required mainly a suitable flow ratio adjustment of the solvent. Between 1964 and 1970, both Hanford and Savannah River Plants produced ^{233}U for research purposes. Some of the ^{233}U comprised fuel for the LWBR experiment at the Shipping port reactor.

The SRP and Hanford adaptations of their respective PUREX facilities for thorium processing were different and have been described in detail elsewhere [14, 15]. The two plants had different contactors, mixer-settlers at SRP and pulse columns at Hanford, a different arrangement of cycles (three cycles at SRP and a four-cycle system at Hanford) that incorporated an initial co-decontamination cycle, and different facilities for product isolation and oxide preparation.

In the five separate thorium processing campaigns conducted at SRP, thorium was processed in equipment and facilities which had been converted in 1959 to recover highly enriched uranium. Two different flow sheets were used and a total of approximately 240 tons of thorium and 580 kg of uranium were processed. In the first two initial campaigns, uranium was recovered with a dilute 3.5 % TBP flow sheet and the thorium was sent to waste. In these two campaigns, the irradiation conditions for the thorium metal resulted in different ^{233}U product batches with ^{232}U concentrations of 40–50 ppm and 200 ppm.

In the first of the last three campaigns (THOREX campaigns), thorium metal and thorium oxide were processed. Only thorium oxide was processed in the remaining two THOREX campaigns. These last three THOREX campaigns used 30 % TBP to recover both uranium and thorium. Irradiation conditions were set to produce a ^{233}U product with 4–7 ppm ^{232}U content.

The final procedure adopted for the thorium oxide dissolution was a two-stage process based on the favorable dissolving rate of thoria at high acid concentrations and the more rapid dissolution of aluminum at low acid concentrations with mercury as catalyst. Concentrated acid (12 M HNO_3, 0.05 M KF) was added to dissolve the substantial amount of the thoria heel leftover in the dissolver from the previous charge. Then freshly irradiated elements were added, the acid was diluted, mercuric nitrate was added, and the aluminum can plus a fraction of the thoria were dissolved, leaving a fresh heel of thoria. As a steady state was reached after few charges, there was no continuous buildup. A small fraction of the oxide, entrained when solution was transferred from the dissolver, was removed by centrifugation. This oxide was routinely slurried back into the dissolver. Specific steps taken for good decontamination from Pa, when deemed necessary, were a scavenging precipitation with MnO_2 and use of dilute phosphate (0.01 M) in the extraction bank scrub streams.

The first production scale thoria campaign at Hanford conducted in 1966, in a suitably modified existing PUREX facility after extensive flushing and decontamination was the culmination of an extensive developmental program. The process flow sheet development work was conducted both at the Hanford Chemical Processing Department and at Battelle Northwest Laboratories. On the basis of information obtained from the development work and the process tests, a series of engineering studies dealing with criticality prevention safety, flow sheet, and equipment capabilities were done for efficient execution of the campaign. Approximately 165 tons of thorium was processed and about 220 kg of ^{233}U was recovered. The thorium and ^{233}U product quality met all targeted specifications except for the fission product content of the thorium product.

The second thoria campaign, conducted in 1970, was targeted to produce a minimum of 360 kg of ^{233}U for use in the light water breeder reactor (LWBR) program. Extensive plant flushing and prior decontamination was done to reduce the ^{238}U contamination levels in the facility so as to have a pure U product. The process flow sheet and operations were quite similar to those employed in the earlier campaign. The success of this second campaign was, to a large extent, attributable to the experience gained in the preceding thorium processing operations. The aluminum cans containing the thoria powder or wafers were dissolved by the conventional sodium nitrate-sodium hydroxide process and the alkaline sodium aluminate waste solution was centrifuged to recover thoria fines. These fines were dissolved in a nitric acid-aluminum nitrate-potassium fluoride solution. A similar solution was used in the dissolvers to dissolve the major portion of the thoria. The solutions were then combined for concentration and distillation to produce an acid-deficient thorium nitrate solution as feed for the first solvent extraction cycle.

The solvent extraction flow sheet was based on the THOREX-2 process developed at ORNL and development work conducted at other sites. The first solvent extraction cycle was used for co-decontamination and partitioning of the thorium and the ^{233}U. Further decontamination of the products was attained in one additional thorium and two additional uranium solvent extraction cycles. A solvent consisting of 30 volume percent TBP diluted with normal paraffin hydrocarbons (n-C_{10}–n-C_{14}) and pulse column contactors were used in all cycles. The Third Uranium Cycle aqueous product stream was also passed through a fixed bed of cation resin for thorium absorption prior to final concentration.

The success of the separate SRP and Hanford thorium programs confirmed that the facilities meant for versatile PUREX-type operations to treat uranium-based fuels could be converted to thorium processing without great difficulty, thus rendering this as a global approach.

4.2 THOREX Process Developments in Europe

UK, AERE, Harwell had initiated the thorium reprocessing programs in the 1950s [7, 62]. The experimental plant erected at Harwell processed kilogram quantities of irradiated thorium metal to recover ^{233}U alone from targets sufficiently cooled to allow for complete decay of ^{233}Pa. The process was similar to the Interim-23 process developed at ORNL. The choice of solvent was between that of Butex and well diluted TBP in an acid-deficient medium. TBP was favored in acid conditions due to its better flexibility. Two-inch pulsed columns were used in the separation. Scrub solution containing slightly acidified 1 M sodium nitrate was used instead of HNO_3. This enhanced the uranium recovery at ambient temperature during stripping. The subsequent recovery of thorium was carried out with 40 % TBP. Operating conditions were optimized to avoid third phase formation. The fluoride present in the feed during dissolution was complexed with aluminum to permit a high recovery of thorium. Even in the presence of aluminum, fluoride improved the zirconium decontamination. The residual traces of Zr as well as ^{233}Pa present in thorium product were removed by passing the product through a silica gel column. Single cycle TBP process was used. The main requirement was the concentration of the uranium product by a factor of up to 100: evaporation was not used as it also concentrated the impurities and leftover phosphates from TBP in the product. Instead, ion exchange processes were used successfully. An anion exchange process based on sulfate was used. In practice, U products of high purity were obtained from such a concentrate by ADU precipitation, followed by a peroxide precipitation to remove traces of unwanted metallic impurities.

High temperature reactor fuel reprocessing was taken up in Europe, especially in Germany due to the high potential attached to these reactors in the early days. The difficulties of the reprocessing of the HTR spent fuel were due to the nature of its constituents. The graphite and the silicon carbide are chemically inert compounds and only a few chemical reagents are capable of dissolving them quantitatively and effectively. The volume fraction of the kernel containing heavy metals (HMs) to recover (fissile and fertile isotopes) is very small. It is lower than 1 % in case of pebbles type concept. The small dimension of these kernels does not allow a simple step like shearing to reach the HMs as practiced while reprocessing LWR spent fuel. Several strategies ranging from no recycle, to mixed recycle, once through recycle, to full recycle were considered for recycling HTGR fuel. Many of the cycles involve in some way the recycle of bred U-233.

A cold engineering-scale pilot plant was installed at General Atomics, USA in 1970 and operated for several years for the development of HTGR reprocessing unit operations (fuel element crush, burn, dissolution, solvent extraction, and off-gas retention). During the mid-1960s, as part of the joint US/federal republic of germany (FRG) program, a small-scale hot head-end treatment of HTGR-type fuels was done at the Idaho Chemical Processing. A pilot plant was constructed in Germany. This Jupiter Experimental Reprocessing Facility was developed in 1970 to experimentally reprocess AVR fuel elements. It was designed as a model to gain experience and data that could be used later in the design of a bigger plant. The plant had a capacity of 2 kg HM/day and included graphite combustion, dissolution of the fuel particles in THOREX reagent, and solvent extraction. In Germany, the development of economic reprocessing methods was initiated in 1966 leading to another variant of THOREX process scheme.

The THOREX process, which used only HNO_3 as a salting agent, was developed by ORNL using a feed solution of 1.1 M $Th(NO_3)_4$, a special feature being an acid deficiency of about −0.15 M. The scrub solution was 1 M HNO_3. The acid deficiency can create undesirable precipitates during feed adjustment. For this reason, a dual cycle THOREX process was developed by Farbwerke Hoechst as part of the German project [63]. This process used a 1 M HNO_3 feed solution in the first extraction cycle and a 0.1 M HNO_3 scrub, thus avoiding hydrolytic precipitates during feed adjustment. After having separated the main volume of fission products in the first cycle, an acid-deficient feed solution was used in the second cycle to reach the desired decontamination factors. In this cycle 1 M HNO_3 was used as scrub. Re-extraction of thorium and uranium was carried out by co-stripping in the first cycle. In the second cycle, separation of thorium and uranium was accomplished by subsequent re-extraction followed by partitioning.

However, it was felt that for a regular thorium fuel cycle utilization such a dual cycle process with high decontamination factors may not be required as the product ^{233}U will always contain up to a few hundred ppm of ^{232}U along with its very intensive γ-emitting daughter radionuclides as contaminants. During refabrication, ^{233}U can only be handled with heavy γ-shielding, so that a gross fission product DF of better than 10^3 may not be necessary. The same applies to reprocessed thorium which contains considerably more ^{228}Th than its equilibrium value. Thorium, if recycled immediately, must thus be processed in hot cell facilities just as ^{233}U.

These considerations lead to the proposal to return to a single cycle THOREX process. An optimized process with acid feed solution (1 M HNO_3) should provide the required decontamination factors of up to about 10^3 for both uranium and thorium. Co-stripping should be avoided as a re-extraction procedure since $Th(DBP)_4$ precipitates and produces a crud at the low acidity (0.1 M HNO_3) necessary for this step. Partitioning of uranium and thorium by a separate re-extraction of thorium can be carried out at an acidity (<0.7 M HNO_3) which reliably avoids crud formation. Further decontamination of thorium and uranium can then be accomplished in separate steps.

Like other countries, France also started early with its thorium fuel cycle program. Rich monazite deposits in the Madagascar sand permitted the separation of thorium in its pilot plant at Le Bouchet. About 2,000 tons of thorium had been produced until 1955. In 1974, France even considered adapting existing facilities (i.e., UP1 at Marcoule) for the reprocessing of HTR fuels.

An advanced prototype plant with remote maintenance, rack mounted reprocessing plant, with a remote fuel fabrication plant, were built with Allis-Chalmers engineering at the South Italy Rotondella ITREC center. The plants were due to operate originally with spent BWR U-Th fuel from Elk River and later on DRAGON-type fuels. These plants were built in the 1960s and cold tested till 1974. The fuel fabrication plant was stopped but the reprocessing plant was used as a pilot plant for hot and cold tests of advanced components. A first campaign on 7 ELK River spent fuel was conducted between 1975 and 1985 to gain operational experience.

4.3 Thoria/Thorium Reprocessing in India

During the early phase of the Indian program on Thorium Fuel Cycle, the recovery of ^{233}U alone from irradiated thorium was contemplated for initiating ^{233}U related R and D tasks. Accordingly, more emphasis was given to a process using 5 % TBP-in odorless kerosene as extractant for selective extraction of ^{233}U. As only long cooled irradiated targets were to be processed, Pa extraction was not considered as a serious interference to the process. High acid flow sheet was preferred because of the higher residual acidity encountered after dissolution of thorium.

4.3.1 First THOREX Process Campaign at BARC, Trombay

For the first campaign at BARC [64], a small pilot plant was set up at the end of the 1960s for the processing of thorium metal/thoria irradiated in the reflector region of research reactor CIRUS. The facility had two main sections: A dissolver cell for rod charging, decladding, and dissolution and a separation area where the extraction process was carried out in a series of shielded glove boxes. Since aluminum was the clad, chemical decladding with NaOH was adopted. The exposed thorium/thoria was dissolved in a thermo siphon dissolver in HNO_3 containing NaF and aluminum nitrate. In the case of thorium metal, the dissolution could be completed in 12 h and the final feed had a thorium concentration of 200 g/L at an acidity of 4 M with about 0.12 % uranium. Since the irradiation was done at the thermalized low flux region of the reactor and the processing was done after sufficient cooling, the ^{232}U content as well as the fission product activities was low.

From the high acid feeds, ^{233}U alone was preferentially extracted and separated from thorium and fission products using 5 % TBP in odorless kerosene as extractant followed by scrubbing of the organic phase with 1–2 M HNO_3 and stripping with demineralized water. The bulk of the thorium remained in the raffinate. Extraction, scrubbing, and stripping were done in continuous air/vacuum-pulsed mixer settler units comprising only 12 stages due to limitation of available space. There were five stages for extraction, three for scrubbing and four for stripping. The stripped product was then concentrated by evaporation.

Further purification of uranium from thorium was carried out by anion exchange procedure in 8 M HCl medium by selective sorption of anionic chloro complexes of uranium, which was loaded on Dowex 1 X 4 anion exchanger and washed free of thorium by 8 M HCl and the uranium was finally eluted with 0.5 M HNO_3. The product obtained contained pure ^{233}U in which the thorium content was a few ppm. The product was converted to U_3O_8 by ADU route and ignition. The final product had about 1–3 ppm of ^{232}U.

4.3.2 THOREX Process-Related Development Studies

After the first campaign, several process development studies were initiated to overcome the major shortcomings encountered in the process. The main areas addressed in the process were: (a) reduction of thorium contamination in the ^{233}U product, (b) tail end purification of the separated ^{233}U using better techniques, and (c) recovery of thorium form HAW using higher percent TBP as extractant.

(i) Solvent Extraction Studies for ^{233}U Recovery

During the operation of pilot facility, because of the various constraints while processing, the final uranium product often had significant amounts of thorium as contaminant. Extraction data relevant to Th/U solvent extraction equilibrium were collected [65] to evaluate the stage requirements for the selective recovery of only ^{233}U using 5 % TBP. Based on the data two extraction schemes were tested with 20 stage air/vacuum pulsed mixer settlers for use in future THOREX campaigns. The feed acidity was kept at 4 M HNO_3 as against a scrub acidity of 2 M. The acid pick-up by 5 % TBP was low and did not affect the uranium stripping pattern in the strip column. Less stress was placed on fission product decontamination under these conditions as the final ^{233}U product would anyway contain ^{232}U and its fast growing daughters. The presence of about 8 g/l Th in the loaded organic leaving the extraction section also dampened the fission products extraction due to saturation effect. The schemes had 12–14 stages for extraction of uranium and 6–8 stages for scrubbing the co-extracted thorium. Using these schemes, it was possible to get loaded organic containing uranium with only 5–10 % thorium contamination, when the aqueous feed contained 200 g/L thorium with 0.12 % U. By increasing the number of scrub stages, it was possible to improve the product purity further. All the subsequent campaigns were conducted using these results as input.

Later, mixer settler studies with 2 and 3 % TBP in n-dodecane were also conducted to test their efficacy for better thorium removal from the product and also to test different feed conditions like initial thorium feed concentration and acidity. These studies also gave products with good DF from thorium [66].

(ii) Solvent Extraction Studies for Thorium Recovery

In order to remove the bulk of thorium left in HAW, data on thorium behavior such as its solubility as a function of nitric acid concentration, its limiting organic concentration in the solvent range of 30–42 % TBP/n-dodecane at higher extraction acidities, etc., were determined and reported by Dhami et al. [67]. Extraction and stripping schemes were developed in the laboratory using 38 % TBP in n-dodcane as extractant for the extraction and recovery of thorium from raffinate generated during the THOREX campaigns. An engineering flow sheet was evolved for large-scale recovery and purification of thorium from accumulated THOREX raffinate.

(iii) Tail End Purification of ^{233}U

(a) *Anion Exchange in Chloride and Acetate Media*

Though the final purification of uranium from thorium by anion exchange in HCl medium gave excellent decontamination for U from thorium, it posed several problems like corrosion, difficulties in filtration and washing of ammonia precipitated U/Th product, gassing of the ion exchange column during loading, poor decontamination from iron (which also forms anionic complex), etc. As an alternate to HCl, acetic acid was tried since it also forms anionic complexes with uranium [68]. This method gave good product purity (50–150 pmm Th in U), but the drawbacks of the earlier method were experienced here as well.

(b) *Cation Exchange Removal of Thorium from ^{233}U Product*

For regular plant application, the cation exchange procedure was studied in detail which involved the preferential sorption of thorium as Th(IV) at 0.5 M HNO_3 from a feed containing ^{233}U and thorium [69]. Compared to Th(IV), the bivalent uranyl ion was weakly absorbed under the chosen acid conditions and the column operation gave an effluent and preferential elution(wash) stream containing more than 99 % ^{233}U which was almost free from thorium. Satisfactory separation of uranium from thorium could be achieved by careful manipulation of feed and elution acidities and also by controlling the thorium loading to within 60–80 % of the total column capacity. Dowex 50 W X 8 cation exchange was used. Feed solutions had 1 g/L thorium and 0.3–3 g/L uranium at an acidity of 0.5 M HNO_3. Nitric acid was used as the preferential eluent (1 or 2 M) depending on the thorium loading. When 60 % of the column capacity was used up by thorium, the elution of uranium with 1 M HNO_3 gave satisfactory product. However, when less than 30 % of the column capacity was used up by thorium, elution with 2 M HNO_3 was necessary to get a concentrated uranium product. After preferential washing out of uranium, the bulk of the thorium from the column was eluted with 3–4 M HNO_3. With Dowex 50 W X 8 resin, complete removal of thorium from the column with nitric acid alone was rather difficult. To alleviate this problem a lesser cross-linked resin Dowex 50 W X 4 was used [70]. The ^{233}U product strip solution with low acidity from the solvent extraction cycle could be directly passed through this column for final purification, the processing rate would depend more on the thorium content of the solution. Routine column operation would require rigid control on thorium loading pattern and acidity which might vary from batch to batch. Because of its adaptability for routine remote operation, this procedure was used in plant-scale applications.

(c) *Removal of Thorium from ^{233}U Product by Oxalate Precipitation*

Another backup laboratory procedure was also developed for the removal of thorium from uranium product based on sequential precipitation technique [71]. It involved the precipitation of thorium as oxalate in nitric acid medium and after filtration, precipitation of uranium as ammonium diuranate from the oxalate supernatant. The method was tested for a wide range of feed conditions, with feeds containing 10 g/l U and 1-50 g/L thorium in 1-6 M HNO_3. For maximum recovery of uranium during thorium oxalate precipitation step, the optimum feed acidity should be about 1 M and the feed uranium concentration should not exceed 10 g/L. The excess oxalic acid concentration was restricted to 0.1 M. However, acidities up to 5 M HNO_3 could be tolerated in the feed with a slight increase in thorium contamination of the U product. The thorium thus separated, could either be converted to oxide or recycled back to process after dissolution. The ADU supernatant could be disposed off as such or after scavenging the residual uranium with Fe(III) as carrier in ammoniacal medium.

This technique was used in the THOREX campaigns at IGCAR Kalpakkam for treating uranium products without going through the ion exchange route. Later this process was further modified at IGCAR by adding one more carbonate precipitation step to remove the iron contamination, if present in the product [72].

4.3.3 THOREX Process Campaigns at IGCAR, Kalpakkam

After the first campaign at BARC, IGCAR, Kalpakkam conducted the next two THOREX campaigns during 1987–1989 and 1998–2000 with plant and equipment facilities that were being developed and installed in hot cells for studies on fast reactor fuel reprocessing [73]. The dissolver and its off-gas system, raffinate storage, feed conditioning systems, etc., were located in a separate cell designed for remote operation. Equipment that needed remote maintenance or remote handling was located in a separate cell. Equipment that needs only remote operation was housed in another cell. Air turbine operated high-speed centrifuge was used for clarification of aqueous feed to extractor. A bank of eight stages of centrifugal extractor was used for the extraction step. Air pulsed mixer settlers were used for scrubbing and stripping. The reconversion of the ^{233}U product was carried out in a train of glove boxes in a separate laboratory.

The process flow sheet employed was similar to the earlier one used in the first campaign at BARC. Thoria fuel irradiated at CIRUS and DHRUVA reactors were processed and the ^{232}U content in ^{233}U product ranged from 3 to 15 ppm. The final ^{233}U purification step used initially the anion exchange technique in chloride medium, which was later substituted by the oxalate precipitation technique described above. During the campaign, equipment such as centrifugal extractors, centrifuge, and air pulsed mixer settlers which were mainly developed for fast reactor reprocessing applications performed well with respect to capacity and efficiency. Good decontamination for uranium form fission products and thorium was obtained. Feedback from these campaigns helped to refine the designs for future programs.

4.3.4 THOREX Process Campaign at UTSF, BARC, Trombay

The fourth THOREX processing campaign was conducted at a dedicated Uranium Thorium Separation Facility THOREX facility at BARC during 2002. Based on the earlier process campaigns and the R&D results, this dedicated facility was built at BARC, Trombay, for conducting THOREX process-related studies for the first time. A vertical limb thermo-siphon dissolver was installed in a separate cell. The extraction flow sheet used was the same as that of earlier campaigns except 3 % TBP in n-dodecane (in place of 5 %) was used as the extractant. The thoria fuel irradiated at CIRUS was processed in this campaign. The Combined Airlift Mixer (CALMIX)-based mixer settlers, specially designed and developed in-house were used as contactors in the solvent extraction cycle. Adequate number of stages for extraction, scrub, and stripping were provided. The cation exchange process using Dowex 50 X 4 resin was used for the final purification of uranium.

A dedicated reconversion laboratory equipped with glove box train was used for the final conversion of product to U_3O_8 through ammonium diuranate route with a Th-oxalate precipitation backup route for products with high thorium contamination. After the completion of the uranium recovery campaign, the same equipments were utilized to recover the thorium in the raffinate employing the thorium extraction flow sheet based on 38 % TBP described earlier. The thorium lean raffinate was sent for waste management to develop and optimize suitable techniques for the waste disposal.

4.3.5 Fuel Cycle Studies Related to Thorium Reprocessing

Several laboratory studies have been reported as part of the ongoing thorium fuel cycle program. They include (i) thoria-based fuel dissolution studies, (ii) post irradiation examination of irradiated thoria, (iii) flow sheet development for three component (Th/U/Pu) separation from irradiated (Th, Pu)O_2, and (iv) third phase formation during extraction of thorium by TBP and evaluation of alternate solvents for the THOREX process.

(i) *Thoria Pellet Dissolution Studies*

Thoria dissolution with HF-HNO_3 mixtures with and without aluminum/zircaloy and the use of standard THOREX dissolvents, 13 M HNO_3 + 0.025 to 0.05 M HF + 0.1 M $Al(NO_3)_3$, for dissolving thoria are well reported [12, 17, 18]. However, the complete dissolution of zircaloy clad thoria by chop leach process with low Th/U losses to the hull waste meant for final disposal (after assay) still remains a challenge.

Attempts have been made [19, 20] to establish the acid and fluoride ion concentrations required for maximum or complete dissolution of thoria in a single step, at an acceptable rate, with minimum feed adjustments for subsequent solvent extraction cycle. The corrosion rates of SS304L and zircaloy-2 have also been assessed under these conditions. The study used two dissolvent mixtures: 13 and

8 M HNO_3 with 0.1 M $Al(NO_3)_3$ and two concentrations of HF/NaF, 0.03 and 0.05 M. The use of NaF was preferred as it was easier to handle and exhibited enhanced dissolution rates, at least in the initial stages of dissolution. The thoria used for these studies comprised sintered pellets (1,680 °C with 95 % TD), green pellets, MgO doped (0.5–2.5 %) sintered pellets, sintered Thoria-4 % PuO_2 pellets, and PHWR irradiated samples. SS304L and zircaloy –2 coupons were also introduced along with thoria. The tests simulated the plant dissolution conditions as closely as possible and were carried out under reflux conditions (at 110–115 °C). The volume of the dissolvent was so adjusted as to have a final thorium concentration of about 1–1.1 M. In the case of fresh sintered thoria, about 92 h were required for 95 % dissolution with 13 M HNO_3 dissolvent mixture, while the time taken for the same in 8 M HNO_3 mixture was more than 172 h (Fig. 6). In the comparison studies, the dissolution rate with NaF was observed to be marginally higher in the initial stages of dissolution as compared to HF, but the time taken for complete dissolution in both cases was the same. Studies with HF and NaF at 0.05 M instead of 0.03 M showed only marginal enhancement in the rate of dissolution. Studies with MgO doped pellets at 8 M HNO_3 indicated better dissolution behavior, and the time taken for dissolution reduced with increasing MgO content, with 1.5 % being optimum as shown in Fig. 7. PHWR irradiated thoria pellet indicated much faster dissolution.

The corrosion rates of plain SS, welded SS, and ring coupons were within acceptable limits at both acidities. The corrosion rate of zircaloy-2 cladding tube was also measured. In the absence of aluminum nitrate, the corrosion rate of SS was very high despite the presence of dissolved zirconium from zircaloy.

In the case of sintered thoria pellet with 4 % PuO_2 (sintered at 1,680 °C, 94 % TD), using 13 M HNO_3 mixture, it took about 32 h for 95 and 90 %dissolution of Th and Pu, respectively, with the dissolution rates as given in Fig. 8 [74]. However, near-quantitative dissolution of the pellet was reported in 44 h probably due

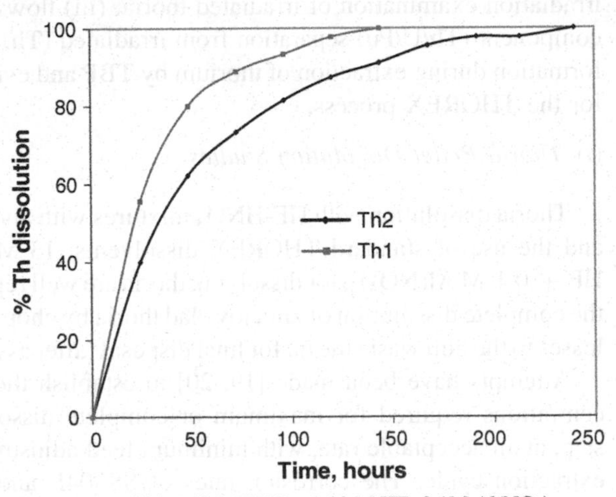

Fig. 6 Thoria dissolution at different nitric acid concentrations

Th1 = 13 M HNO_3+0.03 M HF+0.1M $Al(NO_3)_3$
Th2 = 8 M HNO_3+0.03 M HF+0.1M $Al(NO_3)_3$

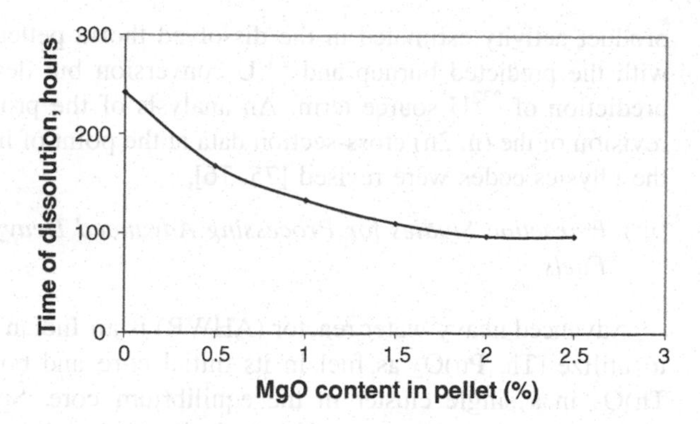

Fig. 7 Thoria dissolution as a function of MgO in the pellet

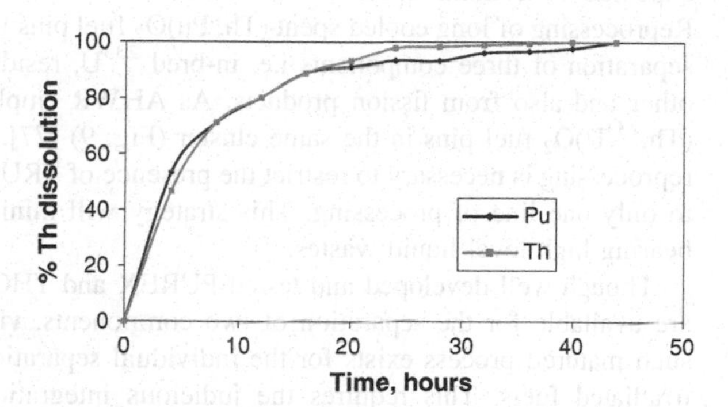

Fig. 8 Dissolution of sintered Thoria- Plutonia (4 % Pu) pellets

to the additives used in preparation of the Pu mixed pellets. The corrosion rate of SS was calculated and was found to be around 14 mpy, within acceptable limits. The zirconium concentration in the final solution was calculated to be 0.002 M.

(ii) *Post Irradiation Examination of Irradiated Thoria*

Since the source term analysis for thorium irradiation is not as matured as that for uranium, the codes developed for the analysis of nuclides generation need experimental validation especially for minor ones like ^{232}U, whose presence will have a direct bearing on the fuel cycle facility design for the final product shielding and handling requirements. Hence, post irradiation examination (PIE) of experimental bundles have been carried out to validate the reactor physics parameters and to generate the necessary database for thorium fuel cycle activities. In the case of thorium irradiation in the reflector region of CIRUS, the evolutions were correctly predicted by the available codes which could easily be verified for actinide and fission product generation during the irradiated rod processing. Irradiation of thoria during the initial PHWR flux flattening could also be achieved as per the predicted physics pattern.

The ^{232}U formation and the ^{233}U conversion were predicted with an existing code and for their validation, PIE and chemical analysis of a PHWR irradiated thoria bundle was carried out. Uranium isotopic composition and the fission product activity estimated in the dissolved thoria pellet showed close agreement with the predicted burnup and ^{233}U conversion but deviated significantly in the prediction of ^{232}U source term. An analysis of the problem revealed a need for revision of the (n, 2n) cross-section data at the point of irradiation and accordingly the physics codes were revised [75, 76].

(iii) *Extraction Studies for Processing Advanced Heavy Water Reactor (AHWR) Fuels*

Advanced heavy water reactor (AHWR) is an Indian Reactor concept designed to utilize (Th, Pu)O_2 as fuel in its initial core and both (^{233}U, Th)O_2 and (Pu, Th)O_2 in a single cluster in the equilibrium core. Spent fuel from AHWR is expected to contain about 2–4 % fissile materials, viz, ^{233}U and plutonium. Reprocessing of long cooled spent (Th, Pu)O_2 fuel pins from AHWR envisages the separation of three components i.e. in-bred ^{233}U, residual Pu and Th from each other and also from fission products. As AHWR employs both (Th, Pu)O_2 and (Th,^{233}U)O_2 fuel pins in the same cluster (Fig. 9) [77], their segregation prior to reprocessing is necessary to restrict the presence of TRU (minor actinide) elements to only one line of processing. This strategy will minimize the volume of TRU bearing high level liquid wastes.

Though well developed and tested PUREX and THOREX process flow-sheets are available for the separation of two components, viz, U–Pu and Th-^{233}U, no such matured process exists for the individual separation of U, Pu, and Th from irradiated fuels. This requires the judicious integration of these two versatile processes.

TBP in n-dodecane was considered for the selective extraction of Pu and U leaving the bulk Th in the aqueous phase. A feed acidity of 3–4 M was envisaged as a higher residual acidity from the dissolution stage inevitably leads to high nitric acid content in the feed which requires adjustment for its subsequent processing by solvent extraction. The feed concentration of thorium was kept around 100 g/L to limit the accompanying fissile content to less than 7 g/L from criticality point of view. The feed adjustment with respect to acidity and thorium content was easier to achieve at this level after dissolution at high acidities. Extraction studies were restricted to the use of only 5 % TBP in dodecane as extractant at 1:1 aqueous to

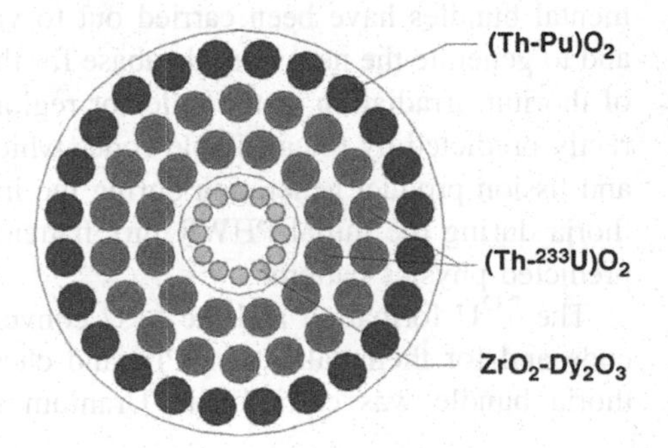

Fig. 9 54 Fuel pin cluster of AHWR

organic phase flow ratio as it had performed well in the earlier THOREX campaigns meant for the recovery of ^{233}U alone and had not posed any third phase formation problems under the normal process operating conditions when the Th content in 5 % TBP organic phase reached a maximum of 8 g/l.

Based on extraction studies with the three-component system (U, Pu, and Th) and 5 % TBP in dodecane as extractant, the optimum extraction, scrubbing, and stripping conditions required for the satisfactory recovery of U and Pu alone from a feed containing 2 % of each of these in a 100 g/l Th feed at 3–4 M HNO_3 have been determined [78]. Effectiveness of the reductant mixture of hydroxyl amine nitrate and hydrazine nitrate as a partitioning agent for Pu was also established in the partitioning cum back extraction studies. Based on these results, an extraction scheme was proposed as shown in Fig. 10 and validated [79] using laboratory mixer settlers with simulated feeds at concentration levels expected in the AHWR spent fuel reprocessing.

Theoretical simulation studies have also been carried out with computer codes for the extraction of Th/U/Pu by TBP and for the partitioning of Pu and U by HAN (the latter mainly for fast reactor fuels) [80, 81].

In view of the favorable extraction characteristics toward Pu(IV), DHOA is being tested as an alternative to TBP for the selective extraction of U and Pu for ternary mixture of Th(IV), U(VI) and Pu(IV) expected from AHWR spent fuel reprocessing. Based on extensive laboratory studies on extraction, partitioning, and

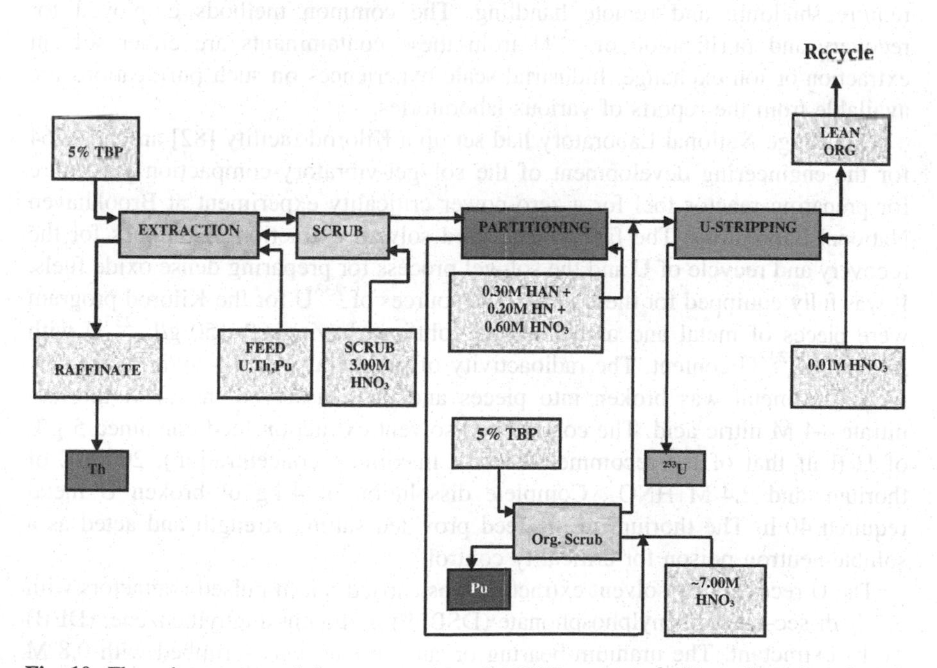

Fig. 10 Flow sheet proposed for reprocessing of AHWR spent fuel, (From Dhami et al. [79]. With permission.)

stripping, a process scheme has been proposed for the reprocessing of spent (Pu, Th)O_2 fuel from AHWR using DHOA [60].

(iv) *Third phase formation during extraction of thorium by TBP and evaluation of alternate solvents for the process*

Extensive investigations on third phase formation tendencies of TBP and other solvents during extraction of Th and Pu have been carried out as they are relevant to both thorium and fast reactor fuel reprocessing domains of the Indian nuclear energy program. As an ongoing effort, alternate solvents to TBP are also being evaluated for thorium processing and these have been covered in the earlier Sects. 3.5 and 3.6.

4.4 ^{233}U Purification from ^{232}U Daughter Products and Recovery from Fabrication Scraps

The production facilities store ^{233}U product either as uranyl nitrate solution or as uranium oxide powder. The radiation dose on ^{233}U bearing materials will grow as a function of time due to decay of ^{232}U and its daughters and a maximum dose level is reached around 10.3 years. This necessitates the fuel fabrication of the finished product within a few days after purification and final assembly would require shielding and remote handling. The common methods employed for recovery and purification of ^{233}U from these contaminants are either solvent extraction or ion exchange. Industrial-scale experiences on such purifications are available from the reports of various laboratories.

Oak Ridge National Laboratory had set up a Kilorod facility [82] around 1964 for the engineering development of the sol–gel-vibratory-compaction procedure for preparing reactor fuel for a zero-power criticality experiment at Brookhaven National Laboratory. The facility had used solvent extraction techniques for the recovery and recycle of U and the sol–gel process for preparing dense oxide fuels. It was fully equipped for these tasks. The sources of ^{233}U for the Kilorod program were pieces of metal and uranyl nitrate solutions having 60–150 g/L ^{233}U with 38 ppm of ^{232}U content. The radioactivity of the metal was 0.4 Sv/hr at the surface. The metal was broken into pieces and then dissolved in 1.4 M thorium nitrate—4 M nitric acid. The conditioned solvent extraction feed contained 5 g/L of U (half that of the recommended safe maximum concentration), 200 g/L of thorium, and 2.4 M HNO_3. Complete dissolution of 4 kg of broken U metal required 40 h. The thorium in the feed provided salting strength and acted as a soluble neutron poison for criticality control.

The U recovery by solvent extraction was carried out in pulsed contactors with 2.5 % di-sec-butyl phenylphosphonate (DSBPP) in diluent diethylbenzene, (DEB) as the extractant. The uranium-bearing organic stream was scrubbed with 0.8 M $Al(NO_3)_3$, which was 0.4 M acid deficient, for removing thorium and other contaminants. The uranium from scrubbed organic phase was stripped with demineralized water in a strip column. The uranium loss in the extraction column was less than 0.01 % of the initial feed. In the aqueous product, the concentration of uranium was 10 g/L. The uranium concentration in the organic streams was controlled by limiting the concentration of DSBPP employed to 2.5 % in DEB, as the saturation of uranium at this extractant concentration was only 10 g/L. This dilute product solution from stripping was diluent washed to remove traces of extractant before concentrating the product to 100–150 g/L to avoid troublesome products during the subsequent sol–gel process.

Ion exchange methods have also been successfully used [83] for satisfactory separation of most of the uranium from the daughters of ^{232}U. Uranyl nitrate solutions with 100 g of ^{233}U, and less than 0.1 g of thorium per liter in less than 0.3 M in HNO_3, were passed sequentially through two resin columns by up flow through the first and down flow through the second. The long-lived daughters of ^{232}U were retained on the resin. The uranium first saturated the resin, and then broke through as a decontaminated product practically free of thorium and radium and partially of lead. More than 99 % of the thorium, about 98 % of the radium, and about 50 % of the lead were retained by the column when 200–400-mesh AG-50W X 12 resin was used at a solution throughput equivalent to 5.3 g of uranium/mL of resin, at 25 °C. The pressure drop across the two columns was about 150 pounds per square inch. After the resin had been washed to displace the residual uranyl nitrate solution, the long-lived daughters that had sorbed on the resin (about 9 % of the total) were eluted with 6 M HNO_3 at 60 °C. The eluate containing radium, lead, about 50 % of thorium and uranium was recycled for uranium recovery by solvent extraction. The resin columns were then ready for reuse. Although solvent extraction was required for the recovery of the residual ^{233}U, associated with the thorium, most of the short-lived ^{232}U daughters were eluted from the resin and decayed within 3–5 days. Purification of ^{233}U by the above process was found to be much simpler than by solvent extraction methods.

In another program [84] during 1973–1976, ORNL under contract with Bettis atomic power laboratory (BAPL), produced several hundred kilograms of ceramic-grade $^{233}UO_2$ for use in the light water breeder reactor (LWBR) demonstration program and also processed 29 tons of UO_2-ThO_2 scrap generated during pellet fabrication for recovery. Though unirradiated, the age of the recycled UO_2-ThO_2 necessitated remote processing. The source material was oxides and nitrate solutions of ^{233}U with 6–10 ppm ^{232}U content. Several kilograms of ^{233}U as its nitrate solution was treated by cation exchange technique immediately prior to $^{233}UO_2$ conversion to remove the ^{232}U daughters as described above. For aged source feeds ($\sim$5 years), it was necessary to have two cycles of ion exchange to obtain the required DF from ^{232}U daughters.

In addition, the scrap materials, $^{233}UO_2$ powder sample residuals, glove box scrap, and unused powder, etc., were dissolved in nitric acid and these along with all the internally generated solution wastes during the above tasks were processed by a solvent extraction flow sheet featuring 5 % DSBPP in DEB. Thorium in the feed solution was discharged to waste.

During the early 1980s, BARC, Trombay had conducted [85] several sub-critical neutron multiplication experiments using beryllium reflected ^{233}uranyl solution systems. Recovery and repurification of ^{233}U from the experimental waste solutions containing significant amounts of boron was carried out after these experiments. Depending on the concentration of boron and other accompanying impurities, either cation exchange or solvent extraction with 30 % TBP in Shell Sol-T was adopted for the purification of ^{233}U in critically safe lots. Solutions containing up to 4 % boron in approximately 0.2 M HNO_3 were purified using cation exchange procedure, when no other impurities were present in the solution. Uranium was loaded on the cation exchange (Dowex 50 X 8) column and the boron impurity was washed off prior to elution with dilute acid. In one cycle the boron impurity could be brought down to 15 ppm and in a repeated cycle, down to 1 ppm. The TBP extraction procedure offered good decontamination when several other cations were present as impurities. ^{233}U was extracted from 3 M HNO_3 solution, scrubbed with 2 M acid and the loaded organic phase was stripped with water. This separation procedure was useful for the removal of many impurities including boron in trace level.

In PURNIMA II reactor experiments at BARC, the ^{233}U as uranyl nitrate in 0.5 M HNO_3 was used as the fissile material. During the experiments, the solution at different ^{233}U concentration levels was pumped into the reactor core vessel to different heights to obtain minimum critical values at each concentration. The solution tank and the pump were located in a glove box adjacent to the core vessel. During the experiments ^{233}U had to be purified at site to remove the ^{228}Th generated from the decay of ^{232}U (1-3 ppm) present as impurity in ^{233}U. The entire solution was passed through a small cation exchange cartridge containing DOWEX 50 X 8 (50–100 mesh) resin. Under the experimental conditions used, the ^{228}Th activity was preferentially held by the column. The cartridge containing ^{228}Th and ^{229}Th (daughter of ^{233}U) was further processed for the recovery of ^{229}Th for other experimental studies.

During the fabrication of alloy fuel for Kamini Reactor, wastes like alloy filings, powder, graphite crucibles with ^{233}U residues, etc., were generated. The technique adopted [85] for recovery and recycle of ^{233}U from these materials was dissolution of alloy scraps (or leaching of residues from crushed crucibles) with nitric acid in presence of mercuric nitrate catalyst. The acid conditions were so chosen as to have only NO_2 and not hydrogen as evolved gas. ^{233}U from this solution was recovered by TBP extraction process with good recovery.

5 Conclusions

Even though the studies on the use of thorium as a nuclear fuel started almost along with uranium, after the initial years, the interest in thorium waned and nuclear energy development based on uranium and plutonium had surged ahead. Thorium did not find favor with the energy developers mainly due to the lack of a fissile component in thorium to initiate the fission in the reactor. This necessitated the use of Pu or highly enriched uranium for the initial start-up. The present global scenario and fissile inventory situation with Th as an efficient Pu burning matrix is expected to boost the thorium fuel cycle from the energy viewpoint.

In spite of the many advantages of using thorium as a fuel in certain reactors, some aspects, especially in its recycling domain need further development prior to its utilization in commercial power production. These include difficulties in complete dissolution of thoria-based fuels, high radioactivity due to the long half-life of ^{233}Pa and ^{234}Th in short cooled spent fuels, handling difficulties of the products due to the decay of ^{232}U and its daughters, etc. However, some of these constraints are currently being viewed as deterrents from proliferation angle in the thorium fuel cycle. The long-lived minor actinide problems are much reduced in this cycle in comparison to that of U–Pu fuel cycle.

These considerations have led to a renewed confidence in the drive to propose thorium-based fuels for two of the Gen IV reactor systems, viz., High Temperature Reactor and the Molten Salt Reactor. Innovative reactor concepts like that of Radkovsky's or the Indian AHWR may boost the thorium utilization in the coming years. For greater public acceptance, Gen IV reactor system designs give due importance to nuclear reactor safety against accidents and better management of the waste left behind in the form of spent fuel or as the high active waste from reprocessing.

Thoria is a good matrix for irradiation and also for confining fission products as a final waste form when compared with urania. Hence, it will be good for reactor operation and for final disposal as such without reprocessing. If recycling is to be adopted, some of the areas which need further attention in the THOREX process to meet these goals are outlined below.

Today the PUREX process has achieved commercial status in the recycling of spent uranium and U–Pu MOX fuel. But THOREX process has still a long way to go prior to its large scale commercial deployment. The recent developments in the PUREX process technology will have their application in THOREX process too for satisfactory resolution of many of the problems faced in the earlier years.

To begin with, at the head-end, complete dissolution of thoria is preferable and this has to be achieved quantitatively in a reasonable time. Laser assisted mechanical systems for the head end operations like chopping and better dissolution techniques may lead to improvements in the head end area of this technology. The method of thoria fuel fabrication and the choice of additives/dopants used during that process seem to influence thoria dissolution and need further evaluation. As of now, fluoride catalyzed dissolution continues to be the best choice till a viable alternative is available. This step leads to fluoride induced corrosion of the equipment during dissolution of the fuel and evaporation of the waste. Even in PUREX process, corrosion of these equipments are encountered often and hence better materials of construction are constantly being developed and evaluated. These developments would help both the processes.

Hull monitoring techniques for verifying complete dissolution of fuel may have to depend on the high energy signal from the ^{232}U decay products and have to be standardized. Hull disposal may adopt the same technique of compaction followed by placing in canister or other innovative techniques under development for PUREX process. Efficient feed clarification systems for both organic and aqueous phases are essential for the success of this process as crud formation tendency is enhanced in presence of bulk amounts of thorium.

All the recent developments in TBP based PUREX technology will also have their impact in THOREX process. For instance, single cycle process envisaged for PUREX may be more relevant to THOREX due to the reduced decontamination requirement of the recovered ^{233}U and Th as remote handling of the products would anyway be necessary due to the gamma activities of the daughters of ^{232}U. Thus development of single cycle processes, maintenance-free short residence contactors and annular pulse columns that are safe with respect to criticality, near real time accountancy of fissile and fertile actinides, use of salt free processes, minimization of losses to waste, improvements in solvent quality, direct denitration of products to oxide, separation of long lived ^{231}Pa and fission products under P and T option with a general reduction of waste volumes are some of the challenging R and D tasks ahead.

Radiation protection protocols and occupational hazard control of workers being simultaneously exposed to various isotopes of Th, U and Pu may need additional techniques to implement regulatory controls on exposure. The criticality values when both the fissile elements $^{233}U/Pu$ are present simultaneously in the streams need to be evaluated to safeguard against criticality hazards. The radiological source terms for relevant long lived actinides generated in this cycle are to be assessed for their long term hazard potential.

A high ^{232}U contamination in the final product will necessitate additional engineering design features to contain ^{220}Rn emanations from ^{232}U contaminated ^{233}U product so as to reduce the operator exposure. Buildup due to the memory effects will necessitate periodic cleanup of processing lines handling the final product. Hence, reactor codes that give a realistic estimate of the expected ^{232}U burden in the ^{233}U product are required to arrive at the requisite optimum design features. Off gas released from the head end contain higher amounts of ^{85}Kr and may have to be effectively trapped to minimize the environmental load. As reactor control of ^{232}U generation during Thoria irradiation to produce ^{233}U in large scale may not be economically viable, methods based on laser may have to be developed for permanent cleanup of ^{233}U from ^{232}U rather than the quick ion exchange or solvent extraction methods whose results are short lived. This will facilitate fuel fabrication step and may help to bring down the fabrication cost. Moreover, laser cleanup system can purify the degraded isotopic composition of the multiple recycled ^{233}U from higher isotopes especially ^{236}U, else this will lead to the formation of higher amounts of minor actinides.

Though TBP is well entrenched as the PUREX extractant for processing U/Pu based fuels, the fuel reprocessing aspects of Thorium fuel cycle are still open and evolving. All the efforts so far to produce ^{233}U have been carried out in modified PUREX facilities after accommodating all the shortcomings of TBP in handling thorium. With the advent of new and specifically synthesized extractants, further investigations and evaluations can certainly lead to better extractants for this still to emerge full fledged Thorium fuel cycle, addressing the specific needs of processing thorium based fuels and the present day environmental safety issues associated with the waste management. As these experiments can be carried out with natural thorium and uranium feed solutions (Unlike U/Pu feeds), full scale tests with the promising extractants can be conducted more easily to arrive at meaningful and reliable data for the ultimate selection of the most suitable solvent. This approach will also ensure that there are no major surprises or unanticipated problems to be faced while processing irradiated thorium fuels on plant scale.

Development of techniques like pyrochemical processes especially for uranium recovery may become attractive for online reprocessing as envisaged for molten salt reactors. Lastly, any developments in robotics, automation and remote handling will have their greatest impact on this fuel cycle and will play a pivotal role in closing the fuel cycle.

References

1. Lung M, Gremm O (1998) Perspectives of the thorium fuel cycle. Nucl Eng Des 180:133–146
2. Greneche D, Szymczak WJ, Buchheit JM, Delpech M, Vasile A, Golfier H (2007) Rethinking the thorium fuel cycle: an industrial point of view. In: Proceedings of ICAPP 2007, Nice, France, 7367
3. Brooksbank RE, McDuffee WT, Rainey RH (1978) A review of thorium fuel reprocessing experience. In: CONF-780223-3, Conference: 84. AIChE national meeting, Atlanta, GA, USA
4. Martin FS, Miles GL (1958) Chemical processing of nuclear fuels. Butterworths Scientific Publication, London
5. Flagg JF (ed) (1961) Chemical processing of reactor fuels. Academic Press, New York and London
6. Staller SM, Richards RB (eds) (1961) Reactor hand book, vol 2—Fuel reprocessing. Interscience Publishers Inc, New York
7. Bruce FR, Fletcher JM, Hyman HH, Katz JJ (eds) (1956) Progress in nuclear energy series III, process chemistry, vol I. Pergamon Press, London
8. Basu TK, Srinivasan M (eds) (1990) Thorium fuel cycle development activities in India. BARC report-1532
9. IAEA-TECDOC-1450 (2005) Thorium fuel cycle—Potential benefits and challenges
10. Raymond GW (1966) Thorium fuel cycle. In: Proceedings of the second international thorium fuel cycle symposium, Gatlinburg
11. Schulz WW, Burger LL, Navaratil JD, Bender KP (eds) (1984) Science and technology of tributylphosphate. CRC Press Inc, Boca Raton
12. Lung M (1997) A present review of the thorium nuclear fuel cycles. EUR17771EN
13. Srinivasan M, Kimura I (eds) (1990) Proceedings on the Indo-Japanese seminar. Bombay, India
14. Walser RL, Jackson RR (eds) (1977) PUREX process operation and performance 1970 Thoria Campaign, ARH-2127
15. Orth DA (1978) SRP thorium processing experience. In: DPSPU-78-30-3; CONF-780622-72, ANS annual meeting, San Diego, CA, USA

16. Ramanujam A (2001) PUREX and Thorex processes (Aqueous Reprocessing). Encyclopedia of Materials: Science and Technology, pp 7918–7924
17. Phillips JF, Huber HD (1968) A process for the dissolution of aluminum-clad thoria fuel elements, BNWL-240
18. Hyder ML, Prout WE, Russell ER (1966) Dissolution of thorium oxide, DP-10084
19. Anantharaman K, Ramanujam A, Kamath HS, Majumdar S, Vaidya VN, Venkataraman M (2000) Thorium based fuel reprocessing and refabrication technologies and strategies. In: Power from thorium: status, strategies and directions, INSAC-2000, Mumbai, India, 1. p 107
20. Vijayan K, Shinde SS, Jambunathan U, Ramanujam A (2000) Studies on the dissolution of sintered thoria pellets in HNO_3+ $Al(NO_3)_3$ mixtures containing HF and NaF. In: Power from thorium: status, strategies and directions, INSAC-2000 Mumbai, India, 2. p 148
21. Gresky AT, Bennett MR, Brandt SS, McDuffee WT, Savolainen JE, Lovelace RC, Talleym JF (1952) Progress report: laboratory development of the Thorex process, ORNL-1367 (Rev)
22. Groenier WS (1991) Technical Manual for Sephis Mod4, Version 2.11, ORNL/TM-11589
23. Breschet C, Miquel P (1971) Improvement of the procedure used to treat highly irradiated fuels (Example of the utilization of complexing and redox agents in a solvent extraction process). In: Proceedings of the international solvent extraction conference ISEC 71, Hague
24. Evans TF (1959) The pilot plant denitration of PUREX wastes with formaldehyde, HW-58587
25. Schulz WW (1970) Macroreticular ion exchange resin cleanup of PUREX process TBP solvent, U.S. atomic energy report ARH-SA-58. Atlantic Richfield Hanford Co, Richland
26. Drain F, Moulin JP, Hugelmann D, Lucas P (1996) Advanced solvent management in reprocessing: five years industrial experience. ISEC 96. p 1789
27. Mailen JC, Tallent OK (1985) Cleanup of savannah river plant solvent using solid sorbents, ORNL/TM-9256. Oak Ridge, TN
28. Manohar Smitha, Srinivas C, Vincent Tessy, Wattal PK (1999) Management of spent solvents by alkaline hydrolysis process. Waste Manag (Oxford) 19:509–517
29. Vasudeva Rao PR, Srinivasan TG, Suresh A (2010) Third phase formation in the extraction of thorium nitrate by trialkyl phosphates. Mater Sci Eng 9:012056
30. Suresh A, Srinivasan TG, Vasudeva Rao PR (2009) Parameters influencing third-phase formation in the extraction of $Th(NO_3)_4$ by some trialkyl phosphates. Solvent Extr Ion Exch 27:132–158
31. Vasudeva Rao PR, Kolarik Z (1996) A review of third phase formation in extraction of actinides by neutral organophosphorus extractants. Solvent Extr Ion Exch 14(6):955–993
32. Healy TV, Mckay HAC (1956) The extraction of nitrates by Tri-n-butyl phosphate (TBP). part 2—The nature of the TBP phase. Trans Faraday Soc 52:633–642
33. Srinivasan TG, Ahmed MK, Shakila AM, Dhamodaran R, Vasudeva Rao PR, Mathews CK (1986) Third phase formation in the extraction of plutonium by tri-n-butyl phosphate. Radiochim Acta 40:151–154
34. Kertes AS (1965) The chemistry of the formation and elimination of a third phase in organophosphorus and amine extraction systems. In: McKay HAC, Healy TV, Jenkins IL, Naylor A (eds) Solvent extraction chemistry of metals. MacMillan, London, pp 377–400
35. Osseo-Asare K (1991) Aggregation, reversed micelles, and microemulsions in liquid–liquid extr: the tri-n-butyl phosphate-diluent-water-electrolyte system. Adv Colloid Interface Sci 37:123–173
36. Chiarizia R, Jensen MP, Borkowski M, Ferraro JR, Thiyagarajan P, Littrell KC (2003) Third phase formation revisited: The U(VI), HNO3-TBP, *n*-dodecane system. Solvent Extr Ion Exch 21:1–27
37. Chiarizia R, Jensen MP, Rickert PG, Kolarik Z, Borkowski M, Thiyagarajan P (2004) Extraction of zirconium nitrate by TBP in n-octane: influence of cation type on third phase formation according to the sticky spheres model. Langmuir 20(25):10798–10808
38. Chiarizia R, Jensen MP, Borkowski M, Thiyagarajan P, Littrell KC (2004) Interpretation of third phase formation in the Th(IV)-HNO3, TBP-n-octane system with Baxter's sticky spheres model. Solvent Extr Ion Exch 22(3):325–351

39. Nave S, Mandin C, Martinet L, Berthon L, Testard F, Madic C, Zemb T (2004) Supramolecular organisation of Tri-n-butyl phosphate in organic diluent on approaching third phase transition. Phys Chem Chem Phys 6(4):799–808
40. Testard F, Zemb Th, Bauduin P, Berthon L (2009) Chapter 7. Third-phase formation in liquid/liquid extraction: a colloidal approach. In: Bruce A Moyer (ed) Ion exchange and solvent extraction-A series of advances, vol 19. CRC Press, Boca Raton, pp 381–428
41. Crouse DJ, Arnold WD, Hurst FJ (1983) Alternate extractants to tributyl phosphate for reactor fuel reprocessing. In: Proceedings international solvent extraction conference ISEC-83, Denver, CO, 90
42. Suresh A, Srinivasan TG, Vasudeva Rao PR (1994) Extraction of U(VI), Pu(IV) and Th(IV) by some trialkyl phosphates. Solvent Extr Ion Exch 12:727–744
43. Venkatesan KA, Robertselvan B, Antony MP, Srinivasan TG, Vasudeva Rao PR (2006) Physiochemical and plutonium retention properties of hydrolytic and radiolytically degraded tri-*n*-amylphosphate. Solvent Extr Ion Exch 24:747–763
44. Siddall TH III (1959) Trialkyl phosphates and dialkyl alkylphosphonates in uranium and thorium extraction. Ind Eng Chem 51:41–44
45. Mason GW, Griffin HE (1980) Demonstration of the potential for designing extractants with preselected extraction properties: possible application to reactor fuel reprocessing. In: Navratil JD, Schulz WW (eds) Actinide separations. American Chemical Society, Washington, DC, pp 89–99
46. Suresh A, Subramaniam S, Srinivasan TG, Vasudeva Rao PR (1995) Studies on U/Th separation using tri-sec-butyl phosphate. Solv Extr Ion Exch 13(3):415–430
47. Suresh A, Srinivasan TG, Vasudeva Rao PR, Rajagopalan CV, Koganti SB (2005) U/Th separation by counter-current liquid-liquid extraction with tri-sec- butyl phosphate by using an ejector mixer-settler. Sep Sci Technol 39:2477–2496
48. Pathak PN, Veeraraghavan R, Manchanda VK (1999) Separation of uranium and thorium using tris(2-ethylhexyl) phosphate as extractant. J Radioanal Nucl Chem 240:15–18
49. Brahammanada Rao CVS, Suresh A, Srinivasan TG, Vasudeva Rao PR (2003) Tricyclohexylphosphate—A unique member in the neutral organophosphate family. Solv Extr Ion Exch 21:221–238
50. Siddall TH III (1960) Effects of structure of N, N-disubstituted amides on their extraction of actinide and zirconium nitrates and of nitric acid. J Phys Chem 64:1863–1866
51. Siddall TH III (1961) USAEC Report DP—541. E I Du Pont de Nemours and Co, Aiken
52. Musikas C, Condamines C, Cuillerdier C, Nigond L (1991) Actinide extraction chemistry with amide type extractants. In: International symposium radiochem rad chem—Plutonium 50, Bombay (India)
53. Musikas C, Condamines N, Cuillerdier C (1990) Advance in actinides separations by solvent extraction. Research and applications. In: International solvent extraction conference ISEC 90, Kyoto (Japan), pp 16–21
54. Musikas C (1988) Potentiality of nonorganophosphorus extractants in chemical separations of actinides. Sep Sci Technol 23:1211–1226
55. Pathak PN, Kumbhare LB, Manchanda VK (2001) Structural effects in N, N-dialkyl amides on their extraction behavior toward uranium and thorium. Solv Extr Ion Exch 19:105–126
56. Pathak PN, Prabhu DR, Manchanda VK (2000) Distribution behaviour of U(VI), Th(IV) and Pa(V) from nitric acid medium using linear and branched chain extractants. Solv Extr Ion Exch 18:821–840
57. Pathak PN, Prabhu DR, Ruikar PB, Manchanda VK (2002) Evaluation of di(2-ethylhexyl)isobutyramide (D2EHIBA) as a process extractant for the recovery of ^{233}U from irradiated Th. Solv Extr Ion Exch 20:293–311
58. Pathak PN, Kumbhare LB, Manchanda VK (2001) Effect of structure of N, N dialkyl amides on the extraction of U(VI) and Th(IV): a thermodynamic study. Radiochim Acta 89:447–452
59. Manchanda VK, Pathak PN (2004) Amides and diamides as promising extractants in the back end of the nuclear fuel cycle: an overview. Sep Pur Technol 35:85–103

60. Pathak PN, Prabhu DR, Kanekar AS, Manchanda VK (2006) Distribution studies on Th(IV), U(VI) and Pu(IV) using tri-n-butylphosphate and N, N-dialkyl amides. Radiochim Acta 94:193–198
61. Rainey RH, Moore JG, Lovelace RC (1962) Laboratory development of the acid Thorex process for recovery of consolidated edison thorium reactor fuel. ORNL-3155
62. Fletcher JM (1961) Aqueous reprocessing of thorium and isolation of ^{233}U. NP-9340 Vol III Sect V
63. Zimmer E, Ganguly C (1987) Reprocessing and refabrication of thorium-based fuel. IAEA-TECDOC-412, p 89
64. Srinivasan N, Nakarni MN, Balasubramaniam GR, Chitnis RT, Siddiqui HR (1972) Pilot plant for the separation of U-233 at Trombay. BARC Report—643
65. Balasubramaniam GR, Chitnis RT, Ramanujam A, Venkatesan M (1977) Laboratory studies on the recovery of uranium233 from irradiated thorium by solvent extraction using 5% TBP Shell Sol-T as solvent. BARC Report—940
66. Chitnis RR, Dhami PS, Ramanujam A, Kansara VP (2001) Recovery of ^{233}U from irradiated thorium fuel in Thorex process using 2 % TBP as extractant. BARC report BARC/2001/E/030
67. Dhami PS, Naik PW, Dudwadkar NL, Poonam Jagasia, Kannan R, Achuthan PV, Dakshinamoorthy A, Munshi SK, Dey PK (2008) Studies on the physico-chemical parameters of extraction and stripping of thorium nitrate from Thorex raffinate using 38 % TBP in n-dodecane. BARC report BARC/2008/E/010
68. Mukherjee A, Kulkarni RT, Rege SG, Achuthan PV (1982) Anion exchange separation of uranium from thorium for the tail-end purification of ^{233}U in Thorex process. In: Proceedings of board of research in nuclear sciences radiochemistry and radiation chemistry symposium: Paper no. SC-12. Pune, India
69. Chitnis RT, Rajappan KG, Kumar SV, Nadkarni MN (1979) Cation exchange separation of uranium from thorium. BARC report –1003
70. Achuthan PV, Janardanan C, Vijayakumar N, Kutty PVE, Mukherjee A, Ramanujam A (1993) Studies on the separation of thorium and uranium on variously cross linked Dowex 50 W Resins. BARC report BARC/1993/E/09
71. Ramanujam A, Dhami PS, Gopalakrishnan V, Mukherjee A, Dhumwad RK (1989) Separation and Purification of uranium product from thorium in Thorex process by precipitation technique. BARC report-1496
72. Govindan P, Palamalai A, Vijayan KS, Raja M, Parthasarathy S, Mohan SV, Subba Rao RV (2000) Purification of ^{233}U from thorium and iron in the reprocessing of irradiated thorium oxide rods. J Radioanal Nucl Chem 246 (2):441
73. Balasubramaniam GR (1990) Reprocessing of irradiated thorium—indian experience. In: Srinivasan M, Kimura I (eds) Thorium utilization, Proc Indo-Japan seminar on thorium utilization: 65. Bombay, India
74. Sinalkar NM, Janardanan C, Vijayan K, Shinde SS, Jambunathan U, Ramanujam A (2001) Studies on the dissolution characteristics of sintered thoria (ThO_2-4% PuO_2) pellet. In: Proceedings of the nuclear radiochemistry symposium: 210. University of Pune
75. Achuthan PV, Dhami PS, Dakshinamoorthy A, Diwakar DS, Vijayan K, Kannan R, Sumankumar Singh, Jambunathan U, Dey PK (2003) Analysis of irradiated thoria dissolved solution. BARC report BARC/2003/I/002
76. Mukherjee S, Ganesan S, Krishnani PD, Jagannathan V, Srivenkatesan R, Anantharaman K, Sahoo KC, Gangotra S, Achuthan PV, Vijayan K, Dakshinamoorthy A, Jambunathan U, Ramanujam A, Aggarwal SK, Govindan R, Jaison PG, Khodade PS, Parab AR, Sant VL, Shah PM, Venugopal V, Ramaswami A, Dange SP, Rattan SS, Manohar SB (2003) Theoretical and experimental analysis of irradiated thorium bundle from KAPS-2. In: Eighth international CNS CANDU fuel conference, Muskoka
77. Sinha RK, Kakodkar A (2006) Design and development of the AHWR—The indian thorium fuelled innovative nuclear reactor. Nucl Eng Des 236:683–700

78. Dhami PS, Poonam Jagasia, Surajit Panja, Achuthan PV, Tripathi SC, Munshi SK, Dey PK (2010) Studies on the development of a flow-sheet for AHWR spent fuel reprocessing using TBP. Sep Sci Technol 45(8):1147–1157
79. Dhami PS, Jagasia P, Panja S, Naik PW, Achuthan PV, Tripathi SC, Munshi SK, Dey PK (2012) Validation of the flow-sheet proposed for reprocessing of AHWR spent fuel: counter-current studies using TBP. Desalin Water Treat 38:184–189
80. Kumar S, Koganti SB (2004) Development and application of computer code SIMPSEX for simulation of FBR fuel reprocessing flow sheets. III: Additional benchmarking results for HAN and U(IV) based U-Pu partitioning and a HAN based alternate partitioning step for FBTR fuel reprocessing. Tech report IGC-257/2004: IGCAR, Kalpakkam, India
81. Kumar S, Koganti SB (2003) Development of a computer code PUThEX for simulation of $^{233}U-^{232}Th$ and ^{233}U-^{232}Th-^{239}Pu separation flow sheets. Tech report IGC-255: IGCAR, Kalpakkam, India
82. Dean OC, Brooksbank RE, Lotts AL (1963) A new process for the remote preparation and fabrication of fuel elements containing uranium-233 oxide-thorium oxide. ORNL-TM- 588
83. Rainey RH (1972) Laboratory development of a pressurized cation exchange process for removing the daughters of ^{232}U from ^{233}U. ORNL-4731
84. Parrott JR Sr, McDuffee WT, Nicol RG, Whitson WR, Krichinsky AM (1979) The preparation of kilogram quantities of ^{233}UO2 for the light water breeder reactor demonstration program. ORNL/CF-79/279
85. Ramanujam A (1990) Experience in final separation, purification and recycling of uranium-233. In: Srinivasan M, Kimura I(eds) Thorium utilization. Proceedings of Indo-Japan seminar on thorium utilization:169. Bombay, India

Radioactive Waste Management in U/Th Fuel Cycles

Kanwar Raj, C. P. Kaushik and R. K. Mishra

Abstract Radioactive waste streams are generated at various stages of uranium and thorium fuel cycles. Treatment and storage/disposal of low and intermediate level wastes (LILW) from U fuel cycle are being practiced in various countries, which have adopted either open or closed fuel cycles for nuclear power generation. The spent nuclear fuel (SNF) from open fuel cycle is being stored. In the case of processing of SNF for closed fuel cycle, the resultant high-level liquid waste (HLW) is solidified by immobilization in an inert matrix like glass, and stored in engineered facilities. The research and development work is being pursued in various countries on disposal of SNF and solidified HLW in deep geological formations. In the case of thorium fuel cycle, since the mining is generally in shallow sea coastal areas, the waste generation is comparatively less. Process and technology for treatment and disposal of the wastes generated during mining and milling of monazite for recovery of thorium and associated rare earth (RE) elements is well established. Waste management practices with respect to the waste generated in fabrication of Th-based fuel and in operation of Th-fuelled reactors will not be much different from similar uranium fuel-based facilities. However, in the immobilization of HLW from the reprocessing of Th-based SNF, some challenges are likely on account of difficulty in dissolution of thorium oxide in concentrated nitric acid which necessitates the use of fluorides and aluminum. The results of R&D on the aspects of reduction of concentration of fluorides and aluminum in the waste, incorporation of thorium oxide in boro-silicate glass matrix, and its impact on the characteristics of vitrified waste product (VWP) as well as future areas of research are discussed in this chapter.

K. Raj (✉) · C. P. Kaushik · R. K. Mishra
Waste Management Division, Bhabha Atomic Research Centre,
Trombay, Mumbai 400085, India
e-mail: rajkanwar50@gmail.com

C. P. Kaushik
e-mail: cpk@barc.gov.in

R. K. Mishra
e-mail: mishrark@barc.gov.in

D. Das and S. R. Bharadwaj (eds.), *Thoria-based Nuclear Fuels*,
Green Energy and Technology, DOI: 10.1007/978-1-4471-5589-8_8,

1 Introduction

Like any other industrial activity, generation of nuclear power also results in different types of waste streams which are characterized by nature and concentration of the radionuclides present in the waste. Each step of nuclear fuel cycle, namely, mining and milling of ore/sand, ore processing to nuclear-grade material, fabrication of nuclear fuel, operation of nuclear power plant, storage/processing of spent nuclear fuel (SNF), radioactive waste management, etc., generates its own characteristic waste. Radioactive waste is also generated during production of isotopes and their application in research, healthcare, and industry. In future, with ageing of present nuclear facilities, substantial quantity of radioactive waste is likely to be generated during their decontamination and/or decommissioning. The nature of waste depends upon various factors like its source of generation, type of fuel and its cladding material, process used in purification/fabrication of fuel, type of nuclear reactor, burn-up of the fuel, off-reactor cooling period, process flow sheet used in SNF reprocessing, techniques adopted in decontamination/decommissioning, etc.

1.1 Radioactive Waste Classification

The waste streams are characterized and categorized on the basis of their radiological, physical, and chemical properties [1]. Most common methods of categorization of solid radioactive waste are on the basis of their physical properties like combustibility and compressibility. This helps to decide their further treatment either by incineration/pyrolysis or by compaction. Solid radioactive wastes are also characterized by the radiation field on the surface of the waste package, which helps to determine their further handling and disposal. Another method of categorization of solid waste is based on the half-lives of the radionuclides present in the waste whether short- or long-lived, which determines the required period for isolation of waste and its disposal concept.

Similarly, liquid radioactive waste streams are categorized based on various characteristics of wastes, e.g., concentration of the radionuclides and their half-lives, aqueous/organic nature, etc. Radioactive liquid wastes are commonly classified as low-level waste ($37–3.7 \times 10^6$ Bq/l), intermediate-level waste ($3.7 \times 106–3.7 \times 10^{11}$ Bq/l) and high-level waste (above 3.7×10^{11} Bq/l). This categorization of waste fulfills many objectives viz. basis of segregation of waste in the waste-generating facilities and in design of relevant storage area/tanks, selection of waste treatment process, selection of matrices for immobilization of waste, choice of volume reduction technique for solid wastes, design of waste disposal modules, selection of disposal site, etc. It also helps in proper documentation and communication with respect to various categories of radioactive waste among waste generators, managers, and regulators.

1.2 Radioactive Waste Management Policy

In view of large variety of radioactive waste streams generated in the nuclear fuel cycle, the processes and technologies used in their treatment and disposal are also very diverse. These are, however, based on the concepts of (a) delay and decay of short-lived radionuclides, (b) concentration and confinement of radioactivity as much as practicable, and (c) dilution and dispersion of the resultant effluents of very low-level radioactivity to the environment as per the permissible levels, set by the national regulatory authority, which need to be in line with the international practices.

The underlying objective governing the management of radioactive waste is protection of human beings and environment, now as well as in future. To meet this objective, the necessary codes and guides have been framed by international bodies like International Commission on Radiation Protection (ICRP) and International Atomic Energy Agency (IAEA). One of the basic requirements is to set up the national regulatory framework in every country. As an example, in India, Atomic Energy Regulatory Board (AERB) has been set up by the act of the Parliament, Atomic Energy Act 1962 [2]. Under this Act, AERB is entrusted with the responsibility of enforcement of safe disposal of radioactive waste in the country. This is to ensure that radioactive wastes are managed and disposed of in a controlled manner with adequate monitoring. This is to further ensure that any unacceptable hazard is not caused to the workers, the public, and the environment.

1.3 Current Radioactive Waste Management Practices

In the countries around the globe with different nuclear power programs, during last more than 50 years, various processes and technologies have been developed and adopted for management of radioactive waste. Based on the extensive R&D efforts on the international scale in various laboratories, processes are developed and are in use for 'conditioning' of radioactive waste. The 'conditioning' of waste includes various processes for decontamination of effluents, pre-treatment, treatment, immobilization in matrices, packaging, etc. Different techniques, selected on the basis of nature of 'waste', are in use for minimization of radioactive waste by (a) recovery and recycle of valuables; (b) segregation of wastes at source of generation; (c) volume reduction of solid waste by compaction, incineration/pyrolysis; (d) volume reduction of liquid wastes by thermal and solar evaporation; (e) decontamination of effluents; (f) concentration of chemical sludges, etc. In general, this applies to nuclear fuel cycles based on uranium fuel. Therefore, vast experience in management of radioactive waste from uranium fuel cycle is available. There are, however, certain challenges still to be satisfactorily tackled like disposal of SNF for countries going ahead with open fuel cycle and disposal of solidified high-level liquid waste (HLW) for countries which have adopted closed fuel cycle.

1.4 Thorium Fuel Cycle Waste Management

The favorable radioactive waste characteristic of thorium fuel cycle is one of the major reason of renewed interest in thorium another being the potential for improving proliferation resistance. A study conducted by Lung [3] indicates that the thorium-^{233}U system will generate less long-lived minor actinides. It is, therefore, believed that the overall radiotoxicity associated with HLW from reprocessing of Th-based fuel will be substantially lower than that of U–Pu waste.

As far as the alternative of open thorium fuel cycle is concerned, the structural stability of thorium oxide is likely to prove helpful in containing the fission products and minor actinides during the interim period of dry storage as well as disposal in final geological repositories [3]. Stability of thoria compared to urania is due to its rigid valency of +4. In the closed fuel cycle being adopted by countries like India, HLW is likely to be managed much in the similar way as in well established uranium fuel cycle except for the presence of fluoride and aluminum introduced during dissolution of thoria fuel in the reprocessing step. Sintered high burn up ThO_2 matrix does not readily dissolve in nitric acid unlike the case of UO_2. Presence of complexing agent like fluoride buffered with Al(III) is necessary to facilitate the dissolution. Another fact that needs attention for management of the waste stream from reprocessing of thorium based fuel is long-term radiological impact from radionuclides such as ^{231}Pa with long half-life of 3.276×10^4 years generated during irradiation of the fuel.

This chapter covers the aspect of management of different types of radioactive waste arising at various stages of the nuclear fuel cycle.

2 Radioactive Waste from Thorium Mining and Milling/Production

Major deposits of thorium are found in India, Canada, the USA, Russia, China, Brazil, and Turkey. Thorium is found in various forms in beach sand, stream placers, vein deposits, precambrian conglomerates, and carbonatites [3].

Like many other mineral recovery operations, mining and milling of uranium and thorium ores lead to generation of various types of wastes. These wastes are characterized by low concentration of materials containing naturally occurring radioactive elements with very long half-lives. In order to have a suitable background with respect to generation and treatment of radioactive wastes, aspect of mining and milling for recovery of thorium will be first reviewed briefly.

2.1 Thorium Mining and Milling

Thorium, in India, occurs in the form of monazite along with other heavy minerals from the beach sand. The monazite content of the beach sand varies from 0.1 to

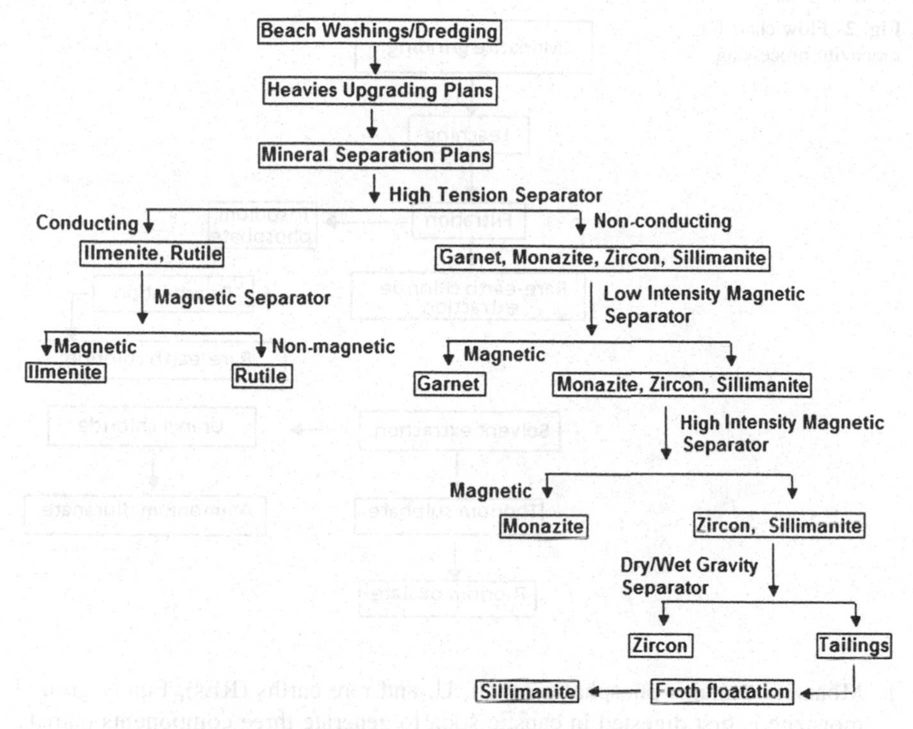

Fig. 1 Flow chart for thorium mining and mineral separation

2 %. Mining of heavy minerals is being carried out in different zones of the coast. In order to separate the heavy minerals, beach sand is mined and processed to obtain thorium, which is used as nuclear fuel and in other industrial applications. Mining of thorium is done by open pit methods involving operation of floating dredge and beach sand collection [4]. The sand contains various heavy minerals, namely ilmenite, rutile, sillimanite, garnet, zircon, monazite, etc. As shown in Fig. 1, the separation process involves gravity separation, application of high voltage and high magnetic fields, operation of dryers, operation of material handling equipment like belt conveyors, bucket elevators, etc.

2.2 Monazite Processing

Monazite is subjected to further processing to produce thorium oxalate/thorium nitrate and ammonium diuranate (ADU) [4]. The details of monazite processing and processing of thorium concentrate are shown in Figs. 2 and 3, respectively. Monazite processing involves three major steps.

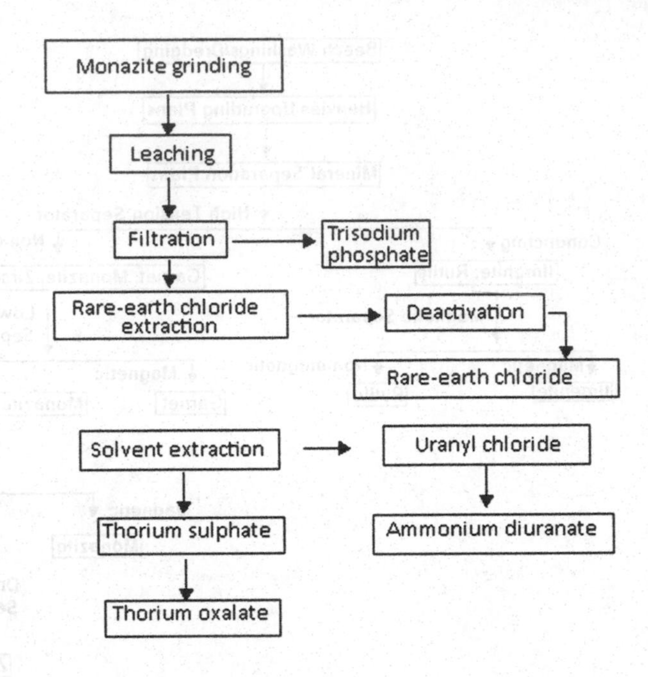

Fig. 2 Flow chart for monazite processing

1. Monazite contains phosphates of Th, U, and rare earths (REs). Finely ground monazite is first digested in caustic soda to generate three components namely tri-sodium phosphate (TSP), mixed hydroxides of RE, thorium and uranium as well as unreacted monazite.
2. Majority of REs from the mixed hydroxides are separated as RE chloride solution, deactivated and converted to flakes. A part of mixed RE is treated to produce products such as RE fluoride and cerium oxide.
3. The mixed hydroxides of Th, U and residual RE are extracted through acid leaching. Solvent extraction is then used to produce Th oxalate and crude uranyl chloride solution besides recycling of residual RE. The crude uranium chloride solution is then refined to produce nuclear grade ADU.

Details of thorium concentrate processing are given in Fig. 3. Thorium concentrate is converted into slurry, dissolved in acid and then subjected to solvent extraction to separate Th, U, and RE compounds. Thorium is precipitated as Th oxalate and then calcined to produce nuclear grade ThO_2 using the process steps shown in Fig. 4.

2.3 Source and Generation of Wastes in Th Production

Various types of waste streams are generated [5, 6] at different stages of Th production, namely,

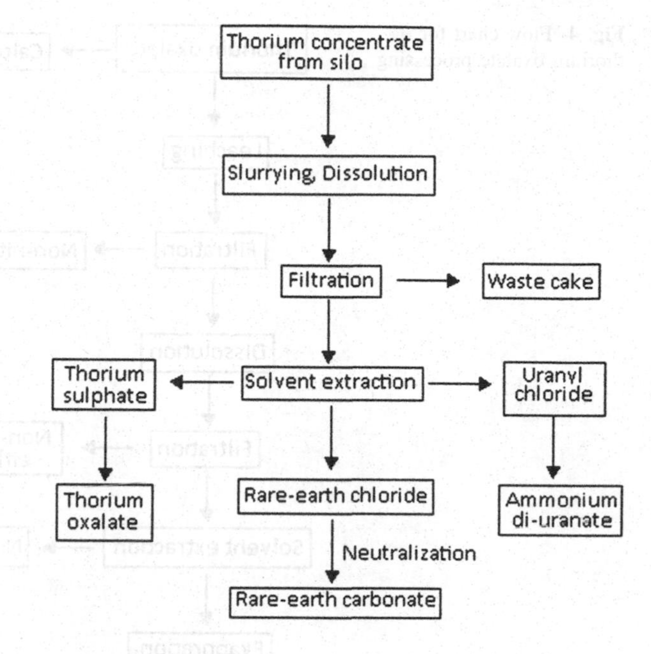

Fig. 3 Flow chart for thorium concentrate processing

1. Mining of sand and pre-concentration,
2. Mineral separation for Th and RE, and
3. Processing of monazite and thorium concentrate.

The characteristics of the generated waste depend upon its source of generation, process chemistry used in production and operational parameters of the manufacturing process.

2.3.1 Waste from Th Mining and Pre-concentration

Large volume of over-burden is generated due to surface mining of beach washings and dredge mining. This over-burden includes top soil, silica sand, clay, organic material like roots, shrubs, vegetation, shells, etc.

Pre-concentration and concentrate up-gradation operations generate solid tailings of silica sand, shells, and organic waste in form of the shreds of plant roots, etc. These two operations account for major part of solid waste from mining and mineral processing steps of entire thorium fuel cycle and contain very low concentration of radioactivity. Table 1 gives the typical radioactivity content of solid waste from mining and mineral separation. The volume of such waste depends essentially on the heavy mineral concentration in the raw sand and efficiency of the recovery processes. Typically 600–700 kg of tailings are generated per ton of raw sand mined.

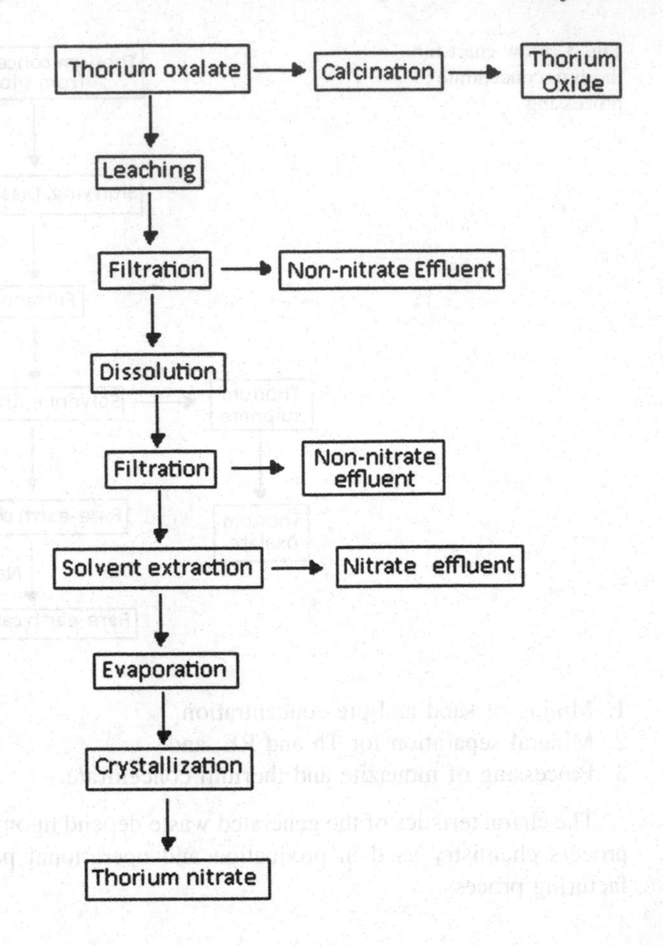

Fig. 4 Flow chart for thorium oxalate processing

The water used in pre-concentration and concentrate upgradation operations contains slime and suspended particulate matter, which are removed by settling. Decant water is recirculated for re-use. The radioactivity concentration in both decanted water and settled mass is very low as indicated by typical values indicated in Table 2.

Table 1 Typical activity in solid waste from mining and mineral separation [5]

Solid waste	Gross alpha (Bq/gm)	Gross beta (Bq/gm)	Radiation field (mGy/h)
Mining tails	0.5	5	0.5
Pre-concentration tails	0.8	4	0.4
Mineral separation plant tails	80	300	50

Table 2 Typical activity content of liquid waste, mining and mineral separation [5]

Activity type	Mining	Pre-concentration	Mill
Gross alpha (Bq/g)	0.04	0.01	0.05
Gross beta (Bq/g)	0.04	0.03	0.09

2.3.2 Waste from Mineral Separation

The mineral separation process generates sand tailing which contain unrecovered minerals and silica sand. The rate of generation of this waste is about 70–100 kg per ton of concentrated feed sand processed. Besides, other waste streams from mineral separation steps are (a) crude monazite concentrate, (b) water used for wet concentration of minerals, and (c) exhaust gases from driers containing mainly suspended particulate matter and SO_2.

2.3.3 Waste from Processing of Monazite and Th Concentrate

Monazite Insolubles

During the dissolution of monazite in sodium hydroxide, solid waste is generated which contains unreacted monazite, traces of Th, U, RE, and Ra. Typical concentration values of radioactivity in monazite insolubles and other solid wastes from chemical processing are given in Table 3. As indicated, concentrations of gross alpha and beta are in the range of 800–2,500 and 800–3,000 Bq/gm, respectively. The quantity of solid waste is about 80–100 kg per ton of monazite processed. Solid waste is also produced during conversion of thorium hydroxide to thorium oxalate at the rate of about 500 kg per ton of thorium concentrate processed.

Calcium Oxalate Sludge

During production of thorium nitrate, two types of solid wastes are generated amounting to 800 kg per ton of thorium concentrate. These solid wastes are (1) calcium oxalate sludge from the non-nitrate removal of process effluents (oxalate removal) and (2) precipitate from RE recovery step containing $NaNO_3$, RE fraction and Th along with traces of Ra and $BaSO_4$. Gross alpha and beta concentrations in Ca Oxalate sludge vary from 500–1,000 and 1,200–1,500 Bq/gm, respectively.

Table 3 Typical activity in solid waste, chemical-processing [5]

Solid waste	Gross alpha (Bq/g)	Gross beta (Bq/g)	^{228}Ra (Bq/g)	Radiation field (mGy/h)
Monazite insolubles	800–2,500	800–3,000	400–1,000	60–100
Mixed cake	2,000–4,000	3,000–7,000	2,000–5,000	400–600
ETP cake	300–500	300–600	25–100	2–3
Ca oxalate cake	500–1,000	1,200–1,500	800–1,000	80–100

Sludge from Effluent Treatment

Acidic and alkaline effluents generated during the processing of monazite are treated in Effluent Treatment Plant (ETP) for removal of residual activity and control of pH before monitored discharge with dilution if required. The rate of generation of chemical sludge from ETP is 100 kg per ton of monazite processed and it contains phosphates besides low levels of radioactivity. This is also named as ETP cake and contains gross alpha and beta radioactivity in the range of 300–500 and 300–600 Bq/gm, respectively.

De-contamination Waste

Chemical processing of monazite leads to re-distribution of radionuclides of thorium chain in the working areas of the facilities manufacturing thorium and RE compounds. Some contamination of plant equipment and structures is unavoidable due to the nature of operations and large-scale handling of thorium and its compounds.

To minimize radiation exposure to the plant workers due to internal and external sources, de-contamination of these nuclides from floors and equipment as well as skin, hand, and clothing is essential. Maximum permissible levels of contamination applicable to the facilities in India involved in thorium processing have been worked out in a study conducted by Haridasan et al. [7]. The salient results of this study are:

1. Derived surface concentration for alpha ($^{232}Th + ^{228}Th$) based on inhalation (1.25 Bq/cm^2) and ingestion (1.0 Bq/cm^2) are comparable and hence a limit of 1 Bq/cm^2 has been recommended. In monazite grinding operation, a limit of 0.3 Bq/cm^2 is found to be appropriate since the observed re-suspension factors are one order of magnitude higher. Similar values of re-suspension factor are reported earlier by Albert [8].
2. Surface contamination limits for beta emitters ($^{228}Ra + ^{228}Ac$) estimated on the basis of skin dose concept as well as ingestion route show that the ingestion route is more restrictive. Hence, a derived concentration limit of 0.4 Bq/cm^2 is found to be appropriate for the beta emitters in the monazite processing plant.
3. Total contamination limit on both hands due to alpha and beta has been set at 275 and 125 Bq, respectively. Beta contamination limits are found to be more restrictive in the case of personnel and clothing and have been set as 4.0 and 1.7 Bq/cm^2, respectively for alpha and beta emitters.

Decontamination of plant area and personnel results in generation of both liquid and solid waste streams. Solid waste due to decontamination consists of mops, used and discarded protective wear, unserviceable plant equipment, etc. Liquid waste from decontamination, which is generally of miscellaneous type, is treated in the ETP.

Miscellaneous Solid Waste

Various types of solid waste are generated during operation and maintenance as well as decommissioning of monazite processing plant. This includes contaminated equipment assemblies and structurals, debris from demolished buildings, etc.

Liquid Waste

Liquid waste generated in monazite processing consists of acidic (pH ~ 2) and alkaline (pH ~12.5) effluents and decontamination solution. About 15 m^3 of effluents are generated per ton of monazite/thorium concentrate processed. During production of thorium nitrate, liquid waste is generated which consists of filtrate from oxalate conversion, aqueous washing from solvent extraction, filtrate from RE recovery, evaporator condensate, and plant washings. Characteristics of these liquid waste streams are given in Table 4. The total volume of such wastes is 10 m^3 containing nitrate and 20 m^3 non-nitrate waste per ton of thorium concentrate processed.

Gaseous Waste

The radioactive gases from thorium fuel cycle contain thorium particulates, ^{220}Rn, ^{222}Rn, and their daughter products. The airborne releases from thorium mining and mineral separation processes are insignificant. The exhaust gases from thorium processing plant consists of suspended particulate matters, Th, ^{220}Rn and its progeny, H_2S, and HCl.

2.4 Treatment and Disposal of Waste

2.4.1 Mining and Mineral Separation Waste

Solid Waste

The solid waste generated in pre-concentration and mineral separation steps does not need processing. The mine tailings are recycled along with fresh feed to the mined area as back-fill. The dry and wet solid tailings are transported to the disposal site. The mechanical means used for transport are pumping, conveyors covered trucks, and dump trucks and bins.

Table 4 Characteristics of liquid waste from processing of monazite and thorium compound pretreatment waste

Effluent	pH	Gross alpha (Bq/g)	Gross beta (Bq/g)	^{228}Ra (Bq/l)
Monazite processing				
Acidic effluent	1.6–2.0	100–300	400–600	150–200
Alkaline effluent	12–13	600–900	900–1,000	300–400
Th-oxalate processing				
Non-nitrate effluent	1–3	23–30	100–110	25–40
Nitrate effluent	1–3	20–70	200–700	25–30

The dry mill waste tailings from mineral separation process are dumped in dredge mining area for recycling along with fresh raw sand being mined. In places where surface mining/beach washing collection only is practiced, the dry mill tailing is stored in earthen trenches. These trenches are covered with soil topping so that radiation level on the top of trench surface is less than 0.2 mGy/h. These trenches are in well-controlled area, properly demarcated by appropriate barriers to prevent public access.

Liquid Waste

The water used in mining and mineral separation steps does not contain enhanced level of radioactivity. Therefore, this water is pumped to settling ponds and re-used. Excess water is discharged to water body (canal or sea) after monitoring.

2.4.2 Chemical Processing of Waste

Solid Waste

The waste sludge containing monazite insolubles and traces of Th and Ra is neutralized and filtered using hydraulically operated filters. The filter cake is filled in suitable containers for disposal.

The slurry (mixed cake) of PbS and $Ba(Ra)SO_4$ is collected in storage tanks and pumped directly to the disposal site. The solid sludge produced (ETP cake) in the ETP is packed in suitable containers for disposal. Similarly, active solid waste produced in the processing of thorium oxalate is packed in suitable container for disposal.

Miscellaneous solid waste is segregated based on physical properties namely compressibility/combustibility or composition namely cellulosic/polymeric. Various volume reduction techniques are in use for reduction of final waste volume, e.g., compaction, incineration, melt-densification selected on the basis of physical characteristics of the waste. Among these, mechanical compaction of metallic waste and melt-densification of polymeric waste (e.g., polythene bags used for transportation, used protective wears, piping, etc.) are simple and eco-friendly technologies for reduction of waste volumes.

The waste filled in containers or packed in bags is disposed in reinforced concrete trenches. The design of these trenches takes into consideration the local conditions like water table, soil properties, rain fall flooding, etc., and radiation shielding like concrete covers and anti-leakage requirement. Inside surface of the trenches are coated with anti-corrosion material if required. After filling with waste, the trenches are closed with proper provision for (a) sealing to eliminate in-leakage of water, (b) drainage to avoid accumulation of rain water, (c) monitoring to check ingress of water, and (d) condition monitoring to assess integrity of structure. Solid waste, such as ETP cake, containing very low levels of radioactivity is disposed in earth trenches.

Liquid Waste

Acidic and alkaline effluents generated during monazite and thorium processing are treated by co-precipitation and allowed to settle in settler. The sludge is filtered and packaged for disposal. The filtrate is segregated in post-treatment tanks, diluted as per the discharge limit requirements and monitored before discharge ensuring compliance with the authorized discharge limits.

Gaseous Waste

The exhaust gases from reaction vessels are scrubbed with alkali in scrubbers which are normally of packed-bed design. After scrubbing, gases are discharged through a stack of height designed on the basis of site meteorological parameters and the permitted ground level concentrations of the discharged radio nuclides.

3 Oxide Fuel Fabrication

The radioactive waste generated during fabrication of oxide fuel from uranium are characterized by low activity concentration of materials containing naturally occurring radionuclides with very long half-lives [9]. In addition, waste streams contain process chemicals and their by-products which have nonradiological characteristics and associated hazards.

3.1 Types of Waste Streams

3.1.1 Solid Waste

During solvent extraction to produce uranium oxide, uranyl nitrate raffinate cake (UNRC) is generated. This has uranium value suitable for recovery by further treatment. Other solid wastes generated during fuel fabrication include contaminate polymer sheets, cotton/rubber gloves, cotton mops, filter element, etc.

3.1.2 Liquid Effluents

During production of UO_2 powder, various process effluents are generated like NH_4NO_3, $NaNO_3$ and waste water containing low level of uranium contamination. Besides, waste water is generated during decontamination of tools, washing of protective wears, floor washings, washing of hands and bathing by plant personnel, and waste water from various operations. These water streams are contaminated with low concentration of uranium.

3.1.3 Gaseous Waste

The off gases from equipment consist of acid vapors like NO_X, ammonia, and uranium particulate.

3.2 Waste Treatment in Fuel Fabrication

Effluents containing ammonium nitrate/sodium nitrate are generated after filtration of ammonium uranate slurry in UO_2 manufacturing process. This effluent is treated with tri-sodium phosphate (TSP) and ferric chloride to remove traces of uranium to generate supernatant liquid suitable for recycle/use. NH_4NO_3 and $NaNO_3$ effluents are monitored to establish their unrestricted use meeting regulatory requirements. Lean active water streams from various sources are collected and treated for removal of uranium. The supernatant is treated to qualify (uranium content less than 0.1 mg/l) for discharge through storm drain.

Gaseous waste from process generating dust and fumes are extracted with exhaust ventilation system. Various devices like filters, scrubbers, or other screening equipment are used to bring down the air-born contaminants.

The off gases mainly consist of acidic/basic gases and particulate matter. These are treated using equipment such as electrostatic precipitators (ESP), scrubber, prefilter, high-efficiency particulate air (HEPA) filter. Gaseous release through air route is monitored both quantitatively and qualitatively (with respect to NO_x, NH_3, uranium, etc.) and discharged through tall stack to meet the prescribed limits set by the regulators.

Solar evaporation ponds are used to concentrate aqueous effluents like laundry wash, sodium nitrate bearing waste. The wet surfaces of the pond are lined with impervious material, design of the pond ensures prevention of overflow even during heavy rain condition, collection of leakages, assessment of integrity during use, etc. Dried salt produced in the ponds is packed and disposed based on radioactivity concentration.

4 Thorium Reactor Waste

In a neutron rich environment, natural Thorium (^{232}Th) would capture neutrons and would become ^{233}Th. ^{233}Th has a half life of 22 min and it decays to become Protactinium (Pa-233), which has a half life of about 28 days to become ^{233}U. With thermal/fast neutrons ^{233}U undergoes fissions quite easily. A thorium reactor is, therefore, called a fast breeder reactor because it breeds fissile ^{233}U from ^{232}Th.

4.1 Characteristics of Waste from Thorium Reactors

In general, the radioactive waste generated from a reactor system may contain activation products, fission products, and transuranic elements.

4.1.1 Activation Products

During the working of a reactor, the reactor components are exposed to neutrons and their stable atoms form activation product, e.g., cobalt (^{59}Co) used as alloying material is converted to ^{60}Co by absorption of neutron and coolant water is converted to radioactive tritium. Besides cobalt; Fe, Mn, etc., are used in metallic components of reactor and are partly converted into radioactive atoms. Due to coolant flow-induced wear and tear as well as corrosion of reactor components, the activation products are detached and are deposited on low-flow parts of the reactor resulting in scale deposits and gradual built-up of radiation field. The coolant is filtered and treated by ion exchange to remove the radioactivity. The resultant chemical regenerate and spent ion-exchange materials are two major waste sources from the reactor operation in liquid and solid forms, respectively. However, half-lives of the activation products are limited to periods of months or years and can be safely managed either by disposal in near surface disposal facility after suitable treatment and immobilization in an inert matrix like cement or cement composites.

It can be concluded that the formation of activation products and their distribution are essentially governed by the neutron flux and the material of construction of the reactor components. Therefore, the radioactive waste generated from conventional nuclear power plants and future thorium reactors will not be different [10].

4.1.2 Fission Products

Fission products are the major contributors of radioactive atoms present in the most of radioactive streams generated during reactor operation and further handling of SNF. On an average, every fission in the reactor results in two fission products. For example, fission of ^{235}U leads to two fission products with atomic masses in the range of 130 and 90. These fission products are normally contained in the fuel on account of the fuel cladding. Wherever breach of fuel cladding takes place, some of the fission products find their way into the coolant. The majority of fission products produced from fission of ^{235}U and ^{233}U are short-lived. There are, however, few fission products like ^{90}Sr, ^{85}Kr, ^{99}Tc, ^{134}Cs, and ^{137}Cs, which have fairly long half lives and are produced in abundance in the reactor. The radioactive waste streams containing these fission products from conventional reactors and thorium reactors are not going to be much different from the point of view of distribution of fission products. The radionuclide content and resultant radioactivity would depend on factors like burn-up of the fuel and off-reactor cooling period.

4.1.3 Transuranic Elements

Thorium reactors have the distinct advantage of generating lesser transuranics. Transuranics are generated by capture of neutrons by uranium atom leading to formation of Pu, Am, etc. ^{232}Th and ^{233}U in thorium reactor lead to formation of far less transuranics compared to ^{235}U in the conventional reactor.

4.2 Wastes from AHWR

As in PHWRs, reactor waste from operation of AHWR will be mainly due to the contamination/leakages of fission, activation and corrosion products.

4.2.1 Liquid Waste

The liquid waste will be similar in nature as those from PHWR except with respect to its higher level of radioactivity on account of high burn-up of the fuel. Based on the process technologies developed for the management of liquid radioactive reactor waste stream, this waste will also be subjected to various decontamination processes employing chemical precipitation, ion-exchange, reverse osmosis, etc., followed by immobilization of concentrates [11].

4.2.2 Solid Waste

Among solid radioactive wastes, spent ion-exchange resin forms major fraction of solid waste with high specific activity. Presently, in Indian Nuclear Power Plants, spent ion-exchange resins are converted into polymerized waste product using polyester styrene matrix. Alternative process using improved cement matrix is undergoing plant scale trials and regulatory clearance review.

5 Fuel Reprocessing Waste

Reprocessing and recycling of both fissile and fertile components back into appropriate reactor systems is an integral part of three stage nuclear energy program of India. Different steps involved in processing of SNF are decladding, dissolution, and recovery of fissile and fertile materials. Reprocessing of SNF is a complex process involving handling of large quantity of radioactive materials and processing chemicals.

Like other industrial activities of nuclear fuel cycle, fuel reprocessing facilities too generate various types of radioactive waste streams. These are generated in all the three physical forms namely solid, liquid, and gas. These waste streams are primarily categorized on the basis of concentration of radionuclides, their half-lives and toxicity. Management of these wastes aims at (a) recovery and recycle of useful materials, (b) concentration and confinement of radioactivity in inert and stable matrices, (c) minimization of final waste volume for disposal, (d) decontamination of effluents following ALARA principle, and (e) minimization of radioactive discharge to the environment.

5.1 Sources and Types of Reprocessing Wastes

As mentioned earlier, in reprocessing plants, wastes in all three physical forms, i.e., solid, liquid, and gaseous, are generated. The radioactivity level of these wastes ranges from low to intermediate to high. In view of use of large variety of processing chemicals in the reprocessing plants, chemical nature of waste too varies from aqueous to organic, neutral to acidic, and presence of low to high total dissolved salts (TDS). A wide spectrum of activities of operation and maintenance lead to generation of solid wastes of various categories. Analytical and production functions of radiochemical laboratories of the plant generate various types of chemical waste containing both short- and long-lived radionuclides. To summarize, the major waste streams generated from the reprocessing plant are listed below:

1. High-level liquid waste (HLW)
2. Intermediate-level liquid waste (ILW)
3. Low-level liquid waste (LLW)
4. Organic liquid waste (OLW)
5. Declad liquid waste or Hulls
6. Laboratory waste
7. Solid radioactive waste.

The origins of major reprocessing waste streams during PUREX process are indicated in Fig. 5. Gaseous wastes are generated during dissolution of SNF and during denitration of high acid stream. ^{85}Kr, ^{131}I, and ^{106}Ru are the main components of gaseous waste. Liquid wastes generated are of both organic as well as aqueous nature. The aqueous wastes, depending upon their radioactivity, are classified into three categories, i.e., LLW, ILW, and HLW. It can be seen from Fig. 5 that HLW is generated during the concentration of raffinate of extraction cycle, whereas ILW is the condensate obtained during evaporation of HLW. In earlier reprocessing facilities, these condensates were neutralized to facilitate storage in carbon steel tanks. In modern reprocessing plants, storage tanks are built

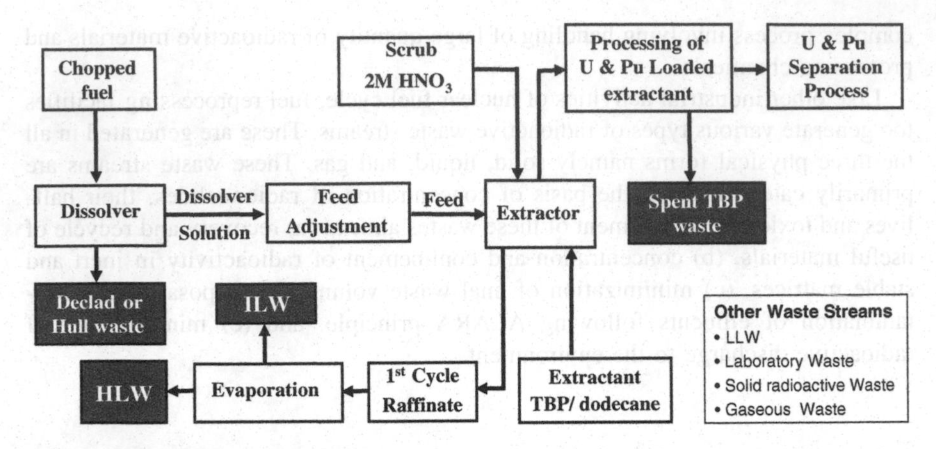

Fig. 5 Major sources of radioactive waste in reprocessing during PUREX process

in stainless construction avoiding necessity of neutralization of ILW. This change in material of construction of storage tanks has two major benefits: (a) overall reduction of generated waste volume and (b) less added salt, hence, low final volume for disposal. The ultimate objective of future flow sheet with respect to ILW management would be to altogether eliminate this stream by splitting the stream into small volume of HLW and major volume of LLW suitable for treatment and discharge. Of course, this requires challenging task of decontamination of LLW with respect to ^{99}Tc, ^{106}Ru, and ^{125}Sb associated with this waste stream.

LLW is generated from off-gas scrubbers of reprocessing plants, active floor drains, decontamination center, laboratories, drain from change room, sampling systems, showers, etc. Majority of radioactivity (more than 99 %) is concentrated in HLW. As a practice, HLW is concentrated by evaporation and stored in specially designed underground stainless steel tanks. These storage tanks require cooling and continuous surveillance. In Indian reprocessing plants, major improvements in reprocessing flow-sheet have been made based on research and development as well as operation experience during last two decades leading to reduction in generation rate of HLW volumes per ton of heavy metal processed and also in quality of the waste. The improvements made are:

1. Avoiding use of ferrous sulphamate, replacing it by uranous stabilized with hydrazine,
2. Avoiding addition of sodium nitrite in process, replacing it by use of NO_2, and
3. Effective use of evaporators for waste concentration.

As a result, HLW generation rate have been reduced from 2,000 to 400 l/ton of fuel reprocessed. These improvements have also led to better waste characteristics resulting in high volume reduction factors during vitrification and lesser volume of final waste product for ultimate disposal.

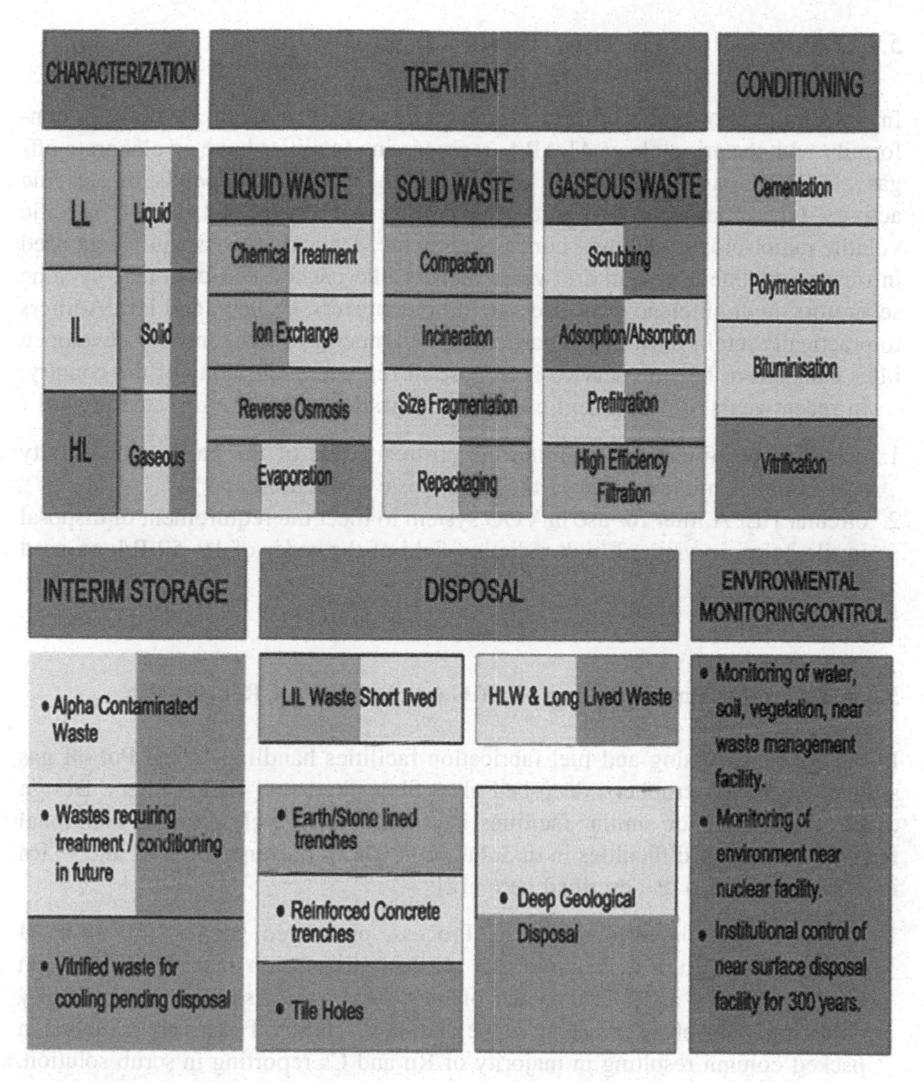

Fig. 6 A schematic for management of different types of radioactive wastes ([11], Copyright Elsevier)

Various processes for management of radioactive waste from reprocessing plants as adopted in India are shown in Fig. 6. The steps involved are: waste characterization, treatment, conditioning, storage, disposal, surveillance/monitoring, etc. Options available for treatment, conditioning, storage, and disposal of these wastes depending on their physical forms are also indicated. The aspects of their management are discussed in following paragraphs.

5.2 Treatment of Gaseous Waste

In order to control and minimize discharge of activity through air route in conformity with the principle of ALARA, reprocessing facilities have an elaborate off-gas cleaning system. Choice of off-gas treatment system depends on specific activity, type of radioactivity, particulate density and its size distribution, specific volatile radioisotopes and their concentration, etc. Typical process equipment used in off gas treatment system are wet scrubbers like packed bed columns, cyclone separators, high-efficiency low-pressure drop demisters, chillers, and HEPA filters to practically retain most of the particulate radionuclides. Indigenously developed filter banks have been in service in almost all reprocessing plants of the country.

In recent years, development efforts in off-gas treatment are directed toward

1. washable pre-filters for retaining the upstream side of the particulate activity and resultant minimization of the solid waste generation and
2. circular HEPA filter for use in VOG system to meet the requirement of disposal in tile holes, in view of high radiation field of the order of 10–50 R/h on used filters.

5.2.1 Treatment and Handling of off Gases in Th Fuel Reprocessing

In present reprocessing and fuel fabrication facilities handling U and Pu, off gas system consist of scrubbers, deep-bed glass fiber filters and HEPA filters. Design of off gas system for similar facilities in thorium fuel cycle requires additional features in view of difficulties in dissolution of ThO_2, presence of ^{232}U and ^{220}Rn in ^{232}U decay chain as described here [12]:

1. In the dissolution step of Thorex Process, prolonged heating (10–18 h) at boiling temperature is required due to slow dissolution rate. This results in escape of ^{137}Cs and ^{106}Ru in the off gas system of dissolver due to highly oxidizing atmosphere prevalent in the dissolver. These off gases are scrubbed in packed column resulting in majority of Ru and Cs reporting in scrub solution. Incorporation of washable glass fiber deep-bed filter results in minimization of solid waste in the form of used off-gas filters as experienced in Advanced Vitrification System (AVS) at BARC, Tarapur.
2. The noble gas ^{220}Rn can pass through HEPA filters and then decay to solid ^{208}Tl. Therefore, a three-stage filtration system is required to prevent the escape of decay products. These are: stage I, a HEPA filter to collect solid including the precursors to ^{220}Rn; stage II, charcoal bed delay lines to enable delay of about 10 min or other special equipment to hold the radon gas in the off-gas system that goes through the first HEPA filter until ^{220}Rn decays to solid material and stage III, a second HEPA filter to remove the solid decay products of ^{220}Rn. Typical off gas systems designed for Pu handling are not suitable for ^{233}U with a high ^{232}U content in view of the fact that they do not contain double

HEPA filters with the time delay between the HEPA filters required to avoid escape of ^{220}Rn. The components used in the three stages of described filtration system would require gamma shielding as well as remote maintenance capability.

5.3 Treatment of Low and Intermediate Level Liquid Waste

LLW generated in the reprocessing plants are normally collected in underground tanks of 200–250 M3 capacity with multiple compartments. Low-level liquid waste streams have low concentration of radioactivity in the range of 0.1–1 micro Curie/liter but volume is comparatively large. These waste streams require special treatment to reduce their activity concentration to a level suitable for monitored discharge according to the national regulations. General philosophy for their treatment is based on concentration of radioactivity and immobilization of the resultant concentrate and sludges. To obtain desired decontamination factor at very low concentration of radionuclide present in the waste is a challenging task. The processes that are employed for treatment of this type of waste are filtration, chemical treatment, ion-exchange, steam evaporation, solar evaporation, and membrane-based processes. Current efforts are toward adoption of processes like ultra-filtration and reverse osmosis in combination with the existing processes to further bring down the release of radioactivity to the environment following "near zero discharge" philosophy. Some of the process systems presently under operation are discussed here.

5.3.1 Ion-Exchange

A variety of sorbents and ion exchangers have been deployed in India for the treatment of diverse types of radioactive aqueous waste streams. Conventional synthetic organic ion exchange resins are used for clean-up of spent fuel storage pool and for polishing of effluents from chemical treatment of low-level waste. Amongst inorganic materials, synthetic zeolites, and the clay mineral vermiculite have found industrial application. While vermiculite is used for decontamination of low-level effluents, a synthetic zeolite is used for reduction of ^{137}Cs activity in spent fuel storage pool water.

A treatment process based on radionuclide separation by selective ion exchange is used for the effective management of alkaline intermediate level reprocessing waste streams, which is characterized by high salt content. In this process, an indigenously developed Resorcinol Formaldehyde Polycondensate Resin (RFPR) is used in repeated loading-elution-regeneration cycles for efficient removal of ^{137}Cs which is the major radionuclide present [13, 14]. Waste processing throughput of 400 l/h is achieved using 100 l columns. As a result of this treatment, the intermediate level waste is split into two streams, viz., a small volume of high-level waste and a large volume of low-level waste which is treated and discharged to the environment. Adoption of ion exchange treatment based on indigenous RFPR has brought very significant change in management of ILW leading to very substantial reduction in final waste disposal volumes compared to earlier designed bituminization and cementation processes. It has also facilitated recovery and use of ^{137}Cs from ILW. In brief, it is a real successful deployment of technology for reducing impact of radioactive waste on the environment. Three examples of this efficient application of these developments are given here:

1. Based on pilot scale experiments, an ion exchange treatment set-up of 30 l/h capacity was first installed at WIP, Tarapur for treatment of ILW generated during reprocessing of SNF from power reactors at PREFRE, Tarapur [15].
2. Subsequently, the system was successfully scaled up to ILW processing rates of 400 lph at Trombay in a unique transportable shielded ion-exchange facility. A chelating imino-diacetic acid resin (IDAR) is used in series mode for the removal of traces of ^{90}Sr [16].
3. Presently, an ion exchange plant is being operated at WIP, Trombay based on the feedback of design and operations of above two facilities. This plant has advanced automation system with the objective of improved flexibilities and minimizing radiation exposure.

Typical waste characteristics and a summary of the results obtained at Tarapur and Trombay are given in Table 5. At Tarapur site, ILW contains carbonates and large amount of salts. It also contains radiation degradation products of tri-butyl-phosphate (TBP) like Na-dibutyle phosphate, Na-monobutyle phosphate, etc. Major radionuclide present in this stream is ^{137}Cs. Other radio nuclides present are ^{90}Sr, ^{106}Ru, ^{125}Sb, ^{99}Tc, and trace concentration of actinides.

5.3.2 Chemical Treatment

Liquid wastes, with low levels (37–3.7 $\times$ 10^6 Bq/l) of activity containing ^{90}Sr and ^{137}Cs as the major radionuclides, are treated by co-precipitation using chemicals

Table 5 Waste characteristics and results of treatment of alkaline intermediate level waste ([11] Copyright Elsevier)

Description of the alkaline intermediate level waste	
Radioactive nucleides (^{137}Cs, ^{90}Sr, ^{106}Ru, etc.)	5–50 Ci/m^3
Total dissolved solids ($NaNO_3$, NaOH, Na_2CO_3, $NaAlO_2$, etc.)	100–300 g/l
pH	9–13
Performance of ion exchange treatment process at Trombay	
Decontamination factor	100–10,000
Total volume treated	1,400 m^3
Total activity removed	34 kCi

Fig. 7 Chemical treatment of low-level radioactive waste at Kalpakkam

like barium chloride, sodium sulfate, potassium ferrocyanide, copper sulfate, etc. Subsequent to precipitation, the resultant sludge from clarifloculator is further concentrated by decantation, filtration and centrifugation. The resulting solids containing bulk of the radioactivity originally present in the liquid waste are immobilized in cement matrix before disposal. Chemical treatment facility at CWMF, Kalpakkam site is shown in Fig. 7, where two clarifloculators are seen along with rectangular tanks for waste storage.

Besides treatment of LLW for decontamination with respect to ^{90}Sr, ^{137}Cs, and traces of alpha activity, the challenges of decontamination of LLW generated after treatment of ILW with respect to ^{99}Tc, ^{106}Ru, and ^{125}Sb has been met successfully both at ETP, Trombay and LWTP, Tarapur. To achieve this, laboratory studies were conducted with actual waste streams to develop process based on nickel sulphide treatment for ^{106}Ru removal, ion exchange for ^{99}Tc and chemical coprecipitation for ^{125}Sb. The results of these studies were subsequently deployed on the plant scale for treatment of large volume of effluents both at Trombay and Tarapur.

The alkaline effluent obtained after ILW treatment by RFPR ion-exchange treatment is LLW in nature with high salt content mainly in the form of sodium nitrate. It comprises of ^{90}Sr, ^{125}Sb, ^{106}Ru, and ^{99}Tc and trace amount of ^{137}Cs, which are not retained by the RFPR column. The concentrations of ^{106}Ru and ^{99}Tc are relatively higher than other radionuclides. Hence a reductive co-precipitation with FeS and $Fe(OH)_2$ was adopted at LWTP, Tarapur for precipitation of Tc and

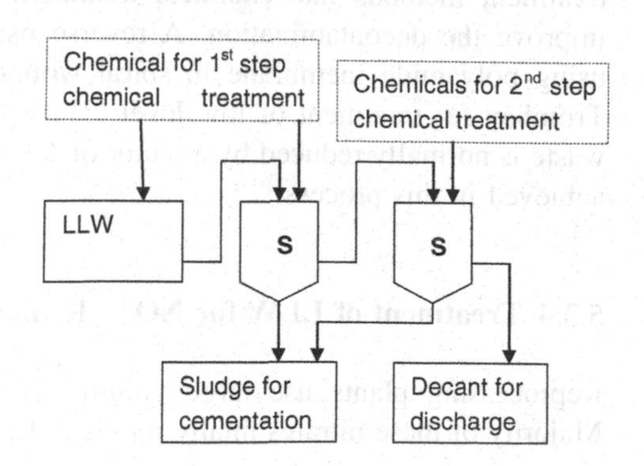

Fig. 8 Flow sheet for two step chemical treatment followed at Tarapur

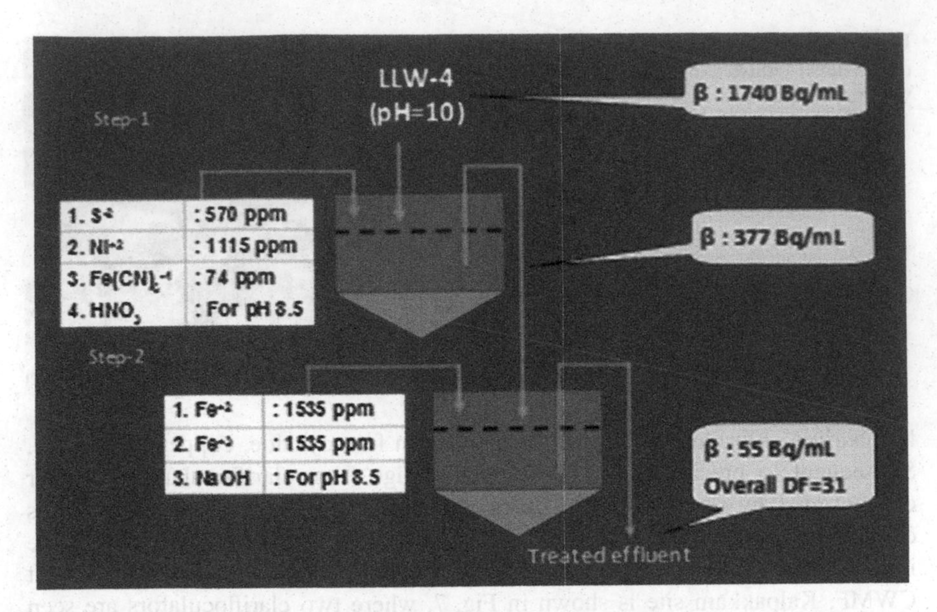

Fig. 9 Flow sheet for chemical treatment of ^{106}Ru at ETP, Trombay

Ru. The reduction of ruthanate and technate ions was carried out using Na_2SO_3 in acidic condition and then precipitated along with FeS and $Fe(OH)_2$. The precipitation of Ru, Tc, and Cs are effective in acidic/neutral pH and that of Sr and Sb in alkaline pH. Therefore, a two-step chemical treatment shown in Fig. 8 was carried out at two different pH conditions.

Large volumes of LLW containing major contribution from ^{106}Ru have been successfully treated at ETP, Trombay by chemical treatment by following the schematic shown in Fig. 9.

5.3.3 Membrane Processes

Membrane-based processes are generally employed in combination with other treatment methods like chemical treatment or ion-exchange process to further improve the decontamination. A reverse osmosis plant of capacity 100 m^3/day using polyamide membrane in spiral wound configuration is installed at WIP, Trombay for treatment of low level (37–3.7 × 10^6 Bq/L) waste. The volume of waste is normally reduced by a factor of 10 and decontamination factor of 8–10 is achieved in this process.

5.3.4 Treatment of LLW for NO_3^- Removal

Reprocessing plants use large volume of nitric acid for dissolution of SNF. Majority of these nitrates finally report in LLW as sodium nitrate. The generation

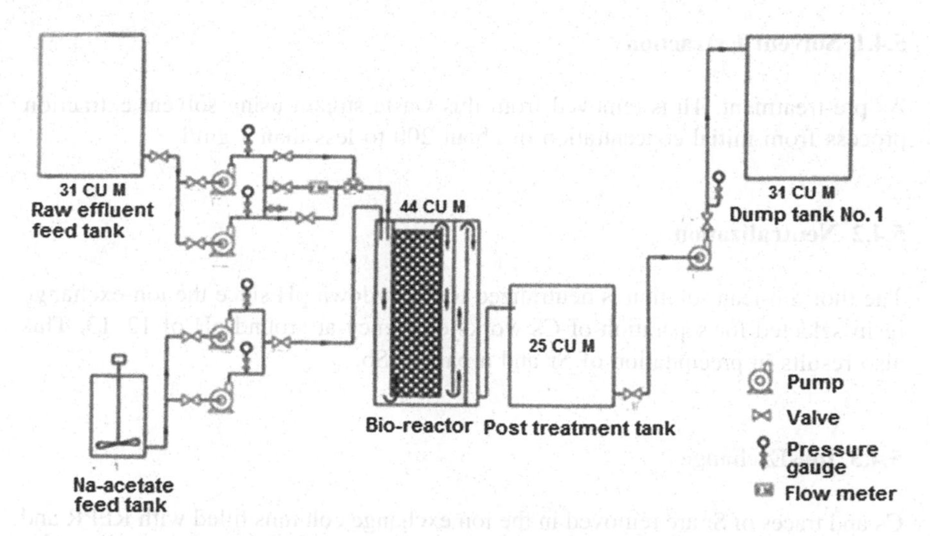

Fig. 10 Nitrate biodegradation pilot plant in operation at ETP, Trombay

of total amount of nitrates in LLW is expected to increase unless recycling and reuse of nitric acid is adopted in all reprocessing facilities. Average concentration of sodium nitrate in LLW is about 2,000 ppm whereas limits prescribed by the regulatory bodies on concentration of nitrate in effluents for discharge is typically 44.2 ppm and limit is met by dilution. However, in India, as a long-term solution, work on denitration of LLW has been taken up involving national institutes/universities as well as by in-house R&D. A promising approach is biodegradation as indicated by pilot studies performed at ETP, Trombay. In this pilot plant, operated at a rate of 2.5 m^3/day with actual LLW, NO_3^- concentration in effluents could be brought down from 2,000 to 30 ppm using locally available consortium of bacteria in a compact equipment assembly as shown in Fig. 10.

5.4 Treatment of Raffinate Waste from Thorium Fuel Reprocessing

The reprocessing of thorium rods irradiated in the research reactor at Trombay has resulted in raffinate waste which is primarily acidic in nature (4–5 M HNO_3) and contains ^{137}Cs and ^{90}Sr along with low concentration of ^{125}Sb as major fission products. It also contains Al (2 gm/L) used for complexing of fluoride ions during dissolution of spent fuel. A process based on ion-exchange separation was developed and adopted on the plant scale for treatment of this waste. The process involved the following steps:

5.4.1 Solvent Extraction

As pre-treatment, Th is removed from this waste stream using solvent extraction process from initial concentration of about 200 to less than 1 gm/l

5.4.2 Neutralization

The thorium-lean solution is neutralized to bring down pH since the ion-exchange resin selected for separation of Cs works efficiency at around pH of 12–13. This also results in precipitation of Sr and a part of Sb.

5.4.3 Ion-Exchange

Cs and traces of Sr are removed in the ion exchange columns filled with RFPR and IDAR, respectively. These ion-exchange media have been extensively used earlier at BARC Tarapur and Trombay for treatment of intermediate level alkaline liquid waste of reprocessing origin resulting in decontamination factor for gross beta activity of the range of 1,000.

5.4.4 Co-precipitation

The resultant effluents are acidified to pH of 8 to facilitate further treatment and discharge. The lowering of pH results in precipitation of Al and co-precipitation of a part of Sb. The precipitate is removed by filtration and immobilized in cement matrix. The effluents from precipitation step comprise of low concentration of ^{125}Sb, ^{137}Cs, and ^{90}Sr and are categorized as low-level waste amenable for further treatment by chemical treatment process which gives a decontamination factor of about 30. Finally, the treated effluents are diluted and monitored for pH and gross alpha and beta activity before discharge.

5.4.5 Vitrification of Cs

The loaded RFPR resin column is eluted with dilute nitric acid, and the elute is vitrified in boro-silicate glass matrix. Alternate process has also been studied where Cs could be further purified and immobilized in special glass formulation of accommodating Cs-rich waste and for deployment as an irradiation source for medical application, e.g., in blood irradiator.

5.5 Immobilization of Intermediate Level Waste

Intermediate level radioactive waste is conditioned depending on the compatibility of the matrix with waste, chemical and mechanical durability of solidified product, cost of processing, throughput, and disposal options. Cementation and polymerization methods are normally adopted in India for conditioning of this type of waste. In early days, bituminization was used as a process for conditioning of ILW at Waste Immobilization Plant, Tarapur. It was discontinued due to high temperature requirement during operation, generation of secondary organic waste during maintenance, etc. Spent ion exchange resins are being immobilized in polymer matrix and their conditioning by cementation is also being evaluated.

5.5.1 Cementation

Cement and cement composites are extensively used for immobilization of low-level radioactive concentrates, chemical sludges, etc. Cementation process offers advantage due to low cost and operational simplicity, higher throughput and product of acceptable quality. Special cement formulations have also been developed by blending cement with suitable additives to improve product characteristics. Cementation facility having in-drum mixing system using re-usable agitator is installed at CWMF, Kalpakkam. Cementation process has also been used for in situ immobilization of intermediate level waste and chemical sludges at WMF, Tarapur. In-situ cementation results in large waste processing rate with extremely low exposure to the radiation workers. At WIP, Trombay; cone mixer located in hot cells with easy remotised operational amenability is in use for conditioning of waste concentrate in cement matrix. Cementation system for chemical sludges and aqueous waste from alkali hydrolysis of spent TBP waste deploying disposable agitator is in use at ETP, Trombay.

5.5.2 Polymerization

Polyester styrene has been used for in situ solidification of low-heat generating liquid waste from reprocessing plant. In resin fixation plant, radioactive spent resins are hydro pneumatically transferred to resin storage tank. A batch of resin is transferred to a specially designed product drum kept on load cell. Excess water is removed by vacuum de-watering. Mixing assembly is then mounted on this product drum. Requisite amount of polyester styrene polymer is premixed with optimized concentration of accelerator and catalyst. This mix is then gradually poured into a product drum with constant stirring. Resin fixation facilities are installed at various NPP sites. Salient features of cement and polymer waste products are presented in Table 6.

Table 6 Salient properties of cement and polymer waste products ([11], Copyright Elsevier)

Properties	Cement waste product	Polymer waste product
Waste loading (%)	60–70	55–60
Compatibility	With alkaline waste	With alkaline waste
Density (g/ml)	1.8–2.0	1.2
Compressive strength (kg/cm^2)	100–150	270
Porosity (ml/g), (total pore volume)	0.18–0.27	Noncontinuous pores
Setting time (minutes)	100–300	NA
Leach rate (g/cm^2.day)	10^{-4}–10^{-5}	10^{-5}
Radiation stability	Up to 10^8 rads	Up to 10^8 rads
Homogeneity	Good	Very Good
Thermal stability	Very good up to 100 °C	Good up to 100 °C

5.6 Treatment of De-clad Waste

Aluminum is in use as a cladding material for the fuel in research reactors. During reprocessing, spent fuel is dissolved in high alkaline medium (NaOH), thereby generating declad waste. Al and Na are the major constituents of this waste in addition to the fission products like ^{137}Cs, ^{90}Sr, ^{106}Ru, ^{134}Cs, ^{99}Tc, etc. Composition of a typical declad waste is given in Table 7. In its proposed treatment schematic, Cs and Sr will be separated using appropriate ion exchange medium. Presence of aluminum adversely affects the performance of treatment processes, necessitating its removal as a pre-treatment step.

5.7 Management of High-Level Radioactive Liquid Waste

The countries involved in reprocessing of nuclear fuel have in general adopted a three step strategy for management of HLW generated during reprocessing [17]. These steps are:

1. Immobilization of HLW in an inert matrix like glass,
2. Interim storage of solidified HLW, and
3. Disposal of solidified waste in deep geological formation.

The challenges involved in vitrification, i.e., immobilization of HLW in vitreous matrix, are due to acidic nature of waste, corrosivity of molten glass and intense radiation environment due to high concentration of radionuclides present in the waste. A brief description of present technology in use, aspects of matrix development, etc., are presented in the following sections. A brief account of treatment of thorium reprocessing waste and matrix development for incorporation of thorium in barium borosilicate is also presented.

Table 7 Detailed characterization of typical declad waste

Property	Value
Alkalinity, M	
OH^-	1.74
CO_3^{2-}	0.3
HCO_3^-	BDL
Density (g/mL)	1.1734
Total solids (g/L)	370.0
Radioactivity (mCi/L)	
	7.9
	3.77×10^{-4}
Principle isotopes (mCi/L)	
^{137}Cs	7.35
^{106}Ru	0.18
^{134}Cs	4.4×10^{-2}
^{90}Sr	8×10^{-3}
^{99}Tc	6.25×10^{-3}
Elemental concentration (mg/L)	
U	<100
Na	39,830
Al	10,320
Ca	BDL
Cr	4.9
Si	14.4
Fe	5.0

5.7.1 Characterization of HLW

The composition of HLW depends on several factors like type of fuel and its cladding material, history in the reactor including burn up, the process used for reprocessing of irradiated fuel and off-reactor cooling period. As an example, the old legacy HLW at Trombay contains significant concentration of U, Na, and SO_4^{-2} in addition to fission products, corrosion products and small amount of other actinides. Presence of sulfate in HLW is because of usage of ferrous sulphamate as a reducing agent for conversion of Pu^{+4} to Pu^{+3} during partitioning stage of reprocessing and is one of the sensitive constituent with respect to its vitrification [18]. Now, uranous stabilized with hydrazine is being used as a reducing agent for reduction of Pu^{+4} to Pu^{+3} instead of ferrous sulphamate. Therefore, fresh waste generated at Trombay does not contain sulfate ions. A comparison of radiochemical properties of HLW generated at three reprocessing plants is presented in Table 8 and distribution of waste oxides of the elements present in HLW is presented in Table 9.

Table 8 Comparison of different types of HLW stored at three sites

Sr. no.	Property	Legacy HLW, Trombay	Fresh HLW, Trombay	HLW, Tarapur	HLW, Kalpakkam
1	Molarity, M	1.34	2.25	4.3	1.8
2	Density, g/ml	1.22	1.10	–	1.13
3	Total solids (g/l)	275	54	–	58
4	Specific activity				
4.1	Gross β (Ci/l)	8.77	60.36	60	40
4.2	Gross α (mCi/l)	4.22	7.3	175	–
5	Principal isotopes identified				
5.1	^{90}Sr (Ci/l)	2.94	6.8	19.5	13.0
5.2	^{134}Cs (mCi/l)	–	0.44	48.9	569
5.3	^{137}Cs (Ci/l)	2.72	8.5	18.5	12
5.4	^{106}Ru (mCi/l)	19	6.8	–	–
5.5	^{144}Ce (mCi/l)	69	11.5	–	–
5.6	^{125}Sb (mCi/l)	–	0.156	83	–

5.7.2 Matrix for Vitrification

The borosilicate glass matrix has been universally adopted for vitrification of HLW [19–21]. One of the limitations of glass matrix is its limited solubility with respect to certain specific waste constituents. Compositional diversity of waste often necessitates modification in the standard basic glass matrix. For instance, vitrification of sulfate containing HLW is a challenging task since solubility of sulfur in the form of sodium sulfate is very less (<1 % wt) in normally deployed borosilicate melts for vitrification of HLW. At higher sulfate concentration, a separate phase of alkali sulfate is formed. Its presence in glass is not desirable as this phase is enriched with ^{137}Cs and has high solubility in water. Suitable modifications to this basic matrix have been developed in order to take care of sulfate and sodium. Barium-based borosilicate glass matrix has been developed at Trombay laboratories to accommodate sulfate in the glass matrix. Barium-based borosilicate glass matrix is able to contain sulfate up to 2.5 wt% without impairing the properties of the conditioned product [18]. Similarly, some modifications in the glass compositions were carried out for vitrification of fresh HLW generated at Trombay and Tarapur. The details of matrix composition and characteristics of VWPs are presented in Tables 10 and 11, respectively.

Before adoption, conditioned products are evaluated in extensive laboratory studies for various properties like product melt temperature, waste loading, homogeneity, thermal stability, radiation stability, and chemical durability. In these studies, advanced analytical instruments are used namely scanning electron microscope, electron microprobe analyzer, X-ray Diffractometer, inductively coupled plasma spectrometer and thermal analysis system. Compositional details of glass matrix are maintained and tried out during inactive plant scale vitrification

Table 9 Distribution of waste oxides of the elements present in HLW

Sr. no.	Oxides	Legacy HLW, Trombay	Fresh HLW, Trombay	HLW, Tarapur	HLW, Kalpakkam
1	Fission product (g/L)				
1.1	SrO	0.01919	0.07096	0.037	0.03974
1.2	Cs_2O	0.131	0.3721	0.682	0.1682
1.3	SeO_2	–		0.072	
1.4	RuO_2	0.002422	0.1448	0.300	0.3555
1.5	PdO			0.123	
1.6	TeO_2			0.030	
1.7	BaO		0.47048	0.071	
1.8	MoO_3		0.3976	0.205	0.135
1.9	CeO_2	0.007034	0.1351	0.155	
1.10	La_2O_3		0.0879	0.065	
1.11	Pr_6O_{11}			0.029	
1.12	Nd_2O_3	0.006567		0.142	
1.13	Sm_2O_3			0.066	
1.14	Y_2O_3			0.043	
1.15	Fe_2O_3	8.4072	2.0657	0.688	0.5004
2.	Corrosion products				
2.1	Cr_2O_3	0.847	0.5336	0.174	0.1016
2.2	NiO	0.9416	0.3436	0.136	0.06363
2.3	MnO		0.7166	0.550	0.1098
3.	Actinides(UO_2)	25.1297	2.0196	8.646	25.245
4.	Added chemicals				
4.1	Na_2O	45.0104	6.3486	8.895	7.0836
4.2	SO_4	9.949			
4.3	Al_2O_3	12.09266	4.1652		
4.4	CaO	5.32	0.6157		0.4016
4.5	K_2O			0.270	
4.6	MgO		0.4063		

runs with simulated waste. Based on the desired product quality, various process parameters are standardized.

5.7.3 Storage and Transfer of HLW

HLW generated during reprocessing of SNF is concentrated by evaporation and is stored in stainless steel tanks in underground vaults. These storage tanks require cooling and continuous surveillance. As a policy, HLW vitrification facility is co-located near reprocessing plant so as to avoid transportation of liquid waste over large distances. Pumping of HLW from reprocessing plant to vitrification plant is an involved job in view of high concentration of radioactive elements present in the waste. Each waste transfer stainless steel pipes is enclosed in secondary

Table 10 Details of various glass compositions in use at Trombay and Tarapur

Oxide (wt%)	VWP Tarapur	VWP Trombay (sulfate-bearing waste)	VWP Trombay (sulfate-free waste)
SiO_2	36.48	30.5	34.0
B_2O_3	19.76	20.0	20.0
Na_2O	8.74	9.5	12.0
BaO	–	19.0	10.0
TiO_2	7.22	–	1.0
MnO	–	–	–
Fe_2O_3	3.8	–	–
Waste oxide	24.0	21.0	23.0
Oxides of fission products (Cs, Sr, Ru, Pd, Ce, Mo, etc.)	2.26	0.03	2.04
Oxides of corrosion products (Fe, Cr, Ni, Mn, etc.)	1.74	1.98	4.46
Uranium oxide	9.7	4.9	2.46
Added chemicals (Na, Al, Ca, Fe, SO_4, etc.)	10.3	14.09	14.04

Table 11 Properties of various vitrified waste products

Properties	VWP Tarapur	VWP Trombay (sulfate-bearing)	VWP Trombay (sulfate-free)
Fusion temperature (°C)	875	850	875
Pouring temperature(°C)	950	925	950
Thermal conductivity at 100 °C (W/m/°K)	0.925	0.92	–
Coefficient of thermal expansion ($°C^{-1}$)	103×10^{-7}	98×10^{-7}	99×10^{-7}
Softening temperature (°C)	540	496	515
Impact strength (RIAJ)[a]	–	0.85	–
Average leach rate on sodium loss basis at 100 °C ($gm/cm^2/day$)	8×10^{-5}	2.32×10^{-6}	1.02×10^{-5}
Density (g/cm^3)	2.8	3.2	2.8

[a] Relative Increase in Area per Joule

stainless steel pipe. The annulus between two pipes is continuously monitored during waste transfer operation. The group of these pipes is isolated from surroundings by a fully-welded stainless steel trough. The stainless steel trough is enclosed in a high integrity underground RCC trench which connects the two facilities.

5.7.4 Vitrification Process

A general flow sheet of the vitrification process is presented in Fig. 11. This involves various processing steps viz. waste transfer and receipt, concentration and acid recovery, vitrification, vitrified waste product (VWP) handling, secondary waste management, and off-gas treatment.

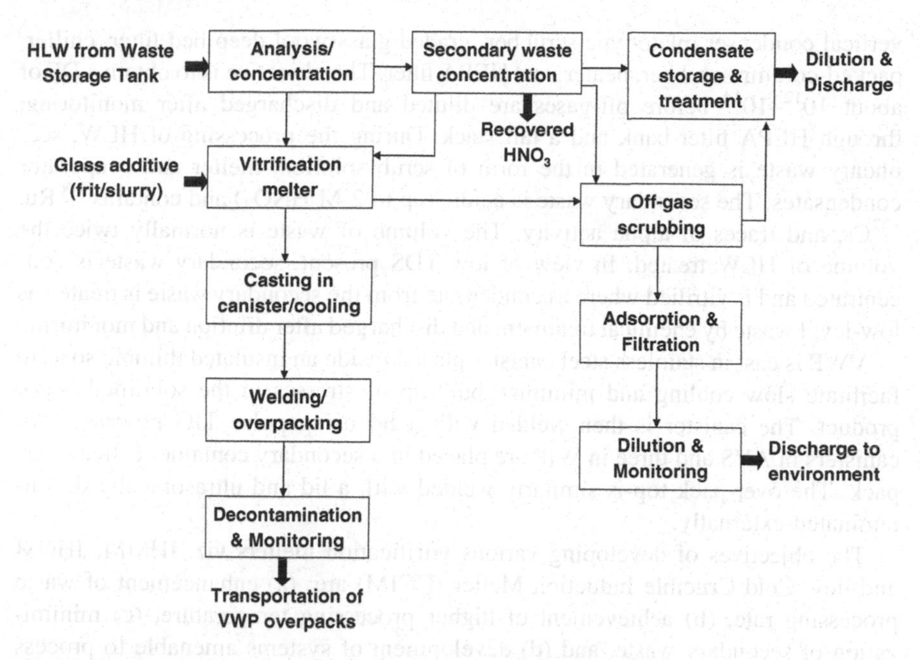

Fig. 11 Process for vitrification of high-level radioactive liquid waste

Depending upon the characteristics, HLW is concentrated with the twin objectives of reducing the evaporative load on the vitrification melter and recovery of nitric acid. The challenges encountered in the waste concentration are: (a) volatilization of Ru due to highly oxidizing environment, (b) choking of air-lift transfer system in view of the presence of salts in the waste, and (c) corrosion on account of boiling 5.5 M HNO_3 in the presence of various corrosion enhancing chemical species.

The processes involved in the vitrification are: evaporation, calcination, fusion, soaking, and casting. In Indian vitrification facilities, all these processes are carried out in the melter in a single step using either Induction Heated Metallic Melters (IHMM) or Joule Heated Ceramic Melter (JHCM). In the two stage process adopted in countries like France and United Kingdom, evaporation and calcination are done in the first step in a rotary calciner followed by fusion and soaking in the second step in an IHMM.

Glass forming additives and concentrated HLW are metered into the vitrification melter in the required ratio. Various safety interlocks are provided so as to avoid pressurization of the melter due to various phenomena like foaming and swelling of the contents of melter. In both IHMM and JHCM processes, temperature measurement of susceptor and process pot and electrode and glass pool is used to determine the stage of vitrification process.

The off-gases from the vitrification melter contain significant concentration of two volatile radionuclides viz. ^{106}Ru and ^{137}Cs besides other radionuclides and oxides of nitrogen. These are cleaned using a series of equipment viz. packed-tube vertical condenser, plate-type scrubber, graded glass-wool deep bed filter, chiller, packed-column scrubber, heater, and HEPA filter. The objective is to obtain a DF of about 10^{12}–10^{14} before off-gases are diluted and discharged after monitoring, through HEPA filter bank and a tall stack. During the processing of HLW, secondary waste is generated in the form of scrub solution, melter and evaporator condensates. The secondary waste is acidic (up to 2 M HNO_3) and contains ^{106}Ru, ^{137}Cs, and traces of alpha activity. The volume of waste is normally twice the volume of HLW treated. In view of low TDS present, secondary waste is concentrated and is vitrified where as condensate from the secondary waste is treated as low-level waste by chemical treatment and discharged after dilution and monitoring

VWP is cast in stainless steel canister placed inside an insulated thimble so as to facilitate slow cooling and minimize built-up of stresses in the solidified waste product. The canister is then welded with a lid using pulse TIG process. Two canisters in AVS and three in WIP are placed in a secondary container called over-pack. The over-pack top is similarly welded with a lid and ultrasonically decontaminated externally.

The objectives of developing various vitrification melters viz. IHMM, JHCM and now Cold Crucible Induction Melter (CCIM) are: (a) enhancement of waste processing rate, (b) achievement of higher processing temperature, (c) minimization of secondary waste, and (d) development of systems amenable to process various types of wastes likely to be generated in the future.

5.7.5 Vitrification Experience in India

The first Indian vitrification facility was set up at Waste Immobilization Plant, Tarapur, which deploys two IHMM operating at 1.2 kHz frequency [22]. Second such facility deploying three IHMMs operating at 3 kHz frequency has been set up at WIP, Trombay [23] where as third facility viz. AVS at Tarapur uses JHCM [24]. Fourth vitrification facility is under commissioning at WIP, Kalpakkam where JHCM-based technology with few improvements has been used. The salient aspects of experience of operation of these vitrification facilities are summarized here:

1. *WIP, Tarapur* was commissioned on February 25, 1984 with simulated waste containing low level of radioactivity and was subsequently operated in campaign mode with maximum radioactivity level of 5 Ci/l. Commissioning and operation of this facility provided very useful experience and inputs for future design and safety related to vitrification facilities. Some of the major gains from WIP, Tarapur are:

   a. Training of a large number of staff in radiological safety procedures and in maintenance of equipment and assemblies in radioactive environment.
   b. Verification of design of process equipment and system with respect to individual and overall decontamination factors.
   c. Establishment of methodology for systematic safety review of systems relevant to HLW processing.

2. *WIP, Trombay* was commissioned on September 5, 2002 with transfer of first batch of 25 m^3 of HLW of about 8 Ci/l. WIP, Trombay; which has many improved design features based on WIP, Tarapur experience; has been operated on a sustained basis in safe manner. The major achievements of WIP, Trombay are:

   a. Successful handling of sulfate-bearing legacy waste on industrial scale demonstrating suitability of barium borosilicate glass matrix.
   b. Establishing usefulness of multicell concept for enhancing plant system availability for operation.
   c. Minimization of generation of secondary waste by acid recovery and recycling.
   d. Adoption of state of art instrumentation and control system for smoother and safer process operation.

3. *AVS, Tarapur* is the third vitrification facility commissioned in August 2006 with HLW of about 30 Ci/l. The operation of AVS has given many positive experiences and the major ones are:

   a. Demonstration of continuous and sustained vitrification of about 250 m^3 of HLW immobilizing 6.0 million Ci ($\beta\gamma$) activity.
   b. Use of washable glass fiber deep bed filters leading to reduction in generation of used filters as solid waste and retention of about 400 Ci of radioactivity.
   c. Concentration of about 500 m^3 of HLW by use of thermo-siphon evaporation achieving a volume reduction factor of 1.8.

4. *Demonstration of JHCM dismantling and de-commissioning*: After successful demonstration of JHCM technology at AVS, Tarapur; task of technology development for decommissioning of used JHCM was taken up. This involved development of procedures, tools, and tackles for remote operations:

   a. Cutting, picking and further handling of ceramic and glass blocks.
   b. Handling of debris generated during de-commissioning, its packaging and transfer.
   c. Cutting of metallic components, cables, supporting structures, pipes, etc.
   d. Decontamination of various components, tools, etc.; devices for localized control of dust and fine particles.
   e. Monitoring and assaying of waste packages to facilitate segregation of waste based on radiation field on contact of package.

The feed-back obtained in analysis of JHCM performance at AVS, Tarapur and from hands-on experience in de-commissioning of used melter system are being utilized for improvements in design of future vitrification facilities with respect to hot-cell layout housing melters, remotely handled components like thermocouples, off-gas connectors, etc.

5.8 *Waste from Reprocessing of Thorium Fuel*

5.8.1 Thorium Solubility in Borosilicate Glass and its Effect on Structural Aspects and Thermophysical Properties

High-level radioactive liquid waste (HLW) from reprocessing of thorium based spent fuel is likely to contain un-recovered thorium, fission products, corrosion products, actinides and added chemicals. In general, solubility of actinides in silicate glass matrix is low and in the case of thorium, a maximum solubility of 2–5 wt% has been reported in borosilicate glasses quenched from near 1,250 °C [25]. In a separate investigation, Sonavane et al. [26]. have reported $\sim$6 wt% thoria solubility in sodium borosilicate glass. Structural aspects of Th^{4+} in silicate glasses have been investigated by Farge [25] using EXAFS technique and found that for 1–3 wt% of ThO_2 contents in the glass, Th^{4+} is octahedrally coordinated with a mean Th–O distance of 2.32 ± 0.02 Å and above this concentration, Th^{4+} is also present in eightfold coordination with mean Th–O distance of 2.40 ± 0.03 Å.

A detailed study of the structural aspects of these glasses using different techniques has been taken up by Mishra [27] to understand the bonding characteristics of Th^{4+} in the borosilicate network and their thermal stability. Solid-state NMR has been used to monitor the changes in the borosilicate network brought about by ThO_2 incorporation using ^{29}Si and ^{11}B as probe nuclei. Barium borosilicate base glass having the general composition (code S-0; Table 12) was prepared and using this base glass a number of glass compositions with varying concentrations of ThO_2 have been prepared and investigated by using X-ray diffraction (XRD), Differential thermal analysis (DTA), ^{29}Si and ^{11}B magic angle spinning nuclear magnetic resonance (MAS NMR), and infrared absorption techniques. Also, to investigate, the particle morphology of ThO_2 in the glass matrix, small angle X-ray scattering (SAXS) technique has been used. This exhaustive work covered the following aspects:

1. Microstructural characterization of glass by X-ray diffraction
2. Microstructural characterization of glass by EPMA
3. NMR studies
4. Infra red studies
5. SAXS studies
6. DTA Studies
7. Dilatometric studies
8. Density measurement
9. Micro-hardness measurement.

The salient results of studies carried out by Mishra [27] are as follows:

Table 12 Composition of BBS glasses containing different concentrations of ThO_2 (wt%)

Code	SiO_2	B_2O_3	Na_2O	BaO	ThO_2
S-0	38.60	25.32	12.03	24.05	0.00
S-1	35.98	23.59	11.21	22.41	6.81
S-2	34.34	22.53	10.70	19.15	13.28
S-3	32.47	21.30	10.13	20.23	15.87
S-4	31.41	20.61	9.79	19.58	18.61
S-5	29.43	19.31	9.18	18.33	23.75

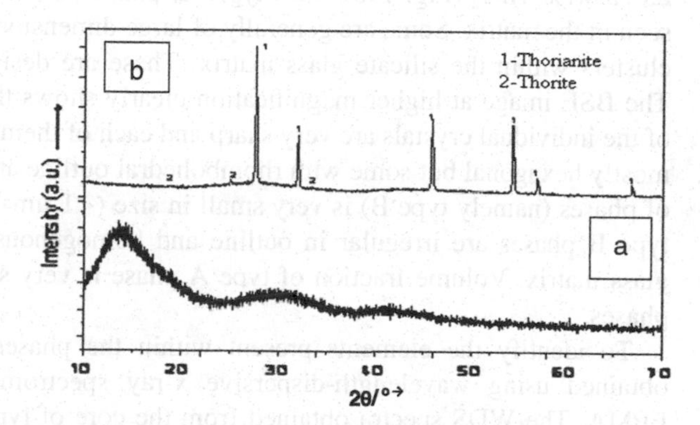

Fig. 12 X-ray diffraction patterns (CuK_α) obtained for BBS base glass (**a**) and BBS glass with 23.75 wt % thoria loading (**b**) ([28], Copyright Elsevier)

Microstructural Characterization of Glass by X-ray Diffraction

XRD pattern of the base glass (S-0) as shown in Fig. 12a [28] indicates the amorphous nature of the glass. Similar feature was also obtained for glass containing up to 15.87 wt% ThO_2. In case of glasses containing more than 15.87 wt% thoria showed the presence of crystalline phases (Fig. 12b) has been observed, identified as thorianite (ThO_2) [29] and thorite ($ThSiO_4$) [30]. X-ray diffraction studies carried out for barium borosilicate glasses containing different amounts of ThO_2 clearly revealed that not more than 15.86 wt% of ThO_2 can be incorporated in such glasses.

Microstructural Characterization of Glass by EPMA

Typical optical and back scattered electron (BSE) image of the Th containing barium borosilicate glass indicate homogenous nature for samples containing up to 15.87 wt% thoria, which is significantly higher than the reported values [25, 26]

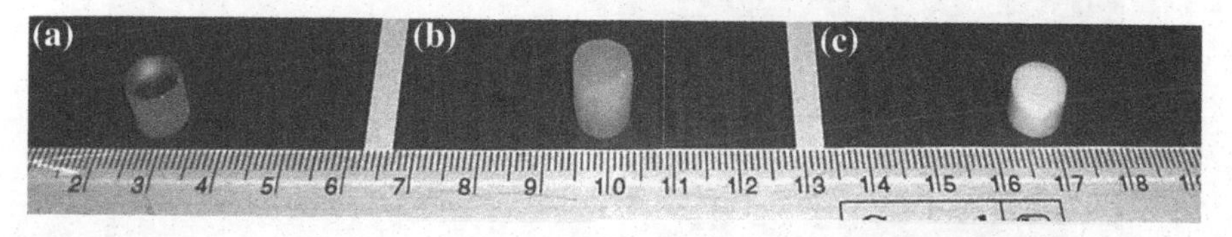

Fig. 13 Hand specimens of thoria glass samples loaded with 15.87, 18.61, and 23.75 wt% thoria, respectively showing, transparent (**a**), translucent (**b**) and opaque optical properties (**c**) ([28], Copyright Elsevier)

indicating the important role played by barium oxide in increasing Th solubility in borosilicate matrix. Beyond 15.87 wt% thoria loading, transparent BBS glass gradually becomes translucent (18.61 wt% thoria) and finally opaque (23.75 wt% thoria) (Fig. 13a–c).

Typical optical and BSE images of base glass (glass without thoria, S-0; Table 12) sample are shown in Fig. 14. Both the images show that the base glass is homogenous in composition. The change in optical property of the glass is due to phase separation as can be seen from BSE images of samples containing 23.75 wt% ThO_2 (Fig. 14b). Two types of phases, having varying dimensions are seen in the matrix. Some are generally of large dimensions 10–30 μm and occur as clusters within the silicate glass matrix. These are designated as Type A phase. The BSE image at higher magnification clearly shows these phases. The outlines of the individual crystals are very sharp and each of them has a well-defined shape, mostly hexagonal but some with rhombohedral outline are also seen. Another type of phases (namely type B) is very small in size (<1 μm) as seen in Fig. 15. These type B phases are irregular in outline and homogenously distributed within the glass matrix. Volume fraction of type A phase is very small compared to type B phases.

To identify the elements present within the phases, X-ray spectrums were obtained using wavelength-dispersive x-ray spectrometer (WDS) attached to EPMA. The WDS spectra obtained from the core of type A phases are shown in Fig. 16a, b. These spectra indicate that type A phase consists of Th, Si and based on XRD studies, this phase has been identified as $ThSiO_4$.

Based on X-ray diffraction pattern obtained from the thoria glass (23.75 wt%) phase B has been identified as thorianite ThO_2.

NMR Studies

^{29}Si MAS NMR patterns of glasses without and with varying amount of ThO_2 are shown in Fig. 17 [31]. An asymmetric peak placed at −89.5 ppm is observed for

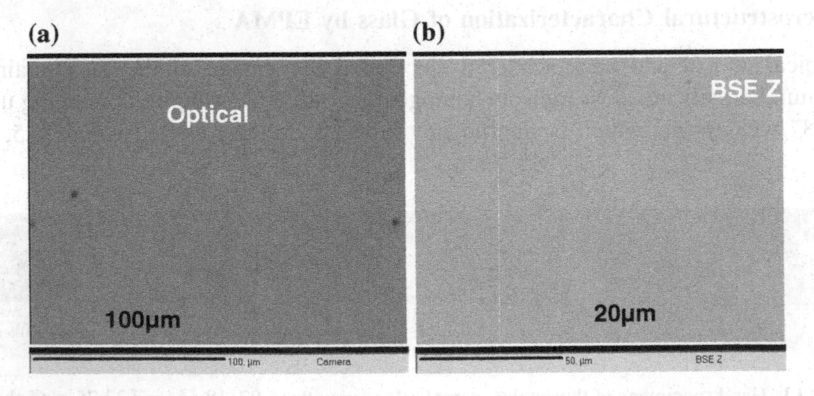

Fig. 14 **a** Optical and **b** BSE images showing homogenous microstructure of the BBS glass sample with 15.87 wt% thoria ([28], Copyright Elsevier)

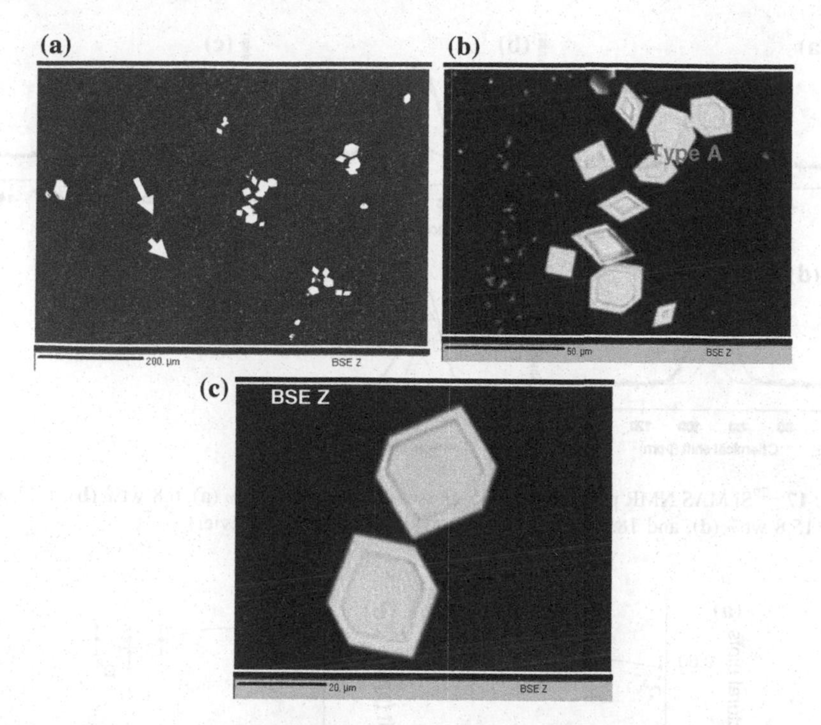

Fig. 15 **a, b** BSE images (different magnifications) showing separation of regular shaped, coarse grained clustered type A and irregular shaped finer grained and homogenously distributed type B phases within the matrix of BBS thoria glass containing 23.75 wt% thoria ([28], Copyright Elsevier). **c** BSE image of thorite (crystallized in BBS glass containing 23.75 wt% thoria) showing hexagonal outline ([28], Copyright Elsevier)

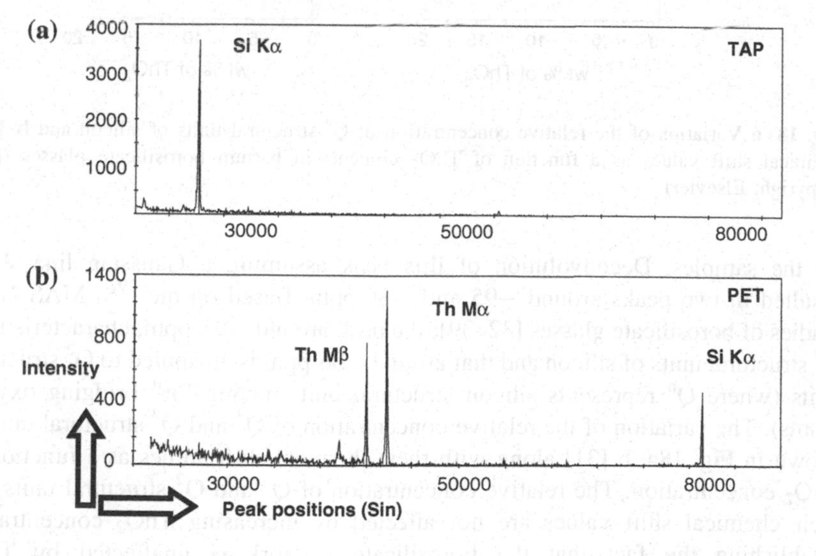

Fig. 16 **a, b** WDS spectrum obtained from the core of thorite crystals using synthetic TAP and PET as diffracting crystals ([28], Copyright Elsevier)

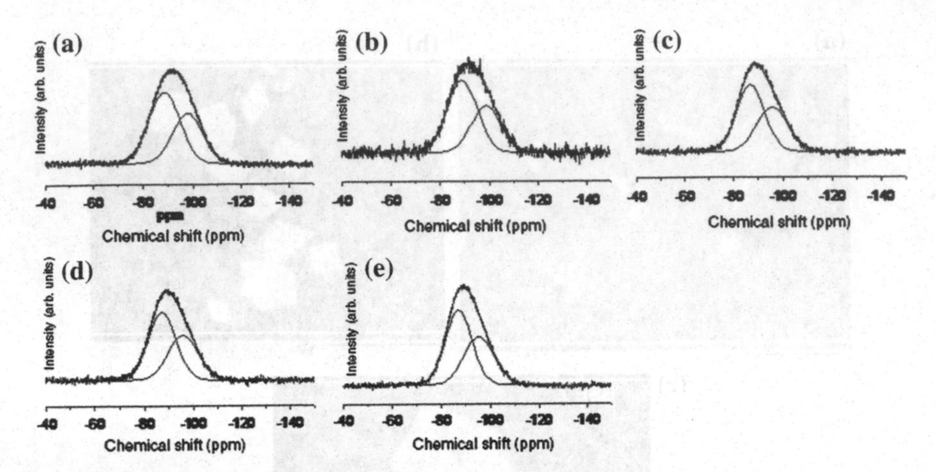

Fig. 17 ^{29}Si MAS NMR patterns for BBS glasses containing, 0 wt% (**a**), 6.8 wt% (**b**), 13.2 wt% (**c**), 15.8 wt% (**d**), and 18.6 wt% of ThO_2 (**e**) ([31], Copyright Elsevier)

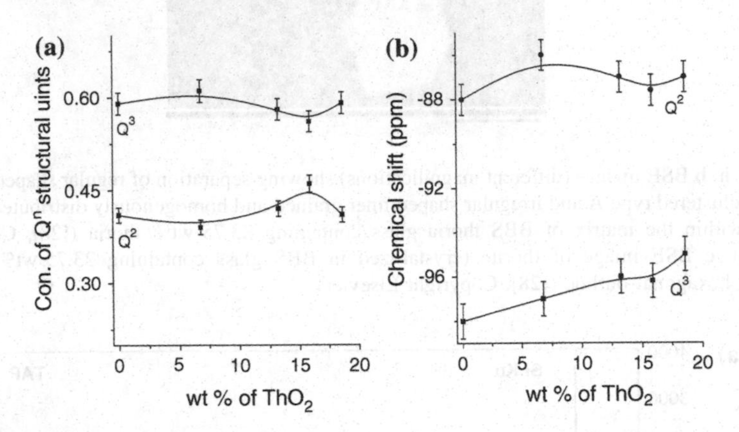

Fig. 18 **a** Variation of the relative concentration of Q^n structural units of silicon and **b** Their chemical shift values as a function of ThO_2 contents in barium borosilicate glasses ([31], Copyright Elsevier)

all the samples. Deconvolution of this peak assuming a Gaussian line shape resulted in two peaks around −95 and −86 ppm. Based on the ^{29}Si MAS NMR studies of borosilicate glasses [32–39], the peak around −95 ppm, characteristic of Q^3 structural units of silicon and that around −86 ppm is attributed to Q^2 structural units (where Q^n represents silicon structural units having "n" bridging oxygen atoms). The variation of the relative concentration of Q^3 and Q^2 structural units is shown in Fig. 18a, b [31] along with their chemical shift values as a function of ThO_2 concentration. The relative concentration of Q^3 and Q^2 structural units and their chemical shift values are not affected by increasing ThO_2 concentration establishing the fact that the borosilicate network is unaffected by ThO_2

incorporation in the glass. This is attributed to the reason that Th^{4+} along with Na^+ ions can get incorporated at the network modifying sites, which are formed by the significant amount of nonbridging oxygen atoms present in these glasses. Thus, there is no detectable interaction between the Th^{4+} ions and the glass network.

In order to find out the role of BaO in ThO_2 containing glasses, barium free borosilicate glasses incorporated with 0, 11, and 18 wt% ThO_2 (composition given in Table 13), were prepared and characterized (keeping all the remaining mole ratios same). XRD patterns for 11 and 18 wt% ThO_2 containing samples revealed that part of the ThO_2 got phase separated during quenching. Figure 19 [31] shows the ^{29}Si MAS NMR patterns for borosilicate glasses without any BaO and containing different amounts of ThO_2. The base glass without any ThO_2, is characterized by an asymmetric peak centered around −100 ppm and deconvolution of which resulted in two peaks around −101 and −90 ppm characteristic of Q^4 and Q^3 structural units of silicon, respectively. With addition of ThO_2 in the glass, Q^4 structural units got partially converted to Q^3 structural units, indicating that the ThO_2 acts as a network modifier in these glasses. Further, it is noted that, the chemical shift values of Q^4 and Q^3 structural units shifted to more negative values with increase in ThO_2 incorporation in the glass. The possible reason for this could be the conversion of Si–O–Si to Si–O–B linkages [32].

Infra Red Studies

The IR absorption spectra for the three representative barium borosilicate glass samples having 0, 13.2, and 18.6 wt% of ThO_2 are shown in Fig. 20 [31]. All the patterns are characterized by peaks at around 987, 707 and 474 cm^{-1}, which are characteristic of the asymmetric, symmetric and bending vibrations of Si–O–Si/Si–

Table 13 Composition of sodium borosilicate glasses containing different concentrations of ThO_2 (wt%)

Code	SiO_2	B_2O_3	Na_2O	ThO_2
S-6	50.83	33.33	15.84	0.00
S-7	45.24	29.66	14.10	11.00
S-8	41.68	27.33	12.99	18.00

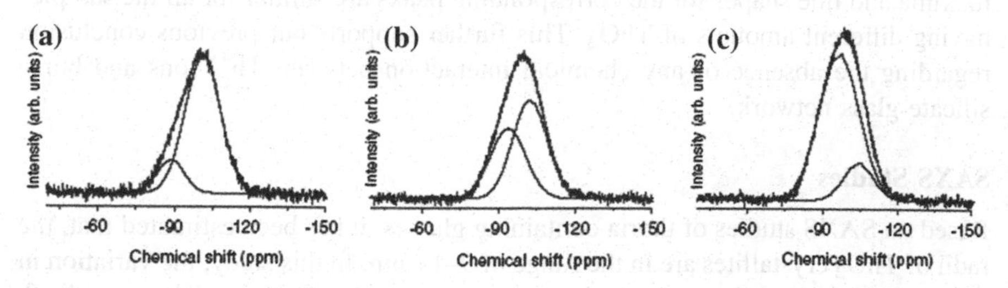

Fig. 19 ^{29}Si MAS NMR patterns for borosilicate glasses without any BaO and containing 0 wt% (**a**), 11 wt% (**b**), and 18 wt% of ThO_2 (**c**) ([31], Copyright Elsevier)

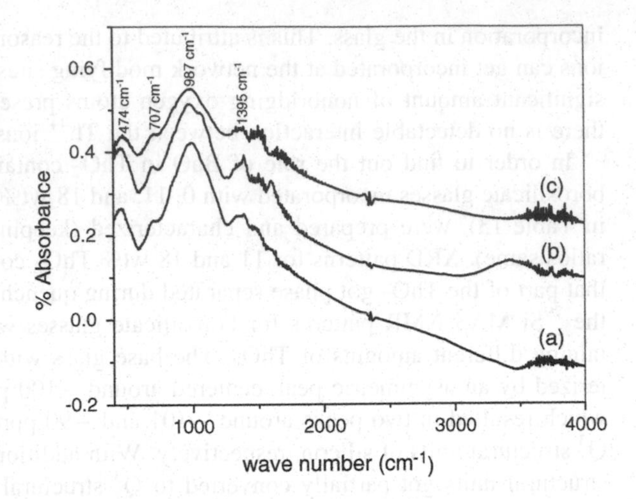

Fig. 20 IR patterns for barium borosilicate glasses containing, 0 wt% (**a**), 13.2 wt% (**b**), and 18.6 wt% of ThO_2 (**c**) ([31], Copyright Elsevier)

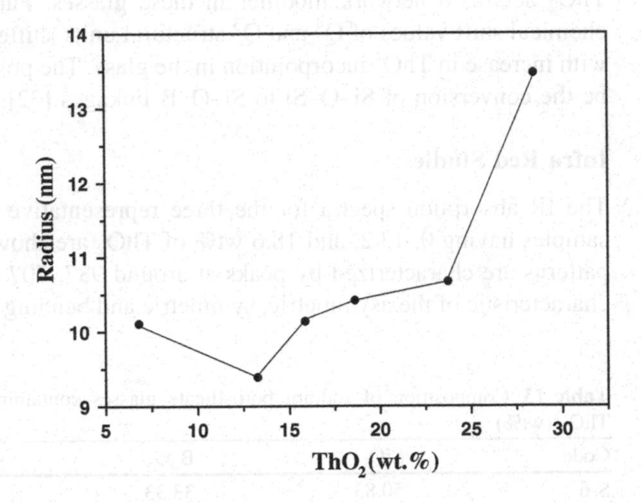

Fig. 21 Variation of average radius of ThO_2 crystallite with content (wt%) ([40], Copyright Elsevier)

O-B structural units [32]. The broad peak around 1,395 cm^{-1} has been attributed to the B–O stretching vibration of the borate structural units [32]. The peak maxima and line shapes for the corresponding peaks are similar for all the samples having different amounts of ThO_2. This further supports our previous conclusion regarding the absence of any chemical interaction between Th^{4+} ions and borosilicate glass network.

SAXS Studies

Based on SAXS studies of thoria containing glasses, it has been estimated that, the radii of ThO_2 crystallites are in the range of 9–14 nm. In this study, the variation in the size of the crystallites with increase in the amount of ThO_2 was also examined. As shown in Fig. 21 [40], the sample S-2 (containing 13.28 wt% of the dopant)

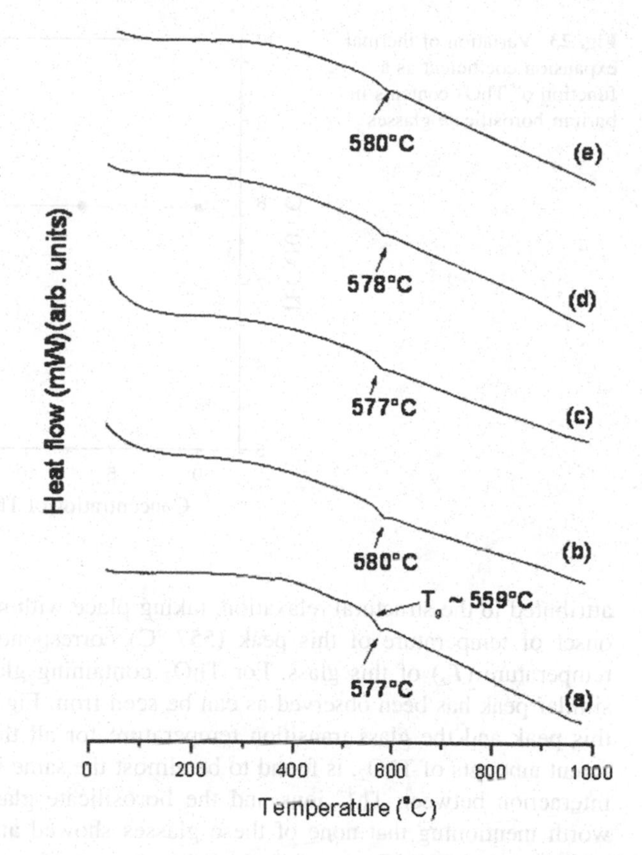

Fig. 22 DTA patterns of barium borosilicate glasses containing, 0 wt% (**a**), 6.8 wt% (**b**), 13.2 wt% (**c**), 18.6 wt% (**d**), and 23.7 wt% of ThO_2 (**e**) ([31], Copyright Elsevier)

has smallest crystallites whereas a steep raise in the size for the samples containing beyond 23 wt% of ThO_2 is clearly evident.

It is noted that, below 16 wt%, the smaller crystallites are accommodated into the void space of the glass-matrix. With incorporation of more ThO_2, at least some of the crystallites probably aggregate to form larger ones resulting in the increase in the average size as shown in Fig. 21 [40]. A maximum in the width of the distribution for sample S-5 also indicates the onset of some rearrangement of crystallites at this concentration (23.75 wt%) of ThO_2. Thus, it may be inferred that these agglomerated larger size crystallites are not able to accommodate in the voids crystallites in S-2 is not known of the matrix and are separated from the matrix. It will be interesting to investigate the significance of the value 16 wt% as the threshold value of the dopant for incorporating into the glass matrix.

DTA Studies

DTA patterns for representative samples having 0, 6.8, 13.2, 18.6, and 23.7 wt% ThO_2, are shown in Fig. 22 [31]. For the glass sample without any ThO_2, DTA pattern (Fig. 22a) is characterized by a broad endothermic peak centered around 577 °C with an onset around 559 °C. The broad endothermic peak has been

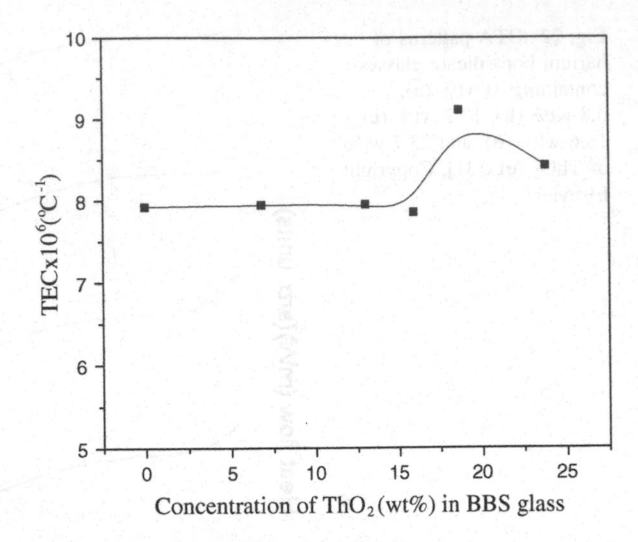

Fig. 23 Variation of thermal expansion coefficient as a function of ThO_2 contents in barium borosilicate glasses

attributed to the structural relaxation, taking place within the glass matrix and the onset of temperature of this peak (557 °C) corresponds to the glass transition temperature (T_g) of this glass. For ThO_2 containing glass samples essentially a similar peak has been observed as can be seen from Fig. 22. The total area under this peak and the glass transition temperature for all these samples, having different amounts of ThO_2, is found to be almost the same indicating that there is no interaction between Th^{4+} ions and the borosilicate glass network. Further it is worth mentioning that none of these glasses showed any tendency for crystallization up to 1,000 °C, as revealed by the absence of any exothermic peaks in the DTA patterns indicating good thermal stability of formed ThO_2 containing glasses.

Dilatometric Studies

Dependence of thermal expansion coefficient (TEC) on thoria content is plotted in Fig. 23. It has been observed that the TEC of these glasses remains same within the experimental errors. However, there is a substantial increase in TEC for the glass sample S-4, where ThO_2 content is 18.6 wt %. Earlier, it was shown by detailed EPMA studies that this composition shows the presence of ThO_2 crystals, unlike other compositions with lower amount of ThO_2. In general, the crystalline material shows higher TEC compared to glassy material. It is known that thermophysical properties of glass ceramics are controlled by the relative concentration and chemical composition of different phases in the system [41]. It was further observed that the next composition S-5, which contains 23.75 wt% of ThO_2 exhibits a slightly lower TEC. The EPMA investigation had shown the presence of $ThSiO_4$ in addition to ThO_2 in this glass. It is well known that $ThSiO_4$ has much lower TEC than ThO_2 [42], which explain the slightly lower TEC of this composition.

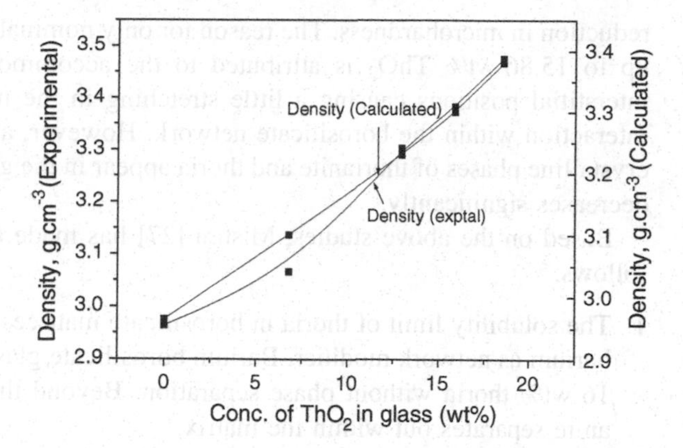

Fig. 24 Variation of density (experimental) and density (calculated) as a function of ThO_2 contents in barium borosilicate glasses

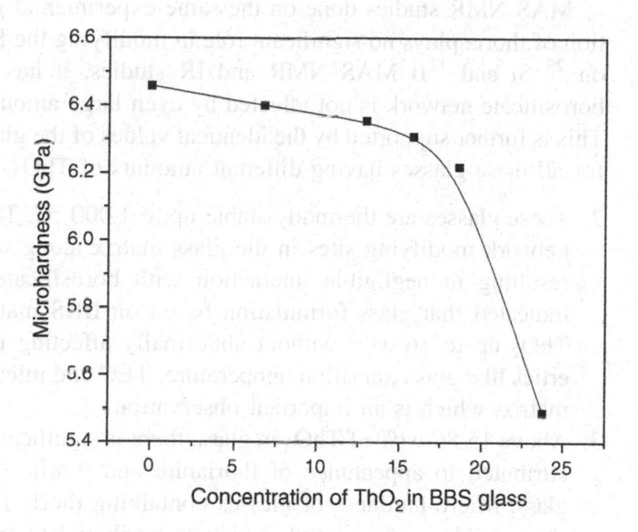

Fig. 25 Variation of microhardness as a function of ThO_2 contents in barium borosilicate glasses

Density Measurement

The compositional dependence of density of different glass samples is shown in Fig. 24. Density of glass samples increases from 2.96 to 3.49 g cm^{-3}with increase in thoria content. Extent of increase in density, calculated theoretically matches well with that of the experimentally observed values (Fig. 24), indicating that ThO_2 does not participate either in the formation or cleavage of the glass network. The increase in density values is understandable as Th^{4+} is a heavier ion compared to all the remaining ions constituting the glass.

Micro-Hardness Measurement

The variation of micro-hardness as a function of ThO_2 content in the glass is shown in Fig. 25. There is a nominal decrease in the value of microhardness up to 18.6 wt% of ThO_2. However, above this concentration of ThO_2, there is a marked reduction in microhardness. The reason for only nominal change in micro-hardness up to 15.86 wt% ThO_2 is attributed to the accommodation of thorium in the interstitial positions causing a little stretching in the matrix and having a weak interaction within the borosilicate network. However, above 15.86 wt% ThO_2, as crystalline phases of thorianite and thoria appear in the glass, micro-hardness value decreases significantly.

Based on the above studies, Mishra [27] has made significant observations as follows:

1. The solubility limit of thoria in borosilicate matrices increases with addition of barium as network modifier. Barium borosilicate glass matrix can contain up to 16 wt% thoria without phase separation. Beyond this limit, thorite and thorianite separates out within the matrix.

MAS NMR studies done on the same experimental glasses indicate that addition of thoria plays no significant role in modifying the BBS glass network. Based on ^{29}Si and ^{11}B MAS NMR and IR studies, it has been concluded that the borosilicate network is not affected by even large amount of ThO_2 incorporation. This is further supported by the identical values of the glass transition temperatures for all these glasses having different amounts of ThO_2.

2. These glasses are thermally stable up to 1,000 °C. Th^{4+} ions are present at the network modifying sites in the glass matrix along with the Na^+ ions, thereby resulting in negligible interaction with borosilicate network. These studies indicated that glass formulation based on BBS matrix is able to incorporate ThO_2 up to 16 wt% without abnormally affecting the thermo-physical properties like glass transition temperature, TEC and micro-hardness of the vitreous matrix, which is an important observation.
3. Above 15.86 wt% of ThO_2 in glass, there is significant change in TEC, which is attributed to appearance of thorianite and thorite crystalline phases in BBS glass. Micro-hardness of glasses containing thoria more than 15.86 wt% also shows a decreasing trend, which is attributed to partial phase separation of ThO_2 and $ThSiO_4$ phases in the glass.
4. Thus based on these studies, it can be inferred that as high as 16 wt% ThO_2 can be incorporated in BBS glass matrix, without any adverse effect on its thermochemical or thermophysical properties. This is a significant result with respect to management of waste from reprocessing of spent fuel from Advance Heavy Water Reactors.

5.8.2 Solubility of U/Th Oxides in Barium Borosilicate Glass Matrix

Thorium and uranium will be present in the waste along with fission products, corrosion products and added chemicals during reprocessing fuel from reactors based on Th–U^{233} MOX fuels [43].

Table 14 Composition of BBS glasses containing 15.86 wt% of ThO_2 and different concentrations of UO_3(wt%)

Code	SiO_2	B_2O_3	Na_2O	BaO	ThO_2	UO_3
ThU-0	32.48	21.30	10.12	20.24	15.86	0.00
ThU-1	30.55	20.04	9.52	19.03	15.86	5.00
ThU-2	29.58	19.41	9.22	18.43	15.86	7.50
ThU-3	28.62	18.77	8.92	17.83	15.86	10.00
ThU-4	27.65	18.14	8.62	17.23	15.86	12.50

Depending upon the oxidation state of the uranium ions, it can have different coordination poly-hedra around them, which is expected to have significant effect on the structure of the glass network as well as on the extent of ThO_2 incorporation in the glass. Hence, it is of interest to study the structural aspects of borosilicate glasses when both uranium and thorium oxides are incorporated in it. Studies have been carried out by Mishra [31] on the aspect of solubility of uranium oxide and thorium oxide together and influence of simultaneous incorporation of thorium and uranium oxide, on the structural network of barium borosilicate glass. In this study, simultaneous incorporation of thorium and uranium in glass, BBS glasses were prepared with a fixed concentration of ThO_2 (15.86 wt%) and incorporated with different amounts of uranium oxide (UO_3) and studied their structural aspects using ^{29}Si and ^{11}B MAS NMR and XRD techniques. ThO_2 amount was kept at 15.86 wt% (Table 14) as it is close to the optimum ThO_2 concentration that can be incorporated into the barium borosilicate glass without any phase separation.

For all the glass samples, the ratio of Na_2O to BaO and SiO_2 to B_2O_3 were maintained constant, and are 0.5 and 1.56, respectively. Further, the constituents of the glass formulations were proportionately changed in such a way that SiO_2/Na_2O, SiO_2/B_2O_3, SiO_2/BaO, B_2O_3/Na_2O, and B_2O_3/BaO are same for all the samples with different amounts of ThO_2 and UO_3.

All these samples were characterized for homogeneity by XRD, determination of glass transition temperature by DTA and structural characterization by ^{29}Si and ^{11}B MAS NMR as described earlier.

Barium Borosilicate Glasses Containing Thorium and Uranium Oxides

The representative XRD patterns for barium borosilicate glasses containing different amounts of thorium and uranium oxides are shown in Fig. 26 [44]. Barium borosilicate glass without any thorium and uranium oxides showed a broad peak around 2θ value of 13° characteristic of amorphous borosilicate network. With incorporation of thorium and uranium oxides, diffraction patterns became too noisy and broad indicating the distortion of the borosilicate network brought about by the uranium and thorium ions. Above 7.5 wt% uranium oxide incorporation, phase separation in the glass took place as revealed by the sharp peaks observed around 2θ values 27.53°, 31.89°, 45.77°, and 54.25°, etc. The sharp peaks have been attributed to the presence of crystalline ThO_2 phase [29] in the glassy matrix.

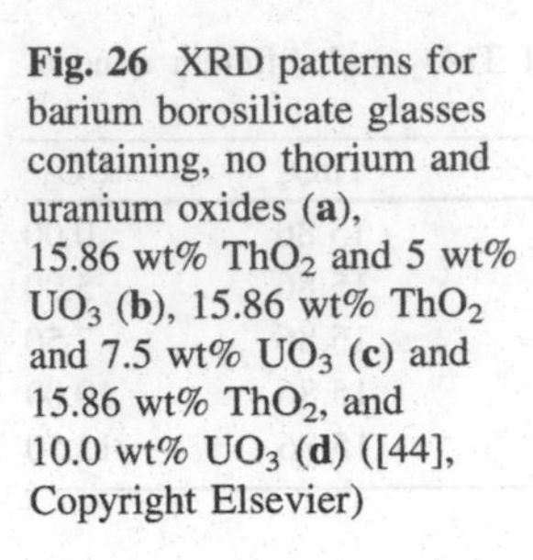

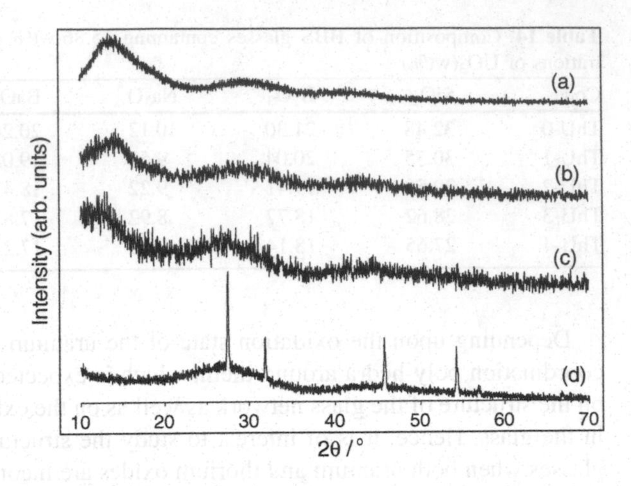

Fig. 26 XRD patterns for barium borosilicate glasses containing, no thorium and uranium oxides (**a**), 15.86 wt% ThO_2 and 5 wt% UO_3 (**b**), 15.86 wt% ThO_2 and 7.5 wt% UO_3 (**c**) and 15.86 wt% ThO_2, and 10.0 wt% UO_3 (**d**) ([44], Copyright Elsevier)

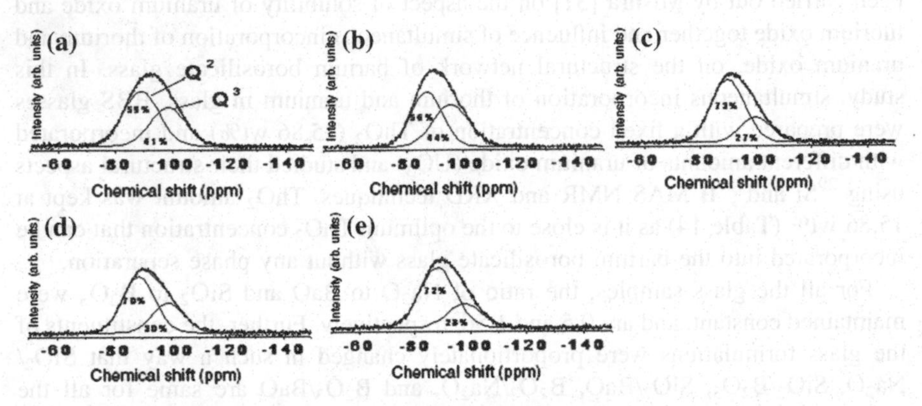

Fig. 27 ^{29}Si MAS NMR Patterns for barium borosilicate glasses containing, no thorium and uranium oxides (**a**), 15.86 wt% ThO_2 (**b**), 15.86 wt% ThO_2 and 5 wt% UO_3 (**c**), 15.86 wt% ThO_2 and 7.5 wt% UO_3 (**d**), 15.86 wt% ThO_2, and 10.0 wt% UO_3 (**e**) (relative concentrations of Q^3 and Q^2 structural units are indicated below each peak) ([44], Copyright Elsevier)

^{29}Si MAS NMR patterns for barium borosilicate glasses containing 15.86 wt% of thorium oxide and incorporated with varying amounts of uranium oxide are presented in Fig. 27 [44]. For the purpose of comparison, the glass sample without any thorium and uranium oxide incorporation is also shown. Barium borosilicate glasses without any thorium and uranium oxide showed a broad asymmetric peak around −92 ppm. De-convolution based on a Gaussian fit resulted in two peaks around −98 and −89 ppm characteristic of the Q^3 and Q^2 structural units of silicon, respectively, (where Q^n represents silicon structural units having "*n*" bridging oxygen atoms) [32]. The relative concentration of both the structural units is indicated below the corresponding peaks.

With 15.86 wt% incorporation of ThO_2 in the glass, the relative concentration of Q^3 and Q^2 structural units remained unaffected (within experimental errors) as can be seen from Fig. 27b, suggesting that thoria incorporation does not result in the breakage of Si–O–Si linkages. Th^{4+} ions must be occupying some of the sites created by nonbridging oxygen atoms (network modifying positions) in the glass, without affecting the borosilicate network. With incorporation of 5 wt% of uranium oxide in the thorium oxide incorporated glass, the ^{29}Si MAS NMR line shape changes and based on the de-convolution of the NMR pattern it can be seen that, the relative concentration of Q^2 structural units increases at the expense of Q^3 structural units. With further increase in uranium oxide concentration, the relative concentration of both Q^2 and Q^3 structural units and their chemical shift values remained unaffected as can be seen from the identical line shapes in Fig. 27c–e. Identical chemical shift values for both Q^2 and Q^3 structural units of silicon with increase in uranium oxide concentration suggest that there is no direct interaction between silicon and uranium structural units and uranium ions probably occupy the sites created by the significant number of nonbridging oxygen atoms present in the glass (network modifying sites). There is no change in the ^{29}Si MAS NMR line shape even after partial phase separation of ThO_2 from the glass matrix.

The following has been concluded based on above studies by Mishra [27]:

1. BBS glasses containing 15.86 wt% of ThO_2 is able to accommodate only up to 7.5 wt% of UO_3, beyond which a crystalline phase of ThO_2 is separated out.
2. Based on ^{29}Si MAS NMR studies on ThO_2 containing barium borosilicate glasses having varying amounts of uranium oxides, it has been concluded that uranium oxide incorporation is associated with the conversion of Q^3 to Q^2structural units of silicon and the increased number of nonbridging oxygen atoms thus produced favors the uranium ions to occupy the sites which are created by the nonbridging oxygen atoms (network modifying sites) in the glass.

5.8.3 Study to Reduce Concentration of F^- and Al^{3+} in HLW

Standard thorex solvent consisting of 13 M nitric acid, 0.03 M sodium fluoride and 0.1 M aluminum nitrate has been used on the engineering scale for dissolution of research reactor irradiated aluminum clad thoria after chemical decladding by using aqueous sodium hydroxide [45]. The corrosion of dissolver vessel made of SS304L is enhanced due to presence of F^- in nitric acid dissolution process. $Al(NO_3)_3$ is added to control the corrosion. Al^{3+} forms complex with free F^- ions present in the solution, thereby, reducing corrosion of stainless steel.

After dissolution, the feed is subjected to thorex solvent extraction process for U, Pu and Th recovery and the resultant HLW is planned to be subjected to concentration by evaporation and vitrification at 950–1,000 °C. It is observed that F^- in the form of HF vapor can escape from HLW during evaporation and vitrification in the range of 2 and 9, respectively [27, 45]. HF in the off-gas could enhance the corrosion of off-gas equipment especially condensers. It is, therefore, desirable to reduce the concentration of F^- that is used during dissolution reaction.

Also, presence of Al^{3+} in such high concentration (0.1 M) has adverse affect on vitrification process namely (a) increase in final volume of VWP due to reduced waste loading, (b) increase in glass melt viscosity, and (c) foaming problem during vitrification process. It is, therefore, beneficial to avoid or minimize the use of Al^{3+} ion during thoria dissolution. A study was undertaken by Srinivas et al. [45] with the twin objectives of (1) avoiding aluminum nitrate addition and (2) lowering the F^- concentration during dissolution study. The salient findings of this work are:

1. Zirconium and thorium ions form stronger complexes with F^- compared to aluminum ion making addition of aluminum nitrate for corrosion mitigation unnecessary for zircaloy-clad thoria dissolution reaction. SS304L corrosion rate studies at 0.03 M NaF concentration clearly showed that if zircaloy is present, the corrosion rate of SS304L in presence and absence of $Al(NO_3)_3$ is quite comparable and is about 10 mpy.
2. F^- concentration can be reduced from presently adopted 0.03 M to lower concentration by following the procedure of adding NaF into the dissolver in two proportions with time gap of about 3–4 h. Reaction times for dissolution reaction were observed to be 48–72, 72–96, and 96–120 h for 0.03, 0.01, and 0.005 M, respectively during the dissolution reaction.
3. Enhancing volume of 13 M nitric acid during the dissolution reaction can reduce the reaction time from 96 to about 67 h at 0.005 M concentration of NaF.

5.8.4 Solubility of F^- in Borosilicate Glass

HLW generated from reprocessing of SNF also contains a good amount of inactive chemicals which are added either during dissolution of spent fuel or partitioning stage of reprocessing flow sheet. As an example hydrofluoric acid, aluminum nitrate and nitric acid are being used for dissolution of thoria based spent fuel [45]. Hence, the high-level waste obtained after reprocessing of the ThO_2-based fuel may contain significant amount of fluoride ions and aluminum. There are a number of investigations on the structural aspects of fluoride ion containing silicate/alumino-silicate glasses investigated using ^{19}F NMR and XPS techniques [46–50]. Based on these studies, it has been concluded that most of the F^- ions in the glass exist as F-M(n) (fluorine with unknown number of network modifying cations like Na^+, Ca^{2+}, Ba^{2+}, etc.) type of linkages. Higher the field strength of the cation, higher will be extent of F-M(n) linkage formation. Further only a negligible amount of F^- ions forms Si–F linkages in the glass. Such linkages in the glass are expected to affect its physico-chemical properties like glass transition temperature, viscosity, TEC and refractive index. Details of study carried out by Mishra [27] are as follows:

Loss of Fluoride during Sample Preparation

In order to check the loss of fluoride ions and silica content from the samples due to the formation of gaseous SiF_4, evolved gas analysis were performed during melting of the samples. It is noticed during glass sample preparation, that there is only around 6–8 wt% loss on the fluoride ions from the glass.

XRD patterns of barium borosilicate glasses containing up to 6 wt% fluoride ions are shown in Fig. 28 [51]. For glasses containing up to 4 wt% a broad hump around a 2θ value of 25° is observed, characteristic of the amorphous borosilicate network. However above 4 wt% addition of F^- ions in the glass resulted in appearance of sharp peaks around 2θ values 24.93°, 28.83°, 41.2°, and 48.75° characteristic of crystalline Frankdicksonite (BaF_2) phase.

Based on the XRD patterns, it has been inferred that barium borosilicate glass can retain up to 4 wt% F^- ions and above that crystalline phase of Frankdicksonite (BaF_2) is separated out (PDF 4–452). These results are in conformity with the previous reports [46–48], which suggests that F^- ions have got tendency to form clusters with network modifying cation with higher z/r values. Here Ba^{2+} has got higher z/r compared to Na^+ and hence it is more favorable to form BaF_2 phase compared to NaF phase.

DTA patterns for barium borosilicate glasses containing different amounts of F^- ions are shown in Fig. 29 [51]. From the onset of slope change the values of glass transition temperatures are obtained for these glasses. Barium borosilicate glass without any F^- content is characterized by a glass transition temperature of 559 °C. With addition of fluoride ions in the glass, the glass transition temperature systematically decrease as can be seen from Fig. 29.

The glass transition temperature values have been found to decrease systematically with incorporation of F^- ions in the glass. Glass transition temperature is nothing but onset of structural relaxation brought about by the heat treatment. Formation of F–Ba(n) or F–Si(n) clusters/structural units disturbs the glass network, reduces the plastic flow and thereby weakening the network. An increased intensity due to the presence of M–F bonds could be another reason for decrease in

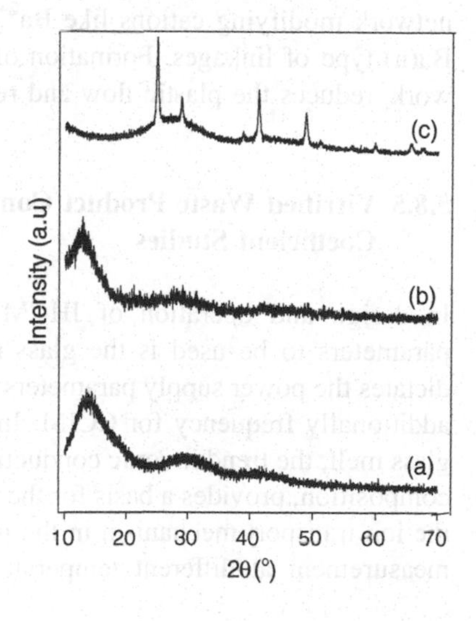

Fig. 28 XRD patterns for barium borosilicate glasses containing, 2 wt% (**a**) 4 wt% (**b**) and 6 wt% F^- ions (**c**) ([51], Copyright Elsevier)

Fig. 29 DTA curves for barium borosilicate glasses containing, 0 wt% (a) 2 wt% (b) and 4 wt% F⁻ ions (c) ([51], Copyright Elsevier)

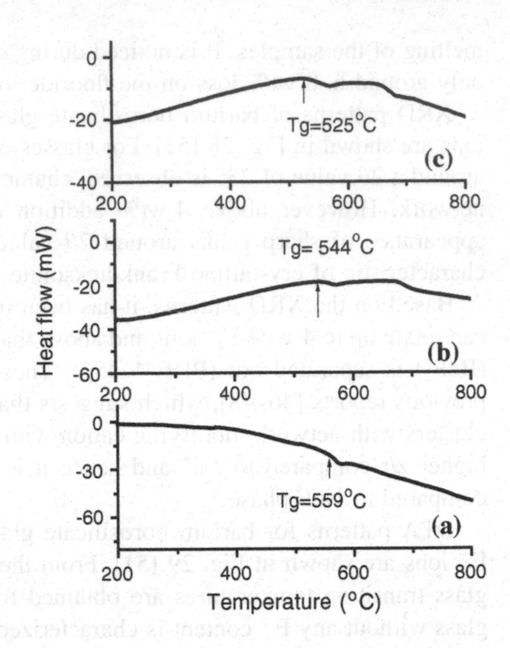

T_g values. This leads to the decrease in the T_g values with increase in F^- ion incorporation the glass.

Based on XRD studies on barium borosilicate glasses containing different amounts of F^- ions, it has been concluded that fluoride ion incorporation above 4 wt% results in phase separation of the glass leading to the formation of crystalline BaF_2 phase. F^- ions in these glasses mainly exist as F–Si(n) or F–Ba(n) type of linkages and their relative extent are unaffected by extent of F^- ion content in the glass. Unlike this, silicon and boron structural units are affected by F^- ion addition in the glass, and this has been explained by the decrease in the number of network modifying cations like Ba^{2+} ions brought about by the formation of F–Ba(n) type of linkages. Formation of such linkages disturbs the borosilicate network, reduces the plastic flow and results in the decrease in the T_g values.

5.8.5 Vitrified Waste Product Conductivity and Diffusion Coefficient Studies

In design and operation of JHCM and CCIM, one of the glass formulation parameters to be used is the glass resistivity [52]. The molten glass resistivity dictates the power supply parameters like voltage and current in case of JHCM and additionally frequency for CCIM. In the absence of the conductivity data of the glass melt, the trend in ionic conductivity below melting point, as function of glass composition, provides a basis for the development of suitable glass. Understanding the ion transport mechanism in the glass matrix is possible from the conductivity measurement at different temperatures. In addition, the electrical conductivity

Table 15 Composition and properties of glasses. ([53] Copyright Elsevier)

Glass Composition (mol%)	SBTh-0	SBTh-2	SBTh-4	SBTh-5
SiO_2	47.36	46.42	45.46	44.99
B_2O_3	26.80	26.26	25.73	25.46
Na_2O	14.31	14.02	13.74	13.59
BaO	11.53	11.30	11.07	10.96
ThO_2	0.00	2.00	4.00	5.00
Glass density ($g.cm^{-3}$)	2.96	3.10	3.23	3.30
($E_a \pm 0.01$) (eV)	1.15(1)	1.13(1)	1.08(1)	1.13(3)
$C_{Na}(\times 10^{21} cm^{-3})$	6.93	6.76	6.58	6.49
Glass transition temperature, T_g(K)	820	822	822	820

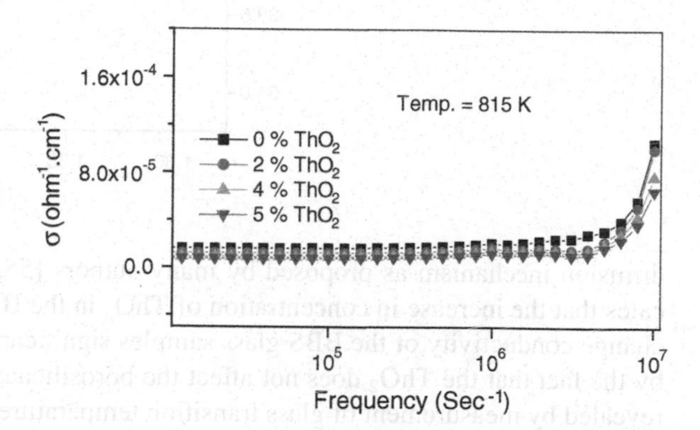

Fig. 30 Conductivity as function of frequency for glass samples at 815 K containing 0, 2, 4, and 5 mol% of ThO_2 ([53] Copyright Elsevier)

measurement is useful in determining the diffusion coefficient of alkali ions in the glass.

In the study conducted by Mishra et al. [53], the effect of ThO_2 on the ionic conductivity of barium borosilicate glasses was studied. Based on the electrical conductivity and density of the glasses, diffusion coefficients of glasses were determined.

Measurement of Electrical Conductivity of Thorium Glasses

Different compositions of glasses were selected having ThO_2 concentration ranges from 0 to 5 mol% while maintaining the SiO_2/Na_2O, SiO_2/B_2O_3, SiO_2/BaO, B_2O_3/Na_2O, B_2O_3/BaO, and BaO/Na_2O ratios constant for all the glass samples (Table 15).

The electrical conductivity $\sigma(\omega)$ for glasses containing 0, 2, 4, and 5 mol% of ThO_2 at the selected temperature of 815 K is given in Fig. 30 [53]. It can be observed from Fig. 30 that the conductivity of the glass samples remains almost constant in the frequency range 1 Hz–1 MHz and increases sharply above this frequency range. The plateau value corresponds to static conductivity for long range ionic displacement, while the increase at high frequency is due to relaxation caused by local motion of Na^+ cations [54] envisaged by single ionic jump

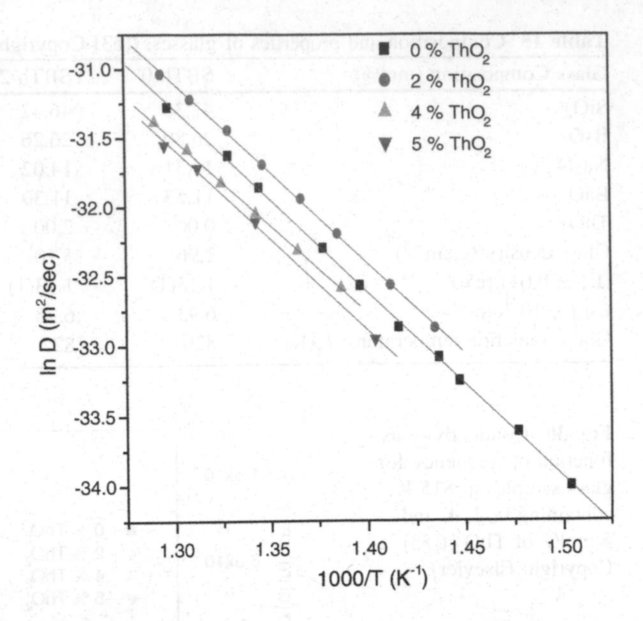

Fig. 31 Plots of ln(D) versus 1/T for glasses containing 0, 2, 4, and 5 mol% of ThO_2 below the glass transition temperature ([53] Copyright Elsevier)

diffusion mechanism as proposed by many authors [55, 56]. Figure 30 also indicates that the increase in concentration of ThO_2 in the BBS glass samples does not change conductivity of the BBS glass samples significantly. This can be explained by the fact that the ThO_2 does not affect the borosilicate network in BBS glass, as revealed by measurement of glass transition temperature by DTA and ^{29}Si and ^{11}B MAS NMR patterns. However, the slight observed decrease in conductivity of glass samples with increase in ThO_2 concentration could be due to the net decrease in Na^+ ion concentration with increase in ThO_2 concentration.

Calculation of Diffusion Coefficient

The diffusion coefficient of the mobile sodium ion has been obtained by relating the electrical conductivity with the diffusion coefficient D using Nernst–Einstein equation [57]. The values of density and C_{Na} obtained are given in Table 15. The diffusion coefficients for the transport of the charge carrier (Na^+ ion) as a function of temperature for glass samples containing different mol% of ThO_2 were calculated by using the above relation. The diffusion coefficient was found to be almost constant with ThO_2 concentration. Figure 31 gives the plots of ln(D) versus 1/T below the glass transition temperature for barium containing glasses. It can be observed from this figure that all the glasses exhibit a linear plot, which is characteristic of a thermally activated transport phenomenon. The activation energy for the glass samples containing 0, 2, 4, and 5 mol% of ThO_2 calculated from the slope of theses curves are found to be 1.15(1), 1.13(1), 1.08(1), and 1.13(3) eV, respectively. The activation energy for the thermally activated transport of Na^+ ion is found to remain by and large same with ThO_2 content in the glass matrix. The constancy in activation energy with increasing ThO_2 content in the glass matrix can be explained on the fact that ThO_2 does not modify the borosilicate network, which makes the number of available sites for ionic movements constant.

Based on above studies, the following has been concluded:

1. The ionic conductivity of the barium borosilicate glasses is independent of ThO_2 content up to 5 mol%. The results of electrical properties, DTA and ^{29}Si and ^{11}B MAS NMR studies suggest that borosilicate network is unaffected by the ThO_2 incorporation in the glass.
2. The diffusion coefficients for migration of Na^+ion in BBS glasses containing different concentrations of ThO_2 have been calculated from the electrical conductivity and density data. The activation energy for ion transport calculated from the ln(D) versus 1/T plot for 0, 2, 4 and 5 mol% of ThO_2 was found to be 1.15(1), 1.13(1), 1.08(1), and 1.13(3) eV, respectively.
3. The electrical conductivity data suggests that addition of small amount of BaO will be beneficial from electrical resistivity point of view, whereas ThO_2 addition will not have any deleterious effect on resistivity.

5.9 Transportation and Interim Storage of Vitrified Waste

After the vitrification, next step in the management of HLW is the interim storage of conditioned waste product. The duration of interim storage varies from 30 to 50 years. Some countries are planning for even longer interim storage periods. The basic objective of interim storage is to allow decay of major fission products like ^{60}Sr and ^{137}Cs which in turn helps in reduction of cooling and shielding requirements. This is helpful in transportation of HLW vitrified packages as well as in optimal design of final underground disposal facility.

5.9.1 Transportation of Vitrified Waste Product Packages

In India, the transportation of VWP overpacks from WIP Trombay to Tarapur is performed in a 18-ton shielding cask with all regulatory requirements of type B(M) package for transportation of high-level radioactive material through public domain. This shielding cask contains three overpacks and is designed with requisite 9 m drop testing of scale down model and validation of analytical model. VWP manufactured at WIP, Trombay, are routinely transferred to Solid Storage Surveillance Facility (SSSF), Tarapur following all radiological safety and security procedures. The experience in transportation of VWP through public domain has been satisfactory and safe.

5.9.2 Interim Storage of VWP

Interim storage facilities for vitrified HLW are based on passive design of natural convection air cooling assisted by induced draught due to stack, not dependent on man or machine [53]. These utilize the decay heat and a suitably designed stack to provide the driving force for the movement of air through the storage vault. In effect, the waste heat increases the air temperature, causing an upward movement of air due to buoyant forces. The driving force due to buoyancy is balanced by the friction effect of air passage through the system to establish equilibrium for the loading condition of each compartment of the vault. Cross flow of air across canisters is adopted in SSSF Tarapur design where as channelized axial flow is adopted for design of Vitrified Waste Storage Facility (VWSF) being set-up at Kalpakkam.

Solid Storage Surveillance Facility, Tarapur

This facility has been designed to ensure optimal heat transfer efficiency. Coolant air distribution, storage unit array and filling pattern are optimized. To facilitate this, the storage vault is divided into two blocks and each of the block is further divided into three compartments [58]. The schematic of the facility is shown in Fig. 32. The structural design of the facility has to take into account certain basic requirement. Tarapur site has a high ground water table, which in the latter days of monsoon season rises up to nearly surface level. Ensuring stability of the structure against uplift at all times and further ensuring water tightness calls for higher wall thickness for the vault. However, the use of ordinary concrete for the structure calls for minimization of thermal effects. This can only be achieved by minimizing the temperature gradients across the concrete wall, which in turn can be achieved by minimizing the vault wall thickness. To reconcile the above requirements, it is necessary to isolate the thermal effects from the load bearing structure and also isolate any water ingress. This is achieved by recourse to a "double vault design." The "inner storage vault" is designed on thermal considerations and the design of the "external vault" is based on structural and biological shielding considerations. The external vault also isolates any seepage of ground water from the immediate environs of the storage units. Seepage water, if any, will be monitored for activity and further processed. The thermal vault is supported on specially designed bearings from the outer vault. This makes possible the free sliding of the thermal vault, thereby, relieving temperature stresses on the thermal vault.

Vitrified Waste Storage Facility, Kalpakkam

Vitrified Waste Storage Facility (VWSF), Kalpakkam is designed to receive vitrified waste in four independent above-ground storage vaults. The schematic view of VWSF is shown in Fig. 33 [59]. An additional safety feature in the form of induced draft air cooling with HEPA filtration is provided in this facility, which is brought in line as and when air born activity, if any, is detected in the cooling air [58].

Radioactivity inventory of the vault in fully filled condition is around one billion curie generating few mega watts of decay heat. The stack design takes into account the above requirement and will provide enough draught to counter

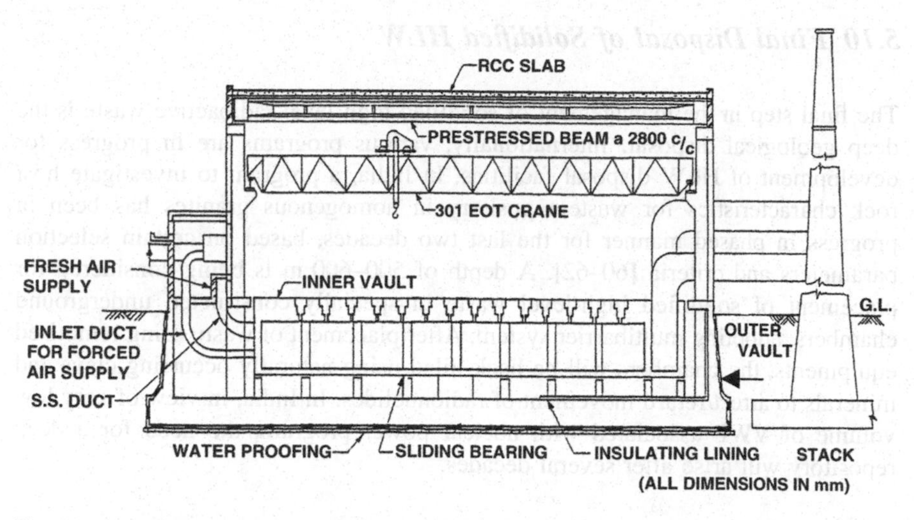

Fig. 32 Schematic view of air-cooled solid storage and surveillance facility, Tarapur ([11] Copyright Elsevier)

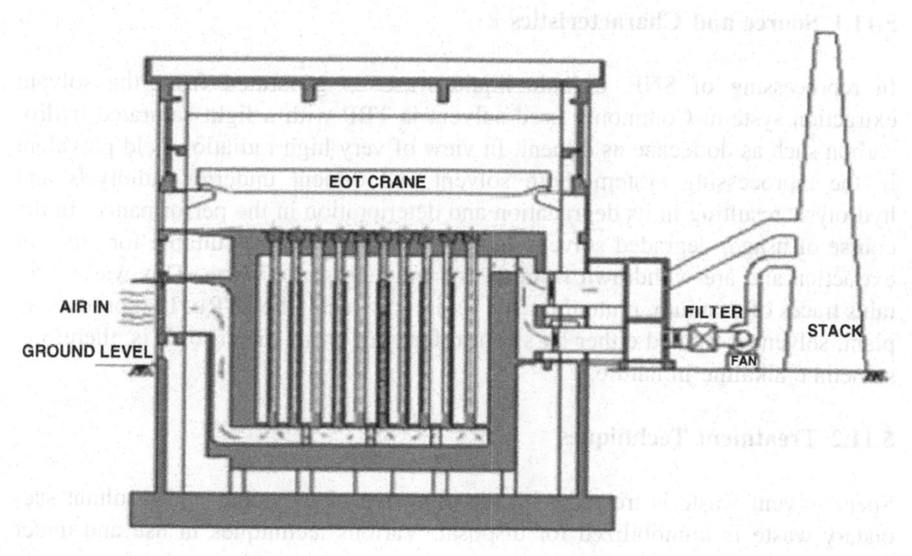

Fig. 33 Schematic view of vitrified waste storage facility being set up at Kalpakkam ([59] Copyright Elsevier)

pressure drop across the vault, inlet port and exhaust plenums and kinetic energy losses at the exit of the stack. The system is self-regulating and can compensate for changes in heat load or seasonal/weather variation conditions. As decay heat reduces, both the air flow and air temperatures decrease.

5.10 Final Disposal of Solidified HLW

The final step in the management of solidified high-level radioactive waste is the deep geological disposal. Internationally, various programs are in progress for development of HLW disposal facilities. In India, a program to investigate host rock characteristics for waste repository in homogenous granites has been in progress in phased manner for the last two decades, based on certain selection parameters and criteria [60–62]. A depth of 500–600 m is being considered for placement of solidified high-level waste in specially constructed underground chambers, adopting multibarrier system. After placement of waste using remotised equipments, the chambers will be back-filled using naturally occurring clays and minerals to arrest/retard movement of radionuclides. In India, in view of very low volume of VWP associated with nuclear power program, the need for a deep repository will arise after several decades.

5.11 Organic Liquid Waste

5.11.1 Source and Characteristics

In reprocessing of SNF, organic liquid waste is generated from the solvent extraction system. Commonly used solvent is TBP with a light saturated hydrocarbon such as dodecane as diluent. In view of very high radiation field prevalent in the reprocessing system, both solvent and diluent undergo radiolysis and hydrolysis resulting in its degradation and deterioration in the performance. In the course of usage, degraded solvent and diluents are no more suitable for efficient extraction and are withdrawn from the system as organic waste. This waste contains traces of uranium, plutonium and fission products like ^{106}Ru. In reprocessing plant, solvent is treated either by steam reforming or alkali scrub. It is, therefore, sometime alkaline in nature.

5.11.2 Treatment Techniques

Spent solvent waste is treated with the objective of recycling and resultant secondary waste is immobilized for disposal. Various techniques in use and under development are discussed here.

Distillation

Distillation under vacuum has been applied for decontamination of spent solvents for re-use [63]. The method is used at the reprocessing facilities in France and the United Kingdom.

Incineration

Incineration is an exothermic process. Organic solvent is destroyed in this process using heat and oxygen. A wide range of technologies have been developed for incineration, namely excess-air, controlled-air, and starved air (pyrolyzing) incinerator.

In incineration of degraded TBP solvent, the products of the reaction with sub-stoichiometric oxygen are phosphoric acid, butanol, butane, etc. In excess air incineration, phosphorus pentoxide is formed which combines with water vapors on cooler parts of the off-gas treatment system to form phosphoric acid. This leads to corrosion of equipment and piping. To tackle this problem, various modifications have been incorporated in incineration system to convert phosphorus into inert phosphates like sodium carbonate and calcium hydroxide. Fixation of phosphorus on alumina/zirconia granules in the fluidized bed incinerator has also been attempted [64].

The ash produced from incineration is usually immobilized in cement matrix for disposal. Besides off gas treatment of gaseous products leads to generation of scrub solution, filter media, etc. This may also lead to furans and dioxins in the off gas system which need to be treated and monitored before discharge.

Pyrolysis

Pyrolysis is based on thermal decomposition of organic waste at comparatively lower temperature of 500–550 °C. The pyrolyzer operates under inert or oxygen deficient environment. The gaseous products of pyrolysis are oxidized in simple combustion chamber, cooled, and decontaminated in off-gas treatment system. An example of pyrolysis of spent TBP and diluents is pebble bed reactor using calcium hydroxide for controlling formation of phosphoric acid [65, 66]. Due to lower temperature of operation, problem of corrosion due to phosphoric oxide is less severe. Besides, due to reduced oxygen level and lower temperature, escape of volatile components of waste like ruthenium and cesium is minimized.

Alkaline Hydrolysis

In alkaline hydrolysis process, organic waste is treated in contact with aqueous alkaline solution. In chemical extraction process, hydrolysis reaction alters the nature of organic solvent and results in transfer of the radioactivity to the aqueous phase. Various phases formed are separated which are diluents with little associated activity suitable for reuse and sodium dibutylphosphate as major by-product which needs further treatment and disposal. In Sellafield Facility, chemical treatment using hydroxide flocculation is employed to remove radioactivity from the aqueous phase and the resultant floc is immobilized in cement matrix [67]. Alkaline hydrolysis process has also been used for treatment of spent organic solvent at the former WAK facility in Germany. A plant based on the alkaline hydrolysis is in operation at BARC, Trombay [68, 69] and another with improvements to reduce secondary waste generation is being commissioned at

BARC, Tarapur. In the process developed at BARC, Trombay; it has been established that temperature during reaction does not exceed 110 °C and pressure is below −400 mm water column. More than 99.6 % conversion of TBP into sodium di-butyl phosphate is achieved in about 5 h in the stirred reactor. As regards the diluents, its quantitative recovery is achieved virtually free of radioactivity and TBP content. The activity associated with the diluents is normally in the range of 0.4–4.0 Bq/ml for alpha and 0.04–10.0 Bq/ml with respect to gross beta. The pre-retention achieved for recovered diluents is 6.23×10^{-3} mg/L for fresh dodecane thereby qualifying it for recycle. Nonrecyclable diluents are incinerated in an excess air incinerator [70].

5.12 Laboratory Waste and its Treatment

Process control analysis, production activities, and development works lead to generation of a variety of waste in the reprocessing laboratories. Some of these waste contain thiocyanate used for uranium analysis, oxalate used for process sample analysis and for rinsing of apparatus, acidic waste containing phosphoric/sulphamic acid, etc., and organic waste from Pu analysis.

In view of small volume and large variation in their composition, laboratory wastes are received in 2.5–5 l carbuoys. The aqueous wastes are neutralized using sodium hydroxide along with ferric nitrate in glovebox. The resultant sludge is immobilized in cement matrix in carbuoy. A number of conditioned waste carbuoys are then embedded in cement matrix in 200 l drum. The small volume of organic waste received from laboratory is also conditioned in cement matrix after treatment.

5.13 Radioactive Solid Waste and its Management

A variety of primary solid waste comprising of tissue materials, glass wares, plastics, protective rubber-wears, used components like filters, piping, structural items, unserviceable equipment, etc., are generated in reprocessing plants. These wastes are categorized depending on the radiation field, concentration and type of radioactivity. They are segregated as compressible or noncompressible and combustible or noncombustible. Major portion of the total solid waste has low activity and is either combustible or compressible. In few facilities, incinerator is used for burning the combustible wastes achieving a volume reduction of about 50. Hydraulically operated baling press is used to compress low active noncombustible waste to obtain volume reduction of 5. Melt densification of polymeric waste has been developed and deployed at CWMF, Kalpakkam, which uses a temperature of about 180 °C to reduce the volume resulting in dense waste product of higher chemical durability [71].

The major portion of waste generated from reprocessing plants is received in standard 200 l carbon steel drums. The waste is categorized with the help of an

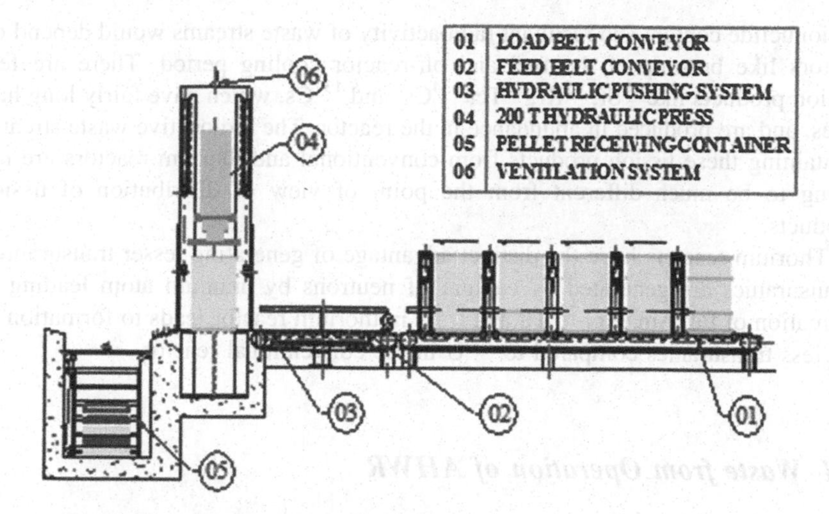

Fig. 34 Schematic of solid waste compaction system

assaying system, based on radioactivity content and radionuclides present. A real-time digital imaging system is also used to segregate compactable and non-compactable low-level waste. The compactable waste packed in drums is pelletized using hydraulic press of about 200 ton capacity. The pelletization system comprises facilities for conveying, indexing, and compacting (Fig. 34). All operations are controlled by programmable logic controller-based control system. Such system is also useful for the compaction of used HEPA filters.

Treatment of primary low and intermediate level wastes (LILW) results in generation of chemical sludge, ash, compacted waste, etc. These are packed in carbon/stainless steel drums and the voids are filled with cement grout. Depending on the category of waste, the waste drums are disposed in near surface disposal facility in engineered modules.

During chemical dissolution of SNF from PHWR, radioactive solid waste in the form of hollow zircalloy tube pieces is generated. The zircalloy hull waste is collected in stainless steel drum and stored for future volume reduction and disposal/recovery. In France, hulls are compacted in a specially designed SS container and packaged in SS over-pack for interim storage prior to final disposal in geological repository.

6 Summary

During the operation of thorium reactors, the radioactive waste generated will not be different from conventional U-based nuclear power plants due to the fact that the formation of activation products and their distribution are essentially governed by the neutron flux and the material of construction of the reactor components. The radionuclide content and resultant radioactivity of waste streams would depend on factors like burn-up of the fuel and off-reactor cooling period. There are few fission products like ^{90}Sr, ^{85}Kr, ^{99}Tc, ^{134}Cs, and ^{137}Cs, which have fairly long half lives, and are produced in abundance in the reactor. The radioactive waste streams containing these fission products from conventional and thorium reactors are not going to be much different from the point of view of distribution of fission products.

Thorium reactors have the distinct advantage of generating lesser transuranics. Transuranics are generated by capture of neutrons by uranium atom leading to formation of Pu, Am, etc. ^{232}Th and ^{233}U in thorium reactor leads to formation of far less transuranics compared to ^{235}U in the conventional reactor.

6.1 Waste from Operation of AHWR

As in PHWRs, reactor waste from operation of AHWR will be mainly due to the contamination/leakages of fission, activation, and corrosion products. The liquid waste will be similar in nature as those from PHWR except with respect to its higher level of radioactivity on account of high burn-up of the fuel. Based on the process technologies developed for the management of liquid radioactive reactor waste stream, this waste will also be subjected to various decontamination processes employing chemical precipitation, ion-exchange, reverse osmosis, etc., followed by immobilization of concentrates. Among solid radioactive wastes, spent ion-exchange resin forms major fraction of solid waste with high specific activity. Presently, in Indian Nuclear Power Plants, spent ion-exchange is converted into polymerized waste product using polyester styrene matrix. Alternative process using improved cement matrix is undergoing plant scale trials and the regulatory review.

6.2 Gaseous Waste in Thorium Fuel Reprocessing

In present reprocessing and fuel fabrication facilities handling U and Pu, off gas system consist of scrubbers, deep-bed glass fiber filters and HEPA filters. Design of off gas system for similar facilities in thorium fuel cycle would require additional features in view of difficulties in dissolution of ThO_2, presence of ^{232}U and ^{220}Rn in ^{232}U decay chain. In the dissolution step of Thorex Process, prolonged heating for 10–18 h at boiling temperature will be required due to slow dissolution rate. This is likely to result in escape of ^{137}Cs and ^{106}Ru in the off gas system due to highly oxidizing atmosphere prevalent in the dissolver. These off gases are normally scrubbed in packed column resulting in majority of Ru and Cs reporting in scrub solution. Incorporation of washable glass fiber deep-bed filter results in minimization of solid waste in the form of used off-gas filters as demonstrated successfully in AVS at BARC, Tarapur.

Typical off gas systems designed for Pu handling are not suitable for ^{233}U with a high ^{232}U content in view of the fact that they do not contain double HEPA filters with the time delay between the HEPA filters required to avoid escape of ^{220}Rn. The noble gas ^{220}Rn present in off-gases can pass through HEPA filters and then decay to solid ^{208}Tl. Therefore, a three stage filtration system will be required to prevent the escape of decay products. These are: stage-I, a HEPA filter to collect solid including the precursors to ^{220}Rn; stage-II, charcoal bed delay lines to enable delay of about 10 min or other special equipment to hold the radon gas in the off-gas system that goes through the first HEPA filter until ^{220}Rn decays to solid material and stage-III, a second HEPA filter to remove the solid decay products of ^{220}Rn. The components used in the three stages of described filtration system would require both gamma shielding as well as remote maintenance capability.

6.3 Liquid Waste from Fuel Reprocessing

Major waste streams generated from the reprocessing plant are high-level liquid waste, intermediate-level liquid waste, low-level liquid waste, organic liquid waste, de-clad liquid waste or hulls, solid radioactive waste, and small volumes of laboratory waste.

6.3.1 Management of LLW and ILW

The processes that are employed for treatment of ILW and LLW are filtration, chemical treatment, ion-exchange, steam evaporation, solar evaporation, and membrane-based processes. Current efforts are toward adoption of processes like ultra-filtration and reverse osmosis in combination with the existing processes to further bring down the release of radioactivity to the environment following "near zero discharge" philosophy.

Reprocessing plants use large volume of nitric acid for dissolution of SNF. Majority of these nitrates finally report in LLW as sodium nitrate. The generation of total amount of nitrates in LLW is expected to increase unless recycling and reuse of nitric acid is adopted in the reprocessing facilities. A promising approach is the biodegradation of nitrates to bring the concentration of nitrates in the effluents to the acceptable values.

The reprocessing of irradiated thoria rods will result in the raffinate waste, which is primarily acidic in nature (4–5 M HNO_3) and contains fission products like ^{137}Cs, ^{90}Sr and low concentration of ^{125}Sb along with Al and fluoride ions added during dissolution of spent fuel. A process based on ion-exchange separation has been developed at Trombay and adopted on the plant scale for treatment of this waste. The process involves various steps viz. solvent extraction to remove residual thorium, neutralization to lower down pH for effective removal of Cs and Sr by ion-exchange, co-precipitation and filtration to remove Al and Sb, concentration and fixation of precipitate in cement matrix for disposal. Finally, the loaded ion-exchange column is eluted with dilute nitric acid and the elute is vitrified in boro-silicate glass matrix. Alternate process have also been studied where Cs could be further purified and immobilized in special glass formulation for accommodating Cs-rich waste and for deployment as an irradiation source for medical application, e.g., in blood irradiator.

Intermediate level radioactive waste is conditioned depending on the compatibility of the matrix with waste, chemical and mechanical durability of solidified product, cost of processing, throughput and disposal options. Cementation and polymerization methods are normally adopted for conditioning of this type of waste. Spent ion exchange resins are immobilized in polymer and cement.

6.3.2 Management of HLW

The countries involved in reprocessing of nuclear fuel have in general adopted a three step strategy for management of HLW. These steps are: (a) immobilization of HLW in an inert matrix like glass, (b) interim storage of solidified HLW, and (c) disposal of solidified waste in deep geological formation. The challenges involved in the first step, i.e., immobilization of HLW in vitreous matrix, are due to acidic nature of waste, corrosivity of molten glass and intense radiation environment due to high concentration of radionuclides present in the waste. Aspects of treatment of thorium reprocessing waste and matrix development for incorporation of thorium in borosilicate have been discussed in this chapter. The salient features are:

- HLW composition depends on factors like type of fuel and its cladding material, history in the reactor including burn up, the process used for reprocessing of irradiated fuel and off-reactor cooling period. The waste is acidic in nature and contains fission products, heavier actinides, unrecovered Pu/U/Th as the case may be, corrosion products, added chemicals, traces of solvent, and its degradation products.
- Borosilicate glass matrix has been universally adopted for vitrification of HLW. Different designs of vitrification system and corresponding type of melter have been developed and deployed. The vitrification involves various processing steps viz. waste transfer and receipt, concentration and acid recovery, vitrification, VWP handling, secondary waste management, and off-gas treatment. The single-step process using liquid-fed JHCM has been adopted in China, India, Japan, and the USA.

Pre-treatment and Immobilizaion of HLW from Thoria Reprocessing

High-level radioactive liquid waste (HLW) from reprocessing of thorium-based spent fuel is likely to contain unrecovered thorium, fission products, corrosion products, actinides and added chemicals. A maximum solubility of ~6 wt% thoria has been observed in sodium borosilicate glass. Based on extensive studies on incorporation of thorium HLW in barium borosilicate glass system, Mishra [27] has made significant observations as follows:

- Barium borosilicate glass matrix can contain up to 16 wt% thoria without phase separation. Beyond this limit, thorite and thorianite separate out within the matrix. Based on ^{29}Si and ^{11}B MAS NMR and IR studies, it has been concluded that the borosilicate network is not affected by even large amount of ThO_2 incorporation. This observation is further supported by identical values of the glass transition temperatures noted for all these glasses with different amounts of ThO_2.
- These studies indicated that glass formulation based on BBS matrix is able to incorporate ThO_2 up to 16 wt% without abnormally affecting the thermo-physical properties like glass transition temperature, TEC and micro-hardness of the vitreous matrix, which is an important observation. These glasses are found to be thermally stable up to 1,000 °C. Above 15.86 wt% of ThO_2 in glass, there is significant change in TEC, which is attributed to appearance of thorianite and thorite crystalline phases in BBS glass. Micro-hardness of glasses containing thoria more than 15.86 wt% also shows a decreasing trend, which is attributed to partial phase separation of ThO_2 and $ThSiO_4$ phases in the glass.
- BBS glasses containing 15.86 wt% of ThO_2 is able to accommodate only up to 7.5 wt% of UO_3, beyond which a crystalline phase of ThO_2 is separated out. Based on ^{29}Si MAS NMR studies on ThO_2 containing barium borosilicate glasses having varying amounts of uranium oxides, it has been concluded that uranium oxide incorporation is associated with the conversion of Q^3 to Q^2structural units of silicon and the increased number of nonbridging oxygen atoms thus produced favor the uranium ions to occupy the sites, which are created by the nonbridging oxygen atoms (network modifying sites) in the glass.

Minimization of F^- and Al^{3+}in Thorium Waste

Another interesting study [45] with the objectives of (1) avoiding aluminum nitrate addition and (2) lowering F^- concentration during dissolution has resulted in the following salient findings:

- Zirconium and thorium ions form stronger complexes with F^- compared to aluminum ion making addition of aluminum nitrate for corrosion mitigation unnecessary for zircaloy-clad thoria dissolution reaction. SS304L corrosion rate studies at 0.03 M NaF concentration clearly showed that if zircaloy is present, the corrosion rate of SS304L in presence and absence of $Al(NO_3)$ is quite comparable and is about 10 mpy.
- F^- concentration can be reduced from presently adopted 0.03 M to lower concentration by following the procedure of adding NaF into the dissolver in two proportions with time gap of about 3–4 h. Reaction times for dissolution reaction were observed to be 48–72, 72–96, and 96–120 h for 0.03, 0.01, and 0.005 M, respectively during the dissolution reaction.

- Enhancing volume of 13 M nitric acid during the dissolution reaction can reduce the reaction time from 96 to about 67 h at 0.005 M concentration of NaF.

Incorporation of F^- and Al^{3+} in Ba-borosilicate Glass

Based on XRD studies conducted by Mishra [53] on barium borosilicate glasses containing different amounts of F^- ions, it has been concluded that fluoride ion incorporation above 4 wt% results in phase separation of the glass leading to the formation of crystalline BaF_2 phase. F^- ions in these glasses mainly exist as F–Si(n) or F–Ba(n) type of linkages and their relative extent are unaffected by extent of F^- ion content in the glass. Unlike this, silicon and boron structural units are affected by F^- ion addition in the glass and this has been explained by the decrease in the number of network modifying cations like Ba^{2+} ions brought about by the formation of F–Ba(n) type of linkages. Formation of such linkages disturbs the borosilicate network, reduces the plastic flow and results in the decrease in the T_g values. Further, in this study, the glass transition temperature values have been found to decrease systematically with incorporation of F^- ions in the glass. Since glass transition temperature is nothing but onset of structural relaxation brought about by the heat treatment, formation of F–Ba(n) or F–Si(n) clusters/structural units disturbs the glass network, reduces the plastic flow and thereby weakening the network.

Characterization of Thorium Containing Glasses

The molten glass resistivity dictates the power supply parameters like voltage and current in case of JHCM and additionally frequency for CCIM. In the absence of the conductivity data of the glass melt, the trend in ionic conductivity below melting point, as function of glass composition, provides a basis for the development of suitable glass. Understanding the ion transport mechanism in the glass matrix is possible from the conductivity measurement at different temperatures. In addition, the electrical conductivity measurement is useful in determining the diffusion coefficient of alkali ions in the glass. In this context, the effect of ThO_2 on the ionic conductivity of barium borosilicate glasses was studied by Mishra et al. [53]. Based on the electrical conductivity and density of the glasses, diffusion coefficients of glasses were also determined in this work. It was observed that the conductivity of the glass samples remains almost constant in the frequency range 1 Hz to 1 MHz and increases sharply above this frequency range. Based on above studies, the following has been concluded:

- The ionic conductivity of the barium borosilicate glasses is independent of ThO_2 content up to 5 mol%. The results of electrical properties, DTA and ^{29}Si and ^{11}B MAS NMR studies suggest that borosilicate network is unaffected by the ThO_2 incorporation in the glass.
- The diffusion coefficients for migration of Na^+ ion in BBS glasses containing different concentrations of ThO_2 have been calculated from the electrical conductivity and density data. The electrical conductivity data suggests that addition of BaO will be beneficial from electrical resistivity point of view, whereas ThO_2 addition will not have any deleterious effect on resistivity.

6.3.3 Interim Storage and Geological Disposal of Vitrified High-Level Waste

After the vitrification, next step in the management of HLW is the interim storage of conditioned waste product. The duration of interim storage varies from 30 to 50 years. Some countries are planning for even longer interim storage periods. The basic objective of interim storage is to allow decay of major fission products like ^{60}Sr and ^{137}Cs which in turn helps in reduction of cooling and shielding requirements. This is helpful in transportation of HLW vitrified packages as well as in optimal design of final underground disposal facility. Interim storage facilities for vitrified HLW based on passive natural convection air cooling design utilize the decay heat and stack to provide the driving force for the movement of air through the storage vault [58]. Third and the final step in the management of solidified HLW is deep geological disposal. Internationally, various programs are in progress for development of HLW disposal facilities. In India, a program to investigate host rock characteristics for waste repository in homogenous granites has been in progress in phased manner for the last two decades, based on certain selection parameters and criteria [60–62]. A depth of 500–600 m is being considered for placement of solidified high-level waste in specially constructed underground chambers, adopting multibarrier system. However, due to less production of Pu and associated minor actinides, global radiotoxic inventory of the solidified HLW for a thorium fuel cycle is likely to be significantly less than that of the standard U–Pu fuel cycle.

6.3.4 Management of Organic Waste

Organic wastes in liquid as well as solid form are generated in nuclear fuel cycle, e.g., from the solvent extraction system in reprocessing of SNF where TBP is commonly used with a light saturated hydrocarbon such as dodecane as diluent. Both solvent and diluent undergo radiolysis and hydrolysis resulting in their degradation and deterioration in Organic waste, thus generated, contains traces of uranium, plutonium and fission products like ^{106}Ru. This is treated with the objective of recycling and resultant secondary waste is immobilized for disposal. Various techniques in use and under development are vacuum distillation, incineration, pyrolysis, alkaline hydrolysis, etc.

6.4 Management of LILW Solid Waste

A variety of primary solid waste comprising of tissue materials, glass wares, plastics, protective rubber-wears, used components like filters, piping, structural items, unserviceable equipment, etc., are generated in operation of fuel fabrication, nuclear reactors, reprocessing plants, and waste treatment facilities. These wastes are categorized depending on the radiation field, concentration, and type of radioactivity. They are segregated as compressible or noncompressible and combustible or noncombustible waste. Different designs of equipment are used to obtain desired volume reduction like incinerator and pyrolyzer for combustible wastes, hydraulically operated baling press for compressible waste, and melt densification of polymeric waste [71]. Treatment of primary LILW results in generation of chemical sludge, ash, compacted waste, etc. These are packed in carbon/stainless steel drums, and the voids are filled with cement grout. Depending on the category of waste, the waste drums are disposed in near surface disposal facility in engineered modules.

6.5 Future Challenges

There are newer challenges to be tackled in future with the adoption of thorium fuel cycle and to fill the gap of conditioning of some of the existing waste categories. In this context, solid waste in the form of zircaloy hulls needs to be taken up for recovery and recycle. Studies on incorporation of thorium in borosilicate matrix have given encouraging results. However, aspects of volatilization of fluoride present in futuristic waste during vitrification as well as estimation of residual fluoride in various glass system need further studies. Another important area of research and development to be continued is minimization of fluoride and aluminum concentrations in the thoria reprocessing waste as well as to study its effect on material of construction of the virtrification melter and associated off-gas system.

To meet the objectives of minimization of final HLW volume and reduce the time frame for isolation of vitrified waste, aspects of reduction of uranium in HLW and separation of minor actinides from it need to be adopted on engineering scales. Separation of ^{90}Sr and ^{137}Cs from HLW would not only result in recovery of valuables from the waste, but will also reduce toxicity and heat generation associated with the waste. Therefore, research and development work in this direction need to be pursued for adoption of these separation steps in future flow-sheets for treatment of waste from U/Th reprocessing. Similarly, work on recovery of nitric acid from HLW and its recycle needs to be pursued to reduce the impact of discharge of nitrates to the environment.

References

1. International Atomic Energy Agency (2009) Classification of radioactive waste. General Safety Guide no. GSG-1, Vienna
2. Atomic Energy Regulatory Board (2009) Management of radioactive waste. Safety Code no. AERB/SC/RW, Mumbai

3. Lung M (1997) A present review of the thorium nuclear fuel cycles. European Commission Final Report no. EUR 17771EN
4. Atomic Energy Regulatory Board (2006) Safety in thorium mining and milling. Guide no. AERB/NF/SG/IS-6, Mumbai, pp 3–9
5. Atomic Energy Regulatory Board (2007) Management of radioactive waste from mining and milling of uranium and thorium. Guide no. AERB/NF/SG/RW-5, Mumbai, pp 42–48
6. International Atomic Energy Agency (2002) Management of radioactive waste from the mining and milling of ores. Safety Guide no. WS-G-1.2, Vienna
7. Haridasan PP, Pillai PMB, Khan AH (2001) Derived contamination limits for the chemical processing of monazite. Radiat Prot Environ 24:219–221
8. Albert RE (1966) Thorium : its industrial hygiene aspects. Academic Press, New York
9. Atomic Energy Regulatory Board (2009) Safety guidelines for uranium oxide fuel fabrication facilities. Guide no. AERB/FE-FCF/SG-3, Mumbai, pp 42–44
10. Driggers FE, Thompson TT (1977) Program plan for R&D in support of thorium fuel cycle technologies. Report DPST-TFCT-77-100, Savannah River Laboratory
11. Raj K, Prasad KK, Bansal NK (2006) Radioactive waste management practices in India. Nucl Eng Design 236:914–930
12. Bhattacharjee B (2003) An overview of R&D in fuel cycle activities of AHWR. INSAC, Kalpakkam
13. Samanta SK, Ramaswamy M, Mishra BM (1992) Studies on cesium uptake by phenolic resins. Sep Sci Technol 27(2):255–267
14. Samanta SK, Theyyunni TK, Mishra BM (1995) Column behavior of a resorcinol formaldehyde polycondensate resin for radio-cesium removal from simulated rad-waste solution. J Nucl Sci Technol 32:425–429
15. Kulkarni Y, Samanta SK, Bakare SY, Raj K, Kumra MS (1996) Process for treatment of intermediate level radioactive waste based on radionuclide separation. In: Proceedings of the waste management 96 symposium, Tucson, Arizona
16. Ozarde PD, Samanta SK, Raj K (2002) Management of intermediate level waste from past reprocessing using cesium specific resorcinol formaldehyde. In: Proceedings of the IAEA international conference on issues and trends in radioactive waste management. IAEA-CN-90/51
17. Raj K, Misra SD (2009) Radioactive waste management in India: present practices and future trends. Energy Secur Insights 4(1):11–15
18. Kaushik CP, Mishra RK, Sengupta P, Kumar A, Das D, Kale GB, Raj K (2006) Barium borosilicate glass – a potential matrix for immobilization of sulfate bearing high-level radioactive liquid waste. J Nucl Mater 358:129–138
19. Ojovan MI, Karlina OK (1992) Synthesis and properties of glass composite materials for solidification of radioactive waste. Radiochemistry 33:97–100
20. Hench LL, Clark DE, Campbell J (1984) High level waste immobilization forms. Nucl Chem Waste Manag 5:149–173
21. Freeman AJ, Lander GH (eds) (1987) Handbook on the physics and chemistry of the actinides. Elsevier, North Holland, Amsterdam, pp 271–312
22. Raj K, Samuel MT (1986) Modified Pot glass process for vitrification of HLW-process engineering aspects. In: Proceedings of the XIV international congress on glass, vol 2, New Delhi, pp 399–407
23. Mehta D, Gangadharan A, Morzaria T, Tomar NS, Verma BB (2008) Industrial scale vitrification of HLW – experience at WIP, Trombay. Indian Nucl Soc News 5(3):5–10
24. Dani U, Kulkarni Y, Banerjee K, Changrani RD (2008) Industrial scale vitrification of HLW – experience at AVS, Tarapur. Indian Nucl Soc News 5(3):11–18
25. Farges F (1991) Structural environment around Th^{4+} in silicate glasses. Implications for the geochemistry of incompatible Me^{4+} elements. Geochim Cosmochim Acta 55:3303–3319
26. Sonavane MS, Yeotikar RG, Ali SS (2003) Indian Nuclear Society symposium proceedings, Mumbai, p 168

27. Mishra RK (2008) Investigations of solubility of different nuclear waste constituents in glasses, their structural and physico-chemical characterization. Ph.D. Dissertation, University of Mumbai
28. Mishra RK et al (2007) Studies on immobilization of thorium in barium borosilicate glass. J Nucl Mater 360:143–150
29. Personal Computer Powder Diffraction Standards (1998) Card No: 42-1462 (Ver. 2.0, 1998)
30. Personal Computer Powder Diffraction Standards. Card No: 18-1371 (Ver. 2.0, 1998)
31. Mishra RK et al (2006) Structural studies of ThO_2 containing barium borosilicate glass. J Noncryst Solids 352:2952–2957
32. Sudarsan V, Shrikhande VK, Kothiyal GP, Kulshreshtha SK (2002). Structural aspects of B_2O_3-substituted (PbO)0.5(SiO_2)0.5 glasses. J Phys Cond Matter 14:6553
33. Bhasin G, Bhatnagar A, Bhowmik S, Stehle C, Affatigato M, Feller S, MacKenzie J, Martin S (1998) Short range order in sodium borosilicate glasses obtained via deconvolution of ^{29}Si MAS NMR. Phys Chem Glasses 39:269–274
34. Martens R, Muller-Warmuth W (2000) Structural groups and their mixing in borosilicate glasses of various compositions – an NMR study. J Non-Cryst Solids 265:167–175
35. Nanba T, Miura Y (2003) Alkali distribution in borosilicate glasses. Phys Chem Glasses 44:244–248
36. Chen D, Miyoshi H, Masui H, Akai T, Yazawa T (2004) NMR study of structural changes of alkali borosilicate glasses with heat treatment. J Non-Cryst Solids 345:104–107
37. Miyoshi H, Chen D, Masui H, Yazawa T, Akai T (2004) Effect of calcium additive on structural changes under heat treatment in sodium borosilicate glasses. J Non-Cryst Solids 345:99–103
38. Nanba T, Nishimura M, Miura Y (2004) A theoretical interpretation of the chemical shift of ^{29}Si NMR peaks in alkali borosilicate glasses. Geochim Cosmochim Acta 68:5103–5111
39. Wood JG, Prabhakar S, Mueller KT, Pantano CG (2004) The effects of antimony oxide on the structure of alkaline-earth alumino borosilicate glasses. J Non-Cryst Solids 349:276–284
40. Mishra RK et al (2008) SAXS study of barium borosilicate glasses containing ThO_2. J Alloys Compd 466:543–545
41. Goswami M, Sarkar A, Sharma BI, Shrikhande VK, Kothiyal GP (2004) Effect of alumina concentration on thermal and structural properties of MAS glass and glass-ceramics. J Thermal Anal Cal 78:699–705
42. Grover V, Tyagi AK (2005) Preparation and bulk thermal expansion studies in $M_{1-x}Ce_x SiO_4$ (M = Th, Zr) system, and stabilization of tetragonal $ThSiO_4$. J Alloys Compd 390:112–114
43. Kakodkar A (1997) Tailorinfg R & D Thrusts in Nuclear Power Technology for Sustainable Development. In: proceedings of international seminar on the role of nuclear energy for sustainable development, New Delhi, pp 62–76
44. Mishra RK et al (2006) Structural aspects of barium borosilicate glasses containing thorium and uranium oxides. J Nucl Mater 359:132–138
45. Srinivas C, Yalmali V, Pente AS, Wattal PK, Misra SD (2012) Thoria/thoria-urania dissolution studies for reprocessing application. Report BARC/2012/E/007, Mumbai
46. Stebbins JF, Zeng Q (2000) Cation ordering at fluoride sites in silicate glasses: a high-resolution 19F NMR study. J Non-Cryst Solids 262:1–5
47. Kiczenski TJ, Stebbins JF (2002) Fluorine sites in calcium and barium oxyfluorides: F-19 NMR on crystalline model compounds and glasses. J Non-Cryst Solids 306:160–168
48. Kiczenski TJ, Du LS, Stebbins JF (2004) F-19 NMR study of the ordering of high field strength cations at fluoride sites in silicate and aluminosilicate glasses. J Non-Cryst Solids 337:142–149
49. Karpukhina NG, Zwanziger UW, Zwanziger JW, Kiprianov AA (2007) Preferential binding of fluorine to aluminium in high peralkaline aluminosilicate glasses. J Phys Chem B 111:10413–10420
50. Hayakawa S, Nakao A, Ohtsutki C, Osaka A (1998) An X-ray photoelectron spectroscopic study of the chemical states of fluorine atoms in calcium silicate glasses. J Mater Res 13:739–743

51. Mishra RK et al (2009) Effect of fluoride incorporation on the structural aspects of barium-sodium borosilicate glasses. J. Noncryst solids 355:414–419
52. Sugilal G, Benny G (2008) Development of cold crucible system for vitrification of HLW. Indian Nucl Soc News 5(3):19–28
53. Mishra RK, Mishra R, Kaushik CP, Tyagi AK, Das D, Raj K (2009) Effect of ThO_2 on ionic transport behavior of barium borosilicate. J Nucl Mater 392:1–5
54. Funke K (1998) Jump relaxation in solid electrolytes. Prog Solid State Chem 22:111–195
55. Souquet JL, Duclot M, Levy M (1998) Ionic transport mechanism in oxide based glasses in the supercooled and glassy states. Solid State Ionics 105:237–242
56. Imre AW, Voss S, Mehrer H (2004) Ionic conduction, diffusion and glass transition in 0.2[xNa2O. (1 − x)Rb2O] 0.8B2O3. J Non-Cryst Solids 333:231–239
57. Grandjean A, Malki M, Simonnet C (2006) Effect of composition on ionic transport in SiO_2–B_2O_3–Na_2O glasses. J Non-Cryst Sol 352:2731–2736
58. Ozarde PD, Haldar KK, Sarkar S (2008) Interim storage of vitrified HLW. Indian Nucl Soc News 5(3):29–32
59. Misra SD (2010) Developments in back end of the fuel cycle of Indian thermal reactors. Energy Procedia 7:474–486
60. Goel RK, Prasad VVR, Swarup A, Dwivedi RD, Mohnot JK, Soni AK, Misra DD (2003) Testing of rock samples and site specific design of underground research laboratory. Technical report GC/MT/R/1: 1–64. Central Mining Research Institute, Roorkee, India
61. Bajpai RK (2008) Characterization of natural barriers of deep geological repository for HLW in India. Indian Nucl Soc News 5(3):40–47
62. Mathur RK, Narayan PK, Arumugam V, Acharya A, Bajpai RK, Balu K (2001) Evaluation of a plutonic granitic rock mass for siting of a geological repository in India. In: Witherspoon PA, Bodvarsson GS (eds) Third world wide review of geological challenges in radioactive waste isolation, LBNL-49767. University of California, Berkeley, pp 153–162
63. IAEA (1989) Option for the treatment and solidification of organic radioactive wastes. Technical Report serial no. 294, Vienna
64. Ford CR (1961) Steam stripping TBP-Amsco solutions from non-volatile contaminants. Idaho operations Report IDO-14546
65. Chrubasik A et al (1987) Process for the Treatment of Spent TBP Kerosene (PUREX) Solvents, vol 2. French Nuclear Society. In: RECOD: Proceedings of the internal conference on nuclear fuel reprocessing and waste management, Paris
66. RWE NUKEM GmbH (2002) Pyrolysis of radioactive organic waste. www.nukem.de/global/downloads/deutch/Pyrolyse.pdf
67. IAEA (2002) Management of low and intermediate level wastes with regard to their chemical toxicity. IAEA-TECDOC-1325, Vienna
68. Smitha M et al (1999) Management of spent solvents of reprocessing origin. In: Technologies for management of radioactive waste from nuclear power plants and back end nuclear fuel cycle facilities Proc Symp Taejon. C &S paper C D series no. 6, IAEA, Vienna (2001)
69. Smitha M, Srinivas C, Tessey V, Wattal PK (1999) Management of spent solvents by alkaline hydrolysis process. Waste Manage (Oxford) 19:509–517
70. Smitha M (2005) Management of spent organic solvents. BARC Highlights. Nucl Fuel Cycle 78–79
71. Rao SVS, Paul B, Shanmugamani AG, Sinha PK (2011) Treatment of plastic waste by melt densification-operational experience at CWMF. Energy Procedia 7:502–506

About the Authors

Dr. D. Das in his 37 years long career worked in the field of high temperature thermochemistry of nuclear materials, and also in solid oxide fuel cell materials. Recently he has retired as senior scientist (Former Head, Chemistry Division, Bhabha Atomic Research Centre, Trombay, Mumbai) in the Department of Atomic Energy, India. He has extensively worked on the thermodynamic evaluation of thoria-based fuels while participating and conducting the programme of a task force for evaluating thermodynamic and transport properties of the oxide fuels for Advanced Heavy Water Reactor (AHWR) technology of thorium utilization. He has published more than 100 papers in reputed international journals.

D. Das and S. R. Bharadwaj (eds.), *Thoria-based Nuclear Fuels*,
Green Energy and Technology, DOI: 10.1007/978-1-4471-5589-8,

Dr. T. R. Govindan Kutty was the former Head, Fuel Property Evaluation Section, Radiometallurgy Division, Bhabha Atomic Research Centre, Mumbai, India, before retiring from service in 2012. He has had a long career in the area of thermophysical property evaluation of fuels and structural materials. He is a well-known expert in the field of thermal conductivity, thermal expansion, sintering especially in densification and sintering kinetics. He graduated in Chemistry from Kerala University and joined Radiometallurgy division of BARC in 1973. He took his M.Sc. (Engg) and Ph.D. degrees in metallurgy from Indian Institute of Science, Bangalore. Dr. Kutty worked about a year as a post-doctoral fellow at McMaster University, Hamilton, Canada. He has authored about 50 peer-reviewed papers and contributed chapters in the field of nuclear fuels.

Mr. Joydipta Banerjee (B.E. Metallurgical Engineering, Calcutta University, India, 1989) is a senior scientist in Fuel Property Evaluation Section, Radiometallurgy Division, Bhabha Atomic Research Centre, Mumbai, India. He is working in the field of characterization of nuclear fuels and structural materials in terms of their thermophysical and thermomechanical properties. His areas of interest include studies related to sintering kinetics and high temperature properties of fuels like thermal conductivity, thermal expansion, specific heat, and creep for advanced fuels for thermal and fast reactors. He has around 40 publications in the national/international symposiums/conferences and 10 publications in the international journals.

Mr. Arun Kumar (B.Sc. Engineering, Metallurgy, BIT Sindri, India, 1971) is presently Director, Nuclear Fuels Group and Head Radiometallurgy Division, Bhabha Atomic Research Centre, Mumbai, India. He has more than four decades of experience in plutonium bearing nuclear fuel fabrication and planning design, installation and commissioning of fuel fabrication plants. He is instrumental in various fuel development and fabrication activities, e.g., $(U,Pu)O_2$, (U,Pu)C for their successful irradiation and testing in Indian nuclear power reactors (BWRs, PHWRs, FBTR, PFBR). His recent contribution has been in development of metallic fuels for future Indian fast reactors and TRISO-coated fuels for Compact High Temperature Reactors (CHTR). He has around 90 technical publications till date.

Dr. Renu Agarwal is a senior scientist working in Bhabha Atomic Research Centre of Department of Atomic Energy, India. She has more than 28 years of experience in the field of nuclear science and technology. Dr. Renu Agarwal has carried out extensive experimental and computational work on thermodynamic behaviour and phase-diagram analysis of different types of nuclear fuels and related materials. She has carried out calculations to predict burn-up behaviour of carbide fuel being used in FBTR, India and thoria-based fuels planned to be used in AHWR, India. She was a member of the committee responsible for development of Minor Actinide Data Base, available at IAEA. She has contributed to IAEA Coordinated Research Project on Thermophysical Properties Database for LWR and HWR. She has more than 110 publications to her credit.

Dr. S. C. Parida joined Bhabha Atomic Research Centre, Mumbai, India in the year 1996. His field of research includes thermodynamics of nuclear materials and development of storage materials for hydrogen isotopes. He has expertise in the field of calorimetry and more than 50 publications in international referred journals. He pursued his post-doctoral research at University of Alabama, USA. He is a recognized Ph.D. guide of Homi Bhabha National Institute, Mumbai for

Chemical Sciences. He has received several awards like TA Instruments-ITAS young scientist award 2008, Extraction and processing division science award 2009, TMS, Warandale, USA, DAE Excellence in Science and Technology Award 2009 and DAE-SRC Outstanding Investigator Award 2012.

Dr. Shyamala Bharadwaj is presently Head, Fuel Cell Materials and Catalysis Section, Chemistry Division, Bhabha Atomic Research Centre, Mumbai, India. Her main area of work during the past 36 years has been determination of thermodynamic properties of nuclear materials using various techniques such as vapour pressure measurements, thermogravimetry, isoperibol calorimetry, etc. She was part of the task force for evaluating thermodynamic and transport properties of the oxide fuels for Advanced Heavy Water Reactor (AHWR) technology for thorium utilization. During 2000–2001, she worked as guest scientist at Juelich Research Centre, Juelich, Germany on Solid Oxide Fuel Cell (SOFC) materials. Her current interests are in the fields of Intermediate Temperature Solid Oxide Fuel Cells (ITSOFC) and Sulphur–Iodine thermochemical cycle for generation of hydrogen from water. She has more than 120 papers in refereed international journals.

Dr. Ratikanta Mishra obtained M.Sc. Degree in Chemistry from Utkal University, Bhubaneswar. He is a senior scientist working in Chemistry Division, Bhabha Atomic Research Centre, Mumbai, India. He obtained his Ph.D. Degree in Chemistry from Mumbai University in the year 1998. His research area of interest includes high temperature chemistry and chemical thermodynamics of nuclear materials and alloys. He is recipient of Alexander von Humboldt and Marie Curie fellowships. He has large number of publications in international journals.

Dr. Manidipa Basu has done her M.Sc. in Physical Chemistry from the University of Burdwan. She is a senior scientist working in Chemistry Division, Bhabha Atomic Research Centre, Mumbai, India. She started research work on high temperature thermochemistry of nuclear fission product compounds and was awarded Ph.D. degree in the same field. Currently, she is engaged in fission gas transport and thermochemical studies on metallic fuels.

Mr. Siddhartha Kolay has done his M.Sc. in Inorganic Chemistry from the University of Calcutta. He is a scientist working in Chemistry Division, Bhabha Atomic Research Centre, Mumbai, India. He started research work on transport and thermochemical studies of nuclear fission products and is currently engaged in similar investigations on metallic fuels.

Mr. Ajit N. Shirsat is a scientist working in Chemistry Division, Bhabha Atomic Research Centre, Mumbai, India. He completed his Master's Degree in Chemistry in 2004. His research areas are kinetic study of fission gas release of Thoria-based fuel and study of thermophysical properties of solid oxide fuel cell materials.

Dr. S. K. Mukerjee is a senior scientist working in Fuel Chemistry Division, Bhabha Atomic Research Centre, Mumbai, India. Since joining the research centre in 1981, he is working in the field of fabrication of ceramic nuclear fuels using sol–gel method in the form of microspheres. He has developed flow sheets for sol–gel preparation of various U, Pu oxides, carbides and for thoria based oxides. His current area of interest is fabrication of thorium carbide, (U,Pu)O_2 microspheres having 50 mole% Pu microspheres (700 μm) for the fabrication of vipac test fuel pin for irradiation in FBTR. He is also working for the preparation of 500 micron UO_2 microspheres having different surface morphology for the carbon/SiC coating studies, fabrication of porous thoria microspheres for uranyl nitrate impregnation studies and development of microwave gelation column and modification of the process for the fabrication of Lithium titanate pebbles to improve the utilization of Li in ITER blanket.

Dr. N. Kumar is a senior scientist working in Fuel Chemistry Division, Bhabha Atomic Research Centre, Mumbai, India. He obtained his Ph.D. from University of Mumbai in 2006. Since his joining, he has been actively involved with the development of sol–gel process for the fabrication of ceramic nuclear fuels of ThO_2, UO_2, mixed (U,Pu)O_2, etc. He has set up a sol–gel demonstration facility for the preparation of (Th,U)O_2 microspheres on a 1 kg/day scale and also engaged in development of sol–gel microsphere pelletization (SGMP) process for the fabrication of AHWR fuel at a laboratory scale. Presently, he is working on the preparation of 500 μm size, high density UO_2 microspheres for studies related to coated particle fuel development program of the department.

Dr. Rajesh V. Pai is a senior scientist working in Fuel Chemistry Division, Bhabha Atomic Research Centre, Mumbai, India. He is mainly associated with the development of sol–gel process for nuclear materials and high technology ceramics. He has developed many flow sheets suitable for fabrication of thoria-urania and urania-plutonia-based fuels by sol–gel process. He also gained experience in synthesis and characterization of nano-size powders of mixed oxides by various techniques. His current area of interest is development of coated fuel kernels suitable for Compact High Temperature Reactor.

Dr. P. V. Achuthan joined Fuel Reprocessing Division of Bhabha Atomic Research Centre, Mumbai, India, in 1978 and was associated with the Trombay Plutonium Plant for more than three decades. He obtained his Ph.D. Degree in Chemistry from

Mumbai University in 1990. He is a specialist in Reprocessing Chemistry, PUREX and THOREX processes, Actinide Chemistry, Fission Product Chemistry, Solvent Extraction, Ion Exchange, Extraction Chromatography, Liquid membrane technology, Gamma Spectrometry, Radiometry, etc. Currently, he is involved in the design of a multipurpose analytical laboratory to cater to the needs of reprocessing, waste management and associated auxiliary systems. He is also involved in the process development of minor actinides separation and analytical automation. He has more than 100 research papers to his credit in national and international journals/symposia.

Dr. A. Ramanujam joined Bhabha Atomic Research Centre, Mumbai, India, in 1963 retired as Head, Fuel Reprocessing Division of the Centre in 2002. He had specialized in reprocessing of high burn-up fuels at CEA, France, as an Indo-French ASTEF scholar in 1970. His was awarded Ph.D in 1980 for his work in the field of solvent extraction and process chemistry of actinides. His specialization and responsibilities in various capacities during his tenure include plutonium and uranium-233 product processing and process control, analytical operations at Purex and Thorex facilities at Trombay, setting up of these laboratory systems for new facilities, development and standardization of the Purex and Thorex-based solvent extraction flow sheets to treat different types of uranium, thorium and thorium/uranium/plutonium-based fuels in these facilities, and towards the end, the overall management of Purex / Thorex reprocessing campaigns conducted by the Division. He was the Chief Investigator from BARC for the IAEA Coordinated Research Programme on P&T (1996) and was a Member of the IAEA-TCM (1996) that reviewed the Safety Series Book on 'Safe Handling of Plutonium'.

Mr. Kanwar Raj, a chemical engineer, from Indian Institute of Technology, Roorkee, India, has 40 years experience in various facets of Radioactive Waste Management. In Department of Atomic Energy in India, he has held the responsibilities of Plant Superintendents at Trombay and Tarapur Complexes . As Head, Waste Management Division, he was responsible for operation RWM Facilities for nuclear fuel fabrication, research and power reactors, spent fuel processing, etc. Kanwar Raj has played a key role in design and commissioning of India's first Vitrification Facility at Waste Immobilization Plant Tarapur. Subsequently, he led the teams for commissioning of second vitrification facility at WIP Trombay and third vitrification facility at Advanced Vitrification System at Tarapur. He has co-authored more than 100 publications in various forums. He has made significant contribution in preparation of safety documents in the areas of RWM both for Atomic Energy Regulatory Board of India and International Atomic Energy Agency, Vienna.

Dr. C. P. Kaushik shouldering the responsibility of Dy. Plant Superintendent (H), Waste Management Facility, Bhabha Atomic Research Centre, Mumbai, India. He is engaged in development and characterization of glass formulations for vitrification of HLW from PHWR, FBR and AHWR. He is also responsible for Plant scale vitrification HLW after finalization of a suitable glass formulation. Other areas of his research cover - compatibility of materials at various stages of vitrification, characterizations of buffer and backfill materials for use in Near Surface Disposal Facilities and Geological Repositories, separation and recovery of valuables from radioactive waste. He has more than 120 publications in journals/international conferences and symposia to his credit.

Dr. R. K. Mishra is associated with Process Control Laboratories of Waste Management Facilities, Trombay for providing analytical support to the waste immobilization plant. He is engaged in preparation and characterization of glass/glass ceramics matrices for immobilization of high-level radioactive liquid waste generated during reprocessing of spent fuel from PHWR, FBR and AHWR. Other areas of his research experience cover separation and recovery of valuables from radioactive waste for industrial application, corrosion study of different alloys with nitric acid as well as molten glass matrix, synthesis and characterization of nano materials for decontamination of low-level radioactive liquid waste. He has more than 75 publications to his credit.

Printed in the United States
By Bookmasters